W9-BQK-976

Phloem Transport in Plants

A Series of Books in Biology

Editors: Donald Kennedy and Roderic B. Park

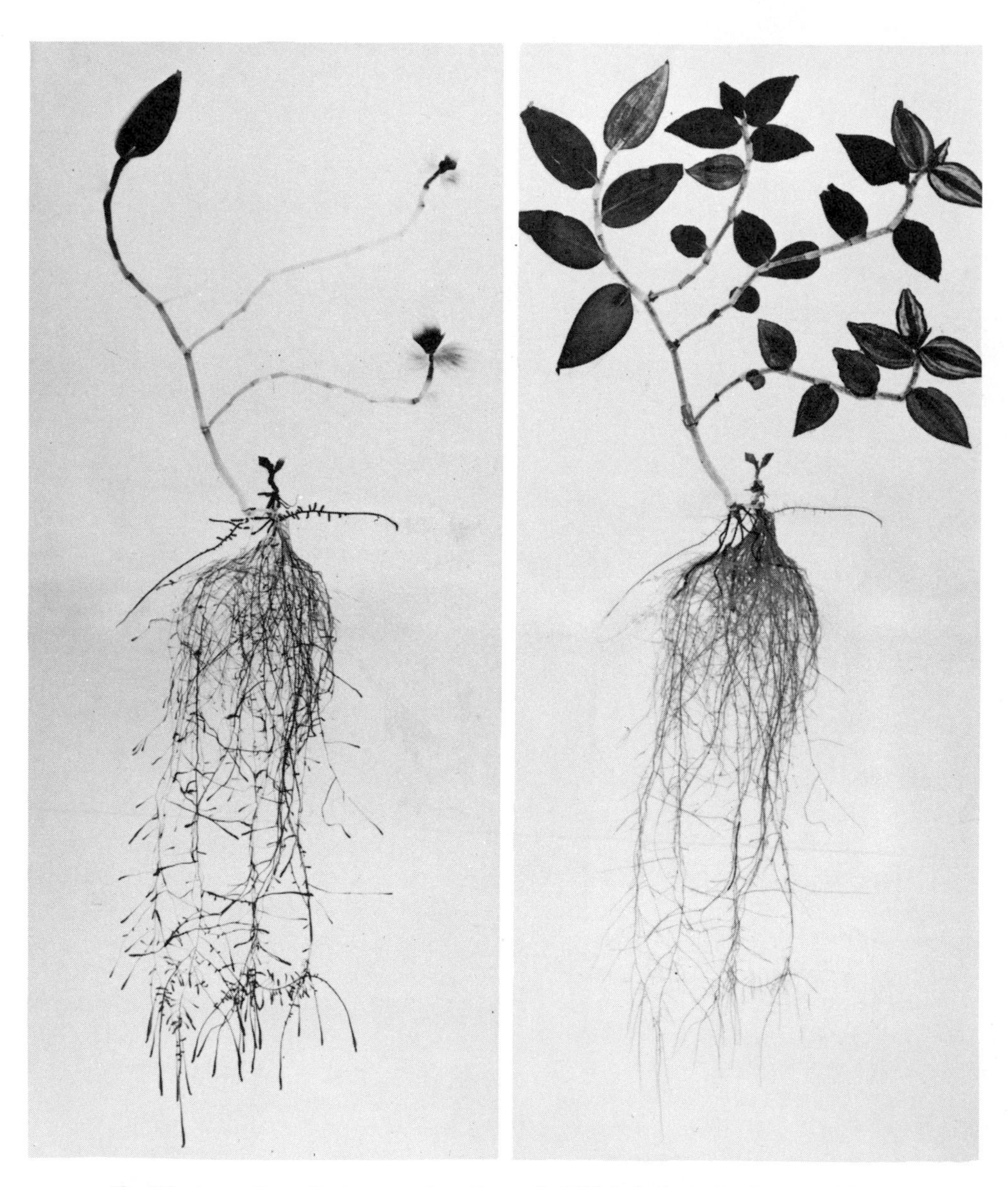

Frontispiece. Symplastic translocation of ^{14}C-labeled amitrole in *Zebrina pendula*. *Left,* Autoradiograph; *right,* photograph. The plant was grown in culture solution with the roots spread on filter paper. Dosage was 0.5 μc. Treatment time was 4 days.

Phloem Transport in Plants

Alden S. Crafts
University of California, Davis

Carl E. Crisp
Pacific Southwest Forest and Range Experiment Station Berkeley, California

W. H. FREEMAN AND COMPANY
San Francisco

Printed in the United States of America

International Standard Book Number: 0-7167-0683-0
Library of Congress Catalog Card Number: 71-125130

10 9 8 7 6 5 4 3 2 1

To Professor Ernst Münch and other pioneers
in the field of plant physiology who,
by hard work, advanced this science to
its present level of sophistication.

Contents

Preface

The flow of sap and the distribution of foods in plants have presented some of the most challenging and baffling problems with which biologists have been faced. These problems were tackled by nearly every prominent plant physiologist of the nineteenth century—Hales, Hartig, von Mohl, Nageli, Sachs, and Strasburger, to mention only a few. More recently, Dixon, Curtis, Mason and Maskell, and Münch have made major contributions to this field.

During the past four decades, research on translocation in plants has gained momentum: the many recent papers cited in the bibliography of this book attest to the increased activity. The greatest progress has resulted from three types of study employing modern methods: (1) studies of phloem anatomy in which the electron microscope has been used to examine material prepared in such a way that sieve-plate plugging has been avoided, which have proven that the protoplasmic connections of the sieve plate are open tubules; (2) studies of phloem exudation, using aphids and excised aphid mouthparts, to determine the undisturbed velocities of flow, and the normal organic and inorganic composition, of the assimilate stream; and (3) studies of tracer movement, which have shown that many endogenous and exogenous materials move from sources to sinks, not along their own individual gradients, but along gradients of assimilates that are undergoing normal distribution in the organic nutrition of the plant.

Scientific progress is slow and painful. It is now more than one-hundred-thirty years since Hartig (1837) first recognized and associated phloem transport with sieve-tube elements, through which the cellular foodstuffs are transported to all parts of the vascular plant. The electron microscopist has studied their structure, conformation, and distribution. The enzymologist has initiated work on the *in situ* biochemical systems and metabolic energy environs of phloem tissue, and the plant physiologist has brought into sharp focus the relationship of exudation and tracer transport to the direction and capacity of phloem translocation.

From these efforts emerges a mechanism to explain assimilate movement at all tissue levels in plants. The aim of this book is to collect and interpret the experimental information available on phloem transport in plants, not only to support the mechanism of assimilate movement, but to provide detailed analyses of the anatomy of the transport path itself, the phenomenon of phloem plugging, the movement of endogenous as well as xenobiotic compounds, the effects of environmental factors, and the complex quantitative aspects of assimilate distribution. More than this, there is now a rapidly growing body of evidence that the companion cell is not just another cell, but that it may, in fact, play a key role in phloem transport as a source of metabolic energy to maintain the integrity of the sieve tube plasmalemma. These interrelationships need considerably more study if we are fully to understand the energy balance between the various tissues and its relationship to long-distance transport in vascular plants. It may be that the answer to the controversy between the proponents of mass flow and those of energy-dependent systems lies in the mechanisms of energy-coupled respiration in the photosynthesizing tissue and the companion cells. In the final analysis we may learn to know and understand a unique molecular biology involving both systems.

We are indebted to our teachers, colleagues, and students, who helped guide our thinking toward the view of phloem transport that is embodied in this book. We are grateful to Rod Park and James Cronshaw for reading and critizing the manuscript. We thank Wanda Winton for skillful and efficient typing of the manuscript, and Mrs. R. Dial for careful collection of the bibliographic citations, and other assistance in preparing the book for publication. We, of course, are responsible for any errors or misinterpretations, and would be grateful if readers would call them to our attention. We would also welcome readers' suggestions for future studies of new principles and mechanisms.

October 1970

ALDEN S. CRAFTS
CARL E. CRISP

Abbreviations and Trade Names

Å Ångstrom unit = 1×10^{-8} cm = 1×10^{-4} μ

abscisic acid 5-(1-hydroxy-2,6,6-trimethyl-4-oxo-2-cyclo-hexen-1-yl)-3-methyl-2,4-pentadienoic acid
(*syn.* dormin; abscisin II; ABA)

actinomycin D *N,N′*-{[2-amino-4,6-dimethyl-3-oxo-3*H*-phenoxazine-1,9-diyl]bis-[carbonylimino(2-hydroxypropylidene)-carbonylimino-isobutylidenecarbonyl-1,2-pyrrolidinediylcarbonyl(methylimino)-methylenecarbonyl]}bis-(*N*-methyl-L-valine)

acrolein 2-propenal

ADP adenosine 5′-(trihydrogen pyrophosphate)

a.e. acid equivalent

Alar mono(2,2-dimethylhydrazide)succinic acid
(*syn.* *N,N*-dimethylaminosuccinamic acid; Alar I; B9; B-995)

AMA methanearsonic acid, amine salts
(*syn.* methyldioxoarsenic acid, amine salts; amine methylarsonate)

amiben 3-amino-2,5-dichlorobenzoic acid

amiton *S*-[2-(diethylamino)ethyl]*O*,*O*-diethyl ester of phosphorothioic acid (*syn.* *O*,*O*-diethyl *S*-2-diethylaminoethyl phosphorothioate; Chipman 6200)

amitrole 3-amino-*s*-triazole
(*syn.* 3-amino-1*H*-1,2,4-triazole; amino triazole; ATA)

AMO 1,618 (5-hydroxycarvacryl)trimethylammonium chloride, 1-piperidinecarboxylate
(*syn.* 2-isopropyl-4-dimethyl-amino-5-methylphenyl 1-piperidinecarboxylate-methochloride)

AMP adenosine 5′-(dihydrogen phosphate)

aniline blue 5,9-dianilino-7-phenyl-4,10-disulfobenzo[*a*]phenazinium hydroxide, inner salt

arthanitin fructosyl-cyclamiretin glucoside
(*syn.* cyclamine; primulin)

D-ascorbic acid D-*threo*-hex-2-enoic acid γ-lactone

ATP adenosine 5′-(tetrahydrogen triphosphate)

atrazine 2-chloro-4-(ethylamino)-6-(isopropylamino)-*s*-triazine

Azodrin® 3-hydroxy-*N*-methyl-*cis*-crotonamide, dimethyl ester with phosphoric acid

bacitracin A 4-{2-[2-(1-amino-2-methylbutyl)-2-thiazoline-4-carboxamido]-4-methylvaleramido}-*N*-(1-⟨{2-[3-aminopropyl]-8-benzyl-5-sec-butyl-14[(1-carbamoyl-2-carboxyethyl)carbamoyl]-11-[imidazol-4-ylmethyl]-3,6,9,12,16,23-hexaoxo-1,4,7,10,13,17-hexaazacyclotricos-22-yl}carbamoyl⟩-2-methylbutyl)glutaramic acid

BAP 6-benzylaminopurine
(*syn.* *N*-benzoyl-adenine; *N*-purin-6-yl-benzamide;BA)

barban *m*-chloro-carbanilic acid, 4-chloro-2-butynyl ester
(*syn.* 4-chloro-2-butynyl *N*-(3-chlorophenyl)carbamate; Carbyne®)

BECT di-(ethylxanthogen) trisulfide
(*syn.* Defoliant 713)

Bidrin® 3-hydroxy-*N*,*N*-dimethyl-*cis*-crotonamide, dimethyl ester with phosphoric acid
(*syn.* 3-(dimethoxyphosphinyloxy)-*N*,*N*-dimethyl-*cis*-crotonamide)

biopterin 1-(2-amino-4-hydroxy-6-pteridinyl)-1,2-propanediol
(*syn.* 2-amino-4-oxy-6-(1,2-dioxypropyl)-pteridine; L-*erythro*-(biopterin); crithidia factor)

romphenol blue 4,4′-(3*H*-2,1-benzoxathiol-3-ylidene)bis-2-bromo-6-chlorophenol

6-(benzylamino)-9-(2-tetrahydropyranyl)-9*H*-purine

-chloroethyl)trimethyl ammonium chloride
Cyocel 1; chlorocholine chloride)

CF1 2-chloro-9-hydroxyfluorene-9-carboxylic acid
(*syn.* Chlorflurenol)

chloramphenicol D-*threo*-(−)-2,2-dichloro-*N*-[β-hydroxy-α-(hydroxymethyl)-*p*-nitrophenylethyl] acetamide
(*syn.* Chloromycetin)

chlorphenol red 4,4′-(3*H*-1,2-benzoxathiol-3-ylidene)-bis-2-chlorophenol

CIPC *m*-chlorocarbanilic acid, isopropyl ester
(*syn.* isopropyl-*N*-(3-chlorophenyl)-carbamate)

count/min counts per minute
(*syn.* CPM)

cP centipoise

CPA 4-chlorophenoxyacetic acid
(*syn.* Tomatotone)

cycloheximide 3-[2-(3,5-dimethyl-2-oxocyclohexyl)-2-hydroxyethyl]-glutarimide
(*syn.* isocycloheximide; Actidione)

D-1991 (1-butylcarbamoyl)-2-benzimidazolecarbamic acid

2,4-D 2,4-dichlorophenoxyacetic acid

dalapon 2,2-dichloropropionic acid, sodium salt

2,4-DB 2,4-dichlorophenoxybutyric acid

DCPA 2,3,5,6-tetrachloroterephthalic acid, dimethyl ester
(*syn.* Dacthal®)

DDS bis(4-aminophenyl)sulfone

DDT 1,1,1-trichloro-2,2-bis(*p*-chlorophenyl)ethane

dehydroascorbic acid D-*threo*-2,3-hexodiulosonic acid, γ-lactone

2,4-DI 2,4-dichloro-5-iodophenoxyacetic acid

di-allate diisopropylthiolcarbamic acid, *S*-(2,3-dichloroallyl)ester
(*syn.* *S*-2,3-dichloroallyl *N,N*-diisopropylthiolcarbamate; DATC; Avadex®)

dicamba 3,6-dichloro-*o*-anisic acid
(*syn.* Banvel D®; mediben)

dihydrostreptomycin *O*-2-deoxy-2-(methylamino)-α-L-glucopyranoxyl-(1→2)-*O*-5-deoxy-3-*C*-(hydroxymethyl)-α-L-lyxo-furanosyl-(1→4)-*N,N*′-diamidino-streptamine
(*syn.* Didromycine®)

diketogulonic acid L-*threo*-2,3-hexodiulosonic acid, γ-lactone

dimefox *N,N,N*′*,N*′-tetramethylphosphorodiamidic fluoride
(*syn.* bis(dimethylamido)phosphoryl fluoride; DMF; Pestox 14)

dimethoate 2-mercapto-*N*-methylacetamide phosphorodithioic acid, *O,O*-dimethyl ester with *S*-ester
(*syn.* Rogor®; Cygon®)

diquat 6,7-dihydrodipyrido[1,2-*a*:2′,1′-*c*]pyrazinediium dibromide
(*syn.* 1:1′-ethylene-2′,2′-dipyridylium dibromide; Regulone®)

disulfoton *O,O*-diethyl-*S*-[2-(ethylthio)ethyl]ester of phosphorodithioic acid
(*syn.* Di-Syston®)

diuron 3-(3,4-dichlorophenyl)-1,1-dimethylurea
(*syn.* DCMU)

DNA deoxyribonucleic acid

DNBP 2,4-dinitrobutylphenol

DNOC 3,5-dinitro-*o*-cresol

DNP 2,4-dinitrophenol

2,4-DP 2-(2,4-dichlorophenoxy)propionic acid
(*syn.* dichlorprop)

DPN 3-carbamoyl-1-ribofuranosyl-pyridinium hydroxide, (5′→5′) ester with adenosine 5′-(trihydrogen pyrophosphate), inner salt
(*syn.* nicotinamide adenine dinucleotide, oxidized; diphosphopyridine nucleotide; NAD)

DPNH adenosine 5′-(trihydrogen pyrophosphate), (5′→5′) ester with 1,4-dihydro-1-α-D-ribofuranosylnicotinamide
(*syn.* nicotinamide adenine dinucleotide, reduced; diphosphopyridine nucleotide, reduced; NADH)

DSMA methanarsonic acid, disodium salt

DTPA diethylenetriaminepentaacetate

Duraset® *N-meta*-tolyl phthalamic acid
(*syn.* 7R5)

EDTA ethylenediaminetetraacetic acid, tetrasodium salt
(*syn.* (ethylenedinitrilo)tetraacetic acid, tetrasodium salt)

EHPG *N,N′*-ethylene-bis-[2-(*O*-hydroxyphenyl)]glycine

endothall 7-oxybicyclo[2.2.1]heptane-2,3-dicarboxylic acid
(*syn.* 3,6-endoxyhexahydrophthalic acid)

endrin 1,2,3,4,10,10-hexachloro-6,7-epoxy-4a,5,6,7,8,8a,9,9-octahydro-1,4-*endo*-5,8-*endo*-dimethanonaphthalene

endrin-ketone 1,2,3,4,10,10-hexachloro-6,7-epoxy-4a,5,6,7,8,8a-hexahydro-1,4-*endo*-5,8-keto-dimethanonaphthalene

eosine Y 2′,4′,5′,7′-tetrabromo-9-(*o*-carboxyphenyl)-6-hydroxy-3*H*-xanthene-3-one,3′,6′-dihydroxyfluoran
(*syn.* 2′,4′,5′,7′-tetrabromofluorescein)

EPTC ethyl-di-*n*-propylthiolcarbamate
(*syn.* Eptam®)

erythrosine 2′,4′,5′,7′-tetraiodo-9-(*o*-carboxyphenyl)-6-hydroxy-3*H*-xanthene-3-one,3′,6′-dihydroxyfluoran
(*syn.* 2′,4′,5′,7′-tetraiodofluorescein)

FAP 6-furfurylaminopurine
(*syn.* *N*-furfuryladenine)

fenac 2,3,6-trichlorophenylacetic acid

FL 9-hydroxyfluorene-9-carboxylic acid
(*syn.* Flurenol)

fluometuron 3-(*m*-trifluoromethylphenyl)-1,1-dimethylurea
(*syn.* Cotoran®)

fluorescein-K 9-(*o*-carboxyphenyl)-6-hydroxy-3*H*-xanthene-3-one,3′,6′-dihydroxyfluoran, potassium salt
(*syn.* resorcinolphthalein, potassium salt; uranine, potassium salt; Uranine Yellow, potassium salt)

ft-c foot-candle

folic acid *N*-⟨*p*-{[(2-amino-4-hydroxy-6-pteridinyl)methyl]amino}benzoyl⟩-L-glutamic acid
(*syn.* pteroylmonoglutamic acid)

gibberellin 2β,4a,7-trihydroxy-1-methyl-8-methylene-4aα,4bβ,-gibb-3-ene-1α,10β-dicarboxylic acid,1,4a-lactone
(*syn.* gibberellic acid; GA_3)

griseofulvin 7-chloro-2′,4,6-trimethoxy-6′β-methyl-spiro[benzofuran-2(3*H*), 1′-[2]cyclohexene]3,4′-dione

GTP guanosine-5′-(tetrahydrogen triphosphate)

^3H_2O $^3H—O—^3H$
(water completely labeled with tritium)

HTO $H—O—^3H$
(water half labeled with tritium)

IAA indol-3-acetic acid
(*syn.* heteroauxin)

INBA 2-iodo-4-nitrobenzoic acid

ioxynil 3,5-diiodo-4-hydroxybenzonitrile

IPC carbanilic acid, isopropyl ester
(*syn.* isopropyl *N*-phenylcarbamate)

Irgafen N^1-(3,4-dimethylbenzoyl)-sulfanilamide

isolan 1-isopropyl-3-methylpyrazol-5-yl ester of dimethylcarbamic acid
(*syn.* dimethyl-5-(1-isopropyl-3-methylpyrazolyl)carbamate)

ITP inosine 5′-(tetrahydrogen triphosphate)

maleic hydrazide 1,2-dihydro-3,6-pyridazinedione
(*syn.* 6-hydroxy-3(2*H*)-pyridazinone; MH)

MCPA 4-chloro-2-methylphenoxyacetic acid
(*syn.* [(4-chloro-*o*-tolyl)oxy]-acetic acid)

MCPB 4-chloro-2-methylphenoxybutyric acid

menazon *S*-[(4,6-diamino-*s*-triazin-2-yl)-methyl],*O*,*O*-dimethyl esters of phosphorodithioic acid

mercurochrome 2′,7′-dibromo-4′-(hydroxymercurio)-9-(*o*-carboxyphenyl)-6-hydroxy-3*H*-xanthen-3-one,3′,6′-dihydroxyfluoran, disodium salt
(*syn.* 2,7-dibromohydroxymercurifluorescein, disodium salt)

methyl demeton mixture of *O,O*-dimethyl-*O*-2(ethylthio)ethyl phosphorothioate and *O,O*-dimethyl-*S*-2(ethylmercaptoethyl)-dimethyl thionophosphate

Mev million electron volts

monuron 3-(*p*-chlorophenyl)-1,1-dimethylurea

MOPA 3-methoxyphenylacetic acid

myoinositol 1,2,3,4,5,6-cyclohexanehexol
(*syn.* *meso*-inositol)

NAA 1-naphthalenacetic acid

NAD *see* DPN

NADH *see* DPNH

NADP *see* TPN

NADPH *see* TPNH

naptalam *N*-1-naphthylphthalamic acid
(*syn.* Alanap)

Nellite® *N,N′*-dimethylphosphorodiamidic acid, phenyl ester
(*syn.* phenyl-*N,N′*-dimethylphosphordiamidate)

neomycin A 2-deoxy-4-*O*-(2,6-diamino-2,6-dideoxy-α-D-glucopyranosyl)-D-streptamine
(*syn.* Neamine)

nicotinic acid 3-pyridinecarboxylic acid

oligomycin Rutamycin (an antibiotic substance)

oxamycin D-4-amino-3-isoxazolidone

P poise

paraquat 1,1′-dimethyl-4,4′-dipyridylium cation

pantothenic acid *N*-(2,4-dihydroxy-3,3-dimethylbutyryl)-β alanine

PCP pentachlorophenol

phlorizin 3,5-dihydroxy-2-(*p*-hydroxyhydrocinamoly)phenyl β-D-glucopyranoside
(*syn.* phloridzin)

Phosdrin® 2-carbomethoxy-1-propen-2-yl-dimethyl phosphate
(*syn.* *O,O*-dimethyl-1-carbomethoxy-1-propen-2-yl phosphate; mevinphos)

phosphamidon 1-chloro-1-diethylcarbamoyl-1-propen-2-yl dimethyl phosphate
(*syn.* dimethyl-diethylamido-1-chlorocrotonyl[2]phosphate)

phosphon 2,4-dichlorobenzyltributylphosphonium chloride
(*syn.* phosphon D)

picloram 4-amino-3,5,6-trichloropicolinic acid
(*syn.* Tordon®)

Pleocidin® an antibiotic substance

polyoxin 1-[5-(2-amino-2-deoxy-L-xylonamido)-6-*C*-(2-carboxy-3-ethylidene-1-azetidinyl)-5-deoxy-β-D-*allo*-hexodialdo-1,4-furanosyl]-5-(hydroxymethyl)-uracil
(*syn.* Polyoxin-A®)

P-protein phloem protein

P1-protein protein component of sieve-tube slime with tubules 231 Å in diameter

P2-protein protein component of sieve-tube slime with tubules 149 Å in diameter

prometryne 2,4-bis(isopropylamino)-6-(methylthio)-*s*-triazine

propanil 3,4-dichloropropionanilide
(*syn.* DPA)

Pyrazoxon® *O,O*-diethyl-*O*-(3-methyl-5-pyrazolyl) phosphate

pyridoxine 5-hydroxy-6-methyl-3,4-pyridinedimethanol

R_f in chromatography, the distance traveled by a given substance divided by the distance traveled by the solvent front

RH relative humidity

rhodamine B [9-(2-carboxy-4-isothiacyanatophenyl)-6-diethylamino-4-3*H*-xanthen-3-ylidene] diethyl ammonium

rhodamine 6G *o*-[6-(ethylamino)-3-(ethylimino)-2,7-dimethyl-3*H*-xanthen-9-yl] benzoic acid

riboflavin 7,8-dimethyl-10-(D-ribo-2,3,4,5-tetrahydroxypentyl)isoalloxazine

RNA ribonucleic acid

RQ respiratory quotient

schradan bis-*N,N,N′,N′*,-tetramethylphosphorodiamidic anhydride
(*syn.* OMPA)

silvex 2-(2,4,5-trichlorophenoxy)propionic acid

simazine 2-chloro-4,6-bis(ethylamino)-*s*-triazine

streptomycin *O*-2-deoxy-2-(methylamino)-α-L-glucopyranosyl-(1→2)-*O*-5-deoxy-3-*C*-formyl-α-L-lyxofuranosyl-(1→4)-*N,N′*-diamidino-D-streptamine

streptothricin 2-{[2-(3-amino-6-{3-amino-6-[3-amino-6-(3,6-diaminohexanamido)hexanamido]-hexanamido}-hexanamido)-2-deoxy-α-D-gulopyranosyl]amino}-3,3a,5,6,7,7a-hexahydro-7-hydroxy-4*H*-imidazol (4,5-*C*)pyridin-4-one

Systox® *O,O*-diethyl-*O*-[2-(ethylthio)ethyl] ester mixed with *O,O*-diethyl-*S*-[2-(ethylthio)ethyl] ester of phosphorothioic acid (*syn.* demeton; demeton O + demeton S)

T-1947 polyoxyethylene-polyoxypropylene-polyol

2,4,5-T 2,4,5-trichlorophenoxyacetic acid

TBA 2,3,6-trichlorobenzoic acid

TCPP 1,3,5-tricyano-3-phenylpentane

Temik® 2-methyl-2-(methylthio)propionaldehyde *O*-(methylcarbamoyl)-oxime

tetramine 2,2,6,6-tetraoxide of 2,6-dithia-1,3,5,7-tetraazaadamantane

thiamine {3-[(4-amino-2-methyl-5-pyrimidinyl)-methyl]-5-(2-hydroxyethyl)-4-methylthiazolium}chloride

thiazamide N^1-2-thiazolylsulfanilamide

Thimet® *O,O*-diethyl-S-[(ethylthio)methyl] ester of phosphorodithioic acid

TIBA 2,3,6-triiodobenzoic acid

TPA 2,2,3-trichloropropionic acid

TPN 3-carbamoyl-1-β-D-ribofuranosylpyridinium hydroxide, (5′→5′) ester with adenosine 2′-(dihydrogen phosphate)5′-(trihydrogen pyrophosphate), inner salt
(*syn.* nicotinamide adenine dinucleotide phosphate, oxidized; triphosphopyridine nucleotide; NADP)

TPNH 2′-(dihydrogenphosphate)5′-(trihydrogen pyrophosphate), (5′→5′) ester with 1,4-dihydro-1-β-D-ribofuranosylnicotinamide
(*syn.* nicotinamide adenine dinucleotide phosphate, reduced; triphosphopyridine nucleotide, reduced; NADPH)

Tween 20 polyoxyethylene sorbitan monolaurate

Tween 80 polyoxyethylene sorbitan monooleate

UDP uridine 5′-(trihydrogen pyrophosphate)

UDPG mono-α-D-glucopyranosyl ester of uridine 5′-(trihydrogen pyrophosphate)

UMP 5′-uridylic acid

UTP uridine 5′-(tetrahydrogen triphosphate)

vitamin B_{12} cobinamide cyanide hydroxide dihydrogen phosphate (ester) inner salt 3′ ester with 5,6-dimethyl-1-α-D-ribofuranosylbenzimidazole

UV ultraviolet

X-77 a blended surfactant containing alkylarylpolyoxyethylene glycols, free fatty acids, and isopropanol

I

Structure–Function Relations

1

Introduction

The growth of plants, their normal functioning, and even their very lives—be they crop plants, weeds, or wild endemics—require the integrated action of a complex variety of processes. The absorption of water, mineral nutrients, and carbon dioxide is elementary to plant function. Transport of a variety of foods from regions of synthesis to regions of utilization is essentail. Basic, too, are a number of metabolic processes, involving assimilation and dissimilation, synthesis of cell walls and cytoplasm, storage of foods, excretion of wastes, and discard of exhausted structures. And throughout this gamut of functions there runs a thread of correlative action. Shall we call it purpose or adaptative response? Whatever we name it, it results in survival: plants thrive, they cover the surface of the earth with a mantle of green, and they provide the basic means of sustenance for man.

This book deals with the processes by which substances (including mineral nutrients, organic foods, hormones, viruses, and various agricultural chemicals, such as pesticides and chemotherapeutants) are distributed throughout the plant body. The term *translocation* has been used to cover this array of processes; alternative terms are *transport, distribution,* and *movement.* These terms will be used more or less interchangeably throughout this volume.

That translocation is an important function is easy to comprehend; the growth of plants to any great size would be impossible without the rapid distribution of inorganic nutrients and organic foods. Less evident is the fact that many correlative phenomena necessary to growth are related to the distribution of minerals and foods and the attendant movement of hormones and inhibitors—factors that are involved in the orderly succession of growth, fruiting, senescense, dormancy, and rejuvenation. Without the functioning of specific translocation mechanisms, it would be impossible for plants to grow and reproduce in the many ways that we observe. Transport of substances, then, includes processes that are essential to the very existence of plants.

The Phenomenon of Solute Distribution

Translocation in plants involves two complex tissue systems, the xylem and the phloem. Each of these consists of a ramifying complex of conduits and parenchyma cells extending from the tips of the leaves to the extremities of the roots. While each is a discrete and unique system, their functions are integrated in such a way that under normal conditions they deliver to each and every cell of the living plant those elements that are essential to life. Primary movement of minerals and water is from the roots to the foliage via the xylem; primary movement of organic nutrients is from the foliage to all non-green cells via the phloem. Minerals may undergo secondary movement in the phloem; foods move secondarily in the xylem (Crafts, 1961b). Some essential elements, such as phosphorus, may circulate and make many cycles via these tissue systems during the lifetime of a plant (Biddulph et al., 1958).

This book will deal mainly with phloem function, and it will attempt to give an overall view of the processes involved in food distribution. Much of the work in this field consists of studies of phloem anatomy, assimilate movement, virus transmission, phloem exudation, tracer transport, and other specifics, with little or no attempt to integrate the various aspects into an overall treatment.

On the mechanism of food movement in plants, opinions have been divided for a long time. Some physiologists favor the mass-flow hypothesis, which holds that the solutes and the solvent water flow as a stream through the mature sieve tubes. Others visualize independent movement by a process analogous to diffusion but greatly accelerated;

in this mechanism, solutes would move independent of each other and of the solvent water. We will attempt to show that, when all of the aspects of phloem structure and phloem function are considered, there can be little question as to the actual mechanism involved. Current research has provided information that clarifies many of the old arguments.

A number of authors have reviewed the literature on translocation, notably Curtis (1935), Swanson (1959), Zimmermann (1960a), Crafts (1961b), Schumacher (1967), Wardlaw (1968a), Weatherley and Johnson (1968), and Milthorpe and Moorby (1969). We will limit our treatment principally to the current decade, and to those papers that seem relevant to the problem of mechanism.

The Symplast–Apoplast Concept

The symplast–apoplast concept of Münch (1930) provides an extremely useful terminology for discussing translocation (Arisz, 1961). According to this concept, the total mass of living cells of a plant constitute a continuum, the individual protoplasts being intimately connected throughout the plant by plasmodesmata. Because the plasmolemma is the outer limiting membrane and extends through the pit pores as the outer surface of the plasmodesmata, and the tonoplast limits the vacuole and does not extend into the plasmodesmata, it follows that the mesoplasm is continuous from cell to cell. Hence, an ion or a molecule that passes either the plasmolemma or the tonoplast and arrives within the mesoplasm is able to migrate from cell to cell throughout the symplast without passing through a permeability barrier. Movement from cell to cell along plasmodesmata must be diffusive; within cells, it is accelerated by protoplasmic streaming, wherever it occurs. Thus, the symplast constitutes the sum total of living protoplasm of a plant: there are no islands of living cells. Ziegler (1964) illustrates a "stretched ray" in *Taxodium distichum*. The terminal cell of the ray becomes separated from its neighbors in the stretching process, and soon dies. Guard cells may prove to be an exception to this rule.

In contrast to the symplast, the apoplast constitutes the total nonliving cell-wall continuum that surrounds and contains the symplast. By its tensile strength, the apoplast counteracts the swelling effect of turgor pressure, and hence enables the plant to grow erect, overcoming the force of gravity. It also constitutes a continuous permeable system through which water and solutes may freely move. All ions and molecules entering the roots do so via the apoplast. And all substances applied to leaves or stems diffuse across the cuticle, pass along the cell walls, and either move with transpiration water to the leaf margins (as do calcium ions, monuron, and simazine), enter the symplast and build up within vacuoles, or migrate to the phloem to translocate to other parts of the plant. Some substances enter leaf cells but do not translocate (dinitro compounds, pentachlorophenol, and diquat, for example); they may be toxic and may soon cause the death of the treated foliage. Some substances (such as phenoxy compounds and PO_4^{---}) enter the leaves and soon translocate to sinks in other parts of the plant via the phloem. Certain other substances move freely in either symplast or apoplast, or both (amitrole, dalapon, TBA, and picloram, for example).

It has been proposed that the sieve-tube system in plants may be looked upon as a highly specialized system of conduits that make up a functional phase of the symplast. The principal features of the functional specialization of sieve tubes are their continuity; their loss of nuclei, tonoplasts, and other plasmatic constituents; and the tubular nature of the sieve-plate connections that accommodate mass flow.

The xylem system may be considered to constitute a specialized phase of the apoplast, adapted to the rapid, long-distance transport of water and salts absorbed by the roots. In its specialization, the protoplasts are completely removed; end walls are perforated in most angiosperms, and the walls develop internal thickenings which enable them to function under reduced pressure (Crafts, 1961b).

Water and salts absorbed by the roots move into the xylem where they are conducted through stems and leaves. Solutes are absorbed into the symplast where they are stored, used in growth, or redistributed to younger tissues via the phloem. Water not stored or utilized in growth is lost from the apoplast walls of the stomatal chambers. Such movement of water and solutes is called *apoplastic translocation.* Solutes synthesized in leaves or chemicals applied to foliage—once they arrive inside the symplast—are moved by diffusion, accelerated by protoplasmic streaming, until they reach the sieve tubes; there, because of their osmotic activity, they lower the water potential; water moves in from the xylem, developing turgor pressure, and the solutes translocate in the assimilate stream to sinks in growing or storing parts of the plant. Such movement of water and solutes is called *symplastic translocation.* This terminology will be used to describe the various aspects of translocation under discussion in this book.

2

Structure of the Phloem

Sieve-cell Characteristics

For a clear understanding of translocation, it is necessary to integrate the structure with the function of phloem. Esau covered the work on the development and structure of phloem up to 1965 in her revised *Plant Anatomy* (Esau, 1965b). Since that time, much new knowledge of phloem structure has been gained as a result of work with the electron microscope (Esau, 1969).

Unique to the sieve cells of the phloem is a series of changes during differentiation, resulting in the degradation—or complete loss—of many of the original organelles, and in the perforation of end walls that adapts these cells to their function in translocation. The young sieve tube resembles the phloem parenchyma cells in many ways; a unique feature is its early division to form the sieve-tube–companion-cell pair. After rapid turgor expansion, during which differentiation of the organelles—including slime bodies—takes place, a series of degenerative changes occur, resulting in the disintegration of nucleus, slime bodies, and tonoplast. As the sieve pores enlarge making the sieve plates perforate, the sieve tube takes on the function of translocation of assimilates; and, from the work that has been done on phloem exudation (see Chapter 5), it seems evident that there takes place, through the sieve tubes, a flow of solution which rapidly displaces the liquid contents of each cell. This soon results in a great loss of the stainable contents.

The mitochondria undergo degenerative changes and assume, with the endoplasmic reticulum, a parietal position, leaving the lumen of the cell relatively clear of obstruction. A filamentous reticulum may develop within the lumen that is continuous through the pores. These features, as described in the literature, will now be covered in more detail.

Loss of nucleus and tonoplast has been confirmed by electron microscopy (Esau and Cheadle, 1962; Esau, 1963, 1966; Engleman, 1965b; Evert and Alfieri, 1965; Evert and Murmanis, 1965). Resch (1961) described in some detail the breakdown of the sieve-cell nucleus and commented on its anatomical-physiological significance. With the tonoplast gone and the parietal cytoplasmic layer continuous through the sieve plate pores, forming an open tubule in each pore, liquid in the lumina is continuous and able to move freely from element to element, as should be the case, judging from the work on phloem exudation.

Fine Structure of Phloem

Methods. For electron microscopy, the methods and materials used in killing, fixing, and embedding the materials are critical. For example, it is widely recognized that permanganate fixation dissolves out most of the cytoplasmic contents of the cell. Potassium dichromate, osmium tetroxide, and other fixing materials are known to render uncritical views of plant-cell fine structure. Glutaraldehyde and acrolein provide more accurate fixation, and new materials and methods are being introduced currently.

With phloem, it is essential, too, that provision be made to prevent rapid surge of the pressurized sap of the sieve tubes to avoid the artifact of sieve plate plugging. Injection of the killing fluid into hollow stems so that killing takes place before the organ is cut, killing of whole plants, and plasmolysis to relieve the turgor of the sieve-tube system have all been used. In the following review of phloem structure, only those results that seem critical will be given attention.

Some Results. The fine structure of differentiating sieve-tube elements of *Pisum sativum* was studied by Bouck and Cronshaw

(1965). Considering only those features relevant to sieve tube function, they described nuclear disintegration. Prior to and after the breakdown of the nucleus, the endoplasmic reticulum aggregated in local patches along the surface of the wall. Eventually, these patches became compressed to form layers of cisternae parallel to the wall. They found these cisternae to be separated by a space which seemed initially to be occupied by layers of fine tubules or granules. Thus, they considered the parietal layer to be composed of membranes derived from the endoplasmic reticulum, and from vesiculation and breakdown of the nuclear envelope. Bouck and Cronshaw termed this system of membranes the "sieve-tube reticulum." They described the cisternae as being oriented in a longitudinal direction with their lateral margins attached perpendicularly to the cell surface. They saw small mitochondria associated with the marginal cisternae. The vacuolar membrane (tonoplast) was lost.

The sieve plate of the mature sieve tube they found to have many pores, each lined by a thin membrane. Under their conditions of fixation, they found the sieve-plate pores to be filled with finely fibrous material, possibly derived from the parietal cisternae. They found that the plugged condition of the sieve plate presented a great obstacle to the mass-flow concept, and they proposed a two-phase system to explain phloem transport: (1) an active pumping through the sieve plate; and (2) possibly a cytoplasm-mediated transport through the sieve cell. They suggested that the free ends of the sieve-tube reticulum may well extend into the sieve tube lumen, and so give rise to transcellular strands along the surfaces of which a streaming transport might occur. They stressed the apparent association of the sieve-tube reticulum and slime to the extent of suggesting that the endoplasmic reticulum might synthesize slime. This matter will be discussed in detail in a later section of this chapter.

Wark and Chambers (1965) also studied phloem fine structure in *Pisum sativum*. They, however, used permanganate fixation, and thus their treatment of the details of cytoplasmic structure is not highly accurate. They stressed the presence of nuclear material in the cytoplasm of mature sieve tubes, the enlargement of cytoplasmic tubules in the connecting strands, and the increase in endoplasmic reticulum running through the cells; these conditions, they concluded, support the view of Kollmann and Schumacher (1964) to the effect that those inclusions are instrumental in assimilate transport; they suggested that the younger stages of sieve-tube ontogeny comprise the functional translocating cells.

Northcote and Wooding (1966) described the development of sieve tubes in *Acer pseudoplatanus*. They categorized five stages in the differentiation process. Stage one starts with the cambial initial and covers its division into another cambial initial and the sieve-tube mother-cell, which gives rise to the young sieve-tube–companion-cell pair. Stage two involves the start of sieve-tube specialization, including cell-wall thickening, slime-body formation, and the close association of endoplasmic reticulum with plastids and the future pore sites. Stage three covers the initiation of callose areas at the pore sites; ribosomes in the cytoplasm are fewer per unit area, the nucleus takes up less electron stain, and the tonoplast leaves the peripheral cytoplasm and enters the vacuolar space. During stage four the perforations of the sieve plate are formed, the peripheral cytoplasm and the slime bodies disperse into the vacuolar space, and ribosomes, endoplasmic reticulum, and nucleus gradually disappear. Stage five represents the mature functioning period in the life of the sieve tube. Its lumen is occupied by a diffuse network of small fibrils which, Northcote and Wooding believe, originate from slime bodies. Remnants of membranous material are associated with the parietal cytoplasmic layer; special pore connections between sieve tube and companion cell are elaborated.

Sieve-plate pore development described by Northcote and Wooding differs in no essential way from that described by Esau, Cheadle, and Risley (1962). The completely perforated pore is a smooth tube, surrounded by callose and lined by a thin layer of plasmolemma. Glucose labeled with ^{3}H and fed to a cut stem was found to be incorporated into the callose around the sieve pores, but not in any other position on the sieve-tube wall; this indicates that the callose cylinders had thickened during the 3-hour incubation period prior to killing and embedding.

Slime bodies, in Northcote and Wooding's material, were forming by stage two. They consisted of closely packed fibers 180 to 240 Å in diameter (their figure 8). In longitudinal view, each fiber appears as two bands with a light region between. By stage three, striated fibrils become prominent and these persist throughout the subsequent stages. In figure 8 of Northcote and Wooding, they appear in the lower part of the cell; in their figures 11 and 12, they are dispersed throughout the lumina; and in their figures 23, 24, and 28 they are shown in detail, both in the lumina and packed in the sieve-plate pores. These striated fibrils or filaments are 90 to 100 Å in diameter, they persist long after the slime has been lost from the elements, and they probably represent the permanent living reticulum of striated filaments that are responsible for sieve-plate plugging in mature sieve elements. The inset in Northcote and Wooding's figure 12 closely resembles figure 8 of Behnke and Dörr (1967); their figure 28 looks like Behnke and Dörr's figure 7. The fibrillar fillings of sieve pores in their figures 22–24 are very much like those of *Cucumis* in Behnke and Dörr's figure 12, and those of *Nicotiana tabacum* in Cronshaw and Esau's figures 10 and 13.

Tucker and Evert (1969) made a 2-year study of phloem development in *Acer negundo*. They described in detail the various stages in sieve-tube differentiation. Mature elements they found to be distinguished by their large size and relatively clear appearance. Whereas, in most woody dicotyledonous species, all of a given season's phloem increment functions as a conducting tissue for that season only, *A. negundo* contains conducting phloem at all times of the year. The greatest amount of functional phloem is present at the end of cambial activity, the smallest amount during winter. Sieve-plate pores measured 3–4 μ in diameter. Sieve plates were both simple and compound, the latter type occurring on long, sloping walls between two elements.

Tamulevich and Evert (1966) studied the sieve-tube constituents of *Primula obconica*, the plant used by Thaine (1961) in his work on strands. In the early stages (their figures 1, 3, 6, and 14) the sieve-tube elements are filled with the microstructures found in phloem parenchyma and companion cells; slime aggregations are prominent (their figures 3, 5, 6, and 14); they consist of tubules, each about 200 Å in diameter. As the elements mature, the nuclei disappear, and the enucleate element is lined with a parietal layer of cytoplasm, one or more cisterna-like layers of endoplasmic reticulum, numerous mitochondria, and plastids with starch grains. Tamulevich and Evert claim that there is also a membrane present, separating the cytoplasm from the central cavity, that is presumably a remnant of the tonoplast. However, their photographs are not convincing, and the fact that plastids can be observed free from such a membrane in their figures 16–18 and 20 would minimize the significance of their claim. Slime is shown disaggregating in their figures 18, 19, and 26. Although they found no strands of the dimensions described by Thaine, they found tubules—which they termed slime—oriented longitudinally in sieve tubes, and they claimed that slime strands less than 0.5 μ in diameter are to be found running the length of the cells and continuous from one cell to the next through the sieve pores; these they claimed to be made up of smaller units, fibrils 100 Å in diameter or tubules 200 Å in diameter, which they called "slime strands."

Careful study of the illustrations of Tamulevich and Evert reveal first a tremendous loss of sieve-tube constituents during maturation of the elements (compare their figures 1, 3, 6, 12, and 14 with their figures 20–26 and 28.) Figure 2.6 reproduces their figures 1 and 21. The large aggregates of slime found in their figures 3, 5, 10, and 14 seem to have disappeared. Meanwhile filamentous material comparable with that of Behnke and Dörr appears in the lumina of the mature elements and this extends through the sieve pores in their figure 17. Possibly, the slime strings described by Tamulevich are actually plasma filaments, and the slime has been carried away in the moving assimilate stream.

The above references describe details of the structure of phloem of dicotyledonous plants. There are fewer papers on the phloem of monocots. Behnke (1965a, 1965b, 1968) and Behnke and Dörr (1967) have noted the loss of nucleus and tonoplast and a universal presence of a filamentous reticulum in sieve tubes of several *Dioscorea* species. Other features of *Dioscorea* phloem will be reviewed in a later section of this chapter. Ie et al. (1966) studied phloem structure of *Yucca flaccida*, a plant which has been extensively used in phloem exudation observations. Attempting to reconcile structure with exudation, they studied sieve plates and found pores not yet opened, pores closed with callose, pores filled with fibrous substance, and open pores with a few filamentous strands running through. These, they concluded, left ample open pore space for flow of liquid between the sieve cells. Parthasarathy and Tomlinson (1967) demonstrated living sieve tubes at the base of a *Sabal palmetto* stem of age 50 years. Tomlinson (1964) suggested that conducting tissues in some arborescent monocots may live for 100 years or more.

The phloem of gymnosperms is unique in that it lacks companion cells but possesses albuminous cells that presumably serve the same purpose. The sieve elements are long and spindle-shaped, and there are few, if any, short cross-walls; communication is by long common walls between adjacent cells, and the interconnecting pores, though of small diameter (0.7 μ, according to Murmanis and Evert, 1966), are great in number.

Wooding (1966) described the differentiation of sieve cells of *Pinus pinea* as seen under the electron microscope. During maturation of sieve cells the dictyosomes break down, the microtubules adjacent to the walls disappear, the large central vacuole phases out, and—according to Wooding—the cytoplasm thins out (is redistributed) to fill the entire cell lumen. Sparse, rough endoplasmic reticulum becomes a part of a complicated meshwork of smooth endoplasmic reticulum. The mitochondria become associated with strands of membranous material that run longitudinally. According to Wooding, when the nucleus undergoes change at sieve-cell maturity the endoplasmic reticulum in the cell collects into vesicular masses that resemble the slime bodies of certain angiosperms. In the most mature sieve cells, these vesicular masses, nuclear remnants, and mitochondria associated with longitudinal strand material, persist along the walls leaving the lumen with a completely empty appearance. Wooding found no strands like those reported by Thaine (1961, 1962), who saw transcellular strands, and strands running along the parietal cytoplasmic layer that Wooding considered to have a cytoskeletal function. Wooding also failed to observe the longitudinally arranged tubule system described for *Metasequoia glyptostroboides* by Kollmann and Schumacher (1964), a fact that he ascribed to different fixation methods. Wooding interpreted the aggregations of membranous or vesicular material found around sieve areas as artifacts of preparation.

Srivastava and O'Brien (1966), in studies on the ultrastructure of the cambium and its derivatives, concluded that in *Pinus strobus* the sieve elements undergo profound modifications in their organelles and membrane systems during differentiation. There occurs vesiculation of the endoplasmic reticulum; loss of ribosomes, dictyosomes, and vacuolar membranes; and loss or degeneration of the nucleus. Mitochondria also undergo some loss of membranes. They found protein crystalloids in the plastids, development of a sieve-element reticulum, and a fibrillar component in the mictoplasm of the mature element.

Work on the phloem of conifers by Evert and Alfieri (1965), Murmanis and Evert (1966), Kollman and Schumacher (1962a, 1962b, 1963), and Kollman (1963) will be presented later.

Contents of Sieve Elements

As indicated above, the contents of sieve elements undergo profound changes during the processes of differentiation, and finally, in the functional stage, the principal assimilate found is sucrose. Meanwhile, the slime bodies, nucleus, and tonoplast—and possibly the dictyosomes, mitochondria and endoplasmic reticulum—have undergone complete dissolution or degenerative changes. As a result, the sieve-tube sap may contain a variety of materials, including proteins, polypeptides, lipoid derivatives of degenerate membranes, and, in addition to sucrose, reducing sugars and oligosaccharides. Table 6.1 (page 89) lists the great number of constituents that have been identified in phloem exudate or by microchemical tests in sieve cells. This great diversity of components is hardly surprising in view of the fact that the sieve cells are conduits through which the total organic nutrient supply of the plant is transported.

The exact relationships between slime, the proteinaceous constituents, the lipoid materials, and the sugars in sieve tubes have not been definitively worked out, and the controversial issues involved have lapped over into the question of mechanism. The literature cited will indicate the nature of some of the anatomical-biochemical-physiological problems that are involved.

Eschrich (1963a, 1963b) described what he termed a lipoprotein network or reticulum in *Cucurbita ficifolia* sieve tubes (his figures 1–4 and 6 of 1963a, and 1a, 1b, and 1c of 1963b). This network resembles very much the plasmatic filament network of Behnke and Dörr; but in Eschrich's second paper (1963b), a view of the same sort of network in the exudate from an apical droplet of sap indicates that the material illustrated comes out in the sap and hence must be slime; this is further confirmed by the fact that a droplet from the basal end of the petiole, having many more mature sieve elements, has little or none of the lipoprotein material. Eschrich's materials were fixed in potassium permanganate, which, according to Johnson (1966), produces a flocculent precipitate in the presence of sucrose. In Johnson's illustrations, a sieve plate of *Nymphoides peltatum* fixed in potassium permanganate has flocculent material on both sides and a dense precipitate in the pores; one fixed in glutaraldehyde has dense contents in the pores, but obviously filamentous material fanning out at both ends of each pore. Apparently, the network shown by Eschrich is made up of slime. The striated filaments that occur in cucurbits were probably destroyed by the potassium permanganate fixation in Eschrich's preparations.

Evert et al. (1966) claimed that strands in sieve plates are all derived from slime. Structures labeled strands in their figures 26, 28, 31, 32, 48, and 49 give every appearance of being composed of slime. If they show plasmatic filaments at all, it is in their figure 21, showing a young sieve tube that has not reached the functional stage of development; yet such material must be present in functioning sieve tubes, or the commonly observed exudation stoppage would not occur. Murmanis and Evert (1966) show slime bodies in *Pinus strobus*. Their series of figures 4–8 give an excellent picture of slime disaggregation. Figure 2.1 presents a reproduction of these. In their figure 7, slime is shown in a stranded form that looks very much like it is adsorbed to filaments of a reticulum. In their figure 8, they show a strand made up mostly of filaments with just a bit of adhering slime. It is strands of this nature that Evert has shown in numerous other publications, stretched from sieve plate to sieve plate in sieve tubes, and from one sieve area to another in sieve cells.

Crafts illustrated strands of this type in *Cucurbita* (1932), *Solanum tuberosum* (1933), *Nicotiana tabacum* (1934), and *Fraxinus americana* (1939b). He discussed their nature in the 1932 paper (p. 188), confirming the earlier descriptions of Hill (1908) and Zimmermann (1922). He discussed this problem again in a recent paper (Crafts, 1968a).

Eschrich (1963b) found that the exudate from *Cucurbita ficifolia* phloem resembles closely the sieve-tube contents; Zimmermann (1961) emphasized the parallelism between the composition of phloem exudate and the organic nutrient requirements of plant growth. Since slime is a normal constituent of the sieve tubes, it must ultimately be consumed in the growth of new cells. Slime may play a more direct role in the plugging of sieve plates following the cutting of the phloem. Such plugging varies widely in timing and effectiveness. Those species in which it is fast and effective (*Cucurbita* spp., for example) seem to have the greatest slime content in their sieve tubes, compared with the giant kelp *Macrocystis pyrifera*, which exudes for 30 minutes or more, *Fraxinus americana*, which exudes for as much as 1 hour from a single cut (Crafts, 1939a), and *Yucca flaccida*, which exudes for hours (Tammes and Die, 1964).

Falk (1964) studied the origin of sieve-tube slime in *Tetragonia expansa* and suggested that it is derived from plastids. He has excellent pictures of the plastids, which resemble those of *Beta vulgaris* in having a dense ring of fibrous material enclosed in a plasma sac, so that, viewed from one angle, the ring on edge looks like a bar with half-round ends, and, from a 90° shift in angle, like a dense ring with a light circular center enclosed in a convoluted envelope. While Falk showed the dispersal of these rings into masses of filamentous material, his suggestion that the slime is derived from degenerating plastids seems hardly justified. Northcote and Wooding (1966) found no such situation in *Acer pseudoplatanus*, nor did Esau (1965a) in *Beta vulgaris*.

In his excellent electron micrographs, Falk produced further evidence for the very great loss of plasmatic constituents from sieve tubes at maturation (his figures 1, 2, and 5). Also, the presence of plasmatic filaments (striated filaments) is indicated and labeled slime in his figure 5.

Huber and Liese (1963) emphasized the lack of stainable contents in the sieve cells of *Larix*. The cytoplasm they found to occupy the cell lumen in a fine stranded arrangement, but the tonoplast, nucleus, and plastids were missing, and the mitochondria were sparse.

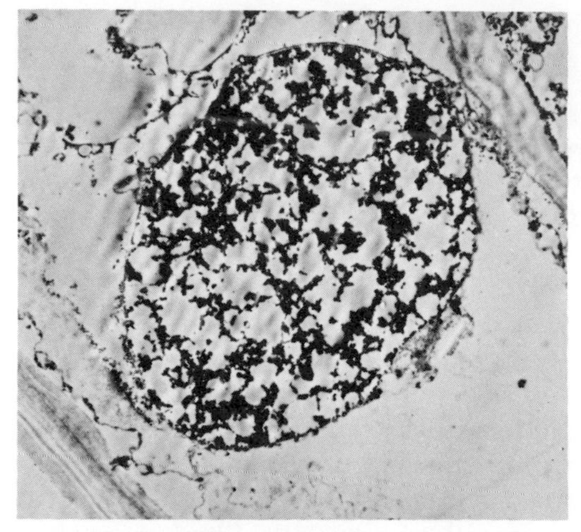

a

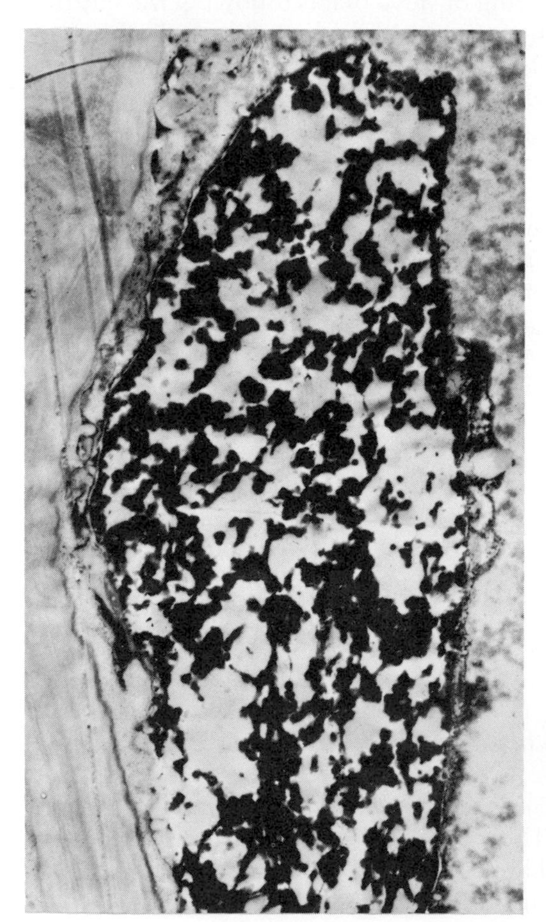

b

Figure 2.1. Five stages in the disintegration of slime bodies of *Pinus strobus:* (*a*) young body × 2543; (*b*) swelling and expansion × 3240; (*c*) disintegration almost complete × 7350; (*d*) slime associated with filaments of the reticulum × 3533; (*e*) strand or fibril made up of associated plasmatic filaments × 8730. From Murmanis and Evert (1966).

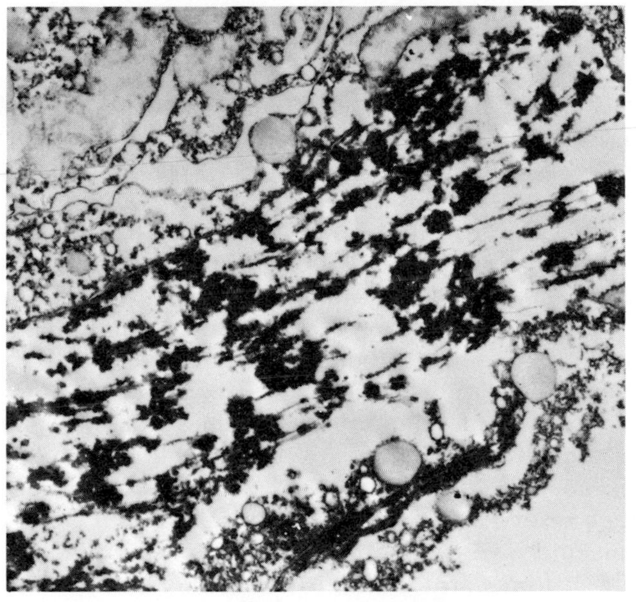

c

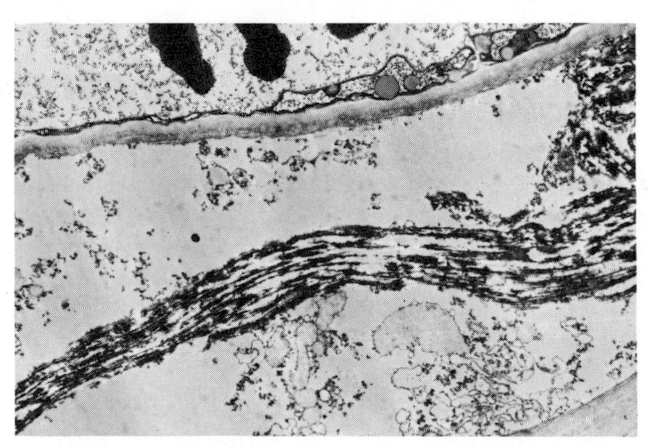

d

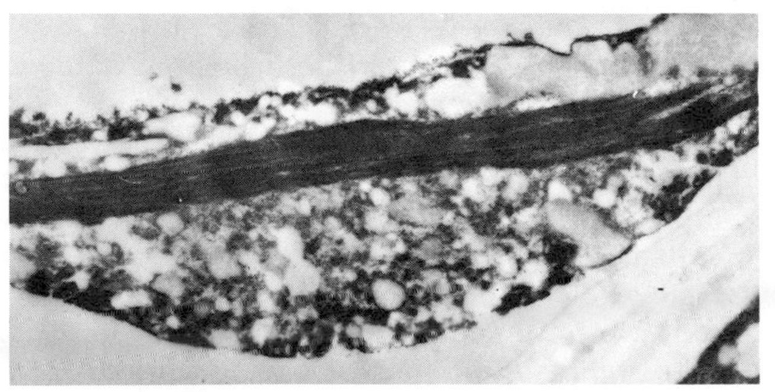

e

In the fifth paper in their series on *Metasequoia glyptostroboides*, Kollmann and Schumacher (1964) described an endoplasmic membrane system which develops toward the end of the period of maturation of the sieve cell. As described, it consists of tubules and cisternae and is oriented mainly parallel with the long axis of the cell. At the stage of maximum development of this system, the tonoplast becomes disorganized, and the endoplasmic system mingles with the parietal cytoplasmic layer as well as occupying the cell lumen. Kollmann and Schumacher expressed the opinion that sieve cells function in translocation during their earliest stages of development. This does not agree with the information on phloem exudation from conifers, with observations on the time of growth of new cells requiring assimilates, nor with the well known phenomenon of fall storage of carbohydrate reserves.

Murmanis and Evert (1966) made electron microscope studies of the phloem of *Pinus strobus*. The young sieve tubes they found to contain all of the protoplasmic components commonly found in young cells. They also found slime bodies with distinct double-layered limiting membranes. The mature sieve cells had the usual thin, parietal cytoplasmic layer containing tubules of endoplasmic reticulum, mitochondria, starch granules, and unidentified lattice-like bodies. Necrotic nuclei were found in the parietal cytoplasm, but ribosomes and dictyosomes were lacking. Strands that Murmanis and Evert considered to be derived from slime were found traversing the central cavity and continuous, through the plasmalemma-lined sieve pores, with strands of adjacent sieve cells. They also observed endoplasmic membranes traversing the pores of mature sieve areas and, in some instances, slime-occupied pores of sieve areas (their figures 18 and 23).

In a series of three excellent papers (Esau and Cronshaw, 1967a; Cronshaw and Esau, 1967; Esau and Cronshaw, 1967b) many of the intricate relations between the various components of the sieve tubes of *Nicotiana tabacum* were clarified. In the first of these, the authors described the two characteristic inclusions of tobacco mosaic virus—striated or crystalline material, and ameboid- or X-bodies—identifiable under the light microscope; and virus particles, which may be identified in electron micrographs. Striated material and X-bodies were found in differentiating tracheary elements and in companion cells in the phloem, but striated material only was found in immature sieve elements. Virus particles were found in all cell types, including mature tracheary and sieve-tube elements. In addition to the constituents mentioned above, Esau and Cronshaw described tubules, larger than virus particles, which they took to be of proteinaceous nature. These tubules were found both within X-bodies and aggregated into discrete masses.

In their second paper (Cronshaw and Esau, 1967) they described a slime, which they name P1-protein, that makes up compact masses of tubules averaging 231 ± 2.5 Å in diameter. At the time of nucleus and tonoplast disintegration, these masses of tubules disaggregate and the tubules disperse throughout the cell. Later, smaller tubules of striated material appear in the cells; these are distinct in appearance and are 149 ± 4.5 Å in diameter; these they designated as P2-protein. In fixed material, the sieve-plate pores are filled with proteinaceous material that frays out into the sieve-tube lumina as striated fibrils of P2-protein. The authors concluded that the contents of contiguous mature sieve-tube elements are continuous through the sieve-plate pores.

To compare the views of young sieve-tube elements in Cronshaw and Esau's paper (their figures 2, 4 and 6) with views of mature elements (their figures 10, 11, 13, and 14) is to become immediately aware that again a large portion of the early contents has been lost; and the arrangement of the filamentous material shown in their figures 10 and 13 resembles very much the plasma filaments of Behnke and Dörr. Again it seems possible that the P1-protein—or slime—has disintegrated and moved out of the cells, that small inconspicuous filaments of the original cytoplasm have become elaborated into a reticulum, and that this reticulum makes up the material clogging the pores of the mature sieve plate shown in their figures 10 and 13.

In their third paper Esau and Cronshaw (1967b) distinguished between P-protein and the tubular protein constituent of X-bodies. The X-body protein particles are straighter, they are arranged in greater order, and they are somewhat larger than the P-protein tubules. The X-component also occurs in other forms; but these may be arranged in a developmental sequence, from groups, to individual tubules assembled in larger, less ordered groups. The writers suggest that the X-component may be the noninfectious protein, or X-protein commonly associated with complete virus particles in *Nicotiana tabacum* infected with tobacco mosaic virus. Thus, the following are to be found in *N. tabacum:* P1-protein (slime), P2-protein, plasma filaments (possibly), X-component, noninfectious protein (possibly), and virus particles.

Currier and Shih (1968) found that there are sieve tubes in the leaves of *Elodea canadensis* and *E. densa.* Using the callose fluorescense technique (Eschrich and Currier, 1964) on incubated leaves, they show pit and plate callose on anticlinal epidermal walls and definite sieve plate callose in midribs. They produced good evidence for open pores in sieve plates in their electron micrographs (their figure 10); slime bodies of a crystalline structure were seen (their figures 11, 12, and 14); and plastids containing crystalline inclusion bodies of probable protein nature were in evidence (their figures 9–12 and 14). They were unable to identify a complete xylem system in these plants; there were isolated spiral and scalariform elements lying about 3 cm from the shoot apex.

Elodea spp. form roots that lack chlorophyll, and it is probably the nutrition of these that requires the formation of phloem tissues. Brown algae, too, possess phloem tissue to nourish the pigmentless cells of their holdfasts, though they have no xylem at all.

Slime

Slime is a term that has come through the literature from the earliest work on phloem anatomy. Slime bodies, or slime drops, are the organelles commonly found in the young sieve tubes of cucurbits. They break down during sieve-tube maturation to produce the slime plugs so sommonly found in these elements. Figure 2.2 shows phloem of *Cucurbita pepo* in various stages of differentiation from the slime-drop stage to maturity. Figure 2.3 shows an electron microscope view of early slime-plug formation, where the identity of the slime drops is still clear. Disintegration of slime bodies to form slime in cucurbit sieve tubes was illustrated by Crafts (1932). Electron microscope pictures of this process were published by Cronshaw and Esau (1968a, 1968b) by Esau and Cheadle (1961), by Evert et al. (1966), and others.

In *Solanum tuberosum* (Crafts, 1933) and *Nicotiana tabacum* (Crafts, 1934; Esau, 1941; Cronshaw and Esau, 1967; Esau and Cronshaw, 1967a, 1967b), slime bodies of various shapes, sizes, and structures have been described. Slime bodies, which are made up of tubular components, break down to form slime masses and, in cut material, slime plugs on the sieve plates. Cronshaw and Esau (1967) coined the term P-protein (phloem protein) to designate this material that is so prominent in *N. tabacum* sieve tubes. They have noted two forms of P-protein in *N. tabacum*, P1-protein tubules of diameter 231 ± 2.5 Å and P2-protein of 149 ± 4.5 Å; the latter is striated, possibly helical, in structure. In plants infected with tobacco mosaic virus, Esau and Cronshaw (1967a, 1967b) identified a third tubule component, the X-component, which may be the non-infectious protein, or X-protein, commonly associated with complete particles; they also pictured the rods of tobacco mosaic virus.

Thus, slime may consist of particulate material (as in the cucurbits), tubular or stranded material of various sorts (as in *Solanum tuberosum* and *Nicotiana tabacum*), and, in its natural state in the living sieve tube, of disintegration products of the nucleus, tonoplast, dictyosomes, ribosomes and mitochondria (Crafts, 1932, 1961b). Hence, it cannot be composed of protein alone, for these various organelles also contain lipoid materials from the numerous membranes involved (Crafts, 1968a).

Eschrich (1963a, 1963b) has spoken of a lipoprotein network in mature sieve tubes, and has illustrated the organization of such material in phloem exudate (his 1963b figures 1c and 1d). The fact that the network is dense in the apical drop and missing in the basal drop emphasized the fact that the coagulable material (slime) is abundant in the young, newly matured sieve tubes where breakdown has recently taken place, and lacking in the exudate of old sieve tubes that have had their contents replaced thousands of times during their functioning life.

Biddulph and Cory (1965) noted the lipoid nature of constituents of the ascending assimilate stream moving from young newly exporting leaves: they termed this material "steroid." During the initial importing period of the leaf, such material must be hydrolized and reabsorbed into the nutrient pool that supplies the food for leaf growth. At compensation, when the stream reverses and is exported, there must remain unresorbed sufficient lipoid material to characterize the stream as "lipoprotein" or "steroid." This material must re-enter the mutrient pool and migrate from the terminal protophloem sieve tubes to the meristems of the shoot; a similar fate must befall the slime in terminal sieve tubes at root tips. Because the terminal, differentiated, protophloem sieve tubes in such organs extend to within the short distance of 260 μ from the root apex in *Nicotiana tabacum* (Esau, 1941), 324 μ from the shoot apex in *N. tabacum* (Esau, 1938), or 400 μ from the shoot apex in *Sequoia sempervirens* (Crafts, 1943), it seems obvious that flow of the assimilate stream through the terminal few protophloem sieve tubes must be rapid to provide nutrients for the vigorous growth in progress. Contents of such sieve tubes must turn over rapidly; it is no wonder that these elements assume a clear, void appearance so rapidly upon maturation.

Figures 2.4 and 2.5 illustrate the clearing phenomenon which takes place with maturation of a protophloem sieve cell in the root of *Nicotiana tabacum* (Esau, 1941). Because it is a diarch root, two of

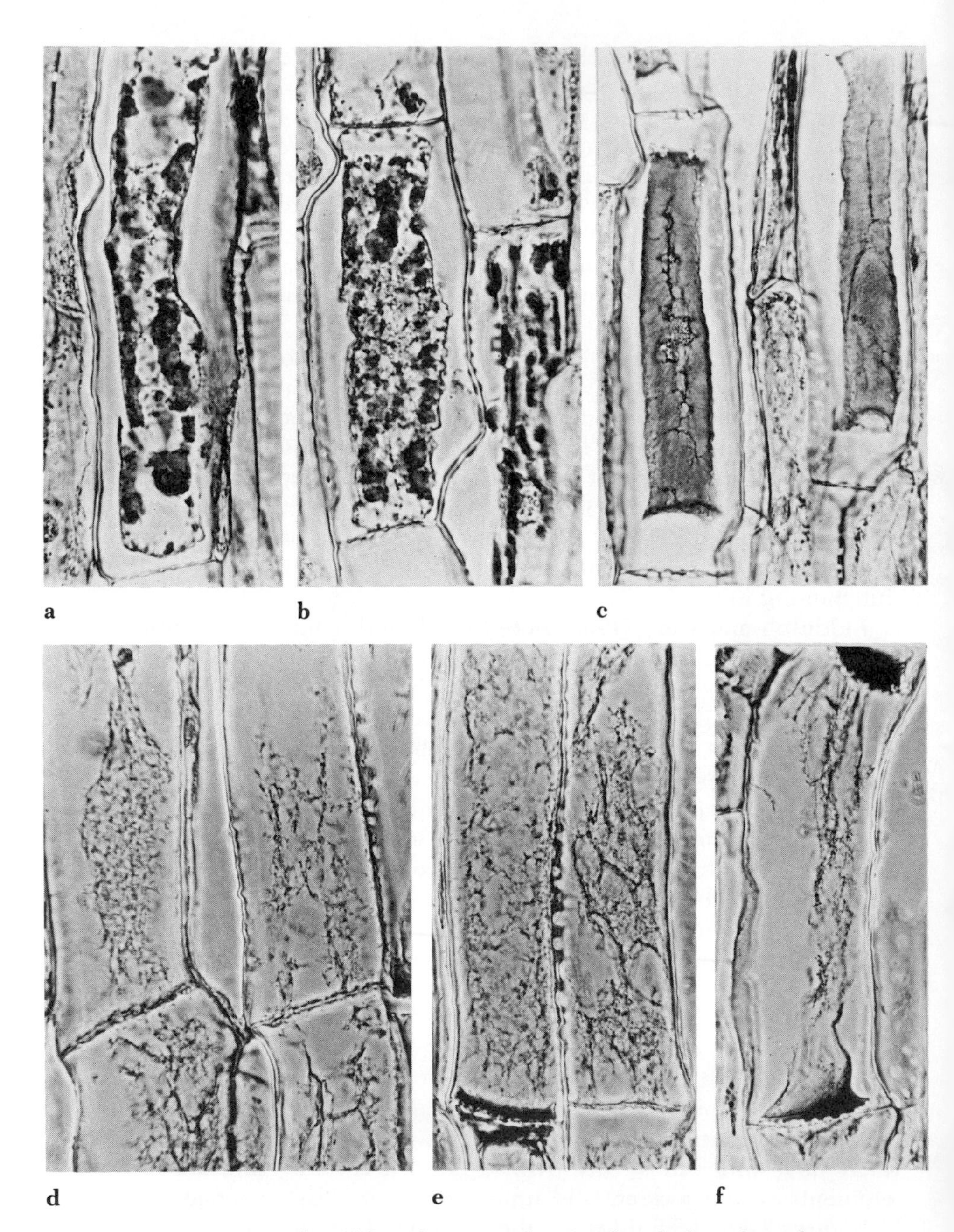

Figure 2.2. Sieve tubes of *Cucurbita pepo* showing slime bodies, slime plugs, the filamentous reticulum with adsorbed slime, and slime sacs: (*a*) slime drops in a young sieve tube; (*b*) disintegrating slime drops, (*c*) amorphous slime, protoplast shrunken; (*d*) filamentous reticulum with adsorbed slime; (*e*, *f*) filamentous reticulum and slime plugs; (*g*) short sieve elements showing slime plugs and

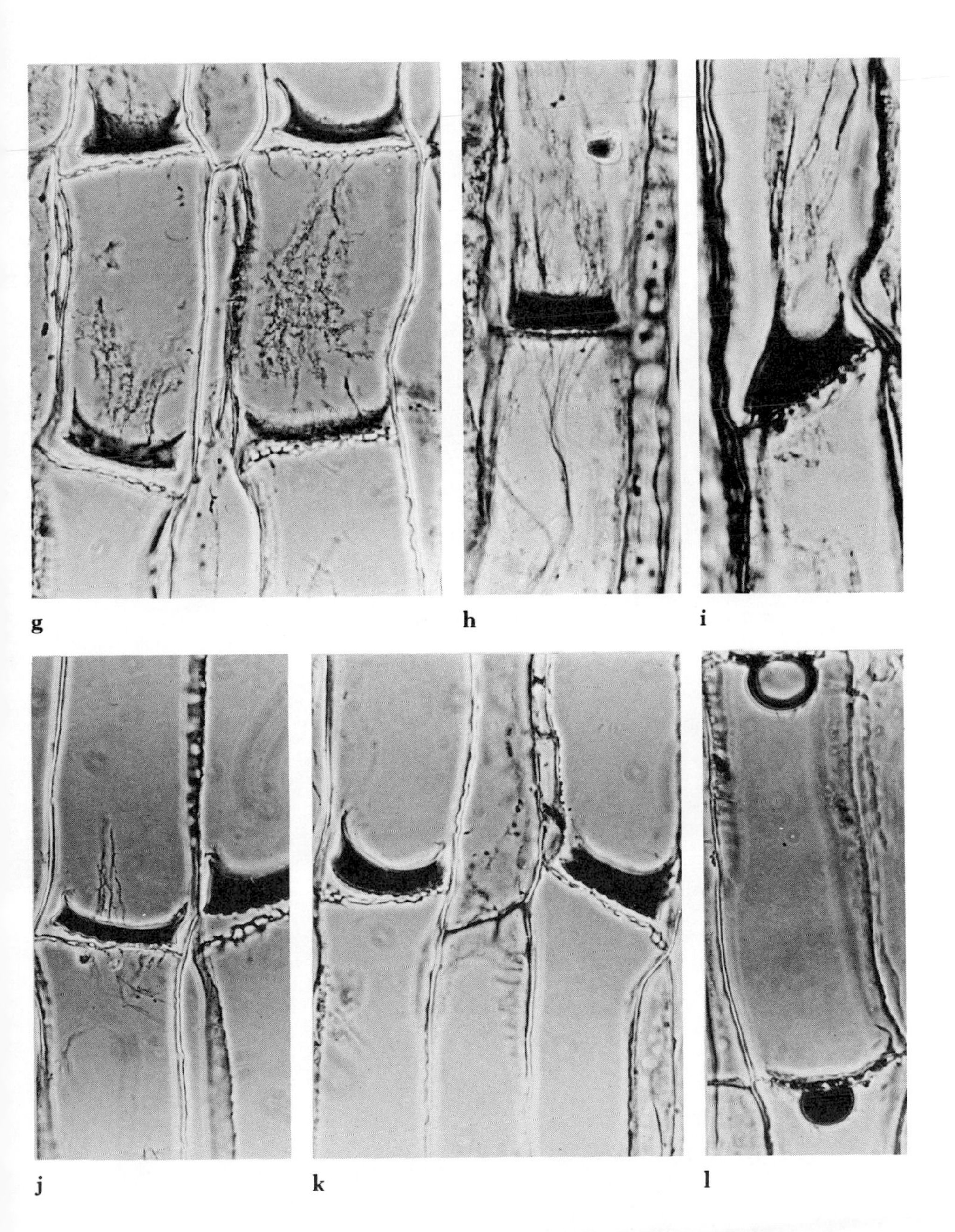

some filamentous reticulum; (*h, i*) slime plugs with fibrils of associated reticulum; (*j*) slime plugs, one with some slime on the reticulum; (*k*) slime plugs—if the reticulum is present, it has no adsorbed slime; (*l*) slime sacs with traces of filaments protruding from their surfaces; all views approximately × 670.

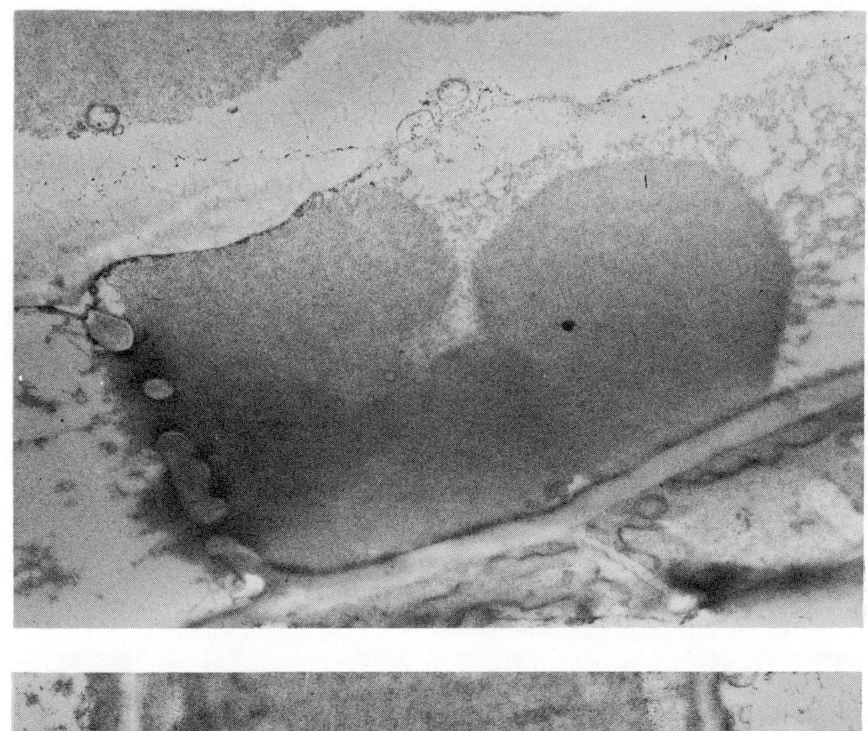

a

b

Figure 2.3. Slime plugs in sieve tubes of *Cucurbita maxima* showing (*a*) disintegrating slime bodies × 16,170, and (*b*) amorphous slime × 6930. Material cut before fixing. From Esau and Cheadle (1961).

the protophloem sieve tubes, each about 8.8 μ in diameter, must deliver to the whole root apex the energy-rich food supply from the leafy part of the plant.

If it can be assumed that the total dry matter content of the root comes from the top of the plant via the phloem, then a simple calculation gives us the velocity of flow through the protophloem elements shown in Figure 2.4. Scaling the diagram on page 453 of Esau's (1941) paper, we find that the diameter of the root is 280 μ, $r = 140\ \mu$, and $\pi r^2 = 61{,}500\ \mu^2$. The diameter of the sieve tube is 8.8 μ, $r = 4.4\ \mu$, and $\pi r^2 = 60.5\ \mu^2$. The two sieve tubes, therefore, have a transverse area of $60.5 \times 2 = 121\ \mu^2$. Because $61{,}500/121 = 508$, roughly, the sieve tubes would have to have their contents replaced 508 times per day, or 21 times per hour, in order to supply the volume required. If the roots grow 1 cm per day, the 21 cm $= 21 \times 10^4\ \mu$ per hour. The longest sieve element in Figure 2.4 is 56 μ long, and $(21 \times 10^4\ \mu)/56\ \mu = 3750$ times per hour, or roughly 1 time per second, that the contents would be replaced at a flow velocity of 21 cm per hour.

This is just a rough calculation, but it should indicate the order of magnitude of the velocity of the assimilate stream required to account for normal root growth. Because the direction of flow through sieve tubes at the apical meristems is constantly acropetal, it follows that the stainable contents of the differentiating elements must be dispersed into the adjacent parenchyma cells, and must move symplastically to supply foods for meristematic activity. And to move laterally, foods must traverse the plasmodesmata, which are very numerous between sieve tubes and companion cells, border parenchyma, or transfer cells. The form in which these materials migrate from sieve tubes to various sink tissues is unknown but it seems that they must be reduced in dimensions to at least the size of the viruses which are known to traverse this pathway. Because the foods consist of sugars, amino acids, and proteins and lipoproteins or their degradation products, it seems logical that they should be of molecular dimensions. Thus slime, which made up the bulk of the stainable contents of the sieve tubes shown in Figures 2.4 and 2.5, must have been rapidly degraded to approximately molecular form, and swept from the cells as they took on the clear empty appearance of the conducting elements. Similar loss of electron-dense materials from the mature sieve tubes of *Primula obconica* is shown in Figure 2.6. This phenomenon can be observed in hundreds of illustrations, where both young and mature sieve tubes are shown.

Evert and his associates (Evert and Derr, 1964b; Evert and Murmanis, 1965; Evert, Murmanis, and Sachs, 1966) introduced a new idea into the concept of slime: namely, that strands often found

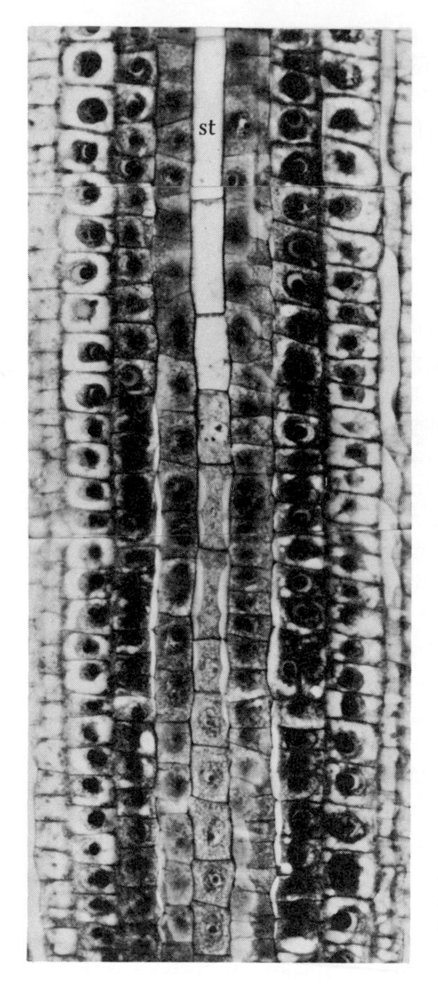

Figure 2.4. Differentiation of a protophloem sieve tube of a root of *Nicotiana tabacum* × 428. The upper three elements have reached the functioning stage. From Esau (1941).

stretched from sieve plate to sieve plate, passing through sieve-plate pores and sometimes attached to side-wall sieve areas, are composed of slime. They claimed that such strands occur not only in dicotyledonous plants, but in monocots and gymnosperms as well. Thus, they pictured a system of strands, mostly oriented in the direction of flow; such strands would offer little resistance to mass flow of assimilates. Evert and Derr (1964b) noted an inverse relation between the presence of strands and the occurrence of slime plugs, and concluded that slime plugs arise through the disruption of strand material and its subsequent accumulation against the sieve plates. Such slime plugs were, in some instances, amorphous; in others, they

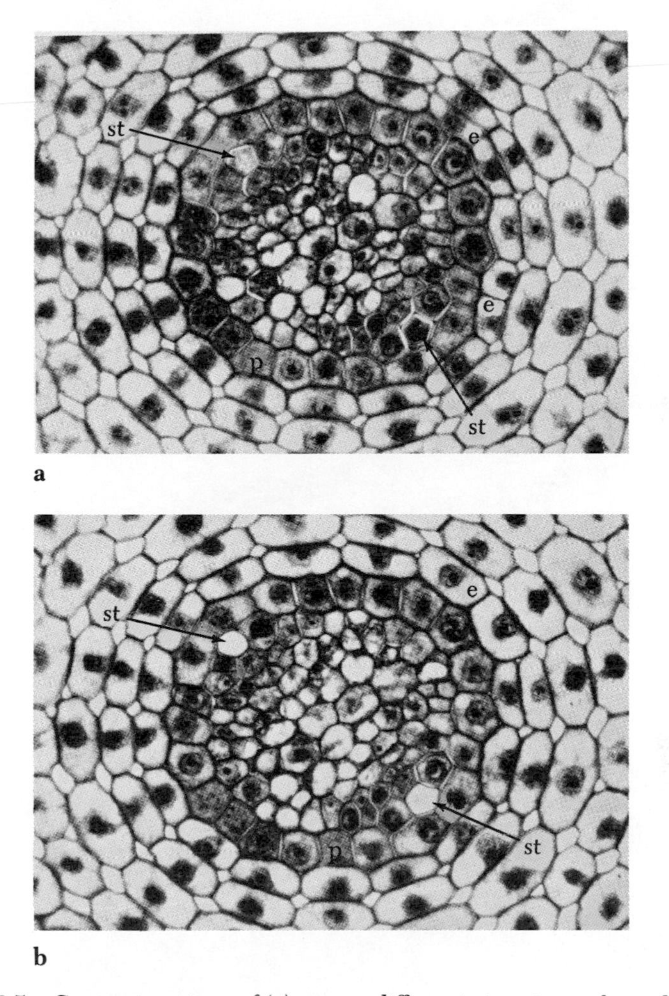

Figure 2.5. Contrasting views of (*a*) young, differentiating sieve tube and (*b*) mature, functioning sieve tube (both in transverse section) of root of *Nicotiana tabacum*, as in Figure 2.4. Both views × 341. *st*, sieve tube; *e*, endodermis; *p*, pericycle. From Esau (1941).

were particulate or reticulate; and in yet others, they were fibrous, a condition that Evert and Derr attributed to preparation methods. They found strands only in those sieve elements in which slime bodies were either dispersing or dispersed. Because they considered that the strands consisted of slime, they reasoned that the large number of internal strands present in mature elements must have resulted from continued slime synthesis, after the initial appearance—and dur-

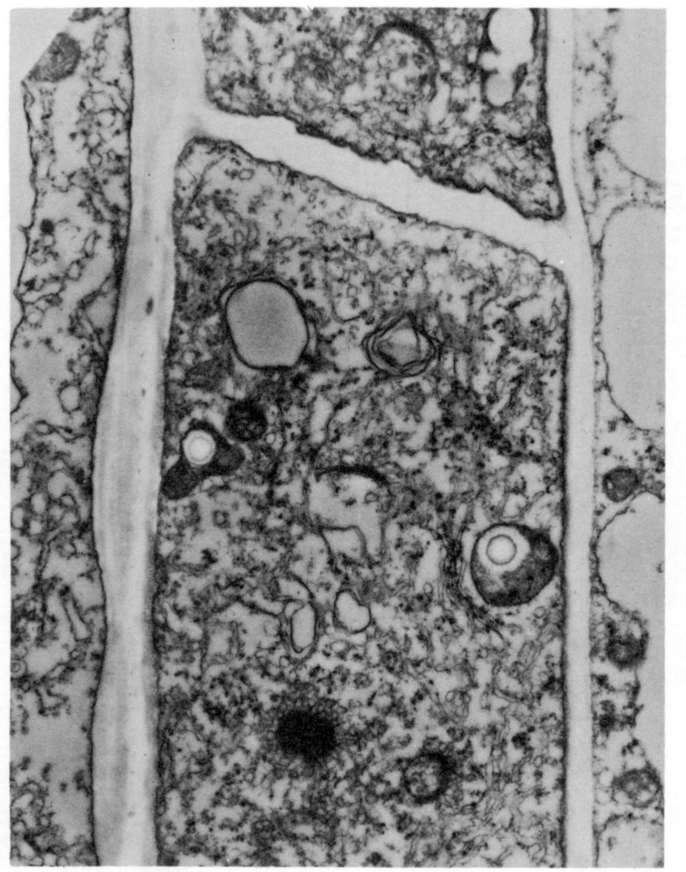

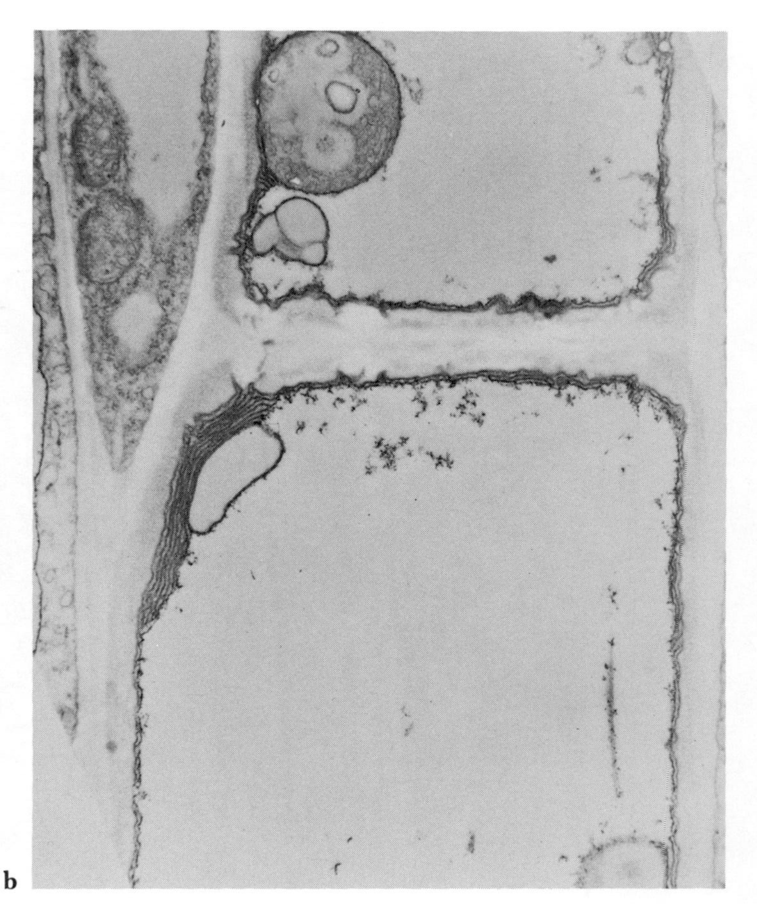

Figure 2.6. Contrasting contents of (*a*) young and (*b*) mature sieve tubes of *Primula obconica*. × 11,620 and × 19,256, respectively. From Tamulevich and Evert (1966).

ing subsequent dispersal—of slime bodies. They found no strands in material treated with potassium permanganate. They concluded that slime is not normally dispersed throughout the sieve-tube lumen, but that it occurs in the form of discrete strands traversing the cell and extending from cell to cell through the sieve pores.

Evert and Murmanis (1965) suggested that the strands play some role in translocation. They denied the existence of endoplasmic reticulum in mature sieve tubes and concluded that many structures found by other workers in sieve-plate pores are actually slime strands. They speculated that enzymes capable of mediating phosphorylation-dephosphorylation processes may be associated with strands, and that movement of assimilates may occur in association with, or along the surfaces of, the strands. Such conclusions fail to give consideration to the phenomenon of phloem exudation, the nature of filamentous sieve-plate plugging, and the quantitative aspects of assimilate movement. The strands that normally occur in sieve tubes of many species must be stationary in normal functioning elements. To the authors of this book, many of the structures labeled slime by Evert appear to be the remnants of slime bodies that are made up of strands of protein, such as those shown in figures 2 and 4 of Cronshaw and Esau (1967). On the other hand, the straight strands stretched between sieve plates—and extending, in some instances, through sieve-plate pores—are parts of the permanent plasmatic filament reticulum that is present in the sieve-tube system of many species, the P2-protein of Cronshaw and Esau's figures 10, 13 and 17.

For excellent descriptions of the development and fate of slime bodies in *Cucurbita*, see Cronshaw and Esau (1968a and 1968b). The dispersed slime (P-protein) from these bodies obviously plays an essential role in the plugging of sieve tubes that results from cutting, heat treatment, or killing by chemicals.

Plasmatic Filaments

The presence, in mature sieve tubes, of an internal system of strands or filaments has been in question for a long time. Crafts (1932) discussed the possible existence of such structures, and pointed out the difficulty of demonstrating or photographing them. This difficulty is now readily understood, in view of the fact that the plasmatic filaments, as viewed under the electron microscope, measure from 90 to 150 Å in diameter, well below the resolving power of the light microscope. Although, under the best optics, a system of linear structures appeared to be present in living or in fixed sieve tubes, they could not be brought into focus in definite form for photography. By using phase optics on some of these old preparations, filaments impregnated with slime have been photographed; these are shown in Figure 2.2.

Bouck and Cronshaw (1965) studied the differentiating sieve-tube elements of stems of *Pisum sativum* under the electron microscope. Of particular interest is their description of the formation, during maturation, of a system of membranes, associated with the parietal cytoplasmic layer, which they designated as "sieve tube reticulum." This system, they said, originates in the cytoplasm and is composed of membranes derived from endoplasmic reticulum and from vesiculation and breakdown of the nuclear envelope. This complex of flattened cisternae is oriented primarily in a longitudinal direction, with its lateral margins attached perpendicularly to the long axis of the cell. Bouck and Cronshaw suggested that the free ends of this reticulum may constitute the transcellular strands of Thaine (1962). Slime is shown in two sieve-tube elements in their figure 11, and the nature and distribution of the contents of the sieve pores in their figure 12 suggest filament plugging—a result of the accelerated sap flow caused by cutting during preparation of the stems for embedding. Because of this sieve-plate plugging, Bouck and Cronshaw favored a metabolic mechanism of assimilate movement.

In a more recent paper, Behnke and Dörr (1967) proposed the term *plasmatic filament* to replace the term *plasmatic fibril* used in Behnke's previous paper. This later paper considers the development and fine structure of the plasmatic filaments. Before the appearance of the filaments, the sieve element of *Dioscorea* is rich in plasmatic constituents; the ground plasm has abundant ribosomes, and the other inclusions common to phloem parenchyma are in evidence; the sieve element can be distinguished by a slight variation in ribosome density; and the plastids early develop osmiophilic inclusions that are unique to sieve cells.

The plasmatic filaments of *Dioscorea* originate as masses of fibrous material (Behnke and Dörr's figure 2) developing in the ground plasm and not confined by membranes. Elements of endoplasmic reticulum and many ribosomes are closely associated with the filaments during all developmental stages. Figure 2.7 shows the filamentous reticulum of Behnke and Dörr.

Figure 2.7. Filamentous reticulum as found in the sieve tubes of *Dioscorea reticulata;* (*a*) × 60,000; (*b*) × 120,000. From Behnke and Dorr (1967).

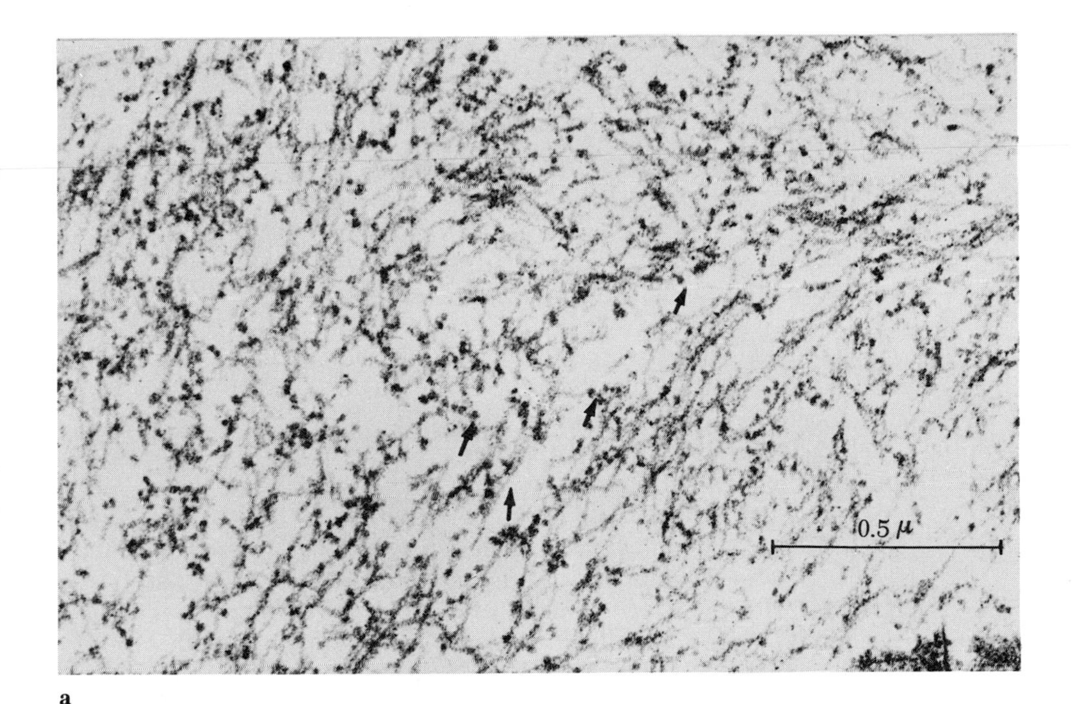

a

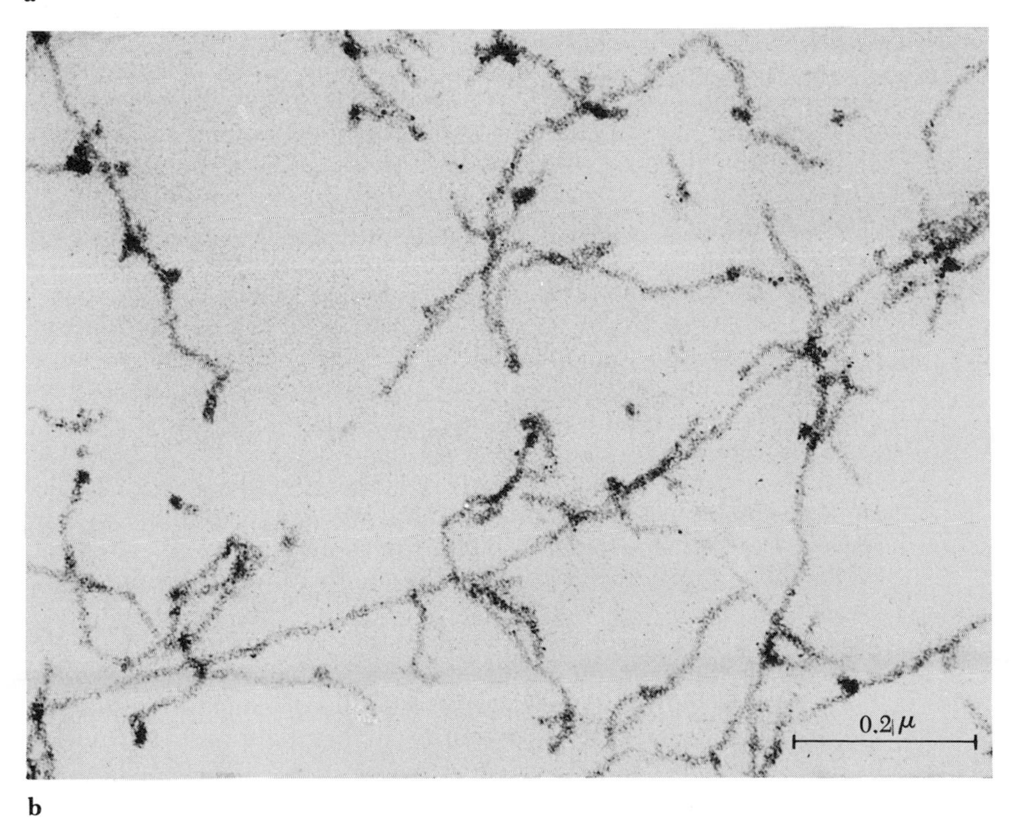

b

Studies on *Dioscorea, Primula, Cuscuta,* and *Cucumis* showed that the filaments in cross section are composed each of an osmiophilic outer ring with a light center, producing a double layered appearance in longitudinal view. An individual filament has an overall diameter of 120–150 Å and an indeterminate length that may be several microns. When the filaments are magnified 120,000 times, it may be seen that they make up a reticulate mesh that occupies the total lumen of the sieve element; Behnke and Dörr showed sieve-plate tubules packed with these strands in a way that has been considered by many to be slime plugging.

Behnke's redefinition of the filaments he found in *Dioscorea* sieve elements focuses attention on the whole subject of slime, slime plugging, and phloem plugging in general. As defined by Crafts (1932, 1961b), slime consists of the breakdown products of slime bodies, nucleus, tonoplast, and, possibly, dictysomes and mitochondria. However, in Behnke and Dörr's illustrations, filaments are shown forming in cells with nuclei (their figure 4), with plastids, mitochondria, dictyosomes, and endoplasmic reticulum (their figure 5), and with tonoplasts (their figure 9). Evidently, plasmatic filaments and slime, as defined above, are different things.

Engleman (1965a) proposed three possible states of aggregation for slime molecules in the sieve tube. His reticular arrangement (his figure 2c) conforms well to Behnke and Dörr's (1967) illustration of plasmatic filaments (their figure 8). Buvat (1963a, 1963b) and Eschrich (1963a, 1963b) have suggested similar arrangements for sieve-tube contents.

Recent phase microscope studies of longitudinal sections of phloem of *Curcurbita pepo* stems have shown that a reticulum is present in mature, functioning sieve tubes of this plant. The filamentous network is most evident in the center of longitudinal sections where slime plugs are lacking; at the ends of the section where slime plugs were present, the reticulum was less evident (Figure 2.2). These observations indicate that, in the phloem of *C. pepo,* some slime may remain lightly bound to the filamentous reticulum after the flow of the assimilate stream has become rapid. This lightly bound slime may be washed free of the reticulum, when the stem is cut, and aggregated on the sieve plates in varying amounts; some of it comes out in the exudate.

Murmanis and Evert (1966) showed an excellent example of the relation of slime to the filamentous reticulum in *Pinus strobus.* Their figures 4–6 show the expansion of the discrete slime body as the sieve element approaches maturity. In their figure 7, the enclosing membrane has broken down and slime particles are seen adhering to the filaments of the reticulum; in their figure 8, the reticulum has contracted to a fibril, similar to that shown in Cronshaw and Esau's (1967) figure 17; the slime is mostly gone (Figure 2.1).

In plants such as *Fraxinus*, *Yucca*, and *Macrocystis*, from which phloem exudation continues for an hour or more, there is probably little slime retained in old sieve tubes. It is possible that callose plugging is responsible for exudate stoppage in these plants. This washing out of slime by exudation may account for the high nitrogen content of the phloem exudate of *Cucurbita* (Colwell, 1942; Crafts and Lorenz, 1944a); it may explain the coagulation of the exudate—rapidly in alcohol, slowly in air; and it may also explain the protein reaction of slime in microchemical tests. When the phloem exudate of *Cucurbita* was hydrolyzed with hydrochloric acid, some 14 amino acids were identified by paper chromatography (Crafts, 1954).

Apparently, the presence of the filamentous reticulum in young sieve tubes is obscured by the presence of a great amount of slime, nuclear material, dictyosomes, mitochondria, ribosomes, and other stainable materials. When these have broken down and left the cell, the reticulum appears as shown in figure 28 of Northcote and Wooding, figures 7 and 8 of Behnke and Dörr (Figure 2.7), and figure 13 of Cronshaw and Esau (1967). The broken appearance of the network results from thin sectioning of the three-demensional structure. Its absence in some preparations may be an artifact of fixation; Evert and Derr state that potassium permanganate destroys strands in sieve tubes. Because the filaments are only around 100 Å in diameter, they are too fine to be seen in the light microscope; in the *Cucurbita* phloem mentioned above, the possible adsorption of slime to the reticulum may well have made it visible at magnification of around 500×. Careful studies will be required to prove the presence of the reticulum under widely differing conditions.

Johnson (1968) found evidence for filaments occurring in the lumina and traversing the sieve-plate pores of *Nymphoides peltata*. Using the conventional method of allowing the leaves to absorb glutaraldehyde fixative through cut petioles, and supplementing this with freeze-etched material, he found sieve plates with open pores. These may be seen in his figures 4 and 5, which also show microfilaments of the filamentous reticulum that resemble those of Figure 2.7. In his figure 8, he shows a sieve pore in longitudinal view with filaments extending through it; and in his figures 16 and 17, he has illustrated plasmatic filaments in lumina and sieve pores of material treated by freezing and etching. This work by critical methods would seem to prove beyond doubt that mature functioning sieve tubes do contain such a reticulum. In *Vitis*, *Tilia*, and other plants whose sieve-plate pores become plugged in autumn and reopened in spring (Esau, 1948; Evert, 1962), the filamentous strands may make up the material that has been termed "slime." If these remain alive through the winter, they could very well contain the enzymes that dissolve the callose and reopen the sieve plates. In *Sabal palmetto* (Parthasarathy and Tomlinson, 1967), sieve tubes function for long periods of time, and slime is lacking; and in the inflorescence of *Arenga pinnata*, which produces phloem exudate for days (Tammes, 1933), callose plugging is probably responsible for stoppage. Palms are limited to climates that do not experience freezing sufficient to solidify the sieve-tube contents. It is possible that their phloem is incapable of sustaining the periods of inactivity that necessarily occur in trees when their bark becomes frozen.

Weatherley and Johnson (1968) gave an excellent discussion of the existence and nature of plasmatic filaments in sieve cells, including an analysis of their possible role in long-distance transport.

Johnson (1969) described crystalline fibrils in mature sieve tubes of *Nymphoides peltata* that resemble the fine fibrils seen by Cronshaw and Esau (1967, figure 17), Tamulevich and Evert (1966, figures 13 and 20), and others. In figures 10 and 11 of his paper, Johnson showed striated filaments associated with the crystalline fibrils. Apparently the crystalline fibrils of Johnson, the fine fibrils of Cronshaw and Esau, the fine strands of Tamulevich and Evert, and the protoplasmic strands of Crafts (1932, figure 27; and 1939b, figure 5) are all aggregates of plasmatic filaments that persist in the mature sieve tubes after the bulk of the slime has been washed away by the assimilate stream. Murmanis and Evert (1966) showed similar structures in *Pinus*. In transverse section, they give the appearance of closely packed tubules. It is difficult to visualize any means by which these structures could serve as "pumps", or could serve to promote or accelerate the longitudinal flow of the assimilate stream in any other way.

The foregoing consideration of the plasmatic filament system of phloem has been recently developed (Crafts, 1968a). It does not conform to the concepts of many electron microscopists. An alternative hypothesis is that the structures which have been identified as plasmatic filaments may actually be fixation artifacts, produced by organization of ordinarily soluble proteins, polypeptides, and amino acids into reticulate structures. This explanation would be more acceptable to many who would derive the filaments from slime (P-protein). In support of this is Eschrich's observation (1963a, 1963b) that a reticulate structure of the type under consideration occurs in phloem exudate from *Cucurbita ficifolia*. It was particularly prominent in sap from the apical part of the stem where slime would have been more plentiful. Left unexplained by this, however, is Figure 2.8 (from Cronshaw and Esau 1967), where the plugging of a mature sieve plate of *Nicotiana tabacum* is shown with no massive deposit of slime. If P2-protein is slime, as Cronshaw and Esau claim, then this phenomenon is a type of slime plugging.

The Nature of Sieve-plate Pores

In the past decade, the greatest advances in our knowledge of phloem have come as a result of studies with the electron microscope. A major breakthrough is the recent accumulation of strong evidence from electron microscopy that the connecting strands traversing the sieve plate are plasmolemma-lined pores capable of

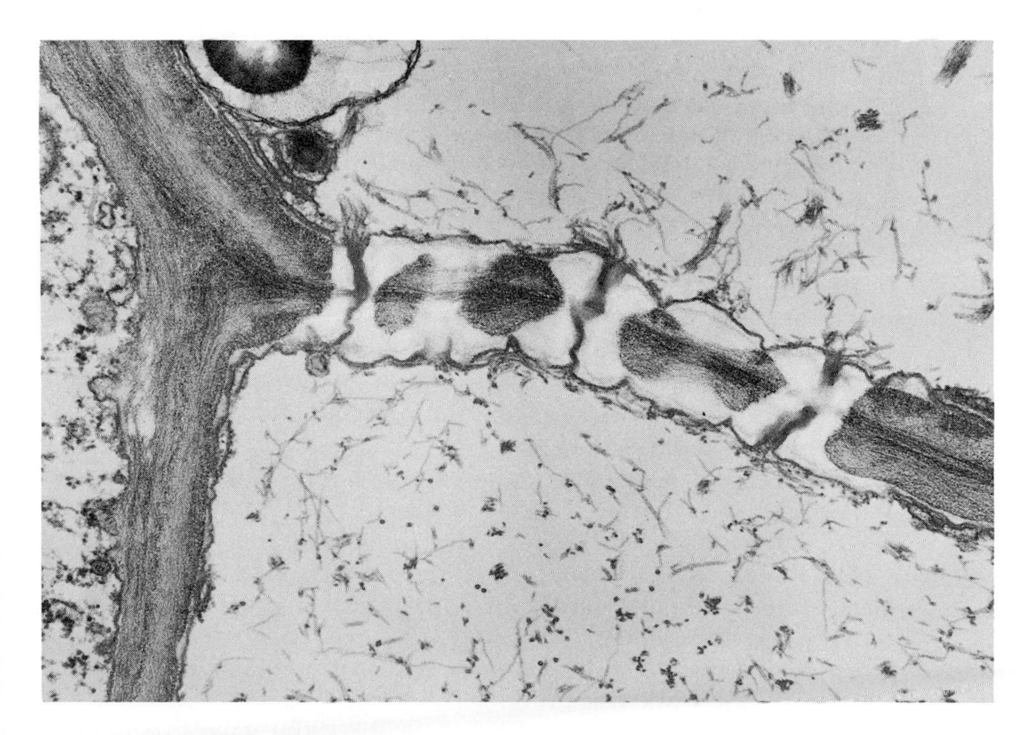

Figure 2.8. Plugged pores of a mature sieve plate of *Nicotiana tabacum*, showing the filamentous reticulum in pores and in cell lumina, × 3813. From Cronshaw and Esau (1967).

accommodating the mass flow so vividly demonstrated by the phenomenon of phloem exudation.

Henry Schneider, in studies of phloem of *Prunus persica* and *P. avium* in 1945, felt certain that the connecting strands of secondary phloem sieve plates are tubular (Schneider, 1945, figure 22). The illustrations of Esau and Cheadle showing sieve plates of four dicotyledonous species appear to show open tubular strands surrounded by callose cylinders (Esau and Cheadle, 1959, figures 10–13). As Esau and Cheadle remarked:

> The role of connecting strands and associated callose in the translocation in sieve elements is a subject of speculation rather than precise knowledge. A common view is that the connecting strands are conduits between sieve elements and that the deposition of callose provides a regulatory mechanism by narrowing the connecting strands or even by blocking the pores completely when the flow through the sieve areas becomes detrimental.

Beer (1959), reporting on electron microscope studies of phloem before the Ninth Botanical Congress in Montreal, stated:

> In mature sieve tubes, the sieve plates appear to be open with no indication of membranes across them. The mature sieve tubes contain no mitochondria, endoplasmic reticulum, or nucleus, but contain some vesicles. The plasmodesmata between the companion cells and the sieve tubes are highly developed.

In stating that cytoplasm penetrates the sieve plates, Beer did not specify whether it does so as tubular connections or as solid strands. Because he stated that the sieve plates appeared to be open, it might be assumed that the sieve plate strands in his preparations were tubular; but because his preparation methods were not described, the question must remain open.

Early electron microscope studies of phloem (Esau and Cheadle, 1961; Engleman, 1963; Hepton et al., 1955; Hepton and Preston, 1960; Kollmann, 1960b; Schumacher and Kollmann, 1959; Ziegler, 1960a; Bouck and Cronshaw, 1965) appeared to show that the sieve-plate strands of dicots are solid with fibrillar cores. Some of the above workers concluded, therefore, that mass flow could not occur across the plates. Most of this work involved permanganate fixation.

More recent work has produced evidence for tubular strands in sieve-plate pores. Eschrich (1963a, 1963b, 1965) gave excellent illustrations of open sieve pores in sieve plates of *Cucurbita ficifolia* and *Macrocystis pyrifera,* and Evert et al. (1966) showed an open sieve plate in *C. maxima.* Esau et al. (1967) illustrated tubular sieve-plate strands in *Beta vulgaris* filled with particles of beet yellows virus. This virus is known to be phloem-limited, similar in its distribution characteristics to the viruses of curly top and aster yellows (Esau et al., 1967). Because these viruses move through the phloem of plants at velocities up to 150 cm per hour (Bennett, 1934), it seems obvious that, as in the case of phloem exudation, there must be conduits structured to accommodate such movement. Demonstration of the existence of truly perforate sieve plates in plants seems finally to have provided an answer to this old question. Esau et al. (1967) also demonstrated the occurrence of virus particles in the plasmodesmata between phloem parenchyma cells, and between companion cells and sieve tubes. These cells surrounding the sieve tubes are the ones that show necrosis, and thus display the first evidence for the presence of virus as it moves into uninfected tissues.

With continued improvement of fixation methods—particularly, the introduction of glutaraldehyde and acrolein—new views showing open, tubular, connecting strands have appeared (Cronshaw, 1969; Cronshaw and Anderson, 1969a; Currier and Shih, 1968; Johnson, 1968; Shih and Currier, 1969; Sjolund, 1968; and Anderson and Cronshaw, 1969, 1970). In two papers, Cronshaw and Anderson (1968b) and Anderson and Cronshaw (1969) illustrated the effects of different fixation methods on the condition of sieve-plate pores. Sieve-pore plugging is evidently a very rapid process in *Nicotiana tabacum.* Both slime (P-protein) and large starch grains are involved, and slime coagulation is avoided only by rapid fixation at low temperatures, or by relieving sieve-tube turgor. The arrangement of protein filaments around sieve pores of sieve plates near a cut end gives strong indications of mass flow of solution immediately after cutting. Anderson and Cronshaw concluded that their work confirms a mass-flow mechanism of the assimilate stream in mature sieve tubes in *N. tabacum.*

Most of the workers mentioned above avoided surging of the phloem sap by injecting the killing fluid, by killing at reduced temperatures, or by using plasmolysing solutions of sucrose to reduce turgor. Thus, by using materials that have not been subjected to any treatment that causes rapid flow of sieve-tube sap, they have been able to show that the sieve-plate strands are normally open tubules occupied by the same solution which fills the lumina of adjacent sieve tube elements.

Mehta and Spanner (1962) made electron microscope studies of the anatomy of the phloem in the petiole of *Nymphoides peltatum,* a species in which Spanner and Prebble (1962) made measurements of ^{137}Cs transport. In spite of the care they took to reduce the turgor of the phloem tissues, Mehta and Spanner found the sieve-plate pores to be filled with a densely staining material, which they interpreted to be cytoplasm (their figures 6, 7, 9 and 10). Their figure 13 appears to show open tubules across a sieve plate. Their figures 4 and 12 appear to show, respectively, partially filled pores and a heavily callosed sieve plate.

Mehta and Spanner felt that their results indicated a flow mechanism, and they suggested the electro-osmotic theory. In view of recent developments, however, it seems more logical to interpret their clogged sieve plates to be the result of plugging, and their single example of open pores as a chance example of the true condition of normal functioning sieve plates. Thus interpreted, the real evidence they present does not controvert the mass-flow concept.

Esau, Cheadle, and Risley (1962) gave a detailed picture of sieve-pore development. They found that the sites of the future pores are at first delimited by small callose platelets that appear in pairs on opposite sides of the sieve plate. These platelets increase in diameter and thickness, and one plasmodesma may be seen in the center of each plate pair extending from one element of the sieve tube to the other. Endoplasmic reticulum adheres to the pore sites and is associated with the callose until the pores are formed.

Perforation occurs in the center of a platelet pair; the middle lamella disappears, and the callose platelets fuse and break through, the fused part forming the pore in the former site of the plasmodesma. Thus, a pore is lined with callose from its inception; callose later spreads laterally to cover the whole sieve plate. Perforation of the sieve plate occurs after the nucleus disintegrates and, in *Cucurbita,* after the slime bodies disaggregate.

In the recent revision of her *Plant Anatomy,* Esau (1965d) stated:

> Electron microscopy based on material prepared with special care to minimize injury . . . indicates that in the sieve plates of dicotyledons with relatively large pores, the contents of the pores resemble those of the cells themselves; that is, they are filled with the mixture of vacuolar sap and disorganized cytoplasmic derivatives delimited by the ectoplast from the pore wall. . . . This kind of structure would be compatible with the concept of mass flow from cell to cell, except that the role of the slime in this system continues to be an enigma.

Electron microscope studies on sieve-cell structure of conifers confirm the previous observation of the great numbers of small-diameter connections across the sloping walls between sieve-cell elements. Kollmann and Schumacher (1962a) studied the phloem of *Metasequoia glyptostroboides* and gave a detailed description of the cytology of the cells. Of particular interest is their description of the sieve pores; these varied from 0.05 μ to 0.5 μ in diameter, and, in their photographs, were very irregular in outline. To see the sieve fields of conifers in their natural, fully hydrated condition, and to compare them with fixed, dehydrated material, is to gain the impression that dehydration greatly shrinks the structures. The connecting strands in the sieve pores in Kollmann and Schumacher's photographs vary in size and appear shrunken. Evert and Alfieri (1965) found the pore size in *Pinus strobus* to be about 0.8 μ and quite uniform; even these could have been shrunken in preparation.

Kollmann and Schumacher showed tubular connecting strands partially filled with filamentous material, which they interpret to be endoplasmic reticulum. Evert and Alfieri (1965) interpret this material to be slime and not endoplasmic reticulum. Because the endoplasmic reticulum in mature sieve cells is—according to Kollmann and Schumacher's later papers (1962b, 1963; Kollmann, 1963)—dilated and vesiculated, presenting a very different appearance from that in young cells, endoplasmic reticulum might not enter sieve pores readily. Slime, on the other hand, gradually disintegrates to molecular dimensions; at an intermediate stage, it might well lodge in the sieve pores of conifers just as it does in those of cucurbits (Murmanis and Evert, 1966). Conifers are known to exhibit phloem exudation and phloem plugging (Crafts, 1939a). Even though Kollmann and Schumacher treated their bark with buffered sugar solution, there is no assurance that this entirely relieved the turgor pressure; and the cutting of the 1-cm bark segments may well have resulted in an aggregation of slime, or of filaments, in the sieve pores. The median nodules in the sieve-area walls give the appearance of being cavities in the wall structure filled with the same stranded material (Kollmann and Schumacher, 1963).

Parker and Philpott (1961) examined the structure of the sieve tubes of *Macrocystis pyrifera* using light and electron microscopes. Their material was cut into pieces and preserved in formaldehyde; it was also treated with potassium permanganate before embedding. From such treatment, it seems quite logical that they should find an accumulation of stranded material on the sieve plates, with filaments penetrating the pores. The sieve plate shown in their figure 1 is very much distorted and shrunken. The enlarged view of a pore shown in figure 1 indicates dimensions of 0.55 $\mu \times 0.15$ μ. The actual dimensions are probably in the range of 1.0 μ to 1.6 μ. Those shown by Esau, Cheadle and Gifford (1953) were 1.2 μ in diameter. Esau and Cheadle (1959) give 1.18 μ as the average sieve pore diameter for 126 dicotyledons.

In his work on the sieve plates of *Macrocystis pyrifera*, Ziegler (1963b) found the sieve pores to be around 1.0 μ in diameter. By scaling the illustrations in Ziegler's paper, the following dimensions were found: figure 11, 1.6 μ diameter; figure 15, 3.1 μ. In the latter, the sieve plate had stood for 6 hours in copper ammonia solution; the callose lining of the sieve pores, however, was not affected; all evidences of a parietal layer, or of the filamentous reticulum, were gone.

Figure 2.9 shows a sieve plate of *Gossypium hirsutum* with open pores in material fixed in glutaraldehyde; Figure 2.10 shows an open sieve plate of *Nicotiana tabacum* from plasmolysed material; Figure 2.11 shows a sieve tube, companion cell, and phloem parenchyma of *Cucurbita maxima;* organelles are abundant in the latter two cell types.

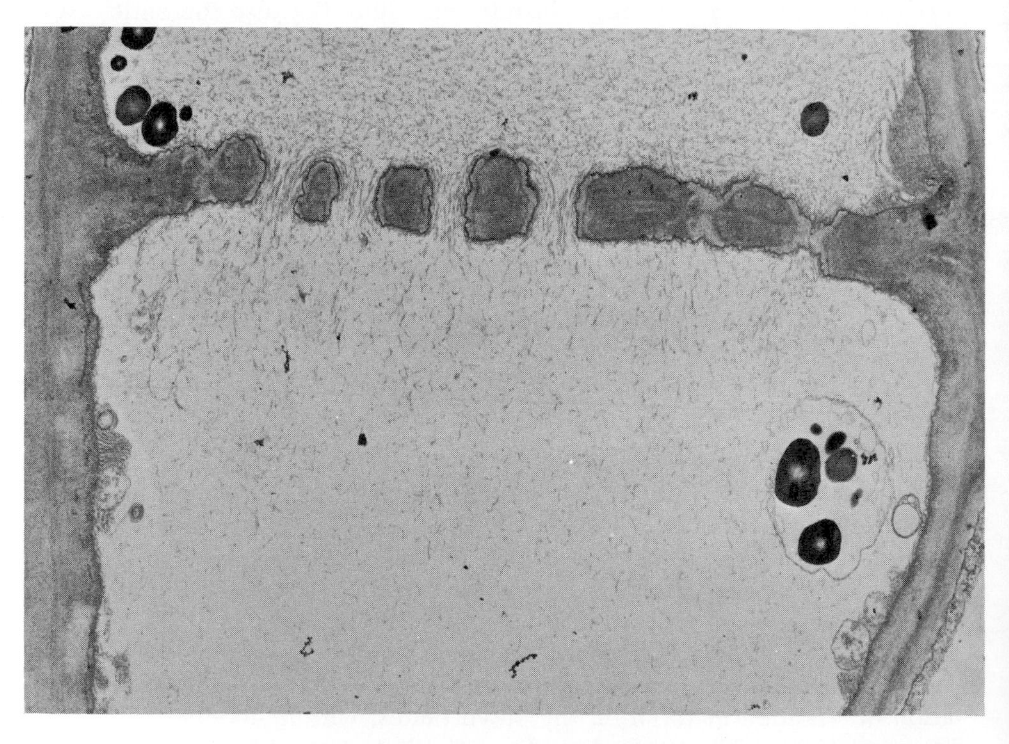

Figure 2.9. Sieve plate of *Gossypium hirsutum* showing open pores, × 6120. Fixed in acrolein-glutaraldehyde mix at 0°C before cutting. From Shih and Currier (1969).

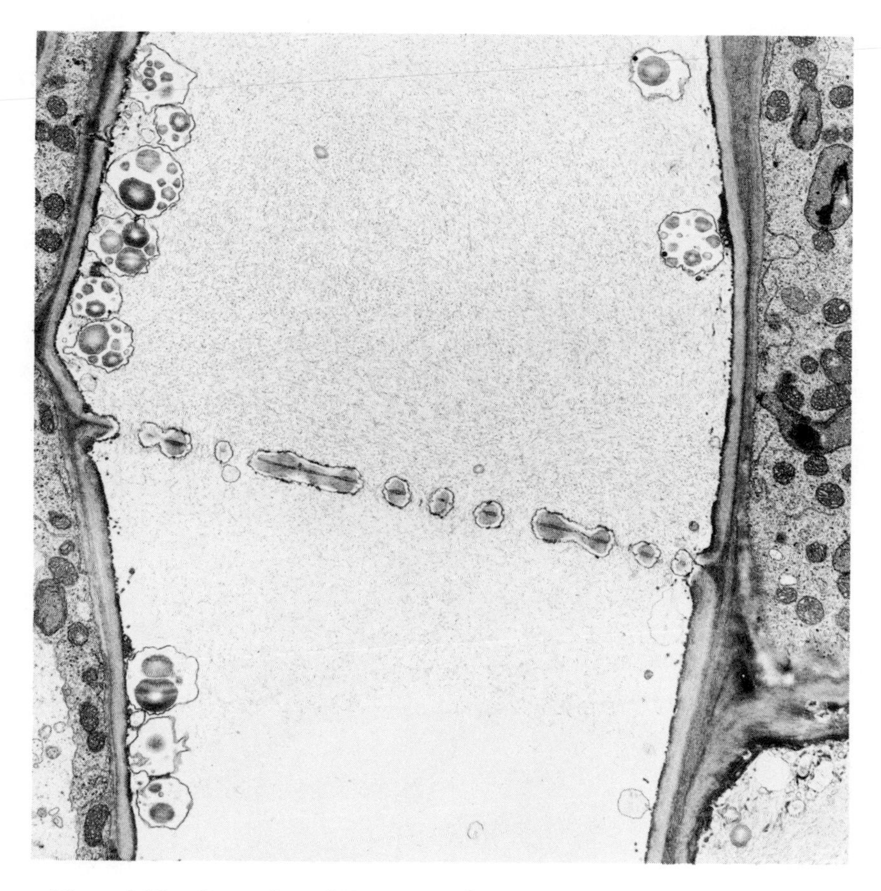

Figure 2.10. Sieve plate of *Nicotiana tabacum* from material plasmolyzed before fixing, × 4400. The pores in this sieve plate are open. From Anderson and Cronshaw (1970).

Special Structures

There are wide variations in the phloem structure of different plants. Various species of *Cucurbita* have been commonly used in the classroom to illustrate phloem structure. The phloem of these plants, however, has many features that make it unique: slime is very abundant, slime plugs are readily demonstrated, the sieve plates have uncommonly wide pores, and the plasmatic features of the cells are readily stained. Further, the phloem of *Cucurbita* spp. differs from that of most other herbaceous plants in occupying the parenchyma tissue of the cortex and pith as strands of ectocyclic, entocyclic, and

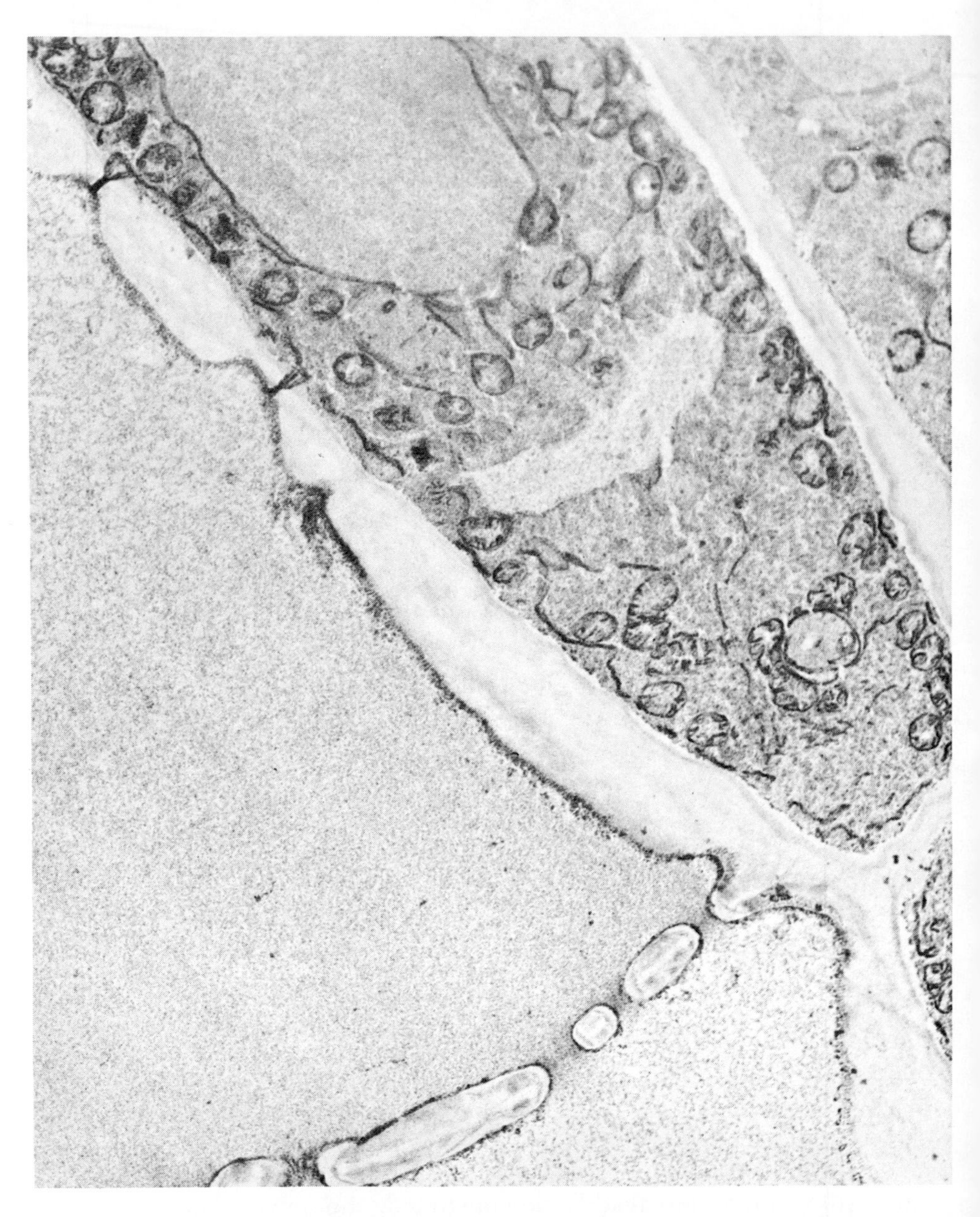

Figure 2.11. Sieve tube, companion cell, and phloem parenchyma cell of *Cucurbita maxima*, × 6500. Organelles in the companion cell are abundant; in the sieve-tube element they are lacking. Electron micrograph by K. Esau.

commissural phloem (Crafts, 1932). These extrafascicular sieve tubes differ from the more normal ones of the bicollateral bundles in having denser contents that are readily coagulated by fixing reagents.

In some leguminous species (in *Robinia* spp., for example) slime bodies do not disperse at sieve-tube maturation (Esau, 1965b). These discrete bodies may be seen suspended in the lumina of mature sieve tubes (Esau, 1965b, plate 43 B; Derr and Evert, 1967, figure 10; Evert and Derr, 1964a, figures 3 and 4). In these plants, the slime body appears to form a slime plug in sectioned material.

Another special phloem structure, the phloem glomerulus, is found in *Dioscorea* spp. Since the work of Mason in 1926, it has been assumed that the bast glomeruli of *Dioscorea* spp. constitute barriers to phloem transport of assimilates. Mason said, "It will be evident that the presence of glomeruli, forming a barrier at every node between the sieve-tubes, must present an insuperable obstacle to the rapid movement of carbohydrates longitudinally in the phloem." The presence of these structures in *Dioscorea* has been cited by several research workers as evidence against a mass-flow mechanism.

Behnke (1965a) studied the nodal anatomy of those species of *Dioscorea* in which glomeruli occur. The nodal ganglia that contain the glomeruli combine some vascular bundles of the two adjacent internodes and the bundles of the axillary bud with those of a leaf of a separate node. Both the xylem and the phloem elements of these bundles are interrupted by networks of short cells. There are distinct parts of the node by which sieve tubes communicate. The phloem network of each nodal ganglion divides into many subunits, the glomeruli. In a glomerulus, a few connecting sieve tubes, and a great number of nodal sieve tubes, unite a single internodal sieve tube having a possible diameter of 100 μ with one hundred or more narrow nodal sieve tubes, each having a diameter of 5–10 μ. Other groups of nodal sieve elements, nodal sieve tubes, and connecting sieve tubes lead to the sieve tube of the next internode. All of these phloem elements of both the node and internode are short sieve-tube-like elements, devoid of nuclei, and containing many small characteristically staining granules, possibly plastids. They are linked by sieve plates having pores (and strands) of varying diameter grading down to connections of the size of plasmodesmata (0.3–0.1 μ).

Every sieve tube enters a glomerulus at least once in the course of three subsequent nodes; nodal parenchyma cells with nuclei are found alternating between the sieve elements of a glomerulus. Phloem anastomoses that combine the phloem parts of two adjoining bundles in the internode show a similar arrangement of small sieve elements like those in a glomerulus.

In a second paper, Behnke (1965b) described the fine structure of *Dioscorea* phloem. The description of the nodal ganglion as seen under the light microscope (Behnke, 1965a) proved to be accurate. Nodal parenchyma cells with nuclei could be readily distinguished from the great numbers of conducting phloem elements of the node; these latter were identified as connecting sieve tubes, nodal sieve tubes, and nodal sieve elements. The cytoplasmic structure of all sieve-tube members and sieve elements in the nodes and internodes proved to be uniform in properties. Each of these cells contained plastids, numerous mitochondria, and an unknown latticelike body, as well as vesicular or cisternal components of the endoplasmic reticulum, and a well developed network of cytoplasmic filaments.

The network of cytoplasmic filaments of each cell extends from the parietal protoplasm throughout the entire cell lumen, including the space occupied by the former vacuole. Each fibril is roughly 100 Å in diameter. Tonoplasts are lacking, and the fibrils run parallel to the long axis of the cells. According to Behnke, these fibrils, accompanied by tubular components of the endoplasmic reticulum, pass through the sieve pores. This, however, could well be an artifact resulting from the preparation method, which included only treatment with sucrose, followed by sectioning into pieces 2–3 cm long, a treatment that by no means prevents exudation from the ends with attendant sieve-plate plugging. Although the sieve-plate strands of these small elements had diameters as low as 0.1 μ, there was protoplasmic continuity from the sieve tubes of one internode, across the connecting strands of the different sieve elements in the nodal ganglion, to the sieve tubes of the next internode: in short, a symplastic continuum. The fact that the cells in the outer layers of the glomeruli had sieve plates, and that the inner elements were connected by sieve-plate-like connections, led Behnke to state that the concept of glomeruli as barriers could no longer be defended; he suggested a new concept of the cells as coupling sieve tubes, nodal sieve tubes, and nodal sieve elements—all of them conducting cells, in contrast with the nucleate parenchyma. Thus, in the phloem anastomoses and in the nodal ganglion, connection by numerous short, narrow elements between two or more large internodal sieve tubes seems likely; evidently, the small sieve elements of the glomeruli, while offering more resistance to flow than the large internodal sieve cells, do not prevent translocation.

Behnke discussed the relations of the above structures to the function of phloem transport. He stressed the significance of the plasma filaments to translocation. Mason had proposed that the glomeruli have a secretory function based on a possible high enzyme content. The more recent demonstration of a high phosphatase activity in glomeruli (Braun and Sauter, 1964), if taken as support for the above theory, must be viewed in relation to the evidence from electron microscopy. The suggestion that this phosphatase activity may be limited to the nucleate phloem parenchyma cells of the glomeruli, which resemble companion cells in being rich in enzymes, cannot be accepted, because the precipitate of lead sulfide was distributed generally over the whole glomerulus phloem. Because the phosphatase must occur in all cells, it should be determined to what structures the phosphatase is bound. For the precipitate in the internodal sieve tubes, Braun and Sauter (1964) pointed out a relation to the plasma filaments, which, from work with the light microscope, they designated as slime. From the distribution of the plasma filaments found, it seems possible that the phosphatase may occur over the whole cell lumen because of its binding to these structures. Whether or not the accumulation of precipitate on the sieve plates signifies an increased enzyme activity at these locations remains, in Behnke's view, an open question; it is possible that the concentration of the enzyme substrate might be an artifact of the preparation.

Behnke questioned the assumption that the particular structure of the phloem glomeruli relates to an active assimilate transport system, but he concluded that, on the basis of the described structural relation between nodal and internodal conduits, a plasma-controlled assimilate transport in the total phloem of *Dioscorea* is a possibility. A barrier for a possible directed stream flow may be assumed, presumably, in the narrow connections between the nodal sieve elements. That the shortening of the conducting elements and the attendant sieve-plate-like wall differentiation might imply an active transport in the sense intended by Spanner (1958), Behnke considered questionable.

If a mass flow is controlled by the plasma content of the sieve tubes, which the narrowing of the sieve plates with their thick plasma-pore content would seem to indicate, then the plasmodesma-like connections of the nodal sieve elements must constitute a resistance. At least it is difficult to imagine how a renewed switch to a plasma transport in each node could exist. This consideration remains if one assumes, for all sieve elements, a uniform metabolic assimilate transport. Behnke reasoned that such transport is possible in angiosperm sieve tubes, as indicated by the anatomical findings of Kollmann (1960a, 1960b) and Buvat (1963a, 1963b), by studies on the respiration dependancy of transport, by analyses of the transported substances, and by the information on enzyme localization and metabolic products of sieve tubes presented by Kursanov (1963).

Behnke was concerned, as had been many before him, by the possible resistance of the fine sieve-plate strands of the glomerulus sieve elements. These, in reality, should not be less permeable than the sieve areas of conifers and ferns, in both of which phloem exudation has been found.

Lawton (1966) showed by normal illumination, UV illumination, and polarized tungsten illumination, the relation between callose–analine blue fluorescence of sieve fields, and birefringence of the cellulose thickenings of sieve plates, of *Dioscorea alata.* In his figure 3, he shows the transverse walls of protophloem elements near the shoot tip. He concluded that callose is a major and universal component of sieve plates from a very early stage, and is not confined to aging sieve plates. Citing his own doctoral thesis at the University of London, Lawton stated that callose around the pores of the sieve plates and on the walls of the phloem-complex cells does not impede the translocation of photosynthates in the Dioscoreaceae.

Gunning, Pate, and Briarty (1968) found a highly specialized form of parenchyma cell, the so-called "transfer cell," in the perivascular tissues of angiosperm leaves. Within these cells occur more or less extended systems of protuberances, made up of loose microfibrils and deposited secondarily on the primary walls. These structures, being covered with plasmolemma, greatly amplify the surface of the protoplast increasing the surface to volume ratio by as much as tenfold. Functionally, they facilitate the exchange processes of the xylem and phloem conductors.

3

The Path of Translocation

The Source

Early work on phloem transport was largely concerned with the mechanism responsible for solute movement. With the researches of Mason and Maskell (1928a, 1928b), however, it became obvious that, regardless of mechanism, movement of assimilates takes place from green, assimilating tissues to non-green, growing and actively respiring tissues. In short, movement follows a source-to-sink pattern. Münch (1930) presented a simple diagram to show how an osmotic system comprised of a source and a sink might drive a mass flow along a turgor gradient in the sieve tubes. Figure 3.1 presents this scheme.

The nature and function of sources, sinks, and their interrelations will be considered in greater detail in Chapter 9. It suffices here to explain that, whereas the green leaves of plants constitute their major source of organic foods, and are hence the major activating organs for translocation, it has been shown that the storage tissues of the phloem and cortex may serve temporarily as sources (Weatherley et al., 1959), and that, under certain circumstances, a single leaf may serve as a source and a sink at the same time (Jones et al., 1959; Peterson, 1968).

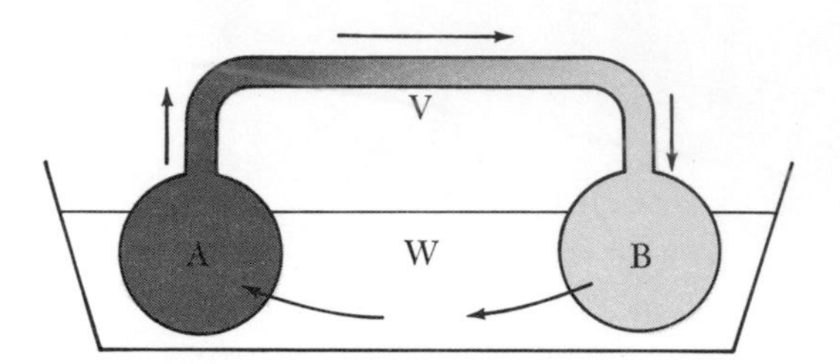

Figure 3.1. The basic mechanism of pressure- or mass-flow. *A* and *B*, osmotic cells; *A* with contents of greater concentration than those of *B*. *V* is a tube that connects the two cells, and *W* is a trough containing water. In the plant, *A* would be the source—the leaves, for example; *B* represents the sinks, and *V* represents the phloem conduits (sieve tubes). From Münch (1930).

The Sink

While it is obvious that photosynthetic organs and storage tissues are the principal sources in plants, a great variety of tissues and tissue systems may act as sinks. All non-green tissues, and those not containing sufficient chlorophyll to meet their own food requirements, are sinks for the import of assimilates. These include apical and vascular meristems; all root tissues; storage organs, such as fruits, seeds, tubers, corms, rhizomes, and so forth; and finally, all parenchyma cells of phloem, xylem, cortex, and pith that constitute living, respiring tissues. The absorbing activity of various sink tissues varies considerably. It is low in pith cells, which need only be maintained on a subsistence level, and quite high in meristems, fruits, and rapidly growing storage organs, which require large amounts of food for their maintainance and growth. Fruits of *Gossypium* and *Curcurbita*, and the tubers of *Solanum tuberosum*, have been favorite organs for studies of velocity of assimilate movement.

Source–Sink Relations

The simple mechanism shown in Figure 3.1 illustrates the prevalent situation in young growing plants where foliage constitutes the major source, and the roots the major sink. To this, however, there are many exceptions. In many young seedlings, for example, cotyledons gorged with starch, protein, or fat may serve as the major source until the leaves expand and take over.

A more complex example is the stem of *Salix,* which Weatherley et al. (1959) showed would produce fairly normal flow of stylet exudate even after the leaves were removed, the phloem parenchyma providing assimilates from stored reserves. A stem length of 16 cm or more was found necessary to maintain the flow of exudate, and the volume dropped off after a few days.

A much more complex situation obtains in a large tree, where every shoot tip with its apical meristem is an active sink, buds in the axils of all leaves are potential growth centers, the vascular cambium represents an immense importing area, the parenchyma of bark, phloem and xylem are active storage tissues, fruits and seeds are organs of accumulation, and the entire root system is an active sink. Source–sink relations of the foliar portion of a tree are exceedingly complex, with each leaf a sink, until compensation, and then a source; as mentioned, some leaves may import and export foods simultaneously. Fruits may serve as sources while young; during their major growth period and on to maturity they are active sinks, drawing on adjacent leaves for assimilates. Apical meristems in shoot tips require great amounts of food during the early growth stages. Later, growth slows and stops as apical buds are formed; these may store starch, proteins, and fats. Compared with all of this complexity of activity in the crown of a tree, translocation of foods below the lowest branch is simply basipetal into trunk and roots. Within the root system, growth initially dominates in activity in the spring; later, storage takes over, and there is competition for foods; soil moisture, temperature, day length, and a host of environmental factors affect the relations between these two processes.

Evidence for Sieve Tubes as Translocation Channels

With the naming of the sieve tube by Hartig in 1837, botanists generally assumed that sieve tubes were the phloem conduits through which assimilates moved rapidly. In their work on *Gossypium hirsutum,* Mason and Maskell (1928a, 1928b) took the position that sugars moved rapidly in sieve tubes. In his book on translocation, Münch (1930) proposed that assimilates moved through sieve tubes by mass flow of solution along a gradient of turgor pressure.

Despite this general agreement among plant physiologists and anatomists that the sieve tubes served as the conduits for rapid transport of foods in the phloem, Mason and Phillis (1937) credit Schumacher (1930) with providing the first experimental demonstration that the sieve tube is the actual channel of transport of both carbohydrates and nitrogen. Schumacher showed that when a leaf of *Pelargonium zonale* is treated with a dilute eosin-Y solution, the sieve pores in petiole and stem are rapidly closed by callose, while protoplasmic streaming in the phloem parenchyma remains unaffected. Under these conditions, export of carbohydrate and nitrogen from the leaf is stopped. Later, Schumacher (1933) reported that fluorescein-K spreads along the phloem in the sieve tubes.

Repeated trials by histoautoradiography (Colwell, 1942; Biddulph, 1956; Eschrich, 1966; Trip and Gorham, 1968a) have indicated that radioisotopes move in the phloem, but they have provided no proof that the sieve tubes constitute the actual conduits. In this sort of test, it is impossible to distinguish between mobile solutes and absorbed or accumulated materials; in fact, when one considers the nature of the procedures required to produce an autoradiograph, it seems that mobile solutes would be largely lost. Thus, the presence of a given tracer at a point removed from its place of application is the only evidence of translocation; its concentration is better evidence for absorption, incorporation, or accumulation than it is for translocation.

In order to identify the transport conduits responsible for movement of assimilates to and from the fruits of *Phaseolus vulgaris*, Bachofen and Wanner (1962a) prepared histoautoradiographs of the peduncle of this organ, the subtending leaf of which was treated with $^{14}CO_2$. In both ringed and unringed peduncles they found an accumulation of silver grains over the xylem, with the highest concentration over the young elements. From this, they concluded that import of assimilates by *P. vulgaris* fruits takes place via xylem, and not via phloem. When they applied $^{14}CO_2$ to the fruits, they found that the concentration of silver grains was localized over the cambium and the young phloem.

In the case of fruit treatment, the location of activity again probably indicated symplastic movement in parenchyma cells and the concentration in cambium and phloem was related to the source–sink condition of the tissues. In the first (ringed) experiment, the xylem constituted the source; in the latter experiment, the inherent nature of the assimilation process channeled the sugars into the phloem, and there, rapid streaming in phloem parenchyma moved the tracer throughout the fruit, and presumably into the peduncle. From the very nature of the distribution function of vascular tissues, and from the weaknesses of histoautoradiography, it seems unlikely that Bachofen and Wanner's interpretations are valid.

More recent attempts at histoautoradiographic identification of phloem conduits have been somewhat more effective. Bieleski (1966a) presented photographs of sections and overlying silver grains showing accumulation of $^{32}PO_4^{---}$ and $^{35}SO_4^{--}$ in both excised and translocating bundles of *Apium graveolens* and in excised bundles of *Malus sylvestris.* In excised bundles of both plants, massed silver grains were formed over recognizable sieve tubes. Although Bieleski was more concerned with accumulation processes than with translocation — and of course, translocation was not going on in his excised phloem tissues — his illustrations show the presence of high concentrations of tracer in sieve tubes.

Confused by the conflicting views of translocation mechanism expressed in the literature, Bieleski concluded that possibly all cells in the phloem, because of similar accumulative capacity, may contribute significantly to the transport of assimilates in plants.

Othlinghaus, Schmitz and Willenbrink (1968) treated leaves of *Phaseolus vulgaris* with $^{14}CO_2$ in light, and studied translocation by autoradiography. They found ^{14}C in the phloem of fruit stalks exclusively; the xylem was always free of ^{14}C.

Even more convincing is work done by Schmitz and Willenbrink (1967), in which $^{14}CO_2$ was fed to leaves of *Cucumis sativus* and *Lycopersicon esculentum.* One to three hours after such exposure, sections of petioles were cut into a petroleum ether–dry ice mixture at −75°C. Assimilate in the frozen sections was counted and chromatographed; freeze-dried sections were autoradiographed. In both species, silver grains were found aggregated above recognizable sieve tubes. To an observer unpersuaded by the evidence from phloem anatomy and phloem exudation studies of the conducting role of functioning sieve tubes, these illustrations by Schmitz and Willenbrink should provide the convincing evidence.

Furthermore, Heyser, Eschrich, and Evert (1969), using gross autoradiography, histoautoradiography, and electron microscopy, provided evidence that it is the phloem tissue that translocates ^{14}C-labeled phenylalanine in *Tradescantia albiflora.* The histoautoradiographs clearly show a close association of the silver grains with the mature sieve tubes.

Other evidence has long indicated that the sieve tubes are the actual conduits of long-distance transport in phloem. In his early work on phloem exudation, Crafts (1936, 1938) noted that if he repeatedly wiped the end of a cut *Cucurbita maxima* stem until the velocity of flow was reduced, he could see, through a binocular microscrope, that small droplets of sap emerged from the individual sieve tubes.

In their work with curly top virus Esau (1933, 1935) and Bennett (1935, 1937, 1940a) found very convincing evidence for movement of the curly top virus in sieve tubes. Bennett found high concentrations of virus in phloem exudate from *Beta vulgaris* (1934) and Esau noted the first symptoms of curly top in phloem parenchyma immediately adjacent to young, newly differentiated sieve tubes in leaves of *B. vulgaris*.

The Age of Functional Sieve Tubes

A reading of the old literature on phloem anatomy reveals that most anatomists assumed that sieve tubes become functional at maturity; that is, when the nucleus and tonoplast disappear, the sieve-plate strands enlarge, and the stainable contents of the lumina become sparse. Schumacher, on the other hand, maintained from his work with eosine Y (1930) and fluorescein-K (1933) that the young sieve tubes are the ones that function in translocation. This conviction has been maintained through his more recent studies with Kollmann on *Metasequoia glyptostroboides* (Kollmann and Schumacher, 1964). It has also been expressed by Wark and Chambers (1965) as a result of their electron microscope work on the phloem of *Pisum sativum*.

Kollmann (1965), using $^{14}CO_2$ as a source of labeled assimilates and employing cryostat sections for microautoradiography, found that the translocation of sugars in *Metasequoia glyptostroboides* branches in October was localized in the sieve cells associated with the inner band of bast fibers. The larch lachnid *Cinara laricicola* was also found to feed upon this inner layer of sieve cells. From studies on the function of conifer phloem it is probable that this layer of sieve cells contains the only functioning elements at that time of year; all older sieve cells are probably nearing obliteration by then; but this inner layer probably functions throughout the winter (Abbe and Crafts, 1939; Alfieri and Evert, 1968).

Kollmann and Dörr (1966) made further studies on the localization of translocation in the phloem of conifers. They used mouth parts of the juniper aphid *Cupressobium juniperi* to obtain phloem exudate from *Juniperus communis* at two different times of year—March and April for greenhouse plants, and August for plants growing outdoors. Exudation was sporadic from the greenhouse plants, although honeydew excretion by intact aphids was copius. In August, the aphids fed on young green shoots, and the exudation was increased over the spring collections. Newly positioned aphids pierced the outermost cells of the metaphloem just inside the crushed protophloem. Insects that had spent time on older branches fed upon the peripheral regions of the functioning secondary phloem. For conifers, Kollmann and Dörr found the rate of sap flow through aphid mouthparts to be slower than that reported for *Salix;* rate values were around one-tenth those for *Salix,* a fact which they explained on the basis of differences in phloem structure.

In further studies on translocation of assimilates in the phloem of *Metasequoia glyptostroboides,* Willenbrink and Kollmann (1966) used 2-year-old branches of intact plants that had been exposed to $^{14}CO_2$. The rate of translocation of labeled assimilates out of the needles was monitored by a Geiger-Müller counting system attached to a fixed position on the branch. Velocities of 48–60 cm per hour were found by counting at two distinct points on the conduit system.

By sectioning, microautoradiography, and chromatography of extracts, it was determined that ^{14}C-labeled sucrose was the main translocated assimilate; that translocation was taking place in the secondary phloem; and that the label was localized in the young sieve cells adjacent to the cambium. Starch labeled with ^{14}C was not stored in these branches in October. This work complements that of Kollmann (1965), and indicates that there is little difference between the sieve cells of conifers and the sieve tubes of angiosperms, as far as translocation rates and the composition of translocated assimilates are concerned. As stated above, the only functional sieve cells in October are the last-formed ones, all older ones having been obliterated.

Kollmann (1967), in continued studies on the translocation of assimilates in *Metasequoia glyptostroboides,* made autoradiographs of sections of 2-year-old branches sectioned in the frozen condition and exposed to stripping film in a cryostat at −25°C. This method excludes any movement of water soluble assimilates during exposure, and should give a critical view of the localization of the conduits within the stem. Radioactivity was found only within the region of the two youngest tangential bands of sieve cells between the cambium and the first band of bast fibers. These are the cells that Kollmann and Dörr (1966) found to be the source of phloem sap to aphids feeding on the bark. Although, as Kollmann stated, these cells are characterized by highly organized cytoplasmic fine structure, they must contain open intercellular pores capable of carrying the assimilate stream at velocities up to 48–60 cm per hour, or they could not have produced the phloem exudate noted by Willenbrink and Kollmann

(1966). Apparently, the endoplasmic reticulum, dictyosomes, mitochondria, plastids, and other cytoplasmic structures assume a parietal position and do not interfere with longitudinal transport in functioning sieve tubes. Were they free to move readily with the assimilate stream, they would block the pores in the sieve fields and prevent phloem exudation (see Wooding, 1966, figures 3, 4, 5, and 7).

In his work on conifers with the mouthparts of aphids, Kollmann has stressed the fact that these insects feed only on the younger sieve cells adjacent to the cambium. As stated above, these are probably the functional elements involved in the conduction of the assimilate stream. Evert et al. (1968), working on *Tilia cordata,* found that the giant bark aphid *Longistigma caryae* may feed not only upon the current season's sieve tubes, but upon the previous season's, as well.

The age of functional sieve tubes in the perennial monocotyledon *Tradescantia albiflora* has been studied by Heyser, Eschrich, and Evert (1969). They found the subcellular anatomy of the sieve tubes of the primary phloem to be classic (that is, the sieve tubes possess plasmalemma, remnants of cytoplasmic membranes, small mitochondria, and plastids, although nuclei are lacking). They found these sieve tubes to be functional for as long as the plant parts in which they occurred remained alive. The plants they used were 15 to 26 months of age.

Alfieri and Evert (1968) studied the seasonal development of the secondary phloem in three species of *Pinus.* They found that there are functional sieve cells present at all times of the year. For a given year's growth increment, all but the last-formed two to four layers of sieve cells cease functioning at the end of the growing season. The last-formed cells, by contrast, overwinter and remain functional until new sieve cells differentiate in the spring. They found all early phloem to be produced by May, before the xylem differentiation starts. Obliteration of sieve cells starts with the over-wintered elements by June, and progresses slowly to sieve cells of the current seasons growth; by mid-December they found all but the last-formed sieve cells to be devoid of contents. (See also Tucker and Evert, 1969.)

These studies on the ages of functioning sieve cells emphasize two facts. The first is that the functioning phloem in trees, and particularly in conifers, is made up of a very narrow layer of cells, at times only a few cells in width. Huber (1942) stressed this in his work on phloem of forest trees. The second fact is that functioning sieve cells are mature elements. Only upon maturity are the physical properties of these cells such that rapid translocation can take place through them.

Before maturation, the most rapid transport that sieve cells can accommodate is that brought about by protoplasmic streaming, a process that proceeds at velocities not above 6 cm per hour in the phloem of higher plants. For a detailed discussion of this subject, see Weatherley and Johnson (1968).

4
Phloem Plugging

The Phenomenon of Plugging

The rapid cessation of phloem exudation in a plant such as a cucurbit, followed by renewed flow after removal of a thin slice of tissue, indicates that there is a mechanism in these plants that plugs the sieve tubes near the cut end of the stem. Although this plugging has been attributed, variously, to slime (Rouschal, 1941) and to callose formation (Currier, 1957; McNairn, 1967), it seems evident from recent work with the electron microscope that there are at least four processes involved; (1) a partial collapse of the sieve plate resulting from reduction of pressure upon cutting, a mechanism suggested by Curtis (1935) and Münch (1930) to explain the small dimension of sieve-plate pores in sectioned material; (2) an accumulation of slime on the sieve plates, a phenomenon very prominent in certain species of *Cucurbita*, *Solanum* and many other genera; (3) a rapid increase in callose that constricts the pore size (see Figure 2.8), suggested by Currier (1957) and Northcote and Wooding (1966); and (4) an agglutination of the filaments or slime in the sieve-plate pores attended by hydration and swelling, triggered by the shock of cutting (Crafts, 1968a). The first of these processes could hardly effect a complete blocking of the pores of active, functioning sieve tubes; but the hydraulic shock involved might serve as a trigger to initiate callose deposition and agglutination of filaments or slime in the sieve-plate pores.

Slime plugging is common particularly of the sieve plates of young sieve tubes of cucurbits, *Solanum tuberosum*, and *Nicotiana tabacum*. Callose plugging has been shown by Currier (1957) and Eschrich (1965) to be very rapid. If callose plugging actually can occur in seconds, it could well account for the plugging phenomenon of cucurbit stems. However, it would have to be limited to the sieve plates close to the cut, because removal of a slice of stem 1 mm in thickness causes exudation to resume.

The sol–gel conversion visualized by Northcote and Wooding, and exemplified by the coagulation of cucurbit sieve-tube sap, could be an additional mechanism. This type of reaction is well illustrated in Figure 2.8, where the filaments inside the sieve pore seem to have lost their discrete structure, whereas, at the ends, they fray out into the lumen. The lack of a typical slime plug characteristic of normal slime plugging, as in Figures 2.2 and 2.3, is obvious.

Possibly, all of these phenomena combine to produce the final effect of cutting or injury. The speed of the reaction might depend upon the maturity of the sieve tubes, the relative space occupied by the filaments in the lumina of the sieve-plate pores, and the capacity of the filaments for swelling. As mentioned previously, exudation in some species stops within a minute or so, although in others it may continue for an hour or more.

It seems evident that the greatly accelerated exudation of sap from a cut stem does not tear the filamentous meshwork free from the parietal layer of cytoplasm, for there is no piling up of the filamentous material against the sieve plates comparable with the accumulations of slime that occur in younger elements. At times, slime plugs may be found with strands extending through and fixed to the sieve plate at the opposite end of the cell. Examples have been figured by Crafts (1932, figures 27 and 28; 1933, figures 13, 16, and 17; 1939b, figure 5) and Evert and Derr (1964b, figures 1, 2, 6–8, 11, and 12).

Callose Plugging

Callose has long been recognized as a sieve-plate constituent, occurring in the form of collars or tubules around the cytoplasmic strands, and capable of controlling movement along the sieve tubes by constricting the size of the tubular strands. There is some question as to the universal occurrence of sieve-plate callose in normal functioning sieve tubes (Zimmermann, 1960a; Esau, 1961; Ziegler, 1963b; Eschrich, 1963a; Evert and Derr, 1964a). Esau et al. (1962) showed callose on sieve plates in very early stages of growth, and heavy platelets of callose at the time of sieve-plate perforation. However, callose deposition soon follows injury, even within minutes (Eschrich, 1956, 1965; Currier, 1957), and definitive callose formation has been recognized as a response to severe injury and to senescence of sieve tubes. There is little question that, following slime or filament plugging of sieve tubes near a cut, near an injury, or near certain disease infestations, callose deposition is soon increased until it effectively closes the phloem to conduction for some distance from the cut (Crafts and Currier, 1963). Also, as the sieve tubes age in the normal course of their development, or in the autumn, definitive callose forms and blocks off the aging elements. In autumn, all of the mature sieve tubes in the bark of trees may develop definitive callose. In many genera, this results in obliteration of the elements; in a few, such as *Vitis, Tilia* and *Liriodendron*, phloem is reopened in the spring, and single sieve tubes may function for two or more years. Parthasarathy and Tomlinson (1967) described sieve tubes in palms that live and function for 50 years or more; even these become blocked with callose when the leaves with which they are associated finally die. Zimmermann (1964a) gives an excellent discussion of definitive callose formation in the autumn in dicotyledonous forest trees. Ziegler (1964) distinguished between ring-porous and diffuse-porous trees with respect to callose function. Engleman (1965a) discussed callose formation in *Impatiens sultanii.*

Callose formation occurs in response not only to physical injury or infectious diseases, but to various other stimuli as well. It can be promoted by exposure to eosin Y (Schumacher, 1930), endothall (Dunning, 1959), boron (Currier, 1957; Dunning, 1959; Eschrich et al., 1965; McNairn and Currier, 1965), TIBA (Lerch, 1960), ultrasound (Currier and Webster, 1964), and heat (McNairn and Currier, 1968).

Schumacher (1930) found that application of dilute eosin Y solution in gelatin droplets to a cucurbit leaf caused callose formation at a distance from the point of application. Currier (1957) found boron to cause callose formations in *Elodea*. Dunning (1959) reported that endothall caused rapid callose formation on the sieve plates of cucurbit petioles when cut petioles were immersed in solution. Dunning found this response to follow treatment with eosin Y, maleic hydrazide, 2,4-D, monuron, boric acid, dalapon, amitrole, and phloridzin. However, these are nearly all phytotoxic compounds, and in a number of instances there was a positive correlation between callose formation and the degree of visible injury to the leaves. High temperature, ultrasonic vibration (Currier and Webster, 1964), and mechanical injury of several forms also induced callose formation. Thus, this response seems to be almost a universal reaction to degenerative influences on the sieve tube protoplast.

Currier and Webster (1964), studying the effects of ultrasound stimulation, found that, where there was no gross injury, callose production by means of locally applied low-intensity ultrasound (850 kc) gave no inhibition of growth; and that the callose eventually decreased to control level. Ultrasound treatment resulted in no irreversible injury.

Eschrich et al. (1965) studied the influence of callose formation on the transport of assimilates in sieve tubes. They used young cucurbit plants treated with calcium chloride solution and 5×10^{-5} M boric acid solution to induce callose formation. They prepared histoautoradiographs of petioles, they studied fluorescein-treated sections, and they used $H^{14}CO_3^-$ solutions and autoradiography to detect the effects of callose formation on assimilate movement. They were successful in inducing callose formation in sieve tubes by injecting the boric acid and calcium chloride solutions into the cavities of cucurbit petioles. Heavy callose deposition was observed in 34.2 percent of the sieve tubes, compared with 5.4 percent in control petioles. Leaf blades absorbed and assimilated ^{14}C, but so far as Eschrich and his co-workers could detect, transport was not slowed down by the callose depositions. Apparently, callose induction by boric acid injection enhanced assimilate movement; both autoradiographs and radioactivity counts showed increased ^{14}C in roots and shoot tips of injected plants. Long-distance transport of fluorescein-K in sieve tubes was not affected by heavy callose deposition. In the region of injection, both fluorescein and ^{14}C-labeled assimilate moved into the phloem parenchyma and starch sheath. These results contrast with those of Schumacher (1930). Evidently, the extent of callose formation brought about by the treatments used by Eschrich et al. was not sufficient to hinder assimilate transport in their plants; probably the younger, most effective sieve tubes were the ones that remained open.

Using heat to induce callose formation in the sieve tubes of the cotyledonary petioles of *Gossypium hirsutum*, Webster and Currier (1965, 1968) found that phloem callose was increased, and that the lateral movement of labeled assimilate from sieve tubes to phloem and cortical parenchyma was inhibited; this decrease was limited to the heated portion. The amounts of phloem callose and lateral movement from sieve tubes returned to normal 1 day after heating. The amount of callose formed by heating could be reduced by storing the plants in darkness for 16 or more hours before heating. There proved to be more lateral assimilate movement in the low-callose (dark stored) plants than in high-callose plants. Webster and Currier concluded that phloem callose is responsible for the decrease in lateral movement, and thus, that lateral movement takes place via protoplasmic connections.

In spite of the substantial increases in sieve-plate callose, autoradiographs showed no restriction in longitudinal movement of assimilates. One explanation for this would be that the conductive capacity of the sieve plates exceeded the demands put upon them, since there were always some sieve plates with little callose. Thus results of these studies agree with those of Eschrich, indicating little effect of chemical callose induction on assimilate transport, and with the ringing experiments of Mason and Maskell (1928a), showing that transport was not seriously hindered until the phloem was reduced to 33 percent, or less, of normal.

Eschrich and Currier (1964) developed techniques for identifying callose by its diachrome and fluorochrome reactions. These techniques should prove valuable in future studies of the role of callose in sieve tubes.

In detailed studies on the petioles of *Gossypium hirsutum*, Webster (1965) found that callose was induced by heat treatment within the temperature range of 40–60°C, being maximum at about 50°C, and slight at or below 43°C. Time saturation at 45°C occurred in less than 15 minutes, with callose formation taking place during heating and being restricted to the heated region. A dark pretreatment of 16 hours prevented callose formation, presumably by limiting callose precursors. The heat treatments in Webster's experiments did not inhibit longitudinal transport of ^{14}C-labeled assimilate through the heated petioles; evidently, the callose formed in these short-term treatments was not sufficient to cause effective plugging. Lateral movement from sieve tubes to parenchyma cells in heated petioles was inhibited, possibly because callose formation restricted movement through the narrow connecting plasmodesmata.

Currier, McNairn, and Webster (1966) reported that short-term longitudinal transport of ^{14}C-assimilate is inhibited by heating the hypocotyl of young *Gossypium hirsutum* plants to 45°C for 15 minutes. Previous failure to demonstrate blockage is attributed to low-level labeling of assimilate.

Maestri (1967), in a comprehensive study on the structural and functional effects of endothall on plants, observed strong callose production in leaf tissues; the callose deposits increased with time and dosage. He found callose deposits on sieve plates and lateral sieve areas. Callose formation was related to visible injury such as wilting of leaves, browning along veins, and desiccation. Detection of callose by the fluorochrome method could be accomplished before visible injury could be detected, but the amount and extent of deposition increased as injury progressed. Endothall inhibited vein loading of leaves.

Continuing the callose investigations, McNairn (1967) used heat treatments on 4-cm lengths of *Gossypium hirsutum* hypocotyl, and found maximum callose production after 15 minutes at 45°C. Translocation, as indicated by movement of ^{14}C from $^{14}CO_2$ treatment of one cotyledon per plant, was inhibited for 3 hours after the heat treatment; but after 6 hours it was greater than normal. Callose was noticeably reduced 6 hours after treatment, and attained normal levels within 2 days. Growth measurements, plasmolytic tests, vital staining, and visual examination gave no evidence of injury to plants heated to 45°C for 15 minutes. These results indicate that callose plugging may be effective, and that callose formation may be reversible, if injury is held to a minimum. Shih and Currier (1969) confirmed these results using *G. hirsutum* seedlings. Their fine-structure studies of sieve-plate blocking are in accord with evidence obtained by tracer tests. Swanson and Geiger (1967) found no evidence for callose formation from their chilling treatments of *Beta vulgaris* and *Phaseolus vulgaris*.

Although the various experimental treatments reviewed here gave variable degrees of translocation inhibition, there can be no doubt that heavy definitive callose formation completely blocks sieve plates to transport.

Slime Plugging

Slime, as defined by the authors of this book, is a viscous, coagulable material present in newly differentiated sieve-tube elements of most dicotyledons. In some species—for example, the cucurbits, *Nicotiana tabacum*, and *Phaseolus* spp.—it is derived from the breakdown of discrete bodies, the "slime bodies." Undoubtedly, in these as well as other species, the breakdown products of the nucleus, tonoplast, plastids, and other inclusions, contribute to the slime content of sieve elements. Evert and Alfieri (1965) and Murmanis and Evert (1966) have illustrated slime bodies and their breakdown in gymnosperms. Breakdown products, as described, make up the material that forms slime plugs lodged against the distal sides of sieve plates when phloem is cut in preparation for microscopic study. Whether this material appears as partially degraded slime bodies (Crafts, 1932, figures 12, 13, 14, and 38; 1933, figures 6, 8, 10, and 11; Esau and Cheadle, 1961, figure 5), or in an amorphous form (Crafts, 1933, figures 12, and 14; Esau and Cheadle, 1961, figure 3; Evert et al., 1966, figures 27, and 28), seems to depend largely upon the age or stage of differentiation of the sieve tubes.

There are three reasons for believing that slime is an ephemeral constituent of the sap of young sieve tubes. First, an examination of an ontogenetic series of sieve tubes reveals that some time after a sieve-tube element attains its functioning condition, it appears to be almost completely devoid of stainable contents. This was illustrated by Esau (1941) in a photograph showing a single differentiating sieve tube in a root tip of *Nicotiana tabacum* (her plate 7; see Figure 2.4). It was illustrated and discussed by Esau, Cheadle, and Gifford (1953). It is apparent in Tamulevich and Evert's (1966) figures 20–25 (see Figure 2.6). This paucity of stainable material indicates that the slime has been reduced to molecular form and absorbed by the surrounding cells. When this state has been reached, cutting results not in the formation of massive slime plugs, but in the clogging of the sieve-plate pores with a filamentous material, namely, the plasmatic filaments. Second, an examination of the rate of displacement of sieve tube sap in the normal flow of the assimilate stream makes evident the fact that the total liquid contents of each element are replaced 1000–5000 times each hour. It is impossible for a molecularly dissolved material to remain within a sieve element under those conditions. Third, if slime were not reduced to molecular form and absorbed by growing cells, it would flow along with the assimilates and accumulate in the last differentiated sieve elements of a root or shoot tip. Figure 2.4 shows that this does not occur.

That slime of *Cucurbita* may coagulate into an elastic gel is shown by the slime sacs illustrated by Nägeli (1861), Crafts (1939a, 1961b) and others (see Figure 2.2). Long considered as evidence for the solid cytoplasmic structure of sieve-plate strands (Crafts, 1939a, 1961b), these structures, it now seems clear, are composed of coagulated slime that has been pushed through the open pores and set in a solid state by the fixation process. In cucurbit sieve tubes, there is evidence that slime may be adsorbed to the filamentous reticulum. This may be washed out in the exudate when the stem is cut (Figure 2.2).

Filament Plugging

That there exists in plants a third plugging mechanism to protect them from complete draining of the phloem system is proved by the fact that exudation from a large, old stem of *Cucurbita pepo* stops within a minute or so. In such stems, the phloem bundles are made up of young, mature, and obliterating sieve tubes with the attendant parenchyma; slime plugs are formed in the newly differentiated sieve tubes, but in the mature ones, strands of filaments clog the sieve-plate pores. Kollman (1960b, figures 20 and 21; 1964, figure 6), Engleman (1963, figure 2), Behnke and Dörr (1967, figure 12), and Cronshaw and Esau (1967, figures 10 and 13) show filamentous plugging of sieve-plate pores in a number of plant species (Figure 2.8).

Although in some species slime is copious and sieve plate plugging is rapid, in others, such as *Fraxinus* spp., *Yucca flaccida*, or *Macrocystis pyrifera*, plugging is slower; in these, true slime may be lacking, and plugging may result from callose, or aggregation of filamentous material in the sieve-plate pores, or both. Renewal of phloem exudation by repeated cutting is physical evidence for the fact that plugging takes place near the cut surface.

The existence of strands within the vacuoles of young sieve tubes, and their persistence after the loss of the tonoplast, has been the subject of some controversy. Crafts (1932) described inner strands of protoplasm in the sieve tubes of *Cucurbita*, and illustrated them in serial order (his plate V, figures 6–21). They are shown again in slides of *Solanum tuberosum*, (Crafts, 1933, figures 8, 10, 13, 16, and 17), of *Nicotiana tabacum* (Crafts, 1934, figures 3, 8, and 10), and of *Fraxinus* sp. (Crafts, 1939b, figure 5).

Engleman (1965a) proposed that slime may be arranged as a stationary reticulum in sieve tubes. This reticulum, as illustrated in his figure 2c, is open enough to allow normal flow of the assimilate stream, but is subject to agglutination and congestion within sieve pores as a result of cutting the phloem, or killing with certain reagents. Buvat (1963a, 1963b) and Eschrich (1963a) reported fibrillar or lipoprotein networks in sieve tubes, and Went and Hull (1949) and Kollmann (1964) characterized the content of the sieve-tube lumen as cytoplasm. Behnke and Dörr (1967) illustrated what they termed plasmatic filaments in *Dioscorea reliculata* (their figures 7 and 8). These are consistent with Engleman's scheme.

Considering the wide array of responses to cutting the phloem, from the very rapid plugging by a coagulable sap in the cucurbits, to the slower stoppage in many tree and palm species having noncoagulable sap, it seems, as mentioned before, that there are four distinct mechanisms of phloem plugging—all of which, possibly, may act in a single plant.

Whereas young, recently matured sieve tubes may be closed by an obvious aggregation of slime, old elements may have the pores of their plates plugged by plasmatic filaments with little or no slime accumulation upon the interpore area. A slower callose plugging is a third mechanism.

That plugging may be localized to the ends of stems or sections of stems, and is complete and effective, is shown by the cessation of exudation within a minute or two after cutting. In the experiments of Currier, Esau, and Cheadle (1955), the cut ends of their sections being tested for plasmolysability must have been plugged, because it is impossible to plasmolyse an open tubule. That the sieve plates in the center of a section remained open was shown by the "surge" phenomenon, which they observed. Rouschal (1941) observed displacement of sieve tube starch grains in *Aesculus* phloem by altering turgor conditions. Crafts (1938) observed deformation of sieve plates from rapid, local increase in pressure. When a plant is cut, mutilated, or injured by insect feeding, it is protected from excessive loss of assimilates. If the wound affects only a part of a stem, wound phloem may be differentiated around the injury and translocation reestablished. If the stem is cut off, an axillary bud at the node below the cut may break into growth and take over the function of apical growth. If injury is to secondary phloem, wound callus may be formed and new phloem developed through or below it to reestablish food distribution. And, as mentioned above, in *Vitis*, *Tilia*, *Liriodendron*, and other genera of perennial plants, definitive callose formed in autumn may be redissolved in spring; and single sieve tubes, therefore, may function two or more seasons. In all of these cases, plugging of sieve tubes is the initial response; recovery may take place in a number of different ways.

5

Phloem Exudation

Incision Method

That the phloem of plants contains a nutritious liquid under positive hydrostatic pressure was discovered by birds (sapsuckers, for example) and insects (aphids and leafhoppers for example) a very long time ago. A view of sapsucker work on *Ceratonia siliqua* was given by Crafts (1961b). A similar picture is Figure 5.1, showing bark perforation on pine. Hartig (1858) described phloem exudation over 100 years ago. Work up to the present decade was reviewed by Crafts (1961b); Zimmermann (1960a, 1961) has reviewed more recent studies.

All early work on phloem exudation was done using an incision technique of some sort. This involved cutting the stem of a cucurbit plant (Hartig, 1958; Zacharias, 1884; Kraus, 1885; Crafts, 1931, 1932, 1936, 1938; Moose, 1938), or puncturing the bark of a tree (Huber, 1937; Huber et al., 1937; Zimmermann, 1957a, 1957b, 1958a). Such exudate, while contaminated with the contents of cut parenchyma cells, gave a fairly accurate picture of the contents of functioning sieve tubes, that is, of the assimilate stream.

Figure 5.1. Sapsucker perforation of the bark of *Pinus halepensis*: (*a*) the holes left in the bark; (*b*) phloem exudate dried on the pavement below the perforations.

Zimmermann, using both bark incision and the aphid stylet technique, has made noteworthy studies on translocation (1960b). He measured successive samples of sieve-tube exudate from cuts in *Fraxinus americana*. He reported a continuous decrease in concentration with time. When one incision was made laterally adjacent to another, the first had no effect on the exudate concentration of the other. Longitudinally, however, incisions into the phloem affected the exudate concentration over great distances. A circular incision at 5 meters height caused, within 1 hour, a decrease of concentration below and an increase above. A concentration maximum develops 50 cm above the incision within about 30 minutes; this concentration peak moves up to about 1 meter above the incision and persists.

Defoliation of one-half of Y-shaped trees showed that sucrose and D-mannitol accumulate in sieve tubes in the absence of leaves. Tangential conduction in the phloem of trees is very slight; phloem conduction from the leafy side fans out at an angle of less than 1°.

Zimmermann was the first to illustrate the location of aphid stylet tips in an individual sieve tube (1961). He has provided essential information on the composition of phloem exudates from trees (1960a, 1961).

Aphid-stylet Technique

Kennedy and Mittler (1953) first reported on the aphid-stylet technique for collecting sieve-tube exudate. When the body of an anaesthetized feeding aphid is cut from the stylets, exudate comes from the cut mouthparts at an average volume of 1 mm^3 per hour. When the aphids fed on *Salix*, the exudate contained sucrose, amides, and amino acids in aqueous solution. Kennedy and Mittler identified some 10 amides and amino acids both in honeydew from aphids and in phloem exudate from severed mouthparts.

Weatherley et al. (1959) used the aphid-stylet technique to study translocation processes in *Salix*. They found the velocity of flow in sieve tubes to be around 100 cm per hour, equivalent to replacement of the lumen contents at a rate of 100 sieve-tube elements per minute. The exudate they found to contain 5–15 percent sucrose, up to 0.4 percent raffinose, no reducing sugars, and about 0.5 percent amino acids. From senescent leaves, the exudate might contain 5.0 percent amino acids.

Zimmermann (1961) was the first to illustrate the location of the tip of an aphid stylet in an individual sieve tube. In 1963, Zimmermann published a beautiful colored picture of aphid stylets inserted into a single sieve tube. This is gratifying evidence that the exudate obtained from severed stylets is pure sieve-tube sap, and that the initial droplets are undiluted samples of the assimilate stream. Coupled with the earlier work of Kennedy and Mittler (1953), Mittler (1953, 1957, 1958), Weatherley, Peel, and Hill (1959), and others, this identification of the source of stylet exudate as the mature, functioning sieve tube lends much strength to the mass-flow hypothesis of phloem transport in plants. More recent work by Evert et al. (1968) confirms the fact that aphids feed in living, functioning sieve tubes. Evert and his co-workers located stylet tips in both current season's and past season's sieve tubes of *Tilia cordata*. They found 20 living sieve elements containing stylet tips, 13 in areas of bark which contained living sieve elements of both the current and the previous season's phloem increments, and 7 containing living sieve elements from the current season only.

Composition of Phloem Exudate

From the reviews cited above, it is evident that phloem exudate may contain roughly 10–25 percent dry matter, of which 90 percent or more is sugar. In many species, sucrose is the only sugar present; stachyose, raffinose, verbascose, or sugar alcohols such as D-mannitol and sorbitol may occur in certain others. Amino acids may make up 0.5 percent of the dry weight of exudate in the summer, and considerably more in the autumn before leaf fall.

More recent studies have added new constituents to the list of materials found in sieve-tube exudate: sugars and organic nitrogen (Tammes and Die, 1964), macronutrients and micronutrients (Tammes and Die, 1966), ATP (Kluge and Ziegler, 1964), and nucleic acids. Braun and Sauter (1964) and Lester and Evert (1965) reported acid phosphatase activity, and Kluge et al. (1964) reported gibberellin activity, but the actual presence of phosphatase and gibberellin was not proven. The general paucity of enzymes and the lack of proportionality between acid phosphatase and sucrose in phloem exudates (Ziegler, 1956) indicate that transformations leading to the high concentration of polysaccharides in phloem exudate are located in the border parenchyma of leaves and phloem parenchyma and companion cells of stems and roots. Zimmermann (1958a) attributes the hydrolysis of oligosaccharides, en route through the phloem, to an α-D-galactosidase located in the parietal layer lining the walls of the sieve tube; D-galactose units are removed as the oligosaccharides pass; raffinose is produced from stachyose, and sucrose from raffinose. Sucrose may also be removed from sieve tubes in the normal nutritional function of phloem; Zimmermann (1958a) proposed that this process is controlled by a hormone. Detailed information on the composition of the assimilate stream is provided in Chapter 9.

Zimmermann (1962) described studies on the effects of double interruption of phloem in *Fraxinus americana*. He found that if double incisions are made, one above the other, exudation takes place from both. This indicates that upward transport toward the lower incision occurs as a reversal of the normal direction of flow. A series of samples taken from each of these loci shows a gradual concentration decrease, which represents an osmotic dilution of the sieve tube sap. The concentration of sap from the upper incision was slightly higher than that from the lower incision, because the supply to the upper incision is from the photosynthesizing source. Sap from the lower incision represents an emptying of the sieve tubes from the trunk, having only the existing sieve-tube contents plus carbohydrate reserves of the stem as a source of osmotically active solutes. Zimmermann tried to determine the possible effect of gravity on exudation rates by inverting living trees. However, the method used was not sensitive enough to discern the gravitational portion of the exudation pressure.

In another paper, Zimmermann (1964a) discussed the relation of transport to growth in dicotyledonous trees. He described and illustrated seasonal phloem activity, and gave, in detail, a picture of the relations between cambial activity, phloem transport, foliage development, storage, and reserve utilization. He reported that aphids feed successfully only on active phloem. When sieve tubes are blocked by callose, feeding fails and the insects die. In some trees, such as *Salix* spp. and *Tilia cordata*, callose formation is incomplete, and aphids can feed on cut branches brought into the laboratory; in *Carya* spp. and *Castonea dentata*, tubes are blocked and insects are unable to feed. Injury, such as cutting, induces callose deposits on sieve plates in both directions from a cut; but only at limited distances, as evidenced by the reestablishment of exudation upon removal of a slice of tissue. In contrast, true dormancy results in complete blocking of the sieve tubes throughout the plant.

Zimmermann discussed the break of dormancy and the reactivation of sieve tubes; and in the discussion reported at the end of the paper, Ziegler pointed out the differences between ring-porous and diffuse-porous species, with respect to the resumption of phloem activity in spring. Zimmermann mentioned bidirectional movement in the intermodes of woody plants; he anticipated Biddulph and Cory's (1965) results by suggesting that ". . . a much simpler interpretation would be movement in one direction in one layer, and movement in the other direction in another layer of the phloem tissue within the bundle."

In a further paper, Zimmermann (1964b) compared the movements of sap in the xylem and in the phloem of trees. Analyzing the various theories that have been proposed, he suggested that there are three requirements to be met, if mass flow in phloem is to operate: (1) there must be a turgor gradient, within the sieve tubes, from sources to sinks; (2) the sieve tubes must be impermeable to outward movement of sugars, the entry of solutes into the sieve tubes and their removal from them being metabolic steps; and (3) the sieve plates—that is, the transverse walls between individual sieve elements—have to be permeable to the flowing solution.

Zimmermann reported studies that substantiate requirements (1) and (2). Concerning the permeability of the sieve plate, he cited the traditional opinion that pores in the sieve plate are closed by cytoplasmic structures. Apparently, he was not impressed with the evidence for open pores through the sieve plates of plants treated by special methods to prevent plugging by slime, callose, or filaments (Esau and Cheadle, 1961; Duloy et al., 1961). Cronshaw (1969), Cronshaw and Anderson (1969), Currier and Shih (1968), Johnson (1968), Shih and Currier (1969), and Sjolund (1968) have further substantiated the open condition of sieve-plate pores.

Yucca flaccida was proven to be excellent material for studies on phloem exudation by Tammes and Die (1964). These workers found that, by cutting the inflorescence of this plant, they could collect a sweet liquid; and by cutting twice daily, the flow continued over periods of several weeks. The exudate has a *p*H of 8.0–8.2, and a dry weight of about 18 percent, of which sucrose constitutes 80–90 percent. Amino acids were found to the level of 0.07 M, based on the glutamine molecule, which proved to be the prevalent nitrogenous constituent. Valine was second in prominence.

Feeding $^{14}CO_2$ to *Yucca flaccida* plants produced exudate containing ^{14}C-sucrose. The ^{14}C content of the exudate exhibited a pronounced daily rhythm (Die and Tammes, 1964). The ^{14}C fraction of the exudate showed a logarithmic decrease in concentration.

In a third paper, Die and Tammes (1966) described experiments on isolated parts of the *Yucca flaccida* inflorescence. Exudate could be seen flowing from the phloem of both the distal and the proximal end of a cut inflorescence. By regularly renewing the cut surfaces, the flow was maintained for at least 48 hours. Comparison of exudates from the two cut ends indicated that cells and tissues adjacent to sieve tubes can reversibly accumulate and secrete carbohydrates, possibly by metabolically maintained equilibria between the solute concentrations inside and outside the sieve tubes. These conclusions resemble those of Weatherley et al. (1959) concerning *Salix*.

Studying the translocation of macronutrients and micronutrients in *Yucca flaccida*, Tammes and Die (1966) found that potassium, nitrogen, and phosphorus are phloem-mobile and occur in relatively high concentrations in exudate. Magnesium was next highest in concentration, followed by calcium, sodium, zinc, boron and iron. Manganese and copper were present in traces.

Electron microscope views of *Yucca flaccida* phloem structure (Ie et al., 1966) show sieve pores constricted by callose and blocked with fibrillar material (pictures 2 and 3) and open sieve pores with a few tubular strands running through them (picture 4). The authors concluded that these allow ample room for continuity of fluid between sieve tube elements. Within the lumina of the elements, and in a sample of exudate first centrifuged and then coagulated with alcohol, is shown a flocculant reticulum that resembles the sieve tube lumina contents of preparations by Esau and Cheadle (1961), described by R. P. C. Johnson (1966) as an artifact of permanganate fixation. Although the exudation from the *Y. flaccida* inflorescences differs from that of the cucurbits, *Fraxinus* spp., *Macrocystis pyrifera*, and many other plants in flowing slowly after the initial cut and accelerating on subsequent days (see Die and Tammes, 1964, figure 5), it resembles phloem exudates in composition, it undoubtedly comes from cut sieve tubes, and these in turn show typical anatomical features under the electron microscope (including sieve pore blocking by callose constriction and fibrillar precipitate, and open pores under proper conditions of preparation). Probably these inflorescences occupy a position somewhere between the bark of *Fraxinus* spp. and the inflorescence stalks of palms (Tammes, 1933, 1951, 1958), which are beaten to induce exudation. Die (1968), studying exudate from both cut ends of an excised inflorescence of *Y. flaccida*, concluded that the source of this material is the secretion product of cells that surround the sieve tubes.

Fife et al. (1962), using lifted roots of *Beta vulgaris* in the laboratory and growing plants of *B. vulgaris* in the field, made a comparative study of phloem and xylem exudates. By careful selection of the roots, they were able to obtain 250 ml of phloem exudate from 100 roots

Table 5.1. Properties of phloem and xylem exudates, and of expressed juice from roots, of *Beta vulgaris*.

Property	Phloem exudate	Xylem exudate	Expressed root juice
Viscosity* (relative to H_2O)	1.58	0.98	2.18
Surface tension* (dyn/cm)	50.3	57.6	43.4
Density*	1.059	1.004	1.082
Total solids (%)	16.0	0.8	20.4
Electrical conductivity (mmho/cm)	10.79	2.30	4.30
Osmotic pressure, by freezing point (atm)	18.04	1.63	21.56
Osmotic pressure, by conductance (atm)	3.86	0.82	1.55
Electrolytes (as % of total solutes)	21.5	50.3	7.7
*p*H	8.02	6.12	6.53
Sucrose (%)	8.8	0.10	16.5
Reducing sugars (%)	0.33	0.03	0.76
Soluble nitrogen, excluding nitrates (g/liter)	5.65	0.10	1.16
Total amino acids, calculated as glutamic acid (%)	1.10		0.44
Protein nitrogen, heat-coagulable (g/liter)	0.48		0.55

Source: Fife et al. (1962).
*At 20°C.

and about 100 ml of xylem exudate from 20 roots in the field. Table 5.1 presents the results of their work, and shows the contrasting properties of these two materials; data on expressed root juice is included for comparative purposes.

Several interesting contrasts are shown in the data of this table. Phloem exudate is higher than xylem exudate in viscosity, density, total solids, conductivity, osmotic pressure, *p*H, sucrose content, reducing sugars, and soluble nitrogen (excluding nitrates). Amino acids were higher in phloem exudate than in expressed sap. The concentration of electrolytes in the phloem exudate was 2.5 times higher than in the expressed juice; and electrolytes in the phloem exudate made up 21 percent of the total solutes present. A good portion of this might have been potassium, which is known to be high in the assimilate stream. (See also Crafts, 1936.)

Parker and Philpott (1961) questioned the collection of phloem exudate from *Macrocystis pyrifera* reported by Crafts (1939a). This collection was made at Pacific Grove, California, in the summer of 1938. The stipes of large, healthy plants were lifted into a boat and cut, leaving the rest of the living plant in the sea water. Plants cast upon the beach did not produce exudate. Volumes of up to 1 ml per hour were obtained from single stipes. Analyzed in the laboratory of the Carnegie Institution's Department of Plant Biology at Stanford, their dry weight proved to consist mainly of sugars. The free flow of phloem exudate from the cut stipe of this giant kelp would seem to be better evidence on the question of mass flow than are the illustrations of Parker and Philpott showing obstructions in sieve-plate pores. The parietal cytoplasmic layer, plus a few strands of plasmatic filament, would account for the contents of these much-distorted pores. In their natural state, the latter inclusions might occupy only around 5–10 percent of the lumen area of the pore.

Exudation in Relation to Assimilate Movement

Phloem exudation continues to provide useful information on translocation processes. Peel and Weatherley (1959) used the aphid-mouthparts technique on *Salix*, to obtain sieve-tube sap, which they analyzed. Their work confirmed earlier analyses showing sucrose to be the principal sugar present, but raffinose also occurred in amounts up to 15 percent of the total carbohydrates; stachyose was present in traces. Mannitol was found, but methylated or phosphorylated sugars were absent. Amino acids varied with the season: from October to April, some ten amino acids could be detected; but during the summer, only aspartic and glutamic acids and asparagine were found.

Stem segments taken from mid-October until March, and sprouted in the laboratory, yielded sap having as much as 25 percent sucrose; sap from summer cuttings ran as low as 4–6 percent. Measuring the osmotic potential of the sap by a vapor-pressure method, they found a large discrepancy between this and the expected value, which was calculated from the sugar concentration. It turned out that as much as 2 percent w/v of potassium might be present, and that the potassium ions were balanced by citric, tartaric, and possibly oxalic acids. When the main constituents of the sieve-tube sap were measured over periods of hours, there was no evidence for a decline in, or a change in ratio between, the concentrations. Peel and Weatherley concluded that potassium normally moves along the sieve tubes with the sugars and amino acids.

Mittler (1959) reported before the Ninth International Botanical Congress that plants may supply aphids with large volumes of sieve-tube sap under pressure. Apparently, aphids are fed physically and chemically by their host plants; an aphid may be regarded as a bona fide sink into which the assimilate stream empties. The distribution of aphids on a host plant suggests that aphids preferentially tap sieve tubes that are supplying existing plant sinks. As aphids require relatively large amounts of nitrogenous matter for growth and reproduction, they prefer sap rich in nitrogenous constituents. A high content of nitrogen in the sap of sieve tubes draining senescent leaves may account for the selection by aphids of these sites for feeding. A heavy aphid infestation may force a plant to mobilize its nitrogenous constituents, which, in turn, may cause its premature senescence. In general, Mittler felt that aphids exploit, and perhaps enhance, the flow to existing sinks, rather than creating new sinks and abnormal sap flow.

Studying phloem exudation from *Tilia americana* by means of the aphid-stylet technique, Hill (1960) found the activity of sieve tubes in detached branches to be dependent upon source–sink relations. During dormancy, branches brought into a warm room and given an extended photoperiod produced exudate for only a short period, indicating a failure of the phloem parenchyma to provide an adequate source under conditions requiring the hydrolysis of starch reserves.

Hill (1962) studied exudation from aphid stylets, during the period from dormancy until budbreak, by taking detached branches of *Tilia americana* to the greenhouse between December and March, and obtaining sieve-tube sap by the method of Kennedy and Mittler (1953). During physiological dormancy, buds on the branches brought to the greenhouse did not break within 2 months. Aphids on such branches were restless, and they produced little exudate. Physiological dormancy ended near the end of December; and after this date buds broke 2–4 weeks after the branches were brought in; by mid-March, this period was reduced to about 12 days. Aphids thrived on these branches, produced copious honeydew, and reproduced rapidly. As the leaves began to expand, the aphids found feeding unsatisfactory, their honeydew production decreased, and they became restless. When buds were removed from branches during the period of active exudation, honeydew secretion persisted for only 5 days, compared with 14–21 days for branches with buds. Determination of sugar concentration, exudation level, and bud dry weight showed that exudation was most intense before sinks in buds became active; it fell off rapidly thereafter. Figure 5.2 shows the

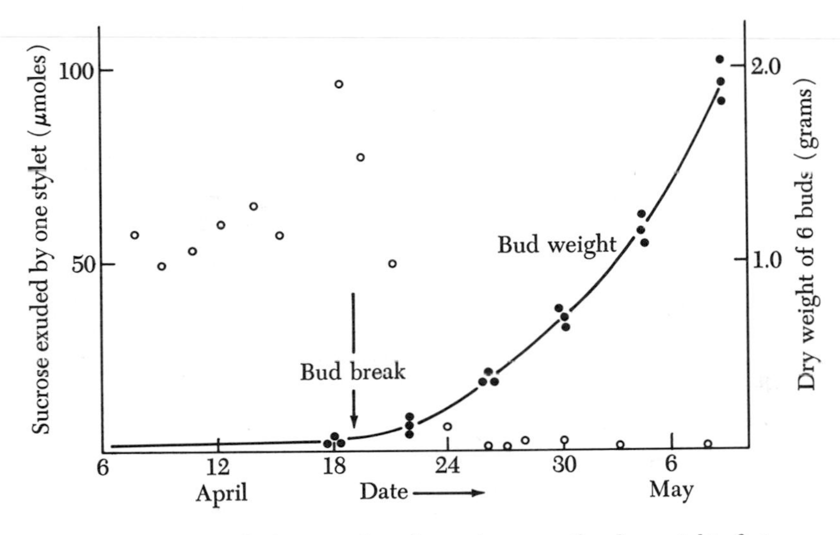

Figure 5.2. Graph showing the relation between the dry weight of six buds of *Tilia americana* and time. *Open circles* show the concentrations of exudate samples from the bark. Concentration of sucrose drops rapidly after budbreak around April 18. From Hill (1962).

relation between exudation level and bud growth during this period. Hill suggested that the stylets act as sinks that compete with the buds; and that as the buds break and go into active growth, they use up the sugar reserves. Hill also considers it possible that a hormone factor produced by buds may activate sieve-tube differentiation after budbreak. He suggests that the drop-off in exudation may reflect a decline in activity of a hormone concerned with hydrolysis and the transport of reserves; or a decline in functional capacity of the old transport system, during the period when leaves are approaching compensation and new sieve tubes are being differentiated.

Peel (1963) made a study of the movement of ions from the xylem sap into the sieve tubes of *Salix*. Using aphids, Peel found that their honeydew contained potassium in concentrations ranging from 5.23 percent to 6.93 percent of the total dry solid content; sodium concentrations ranged from 0.28 percent to 0.42 percent of the total dry solids. On perfusing the xylem with a solution of potassium or sodium chloride, a considerable uptake of the cation took place. This was followed, after several hours, by an increase in the concentration of the particular cation in the honeydew. A relationship was shown to exist between the concentration of these cations in a given stem segment, and their concentration in the honeydew obtained from that segment. Potassium perfused through the xylem resulted in an increased potassium concentration in the bark, but not in the wood. In the case of sodium, storage apparently took place in the wood, not in the bark. However, passage into sieve tubes could go on in either case. This indicates that migration of ions into sieve tubes in these stems takes place by symplastic movement from the parenchyma of either phloem or xylem.

As a sequal to the excellent work of Weatherley et al. (1959) using aphid mouthparts as sources of phloem exudate, Peel and Weatherley (1962) studied the effects of illumination and darkening on translocation in *Salix*. In darkness, the rate of sap and sucrose exudation increased and the concentration fell, relative to values in the light. Similar effects were produced by girdling the stem just beneath the crown of leaves. These effects were shown to be secondary, due to a lessening of sap pressure in the xylem resulting from lowered transpiration during darkness or following girdling. If changes in water tension were eliminated, such fluctuations in phloem sap transport and concentration did not occur.

Use of ^{14}C as a tracer showed that exudation from sieve tubes of intact plants is immediately and intimately connected with events taking place in leaves situated some distance from the point of exudation. Such exudation is not a local phenomenon, nor is it a result of the activities of the cells within the 16-cm contributory length of stem that Weatherley et al. (1959) showed to be responsible for exudation from an isolated stem segment. The fact that when the connection with the leaves is severed by a girdle, the proportion of labeled sugar in the exudate falls, while girdling has no effect on the total quantity of sugar exuded, shows that a switch of source from leaf cells to storage cells in the stem takes place rapidly.

Peel and Weatherley considered that a puncture produces a leak through which a loss of sugar solution takes place. This loss, though strictly local, may simulate a normal sink extending over several or many sieve elements. They point out that such a puncture produces two fluxes; an influx of water and sugar from surrounding cells, and a flow of the assimilate stream from adjacent sieve tubes through sieve plates. These disturbances spread along the sieve tubes in both directions from the point of puncture. Obviously, both the water potential in the phloem and xylem and the sugar potential in the sieve tubes play significant roles in the observed responses. As photosynthesis builds up the sugar potential in leaves, translocation down the sieve tubes of the phloem is induced, and a front of higher sugar potential will pass down the tubes. As this front moves, sugar will move out into the surrounding cells, which were in equilibrium with the pre-front sugar level; this is a natural result of the distribution function of the phloem. Thus, the front itself becomes less steep, and might move forward less rapidly than the original nature of the front might have indicated; the movement of a tracer front may be slower than the actual flow of total assimilate along the sieve tubes. Such differential exchange may explain the different rates of translocation of different solutes observed by Swanson and Whitney (1953) and Biddulph and Cory (1957), in the opinion of Peel and Weatherley. Such exchange, as shown by this work of Peel and Weatherley, indicates that the sieve tube is not isolated with respect to the surrounding parenchyma cells, as postulated by Canny (1960b, 1961).

Using two separate aphid colonies on a single stem of *Salix,* Peel and Weatherley measured the velocity of tracer movement along the stem. Their value was around 30 cm per hour. Since this was only about one-third the volume rate observed by Weatherley et al. (1959), they proposed that exchange with surrounding cells along the route may have been taking place. Canny (1961) obtained a much lower rate, using the same technique; and Peel and Weatherley proposed that their measurements give a more rational picture of the mechanism involved.

Since Mittler (1958) has shown that honeydew from aphids differs little in sugar content from sieve-tube exudate from aphid stylets, honeydew has been collected and used in studies on phloem transport. Hill (1963) used this method to determine the sources of sugars in the sieve-tube sap of *Salix.* Feeding $^{14}CO_2$ to the leaves of *Salix* cuttings, he collected honeydew from groups of aphids situated at different distances from the foliage. He found that the specific activity of honeydew from groups nearest the leaves rose sharply to a peak, then fell—steeply at first, then more gradually. Twenty-four hours after $^{14}CO_2$ assimilation, substantial activity was still present in the honeydew. In groups further from the leaves, the maximum activity was not so high; aphids having other aphid groups between them and the source excreted very little ^{14}C. They did, however, excrete sugar at about the same rate as the other groups, indicating that stem reserves compensated for the loss, to the other aphids, of sugars provided by leaves. Movement of peaks of radioactivity from group to group of aphids indicated a linear velocity of flow of 7.5 cm per hour, a relatively low rate in these *Salix* cuttings.

In an excellent series of papers, Tammes and his associates described their work on phloem exudation from *Yucca flaccida.* Some of this has been described in other sections of Chapter 5. Only the relation of this work to normal food movement in plants will be considered here.

In their first paper, Tammes and Die (1964) described the method for obtaining phloem exudate from the inflorescence of *Yucca flaccida*. Exudation started immediately after cutting, or during the following day; it was necessary to renew the wound by cutting off a slice about 1 mm thick twice each day; exudation was thus induced for a week or more. The exudate had a *p*H around 8.0; it contained about 18 percent total solids, 80–90 percent of which was sucrose; and it had a low content of amino acids. Potassium was the principal inorganic solute; calcium was very low. Thus, this exudate was similar to that obtained by Tammes from several palm species, and the authors give convincing arguments for its phloem origin.

In a second paper, Die and Tammes (1964) described experiments in which $^{14}CO_2$ was fed to single *Yucca flaccida* leaves, and the radioactivity of the exudate was measured over periods of several days. The maximum amount of activity was obtained on the third day after treatment with $^{14}CO_2$. In all exudate fractions, sucrose represented about 90 percent of the radioactivity measured. The ^{14}C-content of the exudate samples exhibited a pronounced daily rhythm; the night samples contained up to four times the concentrations of those exuded during the day. By contrast, the total dry materials and total sugar contents showed only slight variations. Apparently, the various leaves contribute with different rhythms, which tends to damp out their individual rhythms. Export of ^{14}C continued for over 8 days, and the maxima on successive nights decreased along a logarithmic course. Because exposure of the leaf was for only 30 minutes, it must be concluded that an appreciable pool of ^{14}C was assembled in the treated leaf, and that it continued to drain away for more than 8 days.

In the third paper, Die and Tammes (1966) measured exudation from isolated portions of young inflorescences. They found that these isolated organs produced exudate, as did also the cut ends still attached to the plants. They made detailed studies on the origin of the exudate, and found that it proceeded from phloem of the vascular bundles. The bleeding lasted for several hours, and removal of thin slices of tissue from the cut ends would renew the flow. The thin slices, upon test, proved to have heavy deposits of callose. By recutting, bleeding could be maintained for at least 48 hours.

Table 5.2 presents data on the composition of an inflorescence, and of the exudates from the basal part of the inflorescence stalk and from the isolated terminal (apical) portion. It is evident that the two exudate samples are very similar, and that they differ in composition from the total inflorescence.

Table 5.2. A comparison of the composition of an inflorescence of *Yucca flaccida* with the composition of phloem exudates from the basal and apical ends of the severed stalk.

Sample	Composition			pH	Conductivity‡ (mmho/cm)
	Reducing sugar (mg/g)*	Sucrose (mg/g)*	Amino acid (μmole/g)†		
Inflorescence	27.4	18.2	81	5.7	5.2
Basal exudate	0.1–2.0	135–165	85–157	8.2	2.6
Apical exudate	1.1–2.5	55–75	83–115	8.2	

Source: Die and Tammes (1966).
* Calculated as mg/ml for exudates.
† Calculated as μmole/ml for exudates.
‡ At 20°C.

By measuring the cross-sectional area of the phloem, and dividing this into the volume of exudate collected from inflorescence stalks connected with plants, they arrived at a velocity of flow of 44 cm per hour for exudation under their conditions. In 29 hours of bleeding, 4.5 grams of sugar and 0.5 grams of amino compounds moved through 3.0 mm^2 of total phloem area; mass transfer proves to be $(5.0 \text{ g}/0.030 \text{ cm}^2)/29 \text{ hr} = 5.7$ g dry matter per square centimeter of phloem per hour. This is about one-third of the value found by Crafts and Lorenz (1944a) for transport of pure assimilates through phloem into fruits.

From a comparison of the composition of both exudates with the composition of the inflorescence, Die and Tammes concluded that cells and tissues surrounding the sieve tubes in the developing inflorescence can reversibly accumulate and secrete carbohydrates, by metabolically maintained equilibria between solute concentrations inside and outside the sieve tubes. This interpretation is further substantiated by Die (1968) who analyzed exudates from both the acropetal (inflorescence) and basipetal (vegetative portion) ends of cut inflorescence stalks and found the compositions to be quite comparable. More assimilates were exuded from the inflorescences than could possible have been present in the sieve tubes. This resembles the conclusions of Weatherley and his associates with regard to the role of metabolism in the production of exudates from isolated portions of *Salix* stems via aphid stylets. It is a rational explanation for one role of metabolism in the mass-flow mechanism of phloem transport.

Tammes and Die (1966) studied the comparative compositions of phloem exudate, inflorescences, and leaves of *Yucca flaccida* in order to emphasize the role of the assimilate stream in the distribution of macronutrients and micronutrients. Table 5.3 gives the

Table 5.3. A comparison of the mineral composition of phloem exudate, inflorescences, and leaves of *Yucca flaccida*.

Element	Phloem exudate (mg/g dry wt)	Inflorescence (mg/g dry wt)	Leaves (mg/g dry wt)	Concn in exudate / Concn in leaves
K	9.324	17.998	10.489	0.889
Na	0.023	0.182	0.198	0.116
Ca	0.078	5.600	6.665	0.012
Mg	0.283	13.502	10.395	0.027
P	1.720	4.727	2.110	0.815
N	10.656	27.694	17.482	0.609
Zn	0.012	0.041	0.044	0.273
Fe	0.008	0.030	0.066	0.121
Mn	0.003	0.143	0.239	0.012
Cu	0.002	0.004	0.021	0.095
B	0.010	0.020	0.034	0.294

Source: Tammes and Die (1966).

data from their analyses. These data show that potassium, phosphorus, and nitrogen are mobile in the assimilate stream. Nitrogen moves mainly in the form of glutamine and amino acids, and phosphorus partly in organic form. At least six organic phosphorus compounds, as sugar phosphates and adenyl phosphates, were identified. Calcium is quite immobile in the phloem sap stream of *Y. flaccida*, as Tammes had found to be true for calcium in palms.

Exudation as Affected by Applied Hydrostatic or Osmotic Pressure

In a general paper on translocation mechanism, Weatherley (1962) described further experiments with the aphid-mouthpart technique. Although longitudinal cuts through the bark of a ringed *Salix* stem on both sides of an exuding proboscis had no effect on exudation, a transverse cut 10 cm above the point of exudation from the ring caused an immediate reduction in exudation rate; this indicates that flow to the point of exudation proceeds along the sieve tube. When $^{14}CO_2$ was fed to the leaves of *Salix* shoots in the light, honeydew collected from aphids situated 50 cm below soon showed radioactivity; and if the leaves were subjected to alternate two-hour periods of light and dark, the radioactivity of the exudate was higher during the light periods, indicating a direct connection between the leaves and the exuding aphids.

When water was forced through the xylem of exuding *Salix* stem segments, raising the hydrostatic pressure in the xylem, the exudation rate increased from around 1.5 μliters per hour to above 2.0 μliters per hour and lowered again when the pressure was normalized. The concentration of the exudate did not change during a period of 7 hours.

Assuming a pressure-flow mechanism, Weatherley calculated that a pressure gradient of 0.6 atm per meter would be required to drive fluid along sieve tubes at known rates of translocation, if there were no sieve plates; if free flow occurs through the sieve plates, an additional pressure of about 0.4 atm per meter would be required. Hence, for a tree 100 meters in height, 100 atm of pressure would suffice to bring about pressure flow; calculations given in Chapter 11 show that pressures of this magnitude are not required.

Peel and Weatherley (1963) continued their studies on exudation through aphid mouthparts using 15 cm lengths of *Salix* stem. By supplying water under pressure to cut xylem at one end of a segment, and allowing it to run freely from the other, a pressure gradient was set up in the xylem. By closing the free end, the pressure in the xylem could be raised uniformly; this caused the rate of exudation to increase. When pressure was raised by increments of 1 atm every 4 hours, the volume rate of exudation increased with each increment of pressure increase, the sucrose concentration decreased, and sucrose exudation remained constant. This effect of pressure could be balanced by using an equivalent concentration of osmoticum in the xylem. Therefore, Peel and Weatherley interpret this effect as being related to the water potential in the system rather than to any intrinsic effect of pressure.

The effect of an osmotic pressure gradient of 4 atmospheres along the 15-cm stem segment depended upon the position of the stylets in relation to the gradient. If the stylets were near the low-pressure end, exudation from them increased; if near the high-pressure end, exudation decreased; in neither case was the total sugar or nitrogen concentration of the exudate changed. Tests on the effect of xylem pressure on the contributory length showed that such length was decreased with increased pressure, showing that xylem pressure has effects upon the storage parenchyma as well as upon the availability of water to the phloem. Possibly, the above relations of the mouthpart location along a gradient to the exudation rate reflect changes in starch hydrolysis and sugar migration to the sieve tube, more than pressure flow along the sieve tube.

When longer segments were used – 41 cm, for example – the effect of an increase in xylem pressure of 4 atm was greater than with 15-cm segments, although the point of application of the pressure was more remote from the stylets. When the wood at the midpoint of a segment was cut, leaving a narrow bridge of bark joining the two halves, raising the pressure on one half led to an increase in exudation from stylets situated in the other half. Thus, the effect of pressure was transmitted along the phloem, because there was no rise in pressure in the half bearing the mouthparts. A 10^{-4} M solution of DNP applied to abraded bark a few centimeters away from mouthparts reduced exudation after about 6 hours. Applied to the cambial surface, 0.1 percent phloridzin reduced exudation. Exudation was stopped completely by cambial applications of DNP and potassium cyanide, in 10^{-4} and 10^{-6} molal solutions, respectively. Inhibitors had no effect on the concentration of the exudate. Although Peel and Weatherley recognized that the relations of xylem pressure in their studies were complex; and, in 1963, they still believed the sieve-plate pores to be blocked with dense cytoplasm; they felt that their results, in general, lent support to the pressure-flow hypothesis.

From these current studies of phloem exudation through incisions and through aphid stylets, it is apparent that the early conclusions of Münch, Crafts, and others, that the phloem constitutes an osmotic system, operating under positive turgor pressure and capable of transporting the amounts of assimilates normally required for growth and development of plants, are confirmed. With the evidence from phloem anatomy that the plugging of sieve plates is largely an artifact of the preparation processes; and that the sieve plates of normal, functioning sieve tubes have open, tubular strands, capable of conducting fluid at the observed rates of exudation; it seems that much of the controversy surrounding the mechanism of food movement in plants is now settled. Additional evidence that this is true comes from work on the movement of radioactive tracers in plants.

II
Experimental Results

6

Assimilate Movement

The Form in Which Assimilates Move

In most modern research on assimilate movement in plants, the radioactivity of ^{14}C, applied to leaves as $^{14}CO_2$, is measured in plant parts located at some distance from the treated foliage; in this way, translocation may be studied in intact plants. In addition to $^{14}CO_2$, ^{14}C-labeled bicarbonate and urea have proven useful as sources of ^{14}C for synthesis of sugars. Chopowick and Forward (1969) found that when ^{14}C-labeled L-alanine is applied to the leaf surface of *Helianthus annuus*, the bulk of the ^{14}C exported from the absorbing leaf was incorporated in molecules of sucrose. Undoubtedly other sources of this kind will be discovered in the future.

In a much quoted paper, Jones et al. (1959) described experiments on translocation of ^{14}C in *Nicotiana tabacum* following assimilation of $^{14}CO_2$ by a single leaf for periods of 2–3 hours; redistribution of the ^{14}C was determined for periods up to 96 hours. In the treated leaf, there was a complete turnover of sucrose in roughly 24 hours without any change in amount. Starch turned over more slowly, and increased in amount. About 20–30 percent of the ^{14}C was irreversibly incorporated into the treated leaf. About 3 percent of the exported ^{14}C moved acropetally into the young leaves and shoot apex; such movement of labeled assimilate took place for around 6 hours. Mature leaves below the treated leaf did not import labeled assimilate; the balance of activity exported by the treated leaf moved into the stem and roots. Older leaves exported less assimilate to upper leaves; 0.5 percent was so moved by a fully mature leaf; a young leaf only 15 cm in length failed to export assimilate. *Nicotiana tabacum* leaves used in these experiments did not begin to export until they were about one-half their maximum size, and there was a stage when there was simultaneous export and import from a single leaf.

By means of paper chromatography, Mayer and Porter (1960) showed that sucrose, fructose, glucose, and an array of fructose polymers occur in the tillers of *Secale cereale*. Labeled leaves produced radioactive sucrose; all of the fructose polymers were radioactive; in sheaths and internodes, the radioactivity was confined to sucrose. Fractionation of the combined extracts on a charcoal column showed that 13 percent was water-soluble (consisting, presumably, of hexoses); that 82 percent came off in 5 percent ethanol, of which all but 1 percent was sucrose (this 1 percent was trisaccharide); and that some 5 percent of the original extract was composed of oligosaccharides, which did not translocate.

Burley (1961) tested the hypothesis that sugar is the principal form in which carbohydrate is moved in *Rubus occidentalis* and *Glycine max* by analyzing for sugars in stems of these plants that had been fed $^{14}CO_2$. He found that glucose and fructose were radiochemically equal, and that the ratio of sucrose to hexoses increased with the translocation distance. These findings indicate that sucrose is translocated, and that glucose and fructose result from hydrolysis. Small amounts of raffinose were found in both species.

In studies on the translocation of ^{14}C-labeled assimilates in *Solanum tuberosum*, Mokronosov and Bubenshchikova (1962) proposed that there are three types of movement: (1) a slow (20–80 cm/hr) streaming export; (2) a rapid diffuse distribution; and (3) a pulsed passage of assimilates to the stem tip. Radial movement of assimilates from the vascular bundles into cortex and pith was also manifested in *Solanum tuberosum*. The initial flow rate of transportable products from the mesophyll to the conducting elements is higher in young than in old plants. However, the over-all mass flow of assimilates from a leaf, during a 24-hour period, increased with plant age and growth of tubers.

In young plants, the diurnal flow of labeled assimilates is higher under long-day conditions; but for plants 12–20 days old, the diurnal flow rate under short-day conditions exceeds by 20–50 percent that of long-day plants. Thus, tuber formation is enhanced under short-day conditions. Sucrose is the main translocation form of assimilate in *Solanum tuberosum.* In young plants 4–10 days old, considerable amounts of organic acids, amino acids, and hexoses move from leaves to stems. Glucose and fructose are found along with sucrose in stems of plants 30–50 days old. A correlation was found between the translocation rate of assimilates from leaves, and the photosynthetic formation of sucrose in the leaf. No data were presented by Mokronosov and Bubenshchikova that could not be reconciled with a mass-flow mechanism of translocation.

Phloem exudate has proved to be a valuable source of assimilates for studies on the composition of the assimilate streams. Die and Tammes (1964) fed $^{14}CO_2$ to leaves of *Yucca flaccida* and collected exudate from the distal parts of the inflorescense. The maximum amount of radioactivity was collected on the third day after a 30-minute treatment of a single leaf. This indicates that *Yucca flaccida* has an assimilate pool from which export takes place at a steady rate. In all exudate fractions, sucrose accounted for about 90 percent of the radioactivity measured; amino acids accounted for about 6.1 percent, and organic acids, about 3.4 percent. Because the cut infloresсense continues to exude for many days, and produces carbohydrate exceeding the dry weight of the stalk of the infloresсense, Die and Tammes concluded that this phenomenon of continuous phloem bleeding favors the theory of mass movement of assimilates in plants.

Joy (1964), working on assimilate movement in *Beta vulgaris* with $^{14}CO_2$, found that sucrose was the main substance translocated; that glucose and fructose were present in lamina and petiole, but absent in roots; and that amino acids migrated out of leaves during senescence, as proteins were broken down.

Geiger and Swanson (1965a) studied the time-course of the translocation of ^{14}C in small plants of *Beta vulgaris.* The plants had been reduced by pruning to one full-grown source leaf, one small sink leaf, the hypocotyl, and the root. By highly accurate methods, they followed the accumulation of ^{14}C in all of the sinks. When exposure of the source leaf to $^{14}CO_2$ was 7–10 minutes, the rate of arrival of the label attained a maximum value after about 30 minutes, and then diminished to near zero in an additional 50 minutes. The rapid turn over of label in the sucrose pool supported the hypothesis that sucrose is the principal source of solute entering the assimilate stream. A major fraction of ^{14}C at the labeled assimilate front in the petiole was incorporated into sucrose.

In a second paper, Geiger and Swanson (1965b) determined the rate of translocation in *Beta vulgaris* seedlings in units of assimilate-carbon per minute per square decimeter of source leaf. This proved to be 55 μg/min/dm^2, equal to 130 μg/min/dm^2 in sucrose equivalents. A constant rate of ^{14}C accumulation in the monitored sink was attained after about 100 minutes. By evaluating the net rate of sucrose synthesis, the size of the sucrose pool, and its rate of approach to isotopic saturation, Geiger and Swanson obtained, observed, and predicted various values indicating that the total sucrose concentration of the source leaf constituted the active pool for translocation.

Studying translocation and accumulation of sugar in the petioles of *Beta vulgaris*, Geiger and Saunders (1969) reported that a leaf with an area of 0.5 square decimeters exporting at a rate of 11 μg carbon per minute accumulates about 3 percent per centimeter of petiole; accumulation was steady over a 10-hour period of exposure to light. Increasing the specific activity of $^{14}CO_2$ increased the transport velocity to about 1 cm per minute. Measurements on the phloem cross-sectional area indicated the available area to be 6 times that necessary. After 1 hour of steady-state labeling with $^{14}CO_2$, about 70 percent of the ^{14}C in the petiole was in transit. Histoautoradiography of cross sections of petioles demonstrated concentrated silver grains over the young phloem. Probably, the ^{14}C in the mature sieve tubes was lost in preparation.

The few studies cited above are examples of the kinds of experimental work being done on assimilate transport in plants; literally hundreds of such studies could be cited; but, in order to keep the presentation brief, Table 6.1 has been prepared listing the compounds that have been reported to be phloem-mobile.

By the use of both $^{14}CO_2$ and ^{14}C-labeled sugars, Trip et al. (1965), using *Fraxinus americana* and *Syringa vulgaris* plants, determined that verbascose, stachyose, raffinose, sucrose, sorbitol, and mannitol —all nonreducing compounds—were translocated; whereas the reducing sugars melibiose, galactose, glucose, fructose, and pentose were not mobile in the phloem. This selective transport system, in delivering the 6 nonreducing sugars into the sieve tubes for export to other parts of the plant, was not altered by the amount of reducing sugar in the supply leaf. Trip and his co-workers suggested that the nonreducing properties of certain sugars may be related to their function as transport materials. (See also Wanner, 1953a.)

In the rapid transport of solutes along the stem or roots of a plant, one sugar may be preferred over another. This was termed "preferential transport" by Trip et al. (1965). Zimmerman (1958a) referred to this in terms of the enzymes involved in the splitting of the

Table 6.1. Assimilates that have been found to be phloem-mobile in plants.

Group and substance	Plant	Reference
SUGARS		
Fructose	*Beta vulgaris*	Trip et al. (1963); Joy (1964)
	Brassica capitata	Keen and Williams (1969)
	Cucurbita ficifolia	Eschrich and Kating (1964)
	Fraxinus americana	Webb and Burley (1964)
	Glycine max	Vernon and Aronoff (1952); Burley (1961)
	Phaseolus vulgaris	Bachofen (1962b)
	Pinus strobus	Shiroya et al. (1962)
	Rubus occidentalis	Burley (1961)
	Secale cereale	Mayer and Porter (1960)
	Solanum tuberosum	Mokronosov and Bubenshchikova (1962)
	Tilia cordata	Meyer-Mevius (1959)
	Vitis labruscana	Swanson and El Shishiny (1958)
Fructose polymers	*Cucurbita ficifolia*	Eschrich and Kating (1964)
Galactose	*Brassica capitata*	Keen and Williams (1969)
	Brassica oleracea	Trip et al. (1963)
	Fraxinus americana	Webb and Burley (1964)
	Sorbus aucuparia	Meyer-Mevius (1959)
Glucose	*Acer negundo*	Webb and Burley (1964)
	Beta vulgaris	Joy (1964)
	Brassica capitata	Keen and Williams (1969)
	Cucurbita ficifolia	Eschrich and Kating (1964)
	Fraxinus americana	Trip et al. (1963)
	Glycine max	Vernon and Aronoff (1952); Burley (1961)
	Lycopersicon esculentum	Meyer-Mevius (1959)
	Phaseolus vulgaris	Bachofen (1962b)
	Pinus strobus	Shiroya et al. (1962)
	Rubus occidentalis	Burley (1961)
	Secale cereale	Mayer and Porter (1960)
	Solanum tuberosum	Mokronosov and Bubenshchikova (1962)
	Vitis labruscana	Swanson and El Shishiny (1958)
	Yucca flaccida	Die and Tammes (1966)

Table 6.1.—*Continued*

Group and substance	Plant	Reference
SUGARS (contd.)		
Maltose	*Brassica oleracea*	Trip et al. (1963)
Mannose	*Brassica capitata*	Keen and Williams (1969)
Mellibiose	*Apium graveolens*	Trip et al. (1963)
Raffinose	*Asparagus officinalis*	Trip et al. (1963)
	Cucumis sativus	Kluge (1967)
	Cucurbita ficifolia	Eschrich and Kating (1964)
	Cucurbita melopepo torticollis	Webb and Gorham (1964)
	Cucurbita pepo	Hendrix (1968)
	Fraxinus americana	Zimmermann (1958a, 1958b); Trip et al. (1965)
	Malus sylvestris	Webb and Burley (1962)
	Mentha piperita	Cruz-Perez and Durkin (1964)
	Phaseolus vulgaris	Bachofen (1962)
	Pinus strobus	Shiroya et al. (1962)
	Salix viminalis	Peel and Weatherley (1959)
	Syringa vulgaris	Trip et al. (1965)
	Tilia cordata	Meyer-Mevius (1959)
	Tilia tomentosa	Eschrich (1961)
	Verbascum thapsus	Webb and Burley (1964)
Stachyose	*Cucumis sativus*	Trip et al. (1963); Kluge (1967)
	Cucurbita ficifolia	Eschrich and Kating (1964)
	Cucurbita melopepo torticollis	Webb and Gorham (1964)
	Cucurbita pepo	Pristupa (1959); Webb and Burley (1964); Hendrix (1968)
	Echinocystis lobata	Trip et al. (1963)
	Fraxinus americana	Zimmerman (1958a, 1958b); Webb and Burley (1964); Trip et al. (1965)
	Malus sylvestris	Webb and Burley (1962)
	Mentha piperita	Cruz-Perez and Durkin (1964)
	Salix viminalis	Peel and Weatherley (1959)
	Syringa vulgaris	Trip et al. (1965)
	Tilia tomentosa	Eschrich (1961)
	Verbascum thapsus	Webb and Burley (1964)

Group and substance	Plant	Reference
SUGARS (contd.)		
Sucrose	*Acer negundo*	Webb and Burley (1964)
	Beta vulgaris	Fife et al. (1962); Joy (1964)
	Cucumis sativus	Kluge (1967)
	Cucurbita ficifolia	Eschrich (1961); Eschrich and Kating (1964)
	Cucurbita melopepo torticollis	Webb and Gorham (1964)
	Cucurbita pepo	Pristupa (1959)
	Fraxinus americana	Zimmermann (1958a)
	Glycine max	Vernon and Aronoff (1952); Clauss et al. (1964)
	Heracleum mantegazzianum	Ziegler and Mittler (1959)
	Humulus lupulus	Meyer-Mevius (1959)
	Malus sylvestris	Bieleski (1969)
	Ricinus communis	Kriedmann and Beevers (1967a)
	Saccharum officinarum	Hartt and Kortschak (1963)
	Salix viminalis	Peel and Weatherley (1959)
	Secale cereale	Mayer and Porter (1960)
	Sorbus aucuparia	Meyer-Mevius (1959)
	Syringa vulgaris	Trip et al. (1963)
	Tilia cordata	Meyer-Mevius (1959)
	Verbascum thapsus	Webb and Burley (1964)
	Vitis labruscana	Swanson and El Shishiny (1958)
	Yucca flaccida	Tammes and Die (1964); Die and Tammes (1966)
Trehalose	*Brassica capitata*	Keen and Williams (1969)
Trisaccharide	*Cucurbita ficifolia*	Eschrich and Kating (1964)
Verbascose	*Cucumis sativus*	Trip et al. (1963); Kluge (1967)
	Cucurbita pepo	Pristupa (1959); Hendrix (1968)
	Fraxinus americana	Zimmerman (1958a, 1958b); Webb and Burley (1964); Trip et al. (1965)
	Mentha piperita	Cruz-Perez and Durkin (1964)
	Salix viminalis	Peel and Weatherley (1959)
	Syringa vulgaris	Trip et al. (1965)

Table 6.1.—*Continued*

Group and substance	Plant	Reference
SUGAR ALCOHOLS		
Dulcitol	*Cucumis sativus*	Kluge (1967)
Mannitol	*Apium graveolens*	Trip et al. (1963)
	Cucumis sativus	Kluge (1967)
	Fraxinus americana	Trip et al. (1965)
	Salix viminalis	Peel and Weatherley (1959)
	Syringa vulgaris	Trip et al. (1963)
Myoinositol	*Brassica capitata*	Keen and Williams (1969)
	Castanea sativa	Ziegler and Ziegler (1962)
	Cucumis sativus	Kluge (1967)
	Robinia pseudoacacia	Ziegler and Ziegler (1962)
	Salix alba	Ziegler and Ziegler (1962)
Sorbitol	*Cucumis sativus*	Kluge (1967)
	Fraxinus americana	Trip et al. (1965)
	Malus sylvestris	Webb and Burley (1962); Hansen (1967a); Bieleski (1969)
	Salix viminalis	Peel and Weatherley (1959)
	Syringa vulgaris	Trip et al. (1963, 1965)
SUGAR PHOSPHATES		
Fructose-1,6-diphosphate	*Beta vulgaris*	Kursanov (1963)
Fructose-6-phosphate	*Beta vulgaris*	Kursanov (1963)
	Brassica rapa	Bieleski (1969)
	Cucurbita maxima	Bieleski (1969)
Glucose-1-phosphate	*Cucurbita maxima*	Bieleski (1969)
	Tilia tomentosa	Eschrich (1961)
Glucose-6-phosphate	*Beta vulgaris*	Kursanov (1963)
	Brassica rapa	Bieleski (1969)
	Cucurbita maxima	Bieleski (1969)
Hexose diphosphate	*Brassica rapa*	Bieleski (1969)
	Cucurbita maxima	Bieleski (1969)
Mannose-6-phosphate	*Brassica rapa*	Bieleski (1969)
	Cucurbita maxima	Bieleski (1969)

Group and substance	Plant	Reference
ORGANIC ACIDS		
Citric acid	*Cucurbita ficifolia*	Eschrich and Kating (1964); Kating and Eschrich (1964)
	Salix viminalis	Peel and Weatherley (1959)
Diketogulonic acid	*Robinia pseudoacacia*	Ziegler and Ziegler (1962)
α-Ketoglutaric acid	*Cucurbita ficifolia*	Kating and Eschrich (1964)
Maleic acid	*Cucurbita ficifolia*	Kating and Eschrich (1964)
Malic acid	*Glycine max*	Nelson (1962); Clauss et al. (1964)
Oxalic acid	*Cucurbita ficifolia*	Eschrich and Kating (1964); Kating and Eschrich (1964)
	Salix viminalis	Peel and Weatherley (1959)
Succinic acid	*Cucurbita ficifolia*	Eschrich and Kating (1964); Kating and Eschrich (1964)
Tartaric acid	*Cucurbita ficifolia*	Eschrich and Kating (1964); Kating and Eschrich (1964)
	Salix viminalis	Peel and Weatherley (1959)
ORGANIC PHOSPHATES		
ADP	*Brassica rapa*	Bieleski (1969)
	Cucurbita maxima	Bieleski (1969)
AMP	*Beta vulgaris*	Kursanov (1963)*
	Cucurbita maxima	Bieleski (1969)
ATP	*Beta vulgaris*	Kursanov (1963)
	Brassica rapa	Bieleski (1969)
	Fraxinus americana	Kluge and Ziegler (1964)
	Robina pseudoacacia	Kluge and Ziegler (1964)
	Tilia platyphyllos	Kluge and Ziegler (1964); Gardner and Peel (1969)
GTP	*Cucurbita maxima*	Kluge and Ziegler (1964)
Phospho-glycerate	*Brassica rapa*	Bieleski (1969)
	Cucurbita maxima	Bieleski (1969)
Phosphoryl choline	*Brassica rapa*	Bieleski (1969)

*Kursanov identified a number of organic compounds in vascular bundles, but did not prove their presence in the assimilate stream.

Table 6.1.—*Continued*

Group and substance	Plant	Reference
ORGANIC PHOSPHATES (contd.)		
Phosphoryl ethanolamine	*Brassica rapa*	Bieleski (1969)
	Cucurbita maxima	Bieleski (1969)
UDP	*Beta vulgaris*	Kursanov (1963)
UDPG	*Beta vulgaris*	Kursanov (1963)
	Cucurbita maxima	Bieleski (1969)
	Heracleum mantegazzianum	Ziegler (1960b)
UMP	*Brassica rapa*	Bieleski (1969)
UTP	*Brassica rapa*	Bieleski (1969)
	Cucurbita maxima	Bieleski (1969)
AMINO ACIDS		
Alanine	*Cucurbita ficifolia*	Eschrich and Kating (1964); Kating and Eschrich (1964)
	Glycine max	Clauss et al. (1964)
Alanylamino-butyric acid	*Cucurbita ficifolia*	Eschrich and Kating (1964)
γ-Aminobutyric acid	*Cucurbita ficifolia*	Kating and Eschrich (1964)
Arginine	*Cucurbita ficifolia*	Eschrich and Kating (1964); Kating and Eschrich (1964)
Asparagine	*Cucurbita ficifolia*	Eschrich and Kating (1964)
	Glycine max	Clauss et al. (1964)
	Salix viminalis	Peel and Weatherley (1959)
Aspartic acid	*Cucurbita ficifolia*	Eschrich and Kating (1964)
	Glycine max	Clauss et al. (1964)
	Salix viminalis	Peel and Weatherley (1959)
Citrulline	*Cucurbita ficifolia*	Eschrich and Kating (1964)
Glutamic acid	*Cucurbita ficifolia*	Eschrich and Kating (1964)
	Glycine max	Clauss et al. (1964)
	Salix viminalis	Peel and Weatherley (1959)
Glutamine	*Cucurbita ficifolia*	Eschrich and Kating (1964); Kating and Eschrich (1964)
	Yucca flaccida	Tammes and Die (1964)
Glycine	*Cucurbita ficifolia*	Eschrich and Kating (1964); Kating and Eschrich (1964)
	Glycine max	Clauss et al. (1964)
Glycylketo-glutaric acid	*Cucurbita ficifolia*	Eschrich and Kating (1964)

Group and substance	Plant	Reference
AMINO ACIDS (contd)		
Leucine	*Cucurbita ficifolia*	Eschrich and Kating (1964)
Serine	*Glycine max*	Nelson (1962); Clauss et al. (1964)
Valine	*Cucurbita ficifolia*	Eschrich and Kating (1964); Kating and Eschrich (1964)
	Yucca flaccida	Tammes and Die (1964)
GROWTH REGULATORS		
Abscisic Acid	*Salix viminalis*	Hoad (1967)
Gibberellin	*Fagus sylvatica*	Kluge et al. (1964)
	Quercus robur	Kluge et al. (1964)
	Robinia pseudoacacia	Kluge et al. (1964)
	Salix viminalis	Hoad and Bowen (1968)
	Taraxacum officinale	Hoad and Bowen (1968)
	Tilia cordata	Kluge et al. (1964)
	Vicia faba	Hoad and Bowen (1968)
IAA	*Fagus sylvatica*	Huber et al. (1937)
	Phaseolus coccineus	Fletcher and Zalik (1965)
NUCLEIC ACIDS		
DNA	*Robinia pseudoacacia*	Ziegler and Kluge (1962)
RNA	*Cucurbita maxima*	Bieleski (1969)**
	Robinia pseudoacacia	Ziegler and Kluge (1962)
VITAMINS		
Ascorbic acid	*Hippophae salicifolia*	Ziegler and Ziegler (1962)
	Ulmus carpinifolia	Ziegler and Ziegler (1962)
Biotin	*Cornus mas*	Ziegler and Ziegler (1962)
Nicotinic acid	*Robinia pseudoacacia*	Ziegler and Ziegler (1962)
	Ulmus carpinifolia	Ziegler and Ziegler (1962)
Pantothenic acid	*Quercus borealis maxima*	Ziegler and Ziegler (1962)

**Bieleski found that ^{32}P moved down the sieve tubes and rapidly migrated to surrounding cells, and thus concluded that any study involving whole vascular bundles, no matter for how short a translocation time, leads to ambiguous results.

Table 6.1.—*Continued*

Group and substance	Plant	Reference
VITAMINS (contd.)		
Pantothenic acid (contd.)	*Robinia pseudoacacia*	Ziegler and Ziegler (1962)
Folic acid	*Robinia pseudoacacia*	Ziegler and Ziegler (1962)
Pyridoxine	*Fagus sylvatica*	Ziegler and Ziegler (1962)
	Tilia cordata	Ziegler and Ziegler (1962)
Riboflavine	*Populus tremuloides*	Ziegler and Ziegler (1962)
	Tilia europaea	Ziegler and Ziegler (1962)
Thiamine	*Populus regenerata*	Ziegler and Ziegler (1962)
	Robinia pseudoacacia	Ziegler and Ziegler (1962)***
STEROIDS		
Steroids	*Phaseolus vulgaris*	Biddulph and Cory (1965)
ENZYMES		
Acid invertase	*Saccharum officinarum*	Hatch and Glasziou (1964)
Acid phosphatase	*Cucurbita pepo*	Kuo (1964)****
	Tamus communis	Braun and Sauter (1964)
	Tilia americana	Lester and Evert (1965)
Alcohol dehydrogenase	*Robinia pseudoacacia*	Kennecke (1969)
Aldolase	*Beta vulgaris*	Kursanov (1963)
	Robinia pseudoacacia	Wanner (1953b); Kennecke (1969)
Amylase	*Robinia pseudoacacia*	Kennecke (1969)
ATPase	*Cucurbita pepo*	Kuo (1964)
	Robinia pseudoacacia	Wanner (1953b)
ATP diphosphohydrolase	*Beta vulgaris*	Kursanov (1963)

***Ziegler and Ziegler determined the vitamin content of many species, but only a few species were selected for presentation.

****Kuo identified five enzymes in hand sections of three cucurbit species by histochemical tests.

Group and substance	Plant	Reference
ENZYMES (contd.)		
Cytochrome oxidase	*Cucurbita pepo*	Kuo (1964)
	Heracleum mantegazzianum	Ziegler (1958)
Enolase	*Robinia pseudoacacia*	Kennecke (1969)
Fructokinase	*Robinia pseudoacacia*	Kennecke (1969)
Fumarase	*Robinia pseudoacacia*	Kennecke (1969)
α-D-Galactosidase	*Beta vulgaris*	Kursanov (1963)
Glucokinase	*Robinia pseudoacacia*	Kennecke (1969)
Glucose-6-phosphatase	*Robinia pseudoacacia*	Wanner (1953b); Kennecke (1969)
Glutamate dehydrogenase	*Robinia pseudoacacia*	Kennecke (1969)
Glutamate-oxalacetate transaminase	*Robinia pseudoacacia*	Kennecke (1969)
Glutamate-pyruvate transaminase	*Robinia pseudoacacia*	Kennecke (1969)
Glyceraldehyde-3-phosphate dehydrogenase	*Robinia pseudoacacia*	Kennecke (1969)
Glycerolphosphatase	*Robinia pseudoacacia*	Wanner (1953b)
Hexokinase	*Beta vulgaris*	Kursanov (1963)
	Robinia pseudoacacia	Wanner (1953b)
Hexosediphosphatase	*Robinia pseudoacacia*	Wanner (1953)
Inorganic pyrophosphatase	*Robinia pseudoacacia*	Kennecke (1969)

Table 6.1.—*Continued*

Group and substance	Plant	Reference
ENZYMES (contd.)		
Isocitrate dehydrogonase	*Robinia pseudoacacia*	Kennecke (1969)
Malate dehydrogonase	*Robinia pseudoacacia*	Kennecke (1969)
Neutral invertase	*Saccharum officinarum*	Hatch and Glasziou (1964)
Peroxidase	*Cucurbita pepo*	Kuo (1964)
	Passiflora coerulea	Ullrich (1961)
Phosphatase (unspecific)	*Beta vulgaris*	Kursanov (1963)
	Heracleum mantegazzianum	Ziegler (1969)
6-Phosphofructokinase	*Robinia pseudoacacia*	Kennecke (1969)
Phosphoglucomutase	*Robinia pseudoacacia*	Wanner (1953b)
2,3-Phosphoglycerate mutase	*Robinia pseudoacacia*	Kennecke (1969)
Phosphohexose isomerase	*Beta vulgaris*	Kursanov (1963)
Phosphorylase	*Beta vulgaris*	Kursanov (1963)
	Robinia pseudoacacia	Kennecke (1969)
Polyphenol oxidase	*Cucurbita pepo*	Kuo (1964)
Pyruvate decarboxylase	*Robinia pseudoacacia*	Kennecke (1969)
Pyruvate kinase	*Robinia pseudoacacia*	Kennecke (1969)
Succinic dehydrogenase	*Beta vulgaris*	Kursanov (1963)
Sucrose-6-phosphatase	*Heracleum mantegazzianum*	Ziegler (1969)
Triose isomerase	*Robinia pseudoacacia*	Kennecke (1969)

Table 6.1.—*Continued*

Group and substance	Plant	Reference
ENZYMES (contd.)		
UDPG-fructose glucosyltransferase	*Robinia pseudoacacia*	Kennecke (1969)
UTP-glucose-1-phosphate uridylyltransferase	*Robinia pseudoacacia*	Kennecke (1969)
INORGANIC SUBSTANCES		
Boron	*Yucca flaccida*	Tammes and Die (1966)
Calcium	*Yucca flaccida*	Tammes and Die (1966)
Copper	*Yucca flaccida*	Tammes and Die (1966)
Iron	*Yucca flaccida*	Tammes and Die (1966)
Magnesium	*Yucca flacidda*	Tammes and Die (1966)
Manganese	*Yucca flaccida*	Tammes and Die (1966)
Nitrogen	*Yucca flaccida*	Tammes and Die (1966)
Phosphorus	*Yucca flaccida*	Tammes and Die (1966)
Potassium	*Salix viminalis*	Peel and Weatherly (1959); Peel (1963)
	Yucca flaccida	Tammes and Die (1966)
Sodium	*Salix viminalis*	Peel and Weatherley (1959); Peel (1963)
	Yucca flaccida	Tammes and Die (1966)
Zinc	*Yucca flaccida*	Tammes and Die (1966)

oligosaccharides. Accepting the mass-flow hypothesis, it seems unlikely that transport itself would be preferential; it is the transformation of sugars and their uptake into adjoining parenchyma cells that would be preferential. The distribution function that goes on within the total symplast is adapted to provide every living cell of the plant with its nutrient requirements, and the preferences are expressed by those cells that are to receive the nourishment. One additional aspect of the transformation of oligosaccharides is the gain in osmotic activity as the larger molecules are split to simpler sugars. This may well represent an essential feature of the osmoregulatory function of the pressure-flow system.

Citing herbicides as examples, Crafts and Yamaguchi (1964) illustrated the phenomenon of selective transport. In studying the movements of 2,4-D and monuron, they found that the former, upon penetrating the cuticle, is apparently taken up by the symplast and moved to the sieve tubes, in which it is rapidly exported to active sinks. Monuron, under identical conditions, penetrates the cuticle but seems unable to enter the symplast; it follows the flow of transpiration water along the cell walls and forms a wedge-shaped pattern as it moves to the leaf apex. Preferential movement is illustrated by the difference between transport of 2,4-D and amitrole in the phloem. As 2,4-D moves along the sieve tubes, it is preferentially absorbed by phloem parenchyma cells, and it accumulates in the total cross section of petiole, stem, and root. Because of this ready absorption, a given dosage of 2,4-D may not get into the roots, but its autograph fades out as its movement slows and stops within the stem. Amitrole, by contrast, moves rapidly along the phloem, traversing petiole, stem, and roots; it tends to accumulate in root tips, with the autograph of petiole, stem, and old roots showing a light but rather uniform labeling (Figure 6.1).

In a very provocative paper, Arnold (1968) proposed that sucrose, the principal translocated assimilate in so many plants, acts as a protected derivative of glucose, and thus, that it has evolved by dint of certain selective pressures as the preferred form of carbohydrate for translocation. Examining physical parameters, Arnold concludes that solubility of sucrose is not important under conditions which support translocation. The densities, viscosities, and surface tensions of sucrose and glucose solutions are so similar that preference based upon flow properties is not valid.

By a similar analysis, Arnold concluded that a specific role for sucrose in translocation revolving around tonicity factors is not convincing, nor do the differences in osmotic properties between glucose and sucrose provide a logical basis for preference of the latter. From a consideration of energy relations, there is likewise only questionable advantage. A molecule of glucose or fructose can, potentially, yield 38 molecules of ATP by metabolism. Sucrose, acted upon by sucrose phosphorylase to yield glucose phosphate and fructose, can produce 37 and 38 molecules of ATP; thus the advantage amounts to only one-half molecule of ATP. If sucrose is hydrolysed by invertase to yield free glucose and fructose, even this small increment is not realized.

Arnold concluded therefore, that no model based upon the physical properties examined offers more than a superficial advantage. Con-

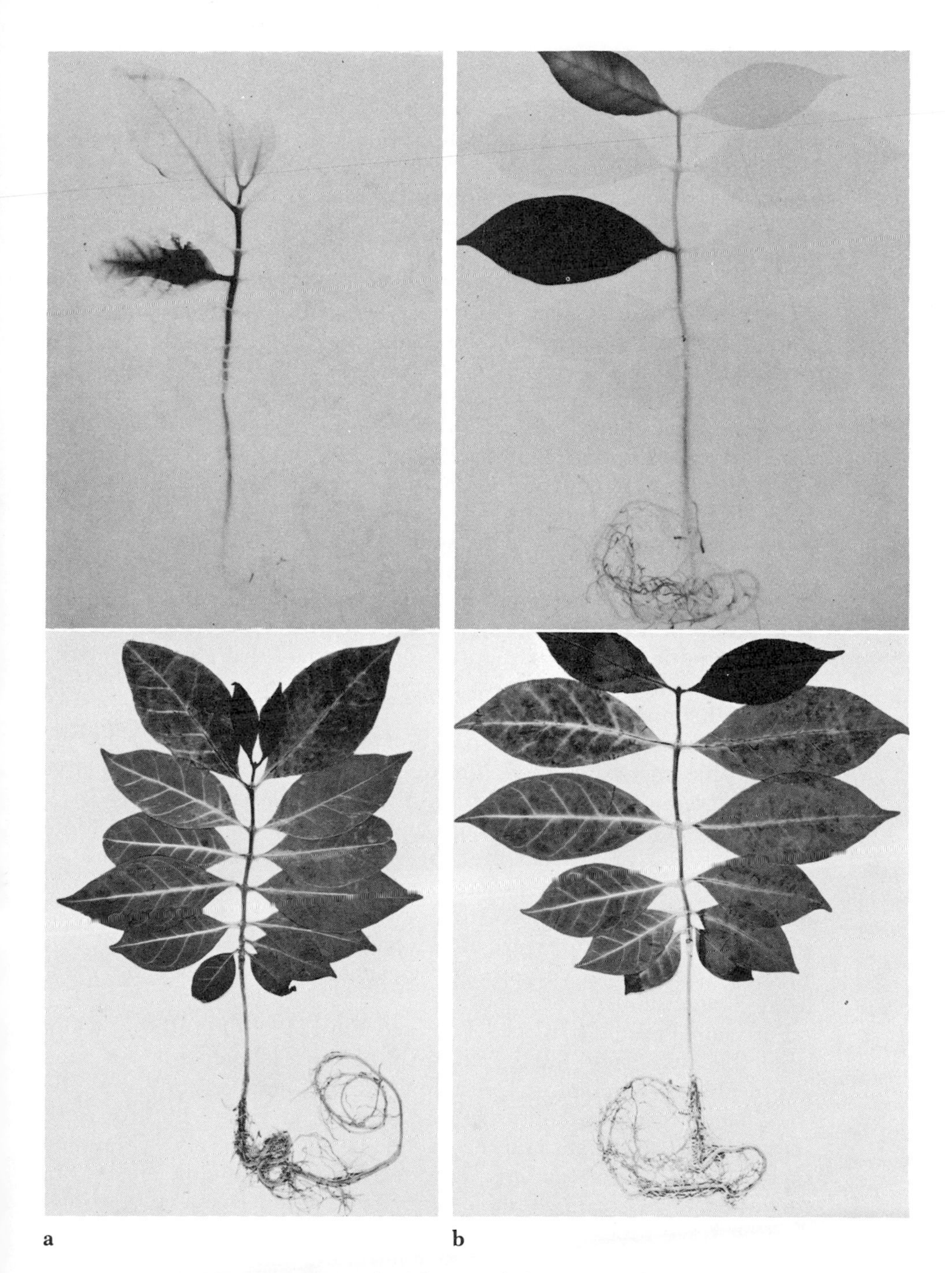

a b

Figure 6.1. Autoradiographs (*above*) and photographs (*below*) showing distribution of ^{14}C in seedlings of *Coffea arabica* treated with (*a*) ^{14}C-labeled 2,4-D, and (*b*) ^{14}C-labeled amitrole. Treated leaf is third from top on the left.

sidering chemical properties, Arnold pointed out that the ubiquitous distribution of enzymes that catalyze the metabolism of glucose would make this carbohydrate extremely vulnerable as a translocated assimilate. Sucrose is much less reactive, and is thus able to move over long distances and to deliver its glucose residues to specific sites of growth or storage. Also, a metabolic constraint is placed upon the acceptor sites, because the subsequent role of sucrose molecules determines the path of carbon in the metabolism of the site tissues.

Examining the protective-derivative hypothesis, Arnold examined the properties of sucrose which equip it for its unique role in translocation. Many glucosides are ruled out: the aglycons that would be released by hydrolysis might be toxic; or they might not be readily metabolized, and hence, be inefficient energy sources. The disaccharides of glucose are the most logical compounds to serve this important role in translocation; Arnold reasoned that the reducing forms would be less desirable because of their reactiveness; Trip et al. (1963) found only the nonreducing sugars to be phloem-mobile. Arnold noted that the relative ease with which sucrose is hydrolyzed is one of its desirable features. He pointed out the fact that young, growing cells may be high in invertase. Mature sieve tubes may have little or none of this enzyme.

Thus, the properties of sucrose, and of the translocation machinery of the phloem, seem fitted, through evolution, for their respective roles in assimilate distribution in plants. As shown in Table 6.1, a good many sugars, sugar alcohols, and organic acids have been identified as being present in the phloem and, presumably, in the assimilate stream. Arnold pointed out that the nonreducing sugars found have 1, 2, or 3 molecules of D-galactose bound to a sucrose residue; their structures are thus consistent with the concept of a protected derivative. Verbascose appears to be the largest of the translocated sugar molecules.

Ford and Peel (1967b) and Peel and Ford (1968) found that when labeled hexoses were fed to bark strips of *Salix*, labeled sucrose came out in phloem exudate or aphid honeydew. Tammes, Vonk, and Die (1967) fed ^{14}C-labeled glucose and fructose to young inflorescence stalks of *Yucca flaccida* and recovered almost exclusively ^{14}C-labeled sucrose in phloem exudate. Evidently these reducing sugars are converted to sucrose in their passage from parenchyma cells to conducting sieve tubes. Arnold pointed out that in the sinks, a few enzymes capable of splitting sucrose control the path of carbon. Shiroya et al. (1962) presented the following scheme to explain their results with distribution of ^{14}C in *Pinus* seedlings.

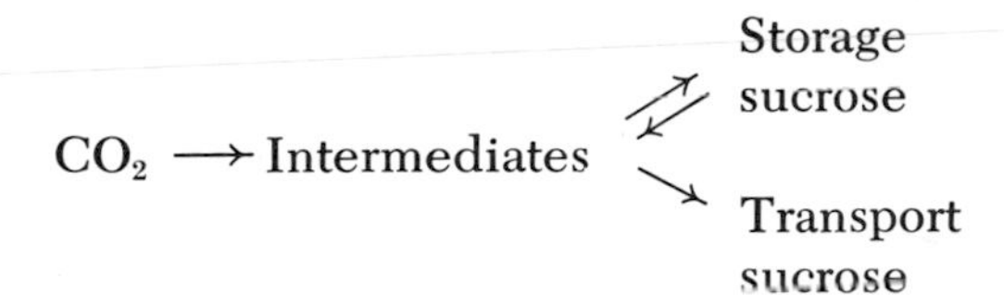

As Arnold concluded, the ultimate fate of assimilates depends upon the presence or absence of one or two key enzymes. In storage tissues, the commonly found glucans, fructans, and galactans are a reflection of the variously evolved systems of utilization for the translocated sugars.

Distribution Patterns of Assimilates

As an integrating force in plant development, translocation is responsible for the organic nutrition of all the living cells of the plant. Thus, the forms in which assimilates move, and the directions of their transport, are critical in determining the ultimate structure and function of a plant, be it a grass, a vine, or a tree. Swanson (1959) reviewed the problems of the distribution of assimilates. This section considers the distribution patterns that have been found, largely by tracing ^{14}C-labeled sugars in plants.

The generalization that states that the lower leaves of a plant export primarily to the roots, that the upper leaves export to the shoot apex, that intermediate leaves export in both directions, that mature leaves only export, and that young, growing leaves only import, is borne out by work of Bel'kov (1955), Crafts (1956b), Crafts and Yamaguchi (1964), Lawton (1967a, 1967b), Quinlan (1965), Thrower (1962), Tuichibaev and Kruzhilin (1965), Wardlaw (1968a), and many others. It is important to realize, however, that this generalized pattern of distribution may vary somewhat with the size, shape, and general organization of the particular plant in question. For example, transport from the cotyledons of *Glycine, Cucurbita,* or *Gossypium* is, initially, almost totally to the roots; from the primary leaves of *Phaseolus* it is the same, until the shoot starts active growth. In contrast, in full-grown plants of *Triticum aestivum* that are producing a heavy crop, very little assimilate moves from leaves to roots during maturation of the seed (Wardlaw and Porter, 1967).

If one prunes a plant, as Geiger and Swanson (1965a) did, one can alter the pattern of distribution. Defoliation, as used by Forde (1966a, 1966b), is another way to change the direction of assimilate flow. And some plants, such as *Lycopersicon esculentum,* may behave differently with respect to assimilate movement (Khan and Sagar, 1966). However, to analyze the results of these many studies from the standpoint of source–sink translocation by pressure flow, and to give proper consideration to the complexities of phloem distribution, phyllotaxis, and growth and accumulation responses, is to discover that there is much evidence to support the simple concept of mass flow, and little or none to refute it.

In studies on translocation of labeled assimilates in *Glycine max*, Thaine et al. (1959) employed $^{14}CO_2$ to provide a source of tracer for determining distribution patterns by autoradiography and counting. They found, as have others, that for the lowermost leaves of a plant having three or more leaves, the major sink for assimilates is the root, for the uppermost leaves it is the apex and young expanding leaves, and that leaves in a median position export to both apex and root; mature leaves are by-passed. This latter pattern characterizes experiments where $^{14}CO_2$ was supplied to a mature, illuminated leaf of a normal plant. They also found translocation to proceed in darkness, following fixation by illuminated leaves; and after 24 hours, the concentration of mobile labeled assimilate in the source leaf was very low. Evidently, during active photosynthesis, a pool of labeled assimilate is built up that continues to move out in the dark. Possibly, this pool is comprised of labeled starch.

Removal of all mature leaves between the source and the apex significantly increased movement to the apex in the experiments of Thaine and his co-workers. Shading of the green apex decreased the import of tracer in some experiments, but not in others. The direction of movement from a given leaf was found to depend upon the age and position of the leaf on the stem. A young, expanding leaf forms part of the major apical sink for upward-moving assimilates, and for its own photosynthetic products. It exports nothing to the rest of the plant. When it has reached about 60 percent of its mature size (approaching compensation), it begins to lose its function as a sink, and starts exporting. Thaine and his co-workers suggested that labeled compounds move along a gradient of the whole assimilate, labeled and unlabeled. While these authors do not commit themselves, and Thaine later adopted an alternative mechanism (Thaine, 1964b), the obvious explanation for the experimental results described here is the mass-flow mechanism.

Hartt and Kortschak (1963) reported on many years research on translocation in *Saccharum officinarum*. Some of their findings were as follows: (1) sucrose is the chief compound translocated in *S. officinarum;* (2) sucrose made in the blade is quickly mobilized into the veins, and immediately proceeds down the blade and sheath into the stem, especially the center of the stem; (3) upon reaching the joint (node and internode), sucrose first turns downward and moves down several joints before any goes upward; (4) some sucrose reaches the roots and goes up into the suckers, but most of the sugar is stored in the millable cane; (5) sucrose moves down at rates generally around 1–2 cm per minute; (6) both the linear rates and the total amount translocated are decreased considerably by deficiencies in potassium, phosphorus, or nitrogen; (7) the linear rates and the total amount translocated increase with higher air and soil temperatures; (8) decreased moisture curtailed translocation to upper joints; (9) a detached blade translocates sugar down the veins at rates of the same order of magnitude as an attached leaf; and (10) the energy for translocation in detached blades comes from light. Newly formed sucrose they found to move into axillary shoots and basal suckers within a few hours; it reached the tops of all stalks in a large stool within 24 hours. Sucrose reaching young stems, leaves, and roots is converted to other compounds used in growth of cell walls and protoplasm. Each day's sucrose proceeds to the stem where most of it is stored in ripening joints; some is added to mature joints.

The range of velocities of translocation they found to be 42–150 cm per hour; the average of 12 determinations was 84 cm per hour. The percentage of counts leaving the treated part of a leaf was 12–54 percent. In studies of injured leaves and parts of leaves, photosynthesis measurements and tissue analyses proved that translocation may be a limiting factor in CO_2 assimilation. Table 6.2 shows the effect of sucrose concentration upon the rate of photosynthesis (Hartt, 1963).

Quinland and Sagar (1962) studied the movement of ^{14}C-labeled assimilates in *Triticum aestivum* plants of different stages of development. Movement was found to be extensive in the early stages, be-

Table 6.2. The effect of sucrose concentration in the blade upon the rate of photosynthesis in *Saccharum officinarum.*

	Time at 6500 ft-c	Sucrose (% dry wt)	Photosynthesis rate (ml $CO_2/dm^2/hr$)
Left lamina, leaf A	10 min in water	3.59	17.25
Left lamina, leaf B	10 min in water	3.84	17.35
Right lamina, leaf A	3 hr in water	6.60	11.40
Right lamina, leaf B	3 hr in 5% sucrose	12.65	1.77

Source: Harrt (1963b).

coming more localized with age until, at a stage after the emergence of the ear, there was little transfer between tillers. In the early stages, ^{14}C from leaves on the main shoot was transported throughout the plant, but accumulation occurred in meristems. Food produced by leaves of newly formed tillers went to meristems of those tillers, with small amounts going to the rest of the plant. In later stages, the ^{14}C was restricted to the treated tiller and to young tillers associated with it. After the emergence of the ear, movement was predominantly towards the ear from the flag leaf, and towards the roots from the leaf below the flag.

In studies on translocation in three varieties of *Triticum aestivum*, Stoy (1963) fed midportions of leaves $^{14}CO_2$ for 30 minutes, and then placed the plants in the dark for 1-hour and 20-hour transport periods. Samples were extracted and counted. Stoy found that translocation of fixed ^{14}C from flag leaves was preferentially to reproductive organs, which were growing rapidly during the translocation period. During flowering, there was no growth, but there was intense export to stem tissues; most of the labeled assimilate was in soluble form in the lower parts of the uppermost stem internode. This is an intercalary region of high metabolic activity. During an 8-week experiment, about 12–15 percent of this accumulation was remobilized and moved into developing grains. There were no essential differences in distribution patterns among the three varieties; the transport periods from flag leaves to ears differed in length; those varieties that can maintain a sufficient rate of photosynthesis in the upper leaves for a long period after flowering may be the high yielding ones. Stoy's study emphasizes the importance of active sinks in assimilate transport.

Carr and Wardlaw (1965) followed the movement of ^{14}C assimilates between the flag leaf, the glumes, and the grain of *Triticum aestivum*. In the cultivar 'Sabre,' 49 percent of the ^{14}C assimilated by the flag-leaf blade, and up to 60 percent of that from the glumes, moved to the grain. After anthesis, photosynthesis by the ear increased for up to 15 days; assimilation by the leaves declined. For the first week after anthesis, the growing top internode acted as a sink for assimilate from both flag leaf and glumes. Seven days after anthesis, this amounted to half of the assimilates leaving the flag leaf. In low light, stem growth may compete with the grain for assimilates during early grain development.

Lupton (1966) treated three varieties of *Triticum aestivum* with $^{14}CO_2$ as they grew in the field. In young plants, translocation proceeded freely, labeling all parts. As the stems elongated, the tillers became autotrophic. The efficiency of translocation to the grain increased during the first four weeks after anthesis, but then decreased as the grain approached ripening. Transport from glumes and flag leaves was toward the grain; that from leaves 2 and 3 was partly toward the grain and partly basipetal.

In an extensive study, Joy (1964) followed the movement of ^{14}C-labeled assimilates in *Beta vulgaris*. Using 8-week-old plants having 13–15 leaves, and roots about 1.5 cm in diameter, he fed individual leaves $^{14}CO_2$ for 4 hours under an illumination of 1000–1200 ft-c. In such plants, the youngest expanding leaf exported no assimilate; older, fully expanded leaves exported to the young leaves and to the roots. He found phyllotaxis very important in the distribution of ^{14}C in *B. vulgaris*, and no labeled assimilate moved into leaves that were older than the source leaf. In normal plants, the label moved into leaves having direct vascular connection with the source. By removing mature leaves from the untreated side of a plant, however, Joy was able to bring about cross-transfer of the label to the young leaves on the pruned side, a result he attributes to the peculiar vascular structure of the crown, which contains complex vascular connections. Translocation was found to be rapid, over 50 percent of the label leaving the source leaf in 24 hours. A small but significant part of the ^{14}C fed to a leaf remained in it in the form of alcohol-insoluble residue; some of this may be released during senescence.

Mortimer (1965) measured the translocation of assimilates moving in *Beta vulgaris* petioles after feeding the blade $^{14}CO_2$ for 1 minute. During the first 5 minutes, activity could be isolated from the midrib, due to direct assimilation by that tissue. After 8 minutes, radioactive sucrose from the blade reached the midrib and began to move down the petiole. This movement, involving about 1 percent of the assimilated ^{14}C per minute, continued at a steady rate. The concentration of ^{14}C in the petiole was a linear function of the distance, and it moved down the petiole at an apparent velocity of 50–135 cm per hour; after 1 hour, the entire length of the petiole contained radioactive sucrose at a relatively uniform concentration. Although the blade still contained 40–50 percent of the ^{14}C as sucrose, export was reduced to a very low rate after 1 hour; ^{14}C-labeled assimilate continued to leave the leaf for an additional hour. Mortimer concluded that the petiole of *B. vulgaris* plays a relatively passive role in the translocation process.

Quinlan (1965) studied the pattern of distribution of ^{14}C-assimilates translocated from single leaves of *Malus sylvestris*. The plants were 1-year-old 'Malling VII' rootstocks pruned to a single shoot. Treatment was with $^{14}CO_2$ in a clear plastic leaf chamber. All of the leaves treated, except the youngest, exported ^{14}C in sufficient quantity to be visible in autoradiographs; the isotope accumulated only in meristematically active regions. The youngest treated leaf assimilated the

$^{14}CO_2$, but none was exported. Labeled assimilates from the 4th and 6th youngest leaves moved predominantly to the stem tip where they accumulated in the apex and the young expanding leaves. Leaves in the same orthostichy as the treated leaf were heavily labeled. From leaf 8, ^{14}C was translocated, both acropetally, and basipetally into the root system. Movement from all leaves below leaf 8 was downward into the roots, with intervening mature leaves being bypassed. In some mature leaves, there were faint traces of ^{14}C, and Quinlan questioned whether this had moved in via the xylem or the phloem. It seems quite possible that this label represented metabolic products of sugar oxidation in the roots. It has been known for many years that the roots of *M. sylvestris* reduce nitrogen compounds to amide and amino forms which move to the foliage via the xylem. Also, under some conditions of mineral nutrition, aliphatic acids are produced, which balance the cations that move upward and provide nourishment for the plant.

Quinlan (1966) continued his studies on translocation of assimilates in *Malus* using partial defoliation as a physiological tool to alter distribution patterns. Using year-old *Malus* sp. rootstocks, Quinlan treated various leaves and defoliated various regions. Using a plant with 15 fully expanded leaves and three expanding leaves, treatment of leaf 14, which had one expanded and three expanding leaves above it, resulted in translocation of ^{14}C-labeled assimilate that was wholly acropetal; leaf 15 was bypassed.

Treating a similar leaf on a plant having all leaves below the treated leaf removed, resulted in acropetal movement into the expanding leaves, plus a strong movement of the assimilate to the roots, and its accumulation there. Treating leaf 11 on a 17-leaved plant resulted in both acropetal and basipetal movement; removing the lower leaves on a similar plant greatly reduced acropetal movement, but the roots were heavily labeled. Treating a mid-leaf on a plant having leaves above the treated leaf removed, eliminated the upper source, and movement was almost totally acropetal to the meristematic shoot tip. These experiments complemented the earlier ones (Quinlan, 1965), and they emphasized the source-to-sink distribution pattern, established for assimilate movement in *Gossypium* by Mason and Maskell in 1928, and established for many other crops in more recent times.

Hansen (1967a) found that in *Malus sylvestris* the presence of fruits promotes the translocation of assimilates from leaves to other organs. Using $^{14}CO_2$ for labelling, Hansen recorded that nearly 90 percent of the ^{14}C taken up by leaves may be moved to adjacent fruits, mostly during the first 4–5 days following labeling. The ^{14}C content of shoots with fruits is reduced more rapidly than that of vegetative shoots. Young leaves incorporate more ^{14}C than do mature ones. Soon after treatment the leaves contained 58–80 percent of the ^{14}C as sorbitol, 7–9 percent as sucrose, and 1–4 percent as glucose. The ^{14}C-labeled sorbitol content of leaves was higher in vegetative shoots than in fruiting ones.

In a second paper on translocation in *Malus sylvestris*, Hansen (1967b) showed that the application of ^{14}C to the base of a long shoot resulted in the downward export of 80 percent or more of the radioactivity. When leaves at the apex were treated, 80 percent of the tracer was retained in the leaves and shoot tip; a high proportion of this was incorporated. As terminal growth decreased, a larger portion of the ^{14}C from apical leaves was exported. The distribution of activity between sorbitol, sucrose, glucose, and fructose, was not very different in young and mature leaves. Hansen (1967c) also studied the influence of season on storage and transport of labeled compounds. Using 1-year-old rootstocks, he applied $^{14}CO_2$ to leaves, and determined the ^{14}C content of different parts of the tree. Translocation of ^{14}C proceeds at the greatest rate during the first few days after application. Distribution of the ^{14}C depended upon source–sink relations; from the time of application until leaf fall, 40–50 percent of the initially absorbed ^{14}C disappeared from the tree. In autumn, most of the tracer moved into the roots, and much of this was lost during winter and spring; 20–25 percent of the absorbed activity disappeared, and only 13–17 percent was recovered in new growth by June.

Khan and Sagar (1966) made a detailed study of the distribution of ^{14}C-labeled assimilate in plants of *Lycopersicon esculentum* grown in the greenhouse under commercial growing conditions. Treating one leaf per plant, they used plants having 6–20 expanded leaves, the oldest plant also having 11 fruiting trusses.

Lycopersicon esculentum is different from many of the plants that have been used in translocation studies. It is highly vegetative; the stems as well as the leaves carry on photosynthesis; the leaves continue to expand for a long period; the basal leaflets grow, and hence import assimilates, while the tip leaflets are exporting; the stems and old leaves act as storage organs; and the fruits are relatively low in solids. Khan and Sagar found that, 24 hours after treatment, the leaves exposed to $^{14}CO_2$ had moved only around 20 percent of the ^{14}C to other parts of the plant. In the vegetative phase, lower leaves exported more carbon up than down, and upper leaves moved a major portion downward. During the development of the crop, upper leaves continued to export more carbon down than up, and lower leaves did likewise. In the young fruiting plants, all leaves supplied all the trusses; but as the number of trusses increased, groups of leaves tended to supply single trusses. At the time of fruit development, the leaves retained more than 80 percent of the carbon they fixed for 24 hours. Mature leaves and stems, both above and below the test leaves, were significant sinks for ^{14}C compounds.

Neither shoot tips nor roots were strong sinks in these plants. (The stems, as well as leaves of the shoot tips of this species, are green; they are probably fairly self-sufficient, so far as carbohydrate nutrition is concerned.) The roots of the 6-leaf plants absorbed appreciable quantities of ^{14}C-labeled assimilates; from the 11-leaf stage on, absorption by roots was low, and the bulk of the assimilates (68 percent) came from basal leaves, a lesser portion (32 percent) came from the middle leaves, and from apical leaves, none.

The vascular system of *Lycopersicon esculentum* is very complex: there is both external and internal phloem, and secondary growth in the stem soon obliterates phyllotactic effects, so far as the main stem is concerned. And because of the highly vegetative nature of the plant (fleshy green stems, indeterminate growth), diurnal variations in photosynthesis, growth, and storage must bring about local adjustments in translocation patterns involving varying velocities, retardations, and even reversals in direction. These conditions, in addition to the complex growth patterns of the leaves, may well explain the exceptions to the general patterns of assimilate distribution which Khan and Sagar experienced.

Crafts and Yamaguchi (1964) used ^{14}C-labeled urea to study assimilate movement in *Zebrina pendula*. Since tests proved that the labeled urea is rapidly hydrolyzed and the $^{14}CO_2$ synthesized to sucrose, this tracer was used as a convenient source of labeled assimilate. Figure 6.2 shows the distribution of ^{14}C in *Z. pendula* after a 4-day translocation time. Labeling of untreated mature leaves, lacking in the case of amitrole transport (frontispiece), probably represents the upward movement of products of sugar assimilation by roots, via the transpiration stream. Figure 6.3 shows the roots shown in the autoradiograph of Figure 6.2 enlarged (left), and the roots of the mounted plant (right). The labeled sugar has been incorporated in the root hairs, labeling them intensely.

In experiments on translocation in *Cucurbita*, Webb and Gorham (1964) found the following pattern of assimilate distribution. About 14 percent of the total of translocated ^{14}C was retained by the petiole of the treated leaf. The remainder was distributed from the node, acropetally and basipetally, following the vascular system. Growing

Figure 6.2. Translocation of ^{14}C-labeled sugar in *Zebrina pendula.* Urea, labeled with ^{14}C, was applied as a droplet to the heavily labeled leaf and, 4 days later, the plant was freeze-dried and autoradiographed: (*a*) autoradiograph; (*b*) mounted plant.

a

Figure 6.3. Enlarged view of the roots of the autoradiographed plant shown in Figure 6.2: (*a*) autoradiograph showing root hairs with incorporated ^{14}C; (*b*) photograph of the roots of the mounted plant.

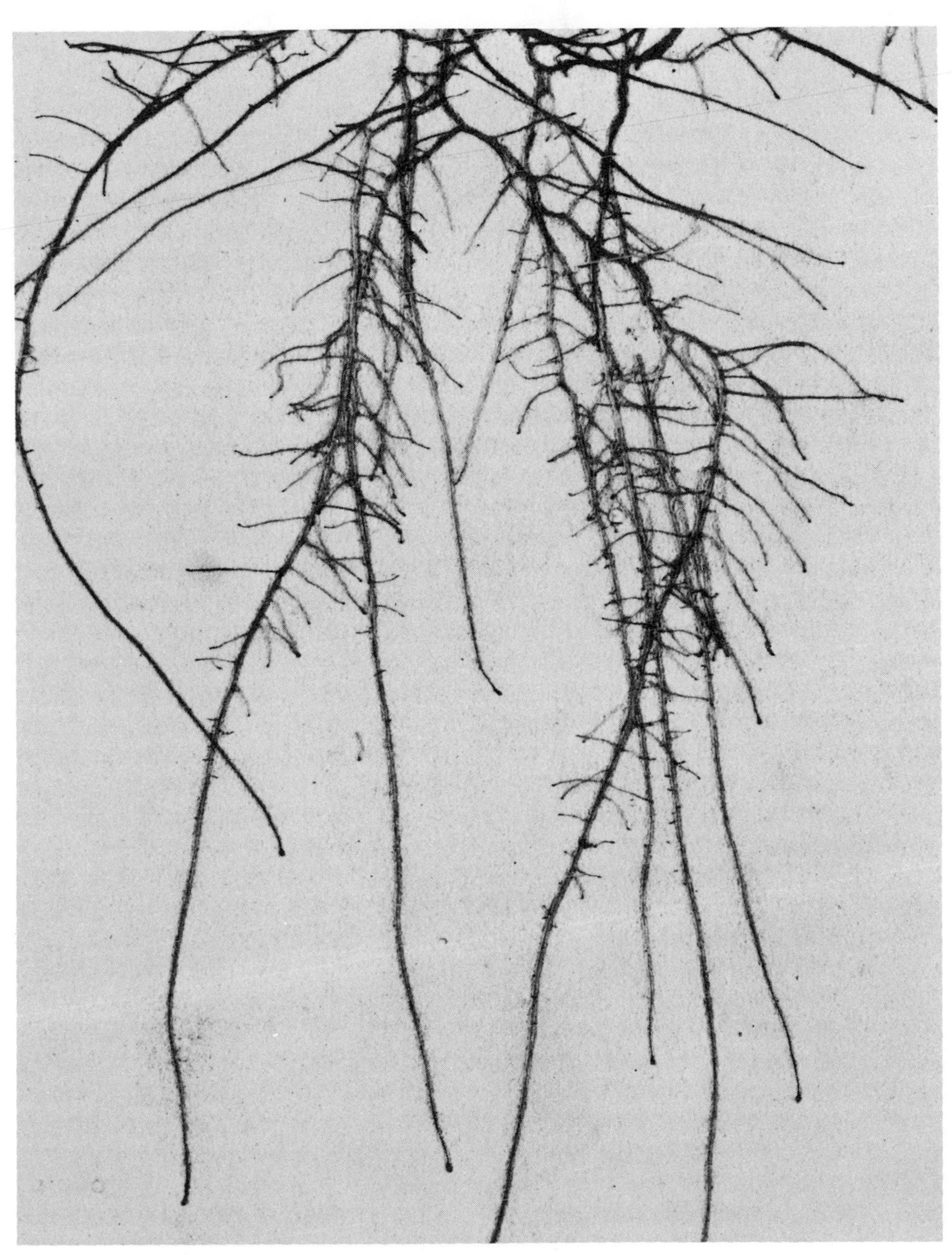

b

leaves and shoot imported ^{14}C-assimilates most actively. An immature leaf could import as much as 50 percent of the total translocate from a mature leaf; there was no export of ^{14}C from such an immature leaf, and no detectable synthesis of ^{14}C-labeled stachyose. As the leaf matured, import declined, stachyose synthesis started, and the export of ^{14}C began; bidirectional movement of ^{14}C in the petiole for 24–36 hours resulted. As the leaf reached maturity, import ceased. Partially yellow, senescent leaves synthesized stachyose and sucrose, and exported them in the assimilate stream.

In mature petioles and stems, transport of ^{14}C took place in the bicollateral bundles and commissural sieve tubes of the phloem accompanied by a slow radial movement to surrounding tissues. Transport of ^{14}C in the phloem of immature petioles was accompanied by rapid movement of ^{14}C into the surrounding tissues. There was no return of ^{14}C from the roots to the foliage in the transpiration stream in these young, active plants.

Continuing their studies on translocation in *Cucurbita,* Webb and Gorham (1965a) made time-sequence studies on the movement of ^{14}C-labeled assimilates at the tissue level by histoautoradiography. Tissue of mature and young petioles, stem, and hypocotyl were examined after 1-minute exposures of the first leaf to $^{14}CO_2$. They found ^{14}C-assimilates to be transported in the phloem from the mature leaf, basipetally to roots, and acropetally to the young leaves and growing point, simultaneously. Movement was rapid, and did not involve gaseous flow of the $^{14}CO_2$. Radial distribution of the ^{14}C-labeled assimilates occurred along the entire length of the phloem, ultimately supplying ^{14}C to every living cell of the plant within 2 hours of assimilation. The velocity and rate of radial movement depended upon the stage of development of each organ. Young, rapidly expanding petioles displayed velocities of about 6 cm per hour; older tissues showed velocities around 1 cm per hour. These values compare well with known rates of protoplasmic streaming in myxomycetes, *Nitella* cells, phloem parenchyma, root hairs, etc.; they contrast sharply with the velocity of translocation of ^{14}C in phloem, estimated at between 250 and 300 cm per hour (Webb and Gorham, 1964). Radial movement of ^{14}C-labeled assimilate seemed to occur in the form of sucrose rather than stachyose, suggesting a differential permeability of the conducting phloem elements toward these two sugars. Because Webb and Gorham seem to be sympathetic to the symplast concept, and to recognize the distributional function of the phloem, they do not term this lateral movement "leakage," as so many do. Because there are no permeability barriers to cell-to-cell movement within the symplast,

Zimmermann's (1958a) suggestion of the enzymatic hydrolysis of stachyose to sucrose within sieve tubes seems a more logical view of this phase of the ^{14}C distribution process.

Combining an autoradiographic technic (for identifying the vascular bundles that conducted labeled assimilates) with a fluorescense method (for locating phloem bundles), Biddulph and Cory (1965) studied metabolite movement from *Phaseolus* leaves at four positions on stems having 1–4 expanded trifoliate leaves. Most of the plants were 18 days old, at which age the third trifoliate leaf had passed compensation and was exporting. Time from exposure to $^{14}CO_2$ to the termination of experiments (translocation time) was $1\frac{1}{2}$ hours. They observed that all mature leaves exported a fraction of their newly synthesized metabolite to the stem apex; the proportion of the total export that moved acropetally in the stem varied with the position of the leaf. It was greatest for the leaf closest to the apex, and decreased for older leaves; for the primary leaves it was a small amount, with almost the entire flow moving downward. This is very similar to the pattern previously found for ^{32}P applied to leaves.

Biddulph and Cory found little tangential movement between adjacent phloem bundles, but in the apical regions as much as 49 percent of the label was found in xylem; although this was termed a loss, it seems much more logical that it was the result of symplastic movement, from sieve tubes to phloem parenchyma to xylem parenchyma, in the normal organic nutrition of the growing shoot. In lower regions, this transfer to xylem was as low as 4 percent.

Metabolite from mature leaves moved downward one node where an anastomosis of bundles permitted a division into upward- and downward-moving components. The upward-moving component was confined to bundles alternating with those conducting the downward-moving component from the next higher leaf; bidirectional movement was in separate phloem bundles. Metabolite from young leaves just past compensation left the leaf traces at the node of insertion and moved directly upward. A downward flow in the leaf traces developed as the leaf matured.

Biddulph and Cory identified two of the major metabolites as steroids, a third as sucrose; exportable metabolite from young leaves was rich in steroids; older leaves exported mainly sucrose. Export from older leaves resembled mass flow; from younger leaves, they thought that a metabolic transport was involved.

In a symposium titled "Isotopes and radiation in soil-plant nutrition studies" held by the International Atomic Energy Agency in Vienna in 1965, Ziegler (1965) described the use of isotopes in studies of translocation in rays. As indicated by their appearance, their spatial arrangement, and their anatomy, rays are radial translocation tissues between the two vertical systems of translocation, the xylem and the phloem. In experiments to elucidate their function, agar cubes containing ^{14}C-labeled sucrose were applied to the bark of 10-year-old *Robinia pseudoacacia* trees, and the radial translocation of the ^{14}C in the wood, internal to the feeding area, was determined by autoradiography. In order to accelerate the radial transport, the trunk was girdled 20 cm below the agar cube. These tests showed that the radial translocation was confined to the rays, and that the velocity of transport was such that it could not be explained by normal diffusion.

That the ray cells transport solutes in a radial direction in the stem is not surprising: they are composed of living cells that display protoplasmic streaming; they are elongated in the radial direction; and they are copiously connected to each other by plasmodesmata. They constitute the only directly connected tissue system between the storage parenchyma of the wood and the functioning phloem. By symplastic movement, the large supply of organic reserves normally stored in the trunks of trees may reach the wood parenchyma during the period of active photosynthesis.

Forde (1966a) used ^{14}C-labeled urea as a precursor of ^{14}C-labeled sucrose for tracing assimilate movement in *Cynodon dactylon.* He found assimilate transport to be strongly polarized toward the growing tip of a large stolon, and less polarized toward the tips of small stolons. Tracer movement to the parent stolon from a treated tiller was as great as, or greater than, to the tip of that tiller. Assimilate movement from one stolen to another occurred only occasionally in intact plants in full light (Figure 6.4).

Marshall (1967) used $^{14}CO_2$ and ^{32}P to investigate the nutritional organization of grass plants. *Lolium multiflorum* seedlings growing under greenhouse conditions were his plant materials, and autoradiography was the method used. He found that the highest amount of radioactivity exported from a mature leaf was located in the younger, expanding leaves and in the roots, especially in the root apices. Thus, assimilates in his plants moved to those regions having the most rapid growth.

In plants having two tillers, the labeled assimilates produced by each tiller were translocated to adjacent tillers and to the roots. This free movement of metabolites was lost as the plants became more complex in their structure; in a plant with five tillers, ^{14}C was not translocated between tillers, but moved solely to the roots. Tillers that were interdependent thus gain independence with development;

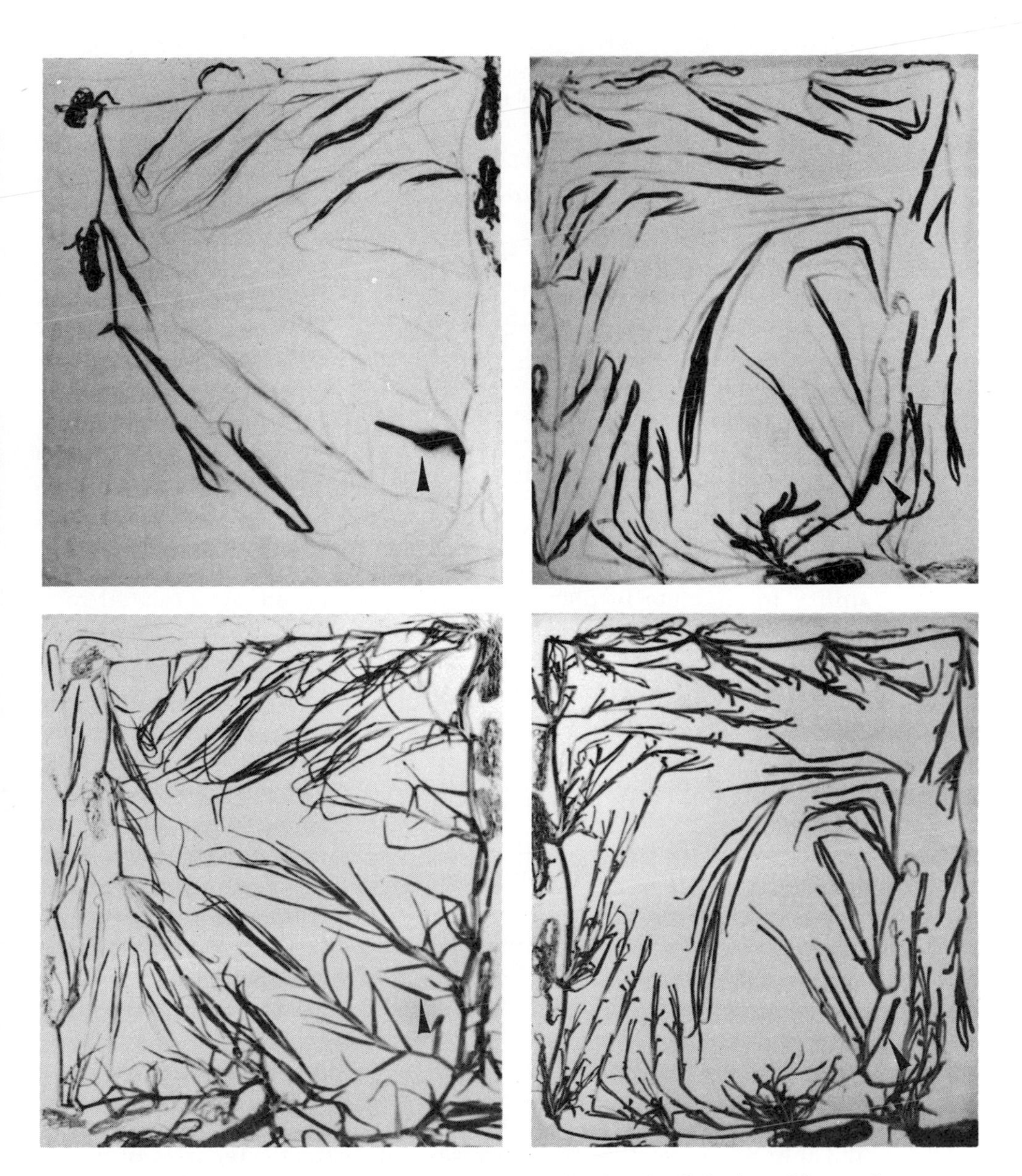

Figure 6.4. Autoradiographs (*above*) and photographs (*below*) used by Forde (1966a) in his studies. The plants, *Cynodon dactylon,* were treated with ^{14}C-labeled urea and allowed a 48-hour translocation time. The plants were covered, except for the treated stolon, for 4 days prior to treatment and during treatment. Arrows indicate the treated leaves.

but, because the root system continues to be fully supplied with assimilates by every tiller, the tiller cannot be regarded as a fully independent unit.

Labeled phosphate moved from a single root to all tillers and to the remaining roots of the treated plant. In the case of a plant with five tillers, ^{32}P moved principally to the tiller to which the root was attached, but also to the other roots. These results show that the grass plant must be considered physiologically as a whole. When all roots but one were removed from the supply of ^{32}P, the isotope moved into all roots, showing that, though they are relatively independent, the tillers can become integrated through vascular interconnection, receiving a supply of phosphate from a single root. Thus, in times of severe stress, the plant may respond positively and overcome its resistance to lateral transport of nutrients, by virtue of the inherent plasticity of its organization.

This response of *Lolium* to the manipulation of the presentation of tracers depends, in the case of phosphate, upon its well-known ability to circulate in plants. Phosphate absorbed by a treated root passes to the shoot, where it enters phosphorylating reactions and serves in the migration of assimilate from mesophyll to phloem. There, according to Wanner (1953a), it is freed from its sugar residues, and then recycles to the mesophyll by symplastic movement. A fraction, however, enters the phloem and is translocated via the assimilate stream to the roots; in this case, to the whole root system. Figure 6.5 shows the distribution of ^{32}P in *Hordeum vulgare*.

The use of ^{45}Ca as a tracer would reveal that calcium has a very different distribution; not being phloem-mobile, it would move from a treated root into the shoot most closely related, but it would not retranslocate to other leaves or roots. In agreement with Evans et al. (1964), Marshall considered the tillers of a grass plant as a hierarchy of sinks competing for translocated assimilates and nutrients. However, as will be described later, the distribution patterns of different molecules—be they assimilates, nutrients, growth regulators or toxic herbicides—depend very much upon the particular properties of each molecular species. As described by Crafts (1967a), and illustrated in Table 8.1 (page 176) there are several distinct categories of molecules, with respect to their translocation properties; and the results of any experiment depend very definitely upon the nature of the

---

Figure 6.5. The distribution of ^{32}P in *Hordeum vulgare* from (*a*) leaf application and (*b*) root application. Dosage was 0.1 μc per plant. Translocation times (*left to right*): 1, 4, and 16 days.

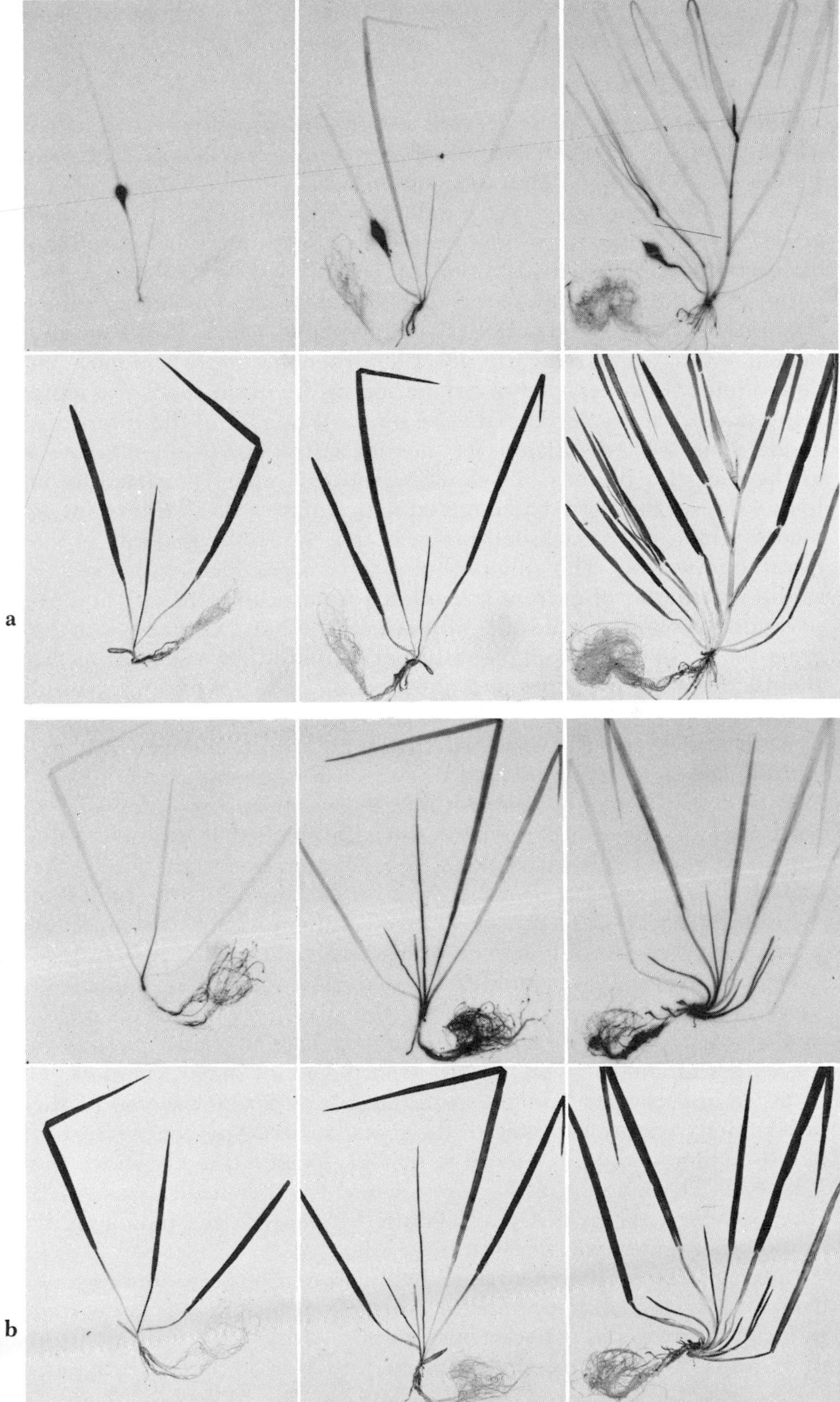

a

b

molecule serving as tracer, as well as upon the growth characteristics of the plants being used, and various environmental factors. This will be discussed in more detail in Chapter 8.

In a subsequent paper on the nature of grasses, Sagar and Marshall (1967), using *Lolium multiflorum* as their plant material, described experiments on the distribution of ^{14}C-labeled assimilates under various conditions. They found that interdependence among tillers continues for prolonged periods in young plants. As the plants age and become complex in structure, the tillers become more and more independent. However, partial defoliation of the tillers leads to a rapid reintegration of the aerial parts of a plant. When all of the tillers of a single plant were defoliated, the current assimilates were not moved to the roots for periods of 1–3 days, and this effect persisted for at least 4 days. The root system proved to be common to all tillers; for no matter which fully expanded leaf was fed $^{14}CO_2$, the majority of the roots acquired ^{14}C. Thus, there seems to be a root pool which serves as the distributor of current assimilates from tiller to tiller when experimental conditions of stress are imposed. When ^{14}C has been in the roots for 24 hours, it is apparently not available for re-export to the shoots; this finding throws into question the value of root reserves in *Lolium*.

Sagar and Marshall suggested that their results indicate that the mobile carbohydrates produced by each tiller become part of a common pool in the root system. Although the root incorporates a large portion of the contents of the pool, those products that are still mobile appear to move readily to regions of utilization. Incorporation by the roots is a fairly rapid process because for 24 hours after the import of ^{14}C into the roots, no re-export occurred, although current assimilates would have been readily moved into active sinks via the root system.

Yamamoto (1967) presented $^{14}CO_2$ to leaf 12 and leaf 17 of plants of *Nicotiana tabacum*, and determined the distribution patterns of ^{14}C in the whole plants. Leaf 12 (the largest leaf on plants having 19 leaves) assimilated actively, and transported about 30 percent of its ^{14}C to other organs after 3 hours. After 21 hours, 20–30 percent of the radioactivity was translocated to the roots, about 20 percent to the upper stem, and about 10 percent to leaf 17, located directly above the 12th leaf. The amount of ^{14}C translocated to other leaves was small after 31 hours. When $^{14}CO_2$ was applied to leaf 17, the radioactivity in the other leaves was negligible.

Judging from the time-course of the incorporation of ^{14}C into organic compounds, sucrose imported into roots from leaf 12 was converted into sugar esters and cationic compounds. Most of the ^{14}C imported into the 17th leaf was incorporated into the 80-percent ethanol-soluble fraction, especially into sucrose. By contrast, ^{14}C fixed photosynthetically by leaf 17 was mostly recovered as starch and protein after 8 hours of $^{14}CO_2$ assimilation.

Whereas Shiroya et al. (1961) found that young leaves of *Nicotiana tabacum* export assimilates more rapidly than old, Yamamoto found that, with plants having 19 leaves, the largest mature leaf was the most active. Yamamoto's plants were only 70 days old, and only 30 cm high; their lower 6 leaves were removed, because they were senescent. The plants used by Shiroya et al. had 8–35 leaves; export was greatest from the young, fully expanded leaves.

Yushkov (1965a, 1965b) studied the distribution patterns of ^{14}C-labeled assimilates in *Pinus* during the vegetative period, using trees 7–8 years old. Branches were enclosed in polythene bags and fed $^{14}CO_2$. At 6 hours, and at 1, 3, 7, and 14 days, the trees were sampled for ^{14}C. The experimental periods were (1) during intensive growth, (2) after intensive growth, and (3) after all growth had ceased. During period (1), the products of assimilation moved into the stems and acropetally to the shoot tips. In period (2), movement was mainly to the roots, with some material still going to shoots. In period (3), ^{14}C-labeled products accumulated only in the roots. In untreated side branches, ^{14}C never appeared. Some labeled compounds moved from bark to wood; the bulk of these products ended up in the wood of stems and roots during all three periods.

From 80 to 90 percent of the ^{14}C in a given branch could be found in the needles, with 8–12 percent in the bark, and 2–12 percent in the wood, especially of 2-year-old organs. Six hours after treatment, ^{14}C-labeled products had moved out of the needles and into the wood; in bark, and in young, importing needles, the concentrations of ^{14}C remained constant. By the end of 1 day, most of the ^{14}C was incorporated into a nonhydrolyzable fraction, presumably cellulose and lignin. After 14 days, 50 percent of the original radioactivity of the needles and bark of 2-year-old branches was still in water- and alcohol-soluble fractions. In the wood, 99 percent of the original radioactivity was in the nonhydrolyzable fraction.

Distribution of photoassimilated $^{14}CO_2$ by seedlings of *Pinus banksiana* was studied by Balatinecz, Forward, and Bidwell (1966). They found the translocated ^{14}C to be incorporated both into phloem and into xylem in the main stem above and below the donor branch. As much as 70 percent was retained within the donor branch; 20 percent moved basipetally, and 10 percent acropetally. By 6 days after presentation, all ^{14}C was in insoluble forms; maximum concentration of ^{14}C in root tips occurred on the second day. The trees used in these experiments were 8 months old.

Shiroya et al. (1966) measured translocation in *Pinus strobus* as the amount of ^{14}C recovered from the roots at the end of a 7-hour period following exposure of the shoot to $^{14}CO_2$. They found such ^{14}C to be high in the spring, dropping to negligible amounts during June and July, increasing again in autumn, and declining after October. Seasonal changes in root respiration paralleled those of translocation. Apparent photosynthesis was low in spring, rising to a maximum during September, and then declining over the winter. The main sugar translocated was sucrose. Raffinose content decreased in the summer.

Ursino, Nelson, and Krotkov (1968) have determined the seasonal changes in the distribution of ^{14}C-labeled assimilates in young *Pinus strobus* plants. Using 1-hour photoassimilation periods at monthly intervals between April and October, they studied subsequent distribution patterns of ^{14}C in the plants, 8 hours, 1 month, 2 months, and 4 months after treatment. In this way, they were able to find the fate of ^{14}C-labeled assimilates during different months of the growing season. They found that, in the spring, old needles make a significant contribution to the photoassimilate supply; they exported this assimilate to the developing new shoots and roots. By July, the new shoots had replaced the old ones, both in assimilation and exportation, especially to the root system.

The roots received current photosynthate from the shoot throughout the entire growing season, a fact that was always shown by the 8-hour analyses. Transport of new photosynthate from shoot to root was very high through August, September, and October. About half of the ^{14}C absorbed by these trees was lost over a 4-month internal, principally through respiration; the bulk of this loss occurred within the first month.

The amount of ^{14}C recovered in the 80-percent ethanol-soluble fraction was initially in the 20–100 percent range, lowering with time as the incorporation of sugars into the protoplasm and cell walls progressed. However, even 4 months after the initial photoassimilation period, as much as 20 percent of the ^{14}C recovered from any part of the plant was in this alcohol-soluble portion. The writers emphasize the great value of using $^{14}CO_2$ in studies on tree growth.

Studying grown *Pinus resinosa* trees for information on the mobilization of ^{14}C-labeled photosynthate from needles of different ages, Dickmann and Kozlowski (1968) found that the uptake and transport of assimilates from leaves to cones in early May (in Wisconsin) was slight, and that it about equalled the transport of assimilates into the bud. During this period, reserves were being mobilized both by cones and by shoots. During June, the mobilization of carbohydrates by cones increased, reaching a maximum in July and August, at a time when the rate of increase of dry-matter in the cones reached a maximum. Thus, carbohydrate reserves are important early in the growing season; but as they become depleted, largely by shoot growth, currently produced carbohydrates—particularly from the 1-year-old needles, and the current-year needles late in the season—become the dominant carbohydrate source for cone development.

Dickmann and Kozlowski found that, as with many other plants, the young leaves of *Pinus* do not export assimilates, but use their photosynthate for their own growth. After leaf expansion slows to some critical level (compensation), flow is reversed; it then becomes increasingly difficult to induce flow into mature leaves. Growing reproductive tissues acted as sinks to preferentially mobilize assimilate; they imported almost three times as much carbohydrate from the old needles as was mobilized by the shoots.

Leonard and Hull (1966) reported work on the translocation of ^{14}C-labeled tracers and $^{32}PO_4^{---}$ in various trees parasitized by mistletoe, and in unparasitized trees of the same species; the mistletoes included species of *Arceuthobium* and *Phoradendron*. Autoradiographic techniques were used to follow the tracers, which were applied to the host foliage, bark, and wood, and to the mistletoe shoots. It was found that phloem-mobile substances were translocated from the hosts into *Arcenthobium* spp., but not into *Phoradendron* spp. When host branches were defoliated, phloem-mobile substances moved into them during the growing season. When dormant, very little transport into defoliated branches occurred, except when they were parasitized by *Arceuthobium;* under this condition, the tracers moved into the parasitized branches and into the tissues of the parasite. Phloem-mobile substances in *Arceuthobium* migrated always in an acropetal direction, accumulating in nodes, flowers, and fruits. In no instance was there any movement in a basipetal direction. By contrast, *Phoradendron* transported tracers in much the same way as other green plants; however, phloem-mobile substances did not migrate significantly out of the endophytic system and into the hosts, even when the branches of the latter were defoliated.

High relative humidity greatly favored the uptake and transport of amitrole in *Abies concolor*; 2,4-D was not so favored, presumably because it penetrates the cuticle via the lipoid route (Crafts, 1961a). Neither *Arceuthobium* nor *Phoradendron* contributed to the organic nutrition of their hosts; *Arceuthobium* was definitely a parasite in the strict sense; and plants of both genera were dependent upon their hosts for water and salts. Herbicides applied to shoots of *Arceuthobium* migrated appreciably in the same manner as assimilates; such treatments were completely ineffective as a method for introducing herbicides into the endophytic system. By contrast, ^{32}P moved in a basipetal direction in *Phoradendron.* The most effective way of introducing substances into plants of both genera was by xylem injection, a method that is being used in Australia to selectively control *Amyema pendula*, a parasite of *Eucalyptus* (Greenham and Brown, 1957; Brown and Greenham, 1965). The herbicides 2,4-D, MCPA, 2,4-DB and MCPB have all proved useful for the control of mistletoes in Australia. Leonard and Hull used 2,4-D, 2,4,5-T, amitrole, atrazine, and paraquat in their experiments, in addition to $^{14}CO_2$ and $^{32}PO_4^{---}$.

In a second paper, Leonard and Hull (1965) gave details of the methods used in the studies described above; they described the results of various girdling experiments, and they illustrated their autoradiographic studies with many photographs. They reported the translocation rates for ^{14}C in parasitized and unparasitized branches of *Abies concolor* in January, February, June, and October, when the trees were dormant, and during July and August, when the trees were actively growing. The highest value was 6.0 cm per hour in a parasitized branch in June; next highest was 5.0 cm per hour in June in an unparasitized branch. The lowest rates occurred in October, when the trees were dormant; three out of nine branches showed no movement; the average rate for all nine branches was 0.24 cm per hour.

Experimental Modification of the Rate and Direction of Translocation

Ringing and shading were early experimental methods for modifying the rate and direction of translocation. More recently, defoliation, deradication, changes in mineral nutrition, local chilling, and the imposition of acute water stress have been used. Where $^{14}CO_2$ is used, tracing the changes in assimilate distribution becomes a convenient method.

In experiments with *Solanum tuberosum*, Anisimov (1964) found that under conditions of culture on grey forest soils, superphosphate accelerated the translocation of assimilates from leaves to tubers when no nitrogenous fertilizer was used. Superphosphate had the opposite effect when nitrogen, as ammonium nitrate, was applied under the drill before planting, or in the furrow with the tubers. Anisimov proposed that nitrogen supply may be delayed either by the high dosage of phosphate (60 kg/ha), by the competition of phosphate and nitrate anions during absorption, or by inhibition of the synthesis of organophosphorus compounds.

Working with *Beta vulgaris*, Anisimov (1965) found, again in grey forest soil, that the application of ammonium sulfate increased the rate of assimilate flow from the leaves to the central rosette. Relative translocation to the roots decreased during the first half of the vegetative growth period when ammonium sulfate was applied, but it increased during the second half. Total translocation (increase in root weight per plant) was higher throughout the entire growth period in the plots receiving ammonium sulfate. Root weight, and yield of sugar per unit of plot area, increased from the use of ammonium sulfate.

Using aphid honeydew, and phloem exudate from severed mouthparts, Hoad and Peel (1965a) studied the movement of solutes between sieve tubes and the surrounding tissues. In order to obtain labeled assimilate, $^{14}CO_2$ was administered to the leaves of rooted cuttings of *Salix*; where perfusion of the xylem was carried out, isolated stem sections were used. With these plant materials, Hoad and Peel showed that interference occurred between sucrose and the cations potassium and rubidium. Upon raising the potassium concentration in the sieve-tube sap by passing a potassium solution through the xylem, the sucrose concentration in the phloem exudate declined. When the sucrose concentration fell over a period of days, as a result of respirational loss in an isolated stem segment, the concentration of potassium or rubidium in the sieve-tube sap rose. When a solution of sodium was perfused through the xylem, the concentration of sodium ions in the sieve-tube sap rose, while that of potassium ions underwent a fall, and later rose to a value higher than the initial one. On introducting solutions of sodium and potassium into the xylem simultaneously, more sodium than potassium was absorbed, but the increase in sodium ions in the sieve-tube sap was less than that of potassium ions. Perfusing the xylem with calcium solution had no effect upon potassium ion concentration in the sieve tube.

The rate of longitudinal translocation between two aphid colonies on a stem of *Salix* has been shown to depend upon the rate of sink activity (removal of the solute from the sieve tube). The amount of such lateral loss is related to the potential gradient for a solute between the sieve tube and the surrounding cells.

Continuing their studies, Hoad and Peel (1965b) found that the time lapse of 2–3 hours, between the introduction of a tracer ion into the xylem and its appearance in the honeydew, is due to a barrier to radial movement of ions; the use of bark strips, where the introduction was to the cambium, reduced the time lapse to less than one hour.

Pretreatment of the bark strips with inactive ions increased the time lapse in the case of $^{32}PO_4^{---}$ and $^{35}SO_4^{--}$, but had no detectable effect upon movement of radioactive rubidium. Possibly, the latter moves directly to the phloem via the apoplast; phosphate and sulfate may have to enter sieve tubes via phloem parenchyma cells. In sieve-tube sap, ^{32}P activity is not limited to inorganic phosphate; three sugar phosphates were identified in this sap.

In 1964 Hartt, Kortschak, and Burr reported experiments with *Saccharum officinarum* involving defoliation, deradication, and darkening. They found that actively growing axillary shoots and basal suckers are active sinks whose presence stimulates the downward translocation of sucrose. They found that the roots of their plants exerted no sink-like action. Defoliation above or below the fed leaf increased translocation to the stem; defoliation of all but the fed leaf did not inhibit its ability to translocate sugar. Cutting off the fed blade above the treated area, or darkening the fed blade, or both, decreased translocation; labeled sucrose accumulated in the sheath of the fed leaf. Hartt and her co-workers concluded that the attachment of the sheath to the stalk constitutes a bottleneck in translocation, and that a major force of translocation is within the blade itself.

Further studies by Hartt related to the effects of temperature (1965a) and light (1965b) upon translocation in *Saccharum officinarum*. Air temperature was found to affect the percentage of ^{14}C-labeled assimilate translocated from the fed blade, as well as the amount moving in the stem. Velocities of transport were 1.40 cm per minute at 20°C, 1.56 cm per minute at 24°C, and 2.00 cm per minute at 33°C. Temperature coefficients for translocation from the fed blade were 1.1–1.5; for translocation down the stem, 1.05–1.7; and for translocation up the stem, 3.9–16.2. Hartt concluded that export and movement down the stem are physico-chemical processes; upward movement is controlled by chemical processes—the metabolism of cell growth, for example. Cold roots decreased translocation in light of high intensity; the plants were under moisture stress. In detached leaves at 250 ft-c of illumination, basipetal translocation at 5°C was zero; at 22°C, it was about 50 percent of the total counts in the blade.

The light experiments (Hartt, 1965b) were performed on detached leaves; both polarity of movement and percentage of translocation were affected by light. Basipetal translocation in blades detached in the morning, compared with blades in the dark, was increased 15.7 times by exposure to light at the rate of 0.1 cal/cm^2/min. In blades detached in the afternoon, translocation was about 75 percent of the morning values, but only 1.9 times as great as from blades in the dark. Darkening the fed section of a leaf and the tip above it resulted in a reversal of polarity after a period of roughly 2 hours. The system controlling polarity was very sensitive; exposure to 2000 ft-c for 5 out of 30 minutes was sufficient to keep acropetal translocation in the normal range; the factor controlling polarity was saturated at 50 ft-c.

Basipetal transport was better in the light than in the dark, and, in light, better in blades detached in the morning than in blades detached in the afternoon. Photosynthetic assimilation of CO_2 was saturated at 6000 ft-c; compensation took place between 120 and 125 ft-c. Because light affects translocation at such low intensities, it is suggested that translocation is under photo-control, which does not involve pressure flow.

Hartt (1966) published a paper reporting the results of her work with lights of different colors. Using detached blades of *Saccharum officinarum* under red, green, blue, and cool-white fluorescent lamps, and under far-red sunlight and incandescent lamps, she found basipetal transport of ^{14}C-labeled assimilate to be stimulated by red or blue more than by green or cool-white illumination. Basipetal translocation took place in both pure red and pure blue light; because the action spectrum for light-induced change in viscosity is of the blue type, Hartt concluded that the effect of light on translocation is not due merely to changes in viscosity of protoplasm. Basipetal transport went on in red light better than in cool white light, suggesting a red light stimulation of movement; far-red was not effective in supporting basipetal translocation; results resembled those in dark.

Because of the wide emission characteristics of the lights used, Hartt could not ascribe light stimulation of phototranslocation to any particular pigment system, but she concluded that phototranslocation of sucrose in the phloem is influenced by the quality of illumination.

Forde (1966a), using ^{14}C-labeled urea as a precursor of ^{14}C-labeled sucrose for tracing assimilate movement in *Cynodon dactylon*, found assimilate movement to be strongly polarized toward the tips of large, vigorous stolons, and less polarized toward the small stolon tips. Darkening, or partial defoliation of a stolon, induced the movement of assimilates to it from another treated stolon on the same plant, but the dependability of such transfer was related to the length of the dark period, or the number of defoliations. Apparently, three defoliations were required to assure movement of assimilates into the branch stolons of this very vigorous grass, and such treatment probably reduced the soluble carbohydrate content; it did not prevent apical growth, and such growth undoubtedly lowered the level of sugars in the phloem (Figure 6.4).

In a second paper, Forde (1966b) described tracer studies on translocation using *Lolium perenne* and *Agropyron repens*. In *L. perenne*, ^{14}C-labeled sucrose, from the precursor ^{14}C-labeled urea, tended to remain in the treated tillers. More mobile were amitrole and maleic hydrazide (both also labeled with ^{14}C), both of which showed some intertiller movement into covered or defoliated tillers.

Movement of ^{14}C in *Agropyron repens* was slower than in *Lolium perenne*. Some tracer moved from the large treated tufts to smaller tufts; the reverse of this did not occur. Covering of *A. repens* had no effect on tracer movement, but partial defoliation caused intertuft transfer of assimilate. Marshall and Sagar (1965) described similar results in tests using $^{14}CO_2$ on *Lolium multiflorum*.

Webb and Gorham (1965b) studied the effect of chilling the primary leaf node of *Cucurbita* seedlings on their assimilation and translocation of ^{14}C. Varying the node temperature from 0–45°C did not alter the rate of either $^{14}CO_2$ assimilation or transpiration of the leaf blade. The rate of export of ^{14}C-labeled sugar from the blade was greatly influenced by the temperature of the node; at 0°C, there was no significant movement of sugars through the node, but a slow movement of ^{14}C-labeled stachyose and ^{14}C-labeled sucrose did take place from the blade into the petiole above the chilled node. This was associated with a loading of the blade with ^{14}C-labeled sugars where the accumulated ^{14}C-labeled stachyose showed little metabolic turnover. Above 0°C, the rate of ^{14}C movement through the node increased to a maximum at 25°C and fell again to zero at 55°C, at which temperature the tissue was killed (Figure 6.6). The exports of ^{14}C-labeled stachyose and ^{14}C-labeled sucrose from the blade were equally affected by the different temperature treatments. The inhibitory effect of chilling on translocation was reversible; when the node was rapidly returned from 0° to 25°C a slow export from the blade started at once; it took 60 minutes for normal export to be restored. Webb and Gorham concluded that their results exclude an explanation on the basis of mass flow, and stated that translocation of sugar through the node of *Cucurbita* ". . . is under direct physiological control." Others (Crafts, 1932; Swanson and Geiger, 1967) concluded that a change in viscosity of a sugar solution of the concentration of the assimilate stream is great enough to block flow, at least temporarily. Furthermore, the sieve-tube sap of *Cucurbita* contains proteinaceous constituents that are capable of setting to a gel; and the filamentous reticulum of the sieve pores is capable of plugging the phloem con-

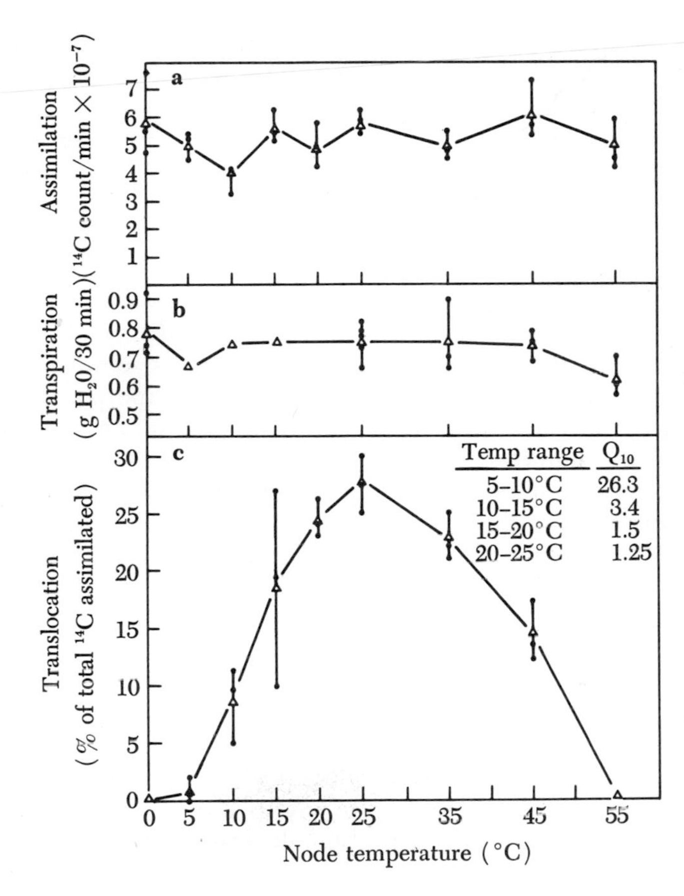

Figure 6.6. Effect of temperature of the primary node of *Cucurbita melopepo torticollis:* (*a*) on the assimilation of $^{14}CO_2$ by the primary leaf blade; (*b*) on transpiration from the primary leaf blade; and (*c*) on the translocation of ^{14}C through the chilled node. From Webb and Gorham (1965b).

duits. If translocation in *Cucurbita* is "under direct physiological control," it is difficult to explain phloem exudation at 1000 cm per hour.

Swanson and Geiger (1967) determined the effect of low-temperature inhibition on sucrose transport in *Beta vulgaris.* Decreasing the temperature of a 2-cm zone of the source-leaf petiole to 1°C decreased the translocation rate 60–95 percent. The reduction, however, was temporary; for when the petiolar zone was maintained at 1°C, the translocation rate gradually recovered, attaining a value of 70 percent or more of the initial rate within 2 hours. If the petiole was prechilled prior to labelling, the translocation rate showed no reduction, and remained unchanged after warm-up to room temperature. The half-time of the chilling response required for 50 percent inhibition was 4–15 minutes, and the half-time of recovery was 30–100 minutes. Maximum inhibition was 68–98 percent. Apparently, inhibition by chilling resulted in little or no impairment of the translocation process, and thermally adapted petiole systems "de-adapted" after periods as short as 1 hour at 25°C. Preliminary studies with *Phaseolus* plants indicated little or no thermal adaptation to chilling, and recovery of transport upon warming to the initial temperature was only to 60–65 percent of normal. Swanson and Geiger suggested that their results are of significance with respect to theories involving high metabolic rates of phloem tissue, or to the functioning of sieve plates as metabolic pumps. Their results would seem to fit, more definitely, a system involving metabolic loading of the sieve tube system, with passive mass-flow accounting for rapid longitudinal transport.

Bowling (1968) observed translocation effects on potassium uptake by *Helianthus annuus* plants growing in water culture. He found that such uptake was closely dependent upon the translocation of sugar to the roots, as shown in Figure 6.7, taken from his paper. This relationship was employed to determine the effect of cooling upon the rate of sugar translocation. Although a Q_{10} of approximately 3 over the range 0–25°C was obtained, tracer tests proved that translocation was not stopped at 0°C. A complete recovery of translocation rate, comparable to that found by Swanson and Geiger (1967), was found in some of the plants after prolonged cooling. Bowling reasoned that the low temperature slows mass flow of the assimilate stream, thus blocking sugar transport, and resulting in an increase in concentration above the chilled region. This, in turn, causes a resumption of flow at a lower velocity, compensated by the high concentration; potassium uptake under these conditions assumes again its initial rate. While Bowling's results seem to be consistent with the Münch mechanism, his abandonment of a hydrostatic driving force for an active mass-flow process on the pattern of that of Spanner (1958) does not seem to be justified. None of the normal components of mechanical or electro-osmotic pumps occur in normal, mature sieve-tube elements; all of the components of osmotic pumps are present.

The effects of water stress on the transport of ^{14}C-labeled sucrose in *Phaseolus* plants were studied by Plaut and Reinhold (1965).

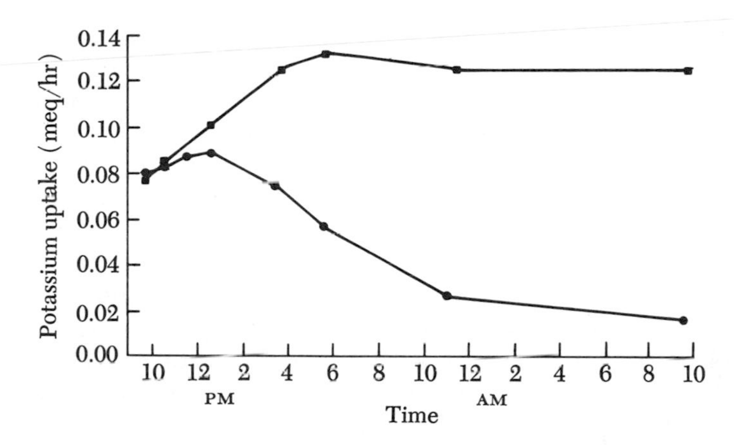

Figure 6.7. The effect of sugar transport, as affected by ringing, (*circles*) on potassium uptake by the roots of *Helianthus annuus* plants over a 24-hour period, beginning at 10 AM. Roots starved of sugar some ceased to absorb potassium. From Bowling (1968).

Sucrose labeled with ^{14}C, in a 0.01 ml droplet of aqueous solution containing surfactant and boric acid, was applied to the underside of a leaf. During the first 45 minutes, the stressed plants absorbed and translocated much more tracer than did normal control plants. Plaut and Reinhold had no explanation for this; but the rapid uptake by stressed leaves implies rapid penetration of the treatment solution. Although the 0.01 ml droplet seems small, the ring in which application was made was covered to prevent evaporation, and the liquid absorbed would account for the total volume of the protophloem sieve tubes in several centimeters of leaf, vein, and petiole. Undoubtedly, the inclusion of surfactant in the treatment solution, bringing about rapid wetting and absorption, would result in accelerated export for so long as the liquid was available for uptake. The experiment recorded in table 3 of Plaut and Reinhold shows that the greatest acceleration of transport to hypocotyl and roots occurred during the first 15 minutes; acceleration was less during the first 30 minutes; and in 45 minutes, the stressed plant showed less tracer in its lower parts than did the controls.

When the interval between the application of ^{14}C and sampling was over 45 minutes, the amount of ^{14}C moved out of a leaf was much reduced by water stress. Stressed plants moved very little ^{14}C in an acropetal direction in the stem, and the disparity in upward transport between stressed and control plants increased with time. The concentration of ethanol-soluble ^{14}C in the lower parts of stressed plants caught up with, and eventually exceeded, that in the control plants. This was due, largely, to inhibition of sugar incorporation by stress. Stress steepened the slope of the logarithmic profile of the distribution of ^{14}C in plants. Irrigation sharply accelerated downward transport; this occurred long before recovery from wilting. Plaut and Reinhold concluded that the pressure-flow hypothesis provides a ready explanation for the inhibition of sugar movement by water stress. While they were unable to account for the rapid acceleration during the first minutes after application, the explanation given above seems logical, and it fits the pressure-flow hypothesis. These observations seem to correlate well with the results of Clor et al. (1962, 1963, 1964), who used polythene bags to create a surplus of moisture around a treated leaf.

In *Triticum aestivum*, Wardlaw (1967) found that water stress reduced the rate of photosynthesis, but did not affect the growth of the grain (Figure 6.8). Movement of assimilates into conducting tissue was prolonged in wilted leaves, but the rate of translocation was little affected. Wardlaw concluded that water stress must act directly on leaf physiology, rather than indirectly through effects on growth or on sugar transport within the conducting tissue. Evidently, *T. aestivum* differs from *Saccharum officinarum* in its response to water deficiency. During ear ripening, carbohydrates are converted from sugars to starch; hence, the pressure gradient within the phloem should be maintained. While the rate of synthesis is lowered in the flag leaf, sugars are diverted from lower plant parts into the grain: the rate of translocation in the top internode is increased. There seems to be no logical reason why this should not support the mass-flow mechanism.

In further studies on the effect of water stress on translocation of assimilates in *Lolium temulentum*, Wardlaw (1969) measured the photosynthetic rate, leaf and root extension, changes in dry weight, and the translocation of labeled assimilates in plants subjected to water stress. While the water deficit was increasing, extension growth was reduced before photosynthesis was lowered. Reduction in the rate of photosynthesis was directly related to the water status of the leaf, as reflected by its relative turgidity, and not to the lowering of the growth rate. Vein loading was delayed in stressed leaves; but the velocity of assimilate movement was only slightly affected at low levels of the relative turgidity of leaves, if sink activity was increased by removal of adjacent leaves. Table 6.3, taken from Wardlaw's paper, shows the effect of defoliation on the distribution of ^{14}C in

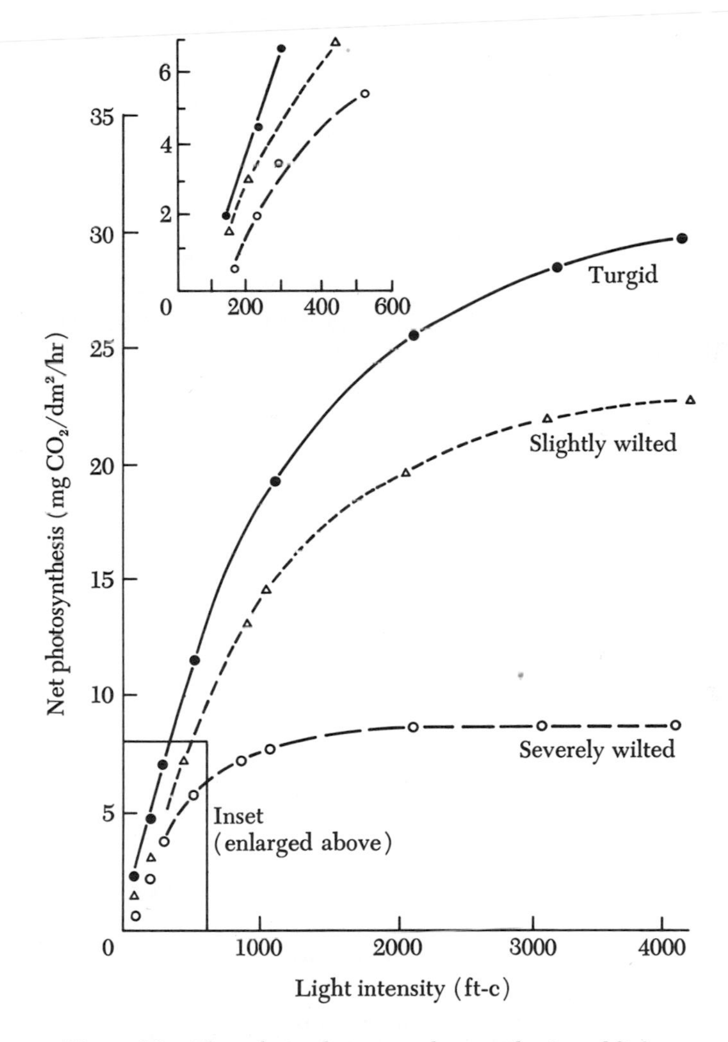

Figure 6.8. The relation between photosynthesis and light intensity in plants of *Triticum aestivum* under three moisture conditions. Wilting severely inhibits photosynthesis. From Wardlaw (1967).

relation to water stress. These results seem logical in terms of a mass-flow mechanism of assimilate movement, where the delivery rate (velocity × concentration) should remain constant over a fairly wide range of water status.

Table 6.3. The effect of defoliation on the distribution of ^{14}C-labeled photosynthate in *Lolium temulentum* as related to water stress. (Each result is the mean of eight replicates ± standard error.)

Plant part	Percentage distribution after 24 hr			
	Controls		Stressed plants	
	Intact	Defoliated	Intact	Defoliated
Leaf 6 (uptake area)	24.5 ± 3.0	20.5 ± 3.2	35.5 ± 2.5	30.5 ± 1.1
Leaf 6 (residual)	3.2 ± 0.3	1.5 ± 0.1	18.6 ± 0.9	11.1 ± 1.2
Young leaves	8.0 ± 0.5	5.7 ± 0.4	13.7 ± 2.2	15.6 ± 1.2
Mature leaves	0.9 ± 0.1	0.6 ± 0.3	2.4 ± 0.2	1.6 ± 0.1
Crown	38.2 ± 4.5	49.2 ± 2.8	12.8 ± 0.8	14.0 ± 0.5
Roots	25.3 ± 1.5	22.7 ± 1.6	17.0 ± 0.9	27.3 ± 3.0
Total plant activity (count/min)	3561 ± 152	3816 ± 131	2667 ± 281	2811 ± 227
Growth rate of leaf 8 (mm/hr)*	2.56	2.38	0.38	0.67

Source: Wardlaw (1969).

*Growth rate calculated from extension over the 24-hr distribution period.

The Role of Sinks in Assimilate Transport

Ever since the work of Mason and Maskell (1928b), it has been recognized that the translocation of assimilates in plants follows a source-to-sink pattern; and that, whereas the leaves form the active sources for growing plants, all non-green, actively growing, storing, or metabolizing tissues are the sinks. Mason and Maskell did ringing and shading experiments to manipulate source–sink relations. Many researchers since their time have done similar studies. Anisimov (1959) used $^{14}CO_2$ and ^{32}P in studies on young seedlings of *Triticum aestivum* and *Helianthus annuus* to determine if nutrient deficiencies in the roots would affect assimilate movement from leaves. In experiments lasting 30–60 minutes, he found that a lack of nitrogen retarded movement of ^{14}C from leaves to roots; potassium deficiency retarded movement to an even greater degree. Replacing potassium chloride with potassium bicarbonate in the culture solution increased the flow of assimilates. Anisimov found that differences in the variables in translocation rates of ^{14}C and ^{32}P had the same character, implying a link between the translocation of phosphorus from leaves and the normal flow of assimilates.

Zimmermann (1964a) reported on the relation of transport to growth in deciduous forest trees at the second Harvard Forest symposium. Starting with the question of whether the annual growth rings of trees are produced from reserve assimilates of the previous year or from the current years product, he posed three possible transport–growth relations: (1) that photosynthates are transported directly from active leaves to sinks via the phloem; (2) that new growth draws on materials retranslocated from places of stored reserves to regions of growth via the phloem or the xylem; or (3) that reserve materials are stored very near the regions of growth, and thus, that mobilization involves only short-distance transport, in which sieve tubes may not be involved. Zimmermann went on to point out that growth rings are formed not only in the xylem, but also in the phloem; and that the new increment of phloem functions during midsummer and autumn until the onset of dormancy. Sugars transported during this functioning season go into late wood formation and into storage as starch in twigs, branches, the main stem, and the roots; most storage is in the xylem and phloem parenchyma and in the rays. In autumn, before dormancy, proteins are broken down in the leaves and exported back into the phloem of branches and stems where they are stored.

Dormancy breaks in the spring, and growth in many temperate regions starts in early summer. In certain species, phloem is reactivated and starts functioning as buds break and the flush of growth is initiated. During this period, reserves are mobilized and moved via the phloem into the young, importing leaves and shoot tips. At compensation, the flow reverses and the mature leaves export: the cycle is complete.

Anatomical studies confirm the functioning period of sieve tubes, and exudation experiments support the conclusions. Only the current season's sieve tubes yield exudate; this point is proved both by controlled incisions by the knife and by investigations on the locus of aphid feeding. Zimmermann (1961, 1963) illustrated the locus of aphid feeding by showing stylet tips in a mature sieve tube; the saliva tracks in the phloem tissue tell the same story.

At leaf senescense, when amino acids are exported (Mittler, 1958; Ziegler, 1956; Zimmermann, 1958a), the onset of dormancy is rapidly followed by callose formation on the sieve plates; in *Fraxinus americana*, dormancy callose is formed about 2 weeks after leaf fall. At this time, exudation from bark incisions stops (Zimmermann, 1958b) and aphids cannot feed on the phloem; greenhouse colonies become restless, and may be lost.

An interesting difference has been found between tree species: in some (in *Salix* spp. and *Tilia* spp., for example), callose formation in excised branches is incomplete, and aphids can be maintained on these in the greenhouse or laboratory; in others (*Carya* spp. and *Castanea* spp., for example), the sieve tubes of excised branches become completely blocked by callose, and although water movement in the xylem continues and the branches stay green, aphids are unable to feed. Release of turgor is possibly the stimulus that triggers callose plugging of this type; leaf abscission and continued utilization of assimilates results in reduced turgor. Injury-induced callose formation may not completely close all sieve plates, but dormancy callose formation provides a fairly complete seal.

Phloem reactivation comes about in the spring, and provides functioning sieve tubes for the translocation of foods during the spring flush of growth. Stylet exudation is resumed soon after the break of dormancy (Mittler, 1958), and the early exudate is rich in nitrogenous substances (presumably those stored during the previous autumn). Hill (1962) found that the exudate concentration dropped rapidly during bud growth of excised branches. Stylet exudation declined further during leaf expansion, and aphid colonies suffered from lack of nutrition. Hill explained this as competition between the leaves and the aphids for the phloem sap. Zimmermann (1964a) suggested that it might represent exhaustion of reserves, but the ability of deciduous trees to releaf after the leaves have been removed mechanically, or by fungus attack, would seem to refute such an explanation.

As the new leaves start exporting, reactivated sieve tubes are obliterated, and new phloem takes on the translocation function. Because phloem reactivation is a requirement only near the shoot tips, Zimmermann (1964a) suggested that dormancy callose formation may not extend all the way into these. He cited *Quercus rubra* as a species in which reactivation seems to extend all the way down the main stem; Schneider (1945) found the collapse of phloem restricted to the old layers of the growth ring, and reactivation taking place in the young sieve tubes near the cambium, in *Prunus avium*.

In species in which growth continues and leaves are produced after all of the leaf primordia of the terminal bud are used up (in *Populus* spp., for example), assimilate movement from median leaves may be both acropetal into the shoot tip, and basipetal toward roots. Thus, in some internodes, bidirectional movement must take place. Work by Biddulph and Cory (1965), done since Zimmermann's (1964a) report, proved that this goes on in separate phloem bundles. Zimmermann questioned whether it is accurate to speak of sieve tubes as functioning for more than one year: although they may be formed in autumn and reactivated in spring, they probably do not function for more than 12 months. He made a plea for simultaneous use of different study methods on phloem of one or more species.

At the second Harvard Forest symposium, Ziegler (1964) discussed the storage, mobilization, and distribution of food reserves in trees. Much of his paper involved the anatomical arrangement of storage cells in relation to the vascular tissues concerned with supply of the storage materials. Much of the anatomy of woody plants seems related to the efficiency of supply and distribution of their reserves.

Tree species may be divided into "starch trees" and "fat trees," on the basis of the nature of the storage materials in their tops. All trees store mainly starch in their roots, probably because the roots are not usually subject to conditions of extreme cold during the food storage period. During extremely cold winters, however, when roots are subject to freezing temperatures, roots may store some fat.

Diffuse-porous species usually store fat, whereas ring-porous ones store starch. Ziegler would have explained this on the basis of his observation that the radial extension of the storage tissue in wood in ring-porous species is limited more than it is in diffuse-porous species; his assumption was that the material with the highest energy content (fat) would be accumulated in ring-porous trees. The fact that an energy capacity of about 8 kcal/cm^3 obtains for fat storage, whereas for starch storage the value is 4.9 kcal/cm^3, would seem to indicate that starch storage in ring-porous species would not be energetically efficient.

Since analyses of phloem and of phloem saps indicate that sucrose is the principal form of mobile carbon in plants, fat and starch must be formed in the storage cells themselves. Reserves used in the formation of young shoots, the thickening of stems, and the formation of fruits and seed, must undergo the reverse process before they can undergo symplastic movement. Growing buds have been shown to be sinks more active than cambium, the latter receiving only the surplus reserves after the needs of the buds are met. Although shoot development in deciduous trees depends almost completely on reserves, a large part of the building materials in evergreens comes from the current activity of photosynthetic organs; these supply the assimilates used in the growth of fruits and seeds in both types of trees.

Rays are recognized as the conducting tissues that carry on the lateral movement of assimilates from the phloem to the wood and bark; they also carry water and salts from the xylem to the wood and bark. Assimilates may move from sieve tubes to sink tissues via companion cells and phloem parenchyma; in conifers, albuminous cells serve in the same capacity as do the companion cells in angiosperms. Cambium is apparently supplied via the rays; ray initials are unique in that they divide to form ray cells, and at the same time that they must serve in translocation. While cohesion of water probably explains the radial movement of xylem sap, the distribution of assimilates must involve some sort of metabolically driven transport, presumably diffusion accelerated by streaming.

In experiments with *Pinus* seedlings, Nelson (1962) found that vigorous roots imported more labeled assimilate than weak roots; roots of plants grown under light of high intensity imported more than the roots of plants grown under low-intensity light. Two 2-year-old *Pinus* seedlings were allowed to assimilate $^{14}CO_2$ for one hour; one of them was then placed in normal 0.03 percent $^{12}CO_2$ for 8 additional hours, whereas the other was kept in 0.006 percent $^{14}CO_2$ for the 8-hour period. The second tree moved 91 μc into its roots, whereas the first moved only 51 μc. Nelson explains this on the basis of storage as transport sucrose and he states:

> If $^{12}CO_2$ is supplied after the initial exposure to ^{14}C, then ^{12}C is translocated and ^{14}C remains stored in the leaf. If no CO_2 is supplied, but the shoot is at compensation point (0.006% $^{14}CO_2$), then some of the storage sucrose is available for translocation. . . . It seems that translocation of organic compounds in plants is regulated by the metabolic activity of all parts of the plant.

Nelson (1964), reporting on production and translocation of ^{14}C-labeled photosynthate in conifers, presented data showing that *Pinus* seedlings with mycorrhizae translocated up to ten times as much assimilate to their roots as did seedlings lacking mycorrhizae, irrespective of the root–shoot ratio. This indicates the effect of the symbiotic fungi on the activity of root sinks.

In experiments on nutrition, plants having no supplementation of nitrogen or phosphorus translocated the greatest percentage of ^{14}C-labeled assimilates from their tops to their roots. Although plants given supplemental nitrogen contained large amounts of amino acids labeled with ^{14}C, the roots contained less ^{14}C than did those of the controls. Although the rate of synthesis of amino acids was high in the roots of plants given extra nitrogen, the root growth of those plants was usually low; hence, translocation would not necessarily be increased by the high level of nitrogen.

In the trials on the effect of light intensity on translocation, plants exposed to light of high intensity transported to their roots 12.1 percent of the ^{14}C-labeled assimilate produced in a 1-hour photosynthesis period; plants grown under high-intensity light, but exposed to low-intensity light during the translocation period, transported 11.1 percent; plants grown under low-intensity light translocated only traces of ^{14}C, whether exposed to high- or low-intensity light during the translocation period. Most of the translocated assimilate was in the form of sucrose in shoot, stem, and roots. In a series of trials running through the year, translocation to the roots was highest in May before shoot extension started, and lowest in June and July when leader extension was at a maximum. This reflects the relative activity of sinks in roots and shoot, and illustrates the profound effects of active growth upon assimilate distribution. Leonard and Crafts (1956) found that growth had a similar effect upon the distribution of 2,4-D and 2,4,5-T in woody plants in California.

Quinlan and Weaver (1969) studied the effect of BAP—a cytokinin—on the translocation of assimilates in *Vitis*. Leaves approaching compensation imported only trace amounts of ^{14}C following the assimilation of $^{14}CO_2$ by a lower leaf on the same shoot. Application of a 4.4×10^{-3} M BAP solution caused a marked increase in ^{14}C import, presumably because of stimulated sink activity. Young, rapidly expanding leaves at the shoot tips imported much ^{14}C, regardless of BAP treatment.

Older leaves near the base, which normally export, were less responsive. BAP treatment alone had little effect on the import of ^{14}C from adjacent fed leaves, but when the BAP-treated leaves were darkened, there was increased import of ^{14}C. When these two treatments were superimposed and applied to branches that were ringed at the base, high levels of radioactivity were found in the BAP-treated leaves; less radioactivity was found in those only darkened. The accumulation of imported ^{14}C was restricted to the area treated with BAP. In the BAP-treated leaves, the percentage of activity present in the sugar fraction of the leaf was lower, and that in the amino acid fraction was higher, than in the untreated leaves. Since cytokinins are known to promote growth and, hence, sink activity, these results are compatible with a mass-flow scheme.

Leonard and King (1968) showed that the ^{14}C-labeled assimilates that result from feeding $^{14}CO_2$ to leaves of *Phaseolus* move into, and accumulate in, the veins and petioles of detached leaves within a 24-hour transport period. Further, ^{32}P accumulated in the petioles of *Phaseolus* leaves that received application of $^{32}PO_4^{---}$; Tween 20 enhanced the uptake of the tracer. These results confirmed the opinion that the parenchyma tissues of the veins may serve as sinks for assimilate, if export into the stem and roots is blocked by excision.

Small and Leonard (1969), using plants of *Pisum sativum* and *Trifolium subterraneum*, studied the effects of nitrogen nutrition on the translocation of photosynthate into roots and into nodules. The presence of nitrate nitrogen in the culture medium increased the proportion of labeled photosynthate translocated to the roots, with a corresponding decrease in translocation of photosynthate to the nodules. Nodules on plants grown in the absence of combined nitrogen had a higher radioactivity than nodules on plants treated with combined nitrogen. These results are most easily rationalized in terms of the relative effects of the applied nitrogen upon root and nodule sinks. Figure 6.9 illustrates the results of Small and Leonard's study.

The Bidirectional Movement Problem

The question of bidirectional movement of assimilates in the phloem has intrigued plant physiologists for years. Curtis (1935) believed that such movement was common, and he adopted the protoplasmic-streaming hypothesis to account for it. Mason, Maskell, and Phillis (1936) designed experiments to elucidate this problem. Crafts (1938) criticized their experiments. Biddulph and Cory (1960) demonstrated bidirectional movement in *Phaseolus;* in subsequent work they proved that it occurred in separate phloem bundles (1965).

Jones and Eagles (1962) studied the translocation of ^{14}C from $^{14}CO_2$ in *Nicotiana tabacum* and in a variegated *Pelargonium* having chlorotic leaf margins. They found evidence, as did Jones et al. (1959), for bidirectional movement in *N. tabacum* leaves, with the apical portions exporting, while the immature basal parts imported assimilates. In variegated *Pelargonium* leaves, they found the chlorotic portions to act like immature areas, in that they continued to import assimilate from outside the leaf to a greater extent than from the adjacent green areas of the same leaf. Using masking, they found that ^{14}C failed to move across the darkened mesophyll of *N. tabacum* leaves; they interpreted this as an indication that the translocation mechanism of leaves is located in the veins. Labeled assimilate moved out of the veins of importing leaves along their entire lengths. Jones and Eagles accepted Thaines' (1961) illustration of transcellular strands as supporting bidirectional movement in a single sieve tube. More probable, however, is the interpretation that the mature leaf tip was exporting ^{14}C-labeled assimilate along the midrib, while the still-growing basal lobes were importing it via the lateral veins that lead into them.

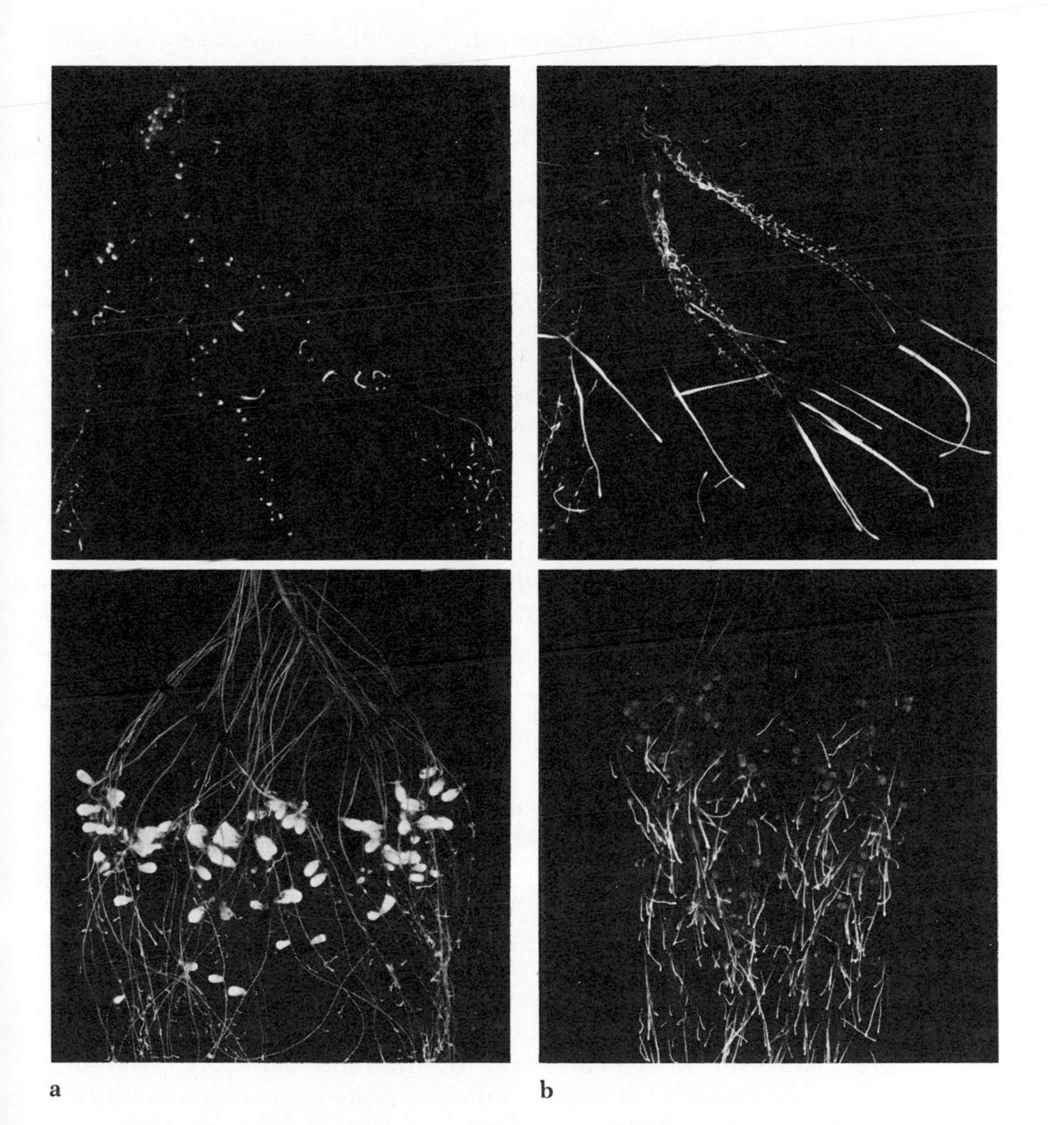

Figure 6.9. Effects of nitrogen nutrition on assimilate distribution in plants of *Trifolium subterraneum* (*above*) and *Pisum sativum* (*below*). Shown are views of roots of plants in (*a*) low-nitrogen culture and (*b*) high-nitrogen culture. The tracer was ^{14}C-labeled assimilate, and the translocation time was 20 hours. From Small and Leonard (1969).

Jones et al. (1959) suggested that simultaneous movement takes place in separate channels in the leaf, and that the ^{14}C-labeled assimilate may possibly migrate to the xylem, and so move in an acropetal direction. It seems just as logical to visualize export via the phloem from the tip of the leaf (which matures early) simultaneous with import via the phloem from lower leaves into the basal portion of the leaf that is still in rapid growth. Such export would probably involve the midrib, although import would occur along lateral bundles of the petiole. Although phloem-to-xylem migration of phosphorus, potassium, maleic hydrazide, and dalapon are well documented (Biddulph and Markle, 1944; Crafts and Yamaguchi, 1964), such transfer of ^{14}C-labeled sucrose has not been reported.

Jones and his co-workers found that their label followed the phylotactic pattern of the *Nicotiana tabacum* plant, and that there was no leakage from the bundles of treated leaves to the neighboring leaves. This would seem to confirm the above interpretation of the simultaneous bidirectional movement which they observed.

Bidirectional movement of three tracers in *Glycine max* seedlings was reported by Crafts in 1967. The herbicides 2,4-D, amitrole, and maleic hydrazide, applied to the primary leaves of these seedlings, moved into the roots and shoot apices. In order to ascend the stem from the node of the primary leaf, it was necessary that the tracer move downward as well as upward in the epicotyl of the plant. Study of the vascular anatomy of the epicotyl of *G. max* proves that the leaf traces from the primary leaves traverse the epicotyl, from the primary leaf node almost to the cotyledonary node, before they give off the bundles leading to the first trifoliate leaf; traces to the second trifoliate leaf emerge a bit higher. This means that tracer applied to a primary leaf must pass along the traces of the treated leaf, and through the epicotyl for a distance of almost four inches in a basipetal direction, before it can ascend to the first trifoliate leaf. Hence, there are two streams within the epicotyl moving simultaneously in opposite directions. However, as shown so well by Biddulph and Cory (1965) this is possible through separate bundles of phloem, and hence, it is not evidence counter to the existence of a mass-flow mechanism. It is incomprehensible that distribution patterns of this type could result from independent movement of solutes in the phloem.

Eschrich (1967) used ^{14}C-labeled assimilate and fluorescein applied to separate leaves of *Vicia faba* to demonstrate bidirectional movement; aphid honeydew collected between the two donor leaves gave evidence that ^{14}C-labeled assimilate was moving basipetally, and that fluorescein was moving acropetally in the stem. Of the honeydew droplets studied, 42 percent were doubly labeled.

Eschrich presented a detailed consideration of the possible mechanisms involved. He considered that either there is two-way movement in a single sieve tube; or that the anatomy of the phloem allows for a "homodromous loop path," involving oppositely directed streams in adjacent sieve tubes that are in contact through anastomoses, so that the aphid-mouthpart sink might allow the exit of both streams in a single droplet of honeydew.

As for simultaneous bidirectional movement in a single sieve tube, there is no known physical mechanism to account for it. Protoplasmic streaming does not take place; and if it did, furthermore, it would be far too slow to account for the commonly observed rates of translocation. As for the "homodromous loop path," the anatomy of the experimental stem would have to be examined very carefully to find the degree and direction of anastomosis. That such loops occur in nodes seems possible; but that they are present in the internodes upon which Eschrich's aphids fed seems questionable.

A third possibility would be that the application of 0.3 ml of fluorescein to the leaf below the aphids allowed uptake into the xylem, and movement acropetally past the feeding area, with lateral spread of the dye into the phloem. A number of compounds move readily in both phloem and xylem (see Table 8.1, page 176); some of these migrate between the two. Repetition of the experiment using ^{45}Ca or ^{14}C-labeled monuron on the lower leaf, instead of fluorescein, should provide a solution to this intriguing problem.

In a Ph.D. dissertation, Carol Peterson (1968) described her studies on bidirectional translocation of fluorescein in the phloem. Testing some ten fluorescent dyes, she found fluorescein to be the only one that moved readily in the phloem without transferring to the xylem. Applied to the middle of a young internode of *Vicia faba* for one hour, the dye was found in bundles 1 cm above and 1 cm below the treated area. Sections examined under the fluorescense microscope proved that the dye was localized in the traces of the larger leaves below the treated area, and in the traces of the younger leaves above the treated area. The tracer was never present in the same bundle both above and below the treated area. This indicates that each bundle translocates dye in only one direction. Peterson concluded further that individual sieve tubes conduct the dye in only one direction. She confirmed her finding in *Phaseolus vulgaris, Vinca rosea,* and *Pelargonium hortorum.* In petioles of young leaves of *Ecballium elaterium,* individual phloem bundles translocated the dye in only one direction. In some bundles, the external phloem carried the dye toward the stem while the internal phloem carried it toward the leaf blade.

When longer time intervals were used in her experiments, Peterson found that the dye appeared in the same phloem bundles both above and below the treated area. This was due to a lateral movement of tracer within the phloem, either through secondary phloem or through bundle-anastomoses at the nodes. Fluorescein was not moved in the xylem when applied to an internode; the distribution of the fluorescein in this way resembled the 2,4-D distribution found by Crafts (1967b). Peterson concluded that her results were in accord with the pressure-flow mechanism of phloem transport. Perhaps the most outstanding accomplishment of her work was the recognition that bidirectional movement is a common phenomenon that goes on in many plants, that it occurs commonly in leaves (and possibly in all large leaves), and that it may possibly go on in all plants at certain stages of their development.

In contrast to the work of Peterson (1968), Trip and Gorham (1968a) claim to have demonstrated that two streams of sugars move in opposite directions in the petiole of a half-grown leaf of *Cucurbita melopepo*, and that movement of both streams occurred within the same sieve tubes. They stated that this demonstration does not fit the mass-flow theory of translocation, but that transcellular strands (Thaine, 1961; Evert et al., 1966) may provide a possible mechanism. Trip and Gorham's experiments lasted 3 hours; Peterson found that in trials lasting for more than one hour, fluorescein appeared in the same phloem bundles both above and below the point of application. This she explained by possible lateral movement of the tracer within the phloem, either through the secondary phloem or through bundle-anastomoses at the nodes. There was no secondary phloem in Trip and Gorham's plants, but there could have been anastomoses at the nodes—or even between the nodes, or within the petiole by extrafascicular bundles (Crafts, 1932). Considering the stage of development of the leaf treated with $^{14}CO_2$ in Trip and Gorham's experiments, there could possibly have actually been a reversal of flow within the bundles of the petiole within the 3-hour treatment period. After all, the direction of flow is eventually reversed in every petiole bundle during development, and it could well have taken place in this case.

Although the above criticism of Trip and Gorham's results may well invalidate their work, an even more serious consideration is that of mechanism. From our present knowledge of the physicochemical properties of aqueous solutions, it seems impossible to rationalize the independent movement of different molecules in static water at rates at which assimilates are known to move. And, as explained elsewhere, protoplasmic streaming at the rate of 5 cm per hour (Thaine, 1961) is entirely inadequate. As for transcellular strands, these may occupy, in mature sieve tubes, only around 5 percent of the total transverse area of the lumen, whereas sucrose alone may occupy up to 20 percent or more. Further, only one-half of the streaming strands could be moving in a unidirectional flow pattern.

If bidirectional movement takes place in separate bundles, or, possibly, in separate sieve tubes of the phloem (Biddulph and Cory, 1965; Crafts, 1967b); and if the distribution of labeled assimilates, tracers, and phloem-limited viruses reflects the relative strengths of sources and sinks; then there is much evidence that mass flow serves as a mechanism to bring about rapid longitudinal movement in mature, functioning sieve tubes. In a recent paper, Trip and Gorham (1968c) accepted solution, or mass flow, as a possible mechanism of translocation in *Cucurbita*.

Radioactivity Profiles and Gradients

With the introduction of ^{14}C for use in translocation studies, great importance became attached to the approximately logarithmic form of the profiles, or fronts, found in transport channels (Vernon and Aronoff, 1952; Horwitz, 1958; Canny, 1961, 1962a; Spanner and Prebble 1962; Peel and Weatherley, 1962). In studies on translocation, Canny (1961) used aphid stylets to determine the form of the profile of advancing radioactive sucrose in the bark of *Salix* with respect both to distance and to time. The time profile of the advancing radioactive sucrose was the same shape as, and coincident with, the distance profile of radioactivity extracted from a series of sections of the bark. From the slopes of comparable parts of the two curves, estimates of the velocity of movement of the distance profile through the bark were obtained, and these Canny checked, by measuring the distance of the stylet from the application leaf, and by measuring the time for the peak to reach the stylet; the quotients of these values gave a second set of velocity estimates. Velocities found by both methods varied from 1.5 cm per hour to 4.1 cm per hour; Canny gave an average value of 2 cm per hour.

Canny used the equation $V = z/b$ to estimate the velocity of tracer movement, where V is the velocity of the moving profile, z is the slope constant of the time curve (where log radioactivity, expressed in counts per minute, is plotted against the time elapsed since the application of $^{14}CO_2$), and b is the slope constant of the distance curve (where log radioactivity is plotted against the distance along the bark from the location of the stylet). It should be noted that the time values were not corrected for the time between the presentation of $^{14}CO_2$ and the arrival of ^{14}C-labeled sucrose in the phloem; according to Canny, these times were in the order of 15–30 minutes.

The values for b, the distance-slope constant, would seem to be of more doubtful significance. Since Canny ruled out pressure-flow as a translocation mechanism, he defined velocity as ". . . rate of a disturbance among moving molecules of sugar or as some kind of average of the velocities of individual sugar molecules." With these ideas in mind, he measured the radioactivity of successive 0.5-cm sections of *Salix* bark, and plotted these values against distance to obtain his slope constant b. If sugar moves in solution along a turgor gradient in the sieve tubes, and moves laterally via plasmodesmata into phloem parenchyma cells (the normal distribution pattern), then the profile peak of counts per minute in bark might indicate the storage function of the plant, more nearly than translocation per se. Because b is in the denominator, the low values for translocation velocity that Canny obtained may well be the result of the slower process of accumulation, rather than the true concentration equilibration in sieve tubes. Weatherley et al. (1959) obtained much higher values.

Because of the values that Canny obtained for transport velocity, he concluded that the cross-section, across which sugar arrives at an aphid stylet, is greater than one sieve tube; that is, that a single stylet may tap many sieve tubes. Although lateral movement of assimilates in the phloem is possible, and the opening of one sieve-tube element to atmospheric pressure might set up an extraordinarily active sink, the fact that exudation may continue for 24 hours or more is evidence that plugging of the sieve plates does not occur, and hence, that the change in pressure is not great. Probably, the narrowness of the capillary passageway (0.2 μ) is such that the exudation from a stylet does not represent a great acceleration of flow along the tapped sieve tube; certainly no great number of sieve tubes are emptied in this process. Zimmermann's (1963) photograph would seem to indicate that the stylets are located in a single sieve tube. Work by Evert et al. (1968) substantiates this view.

Determined that the profile of the advancing front of radioactivity along the sieve-tube conduits was a key to velocity measurements, Canny (1962a) studied the profiles that are reflected in radioactivity intensity along stems, and the changes in the rates of evolution of $^{14}CO_2$ from local sites on stems. Plotting these results, he found, for example, that when the points on a chart plotting log radioactivity against time for $^{14}CO_2$ were connected with a line, the slope of that line had a value $z = 0.2$/hr. A line through points on a chart plotting log radioactivity against distance for the ^{14}C content of a stem had a slope with a value $b = 0.84$/cm. Dividing z by b, he derived a value $V = 0.24$ cm/hr for the velocity of transport of ^{14}C in the stem. From seven such experiments, he obtained values running from 0.24 to 1.6 for the transport velocity in *Salix*, and a value of 0.56 cm per hour for transport in *Vicia faba*.

Canny found that the profile characteristics of his CO_2-time curves were closely similar to the profile of ^{14}C-labeled sucrose collected in successive samples of sieve-tube sap from cut aphid stylets. Remarking on the low values for translocation velocities calculated from these experiments, Canny suggested the possibility that part of the labeled sucrose in the sieve tubes is not readily mobile. A more obvious explanation might be that his calculations were based upon erroneous supposition. His value z for the slope of log radioactivity against time would seem to be a value for the rate of respiration of ^{14}C-labeled sucrose by the phloem parenchyma and cortex parenchyma, or for the rate of the subsequent diffusion of $^{14}CO_2$ into the ambient atmosphere, both of which would certainly show an appreciable lag from the arrival of ^{14}C at a given site in the stem. Canny's value b for radioactive-sucrose differences along the stem would seem, in turn, to represent the rate of, and the capacity for, the accumulation of ^{14}C-labeled sucrose by cells other than the sieve tubes (Webb and Gorham, 1965a).

Peel and Weatherley (1962), recognizing the distributional function of the sieve-tube system, suggested that because of loss of assimilates from the moving stream, the front may become less steep, and may move forward less rapidly, than suggested by the original supply. Nevertheless, using two colonies of four aphids each located 34–65 cm apart, they recorded translocation velocities of 25, 27, and 33 cm per hour—as measured by the times of arrival of radioactivity at the proximal site and the distal site from a common source. This indicates, according to Peel and Weatherley, that tracer can move toward the stylet sink at a rate of roughly 30 cm per hour. The discrepancy between this and the rate of 100 cm per hour recorded by Weatherley et al. (1959) is not surprising, if exchange with surrounding cells occurs en route, and if the volume of the exudate comes partly from storage parenchyma as well as from the leaves.

These rates are an order of magnitude greater than those of Canny (1961, 1962a). Peel and Weatherley questioned Canny's estimate of rates: it was Canny's belief that the stylet gives a good sample of sieve-tube sap only in the locality of the puncture, whereas they believed that the stylet creates a sink in the sieve element, which induces movement from a distance. They felt that their interpretation invalidated the bases upon which Canny's calculations were made.

Saccharum officinarum is different from many plants, in that translocation will go on in detached leaves, but at a slower rate than in leaves still attached to the plant (Hartt and Kortschak, 1964). Sucrose is the principal compound that moves in detached leaves, and Hartt and Kortschak concluded that its movement depends upon a strong basipetal polarity within the phloem of the blade. The rate of basipetal translocation in darkness depends upon the time of day when the blade is cut: it is low in morning-cut blades, but it increases during the day. Blades cut in the afternoon translocate very well in total darkness; such blades have a strong, positive gradient of sucrose, with two or more times the concentration in the apex as in the base. As translocation proceeds, sugar accumulates in the basal region, and the final sucrose gradient is the reverse of the initial. Because afternoon-cut blades in the dark—just as morning-cut blades in the light—can translocate against a sucrose gradient in whole sections, Hartt and Kortschak proposed that there must be a regulating factor, other than sucrose synthesis, that determines this polar movement.

In discussing sugar gradients in detached leaves, Hartt and Kortschak stated that accumulation takes place in the veins and not in the parenchyma. Since much of the non-green tissue of the veins is made up of phloem and cortex parenchyma, the distinction they were making must be between the non-green tissue of the veins and the chlorenchyma of the leaf. Apparently, this non-green parenchyma of the veins is like the storage parenchyma of the stem, in that it can accumulate sucrose from sieve tubes (a phenomenon known as vein loading), even against a concentration gradient. If this is true, even such venous tissue must constitute a sink in the same way that the stem does, and Hartt and Kortschak were in error in stating that detached blades exhibit translocation without a sink.

Furthermore, conclusions about gradients determined from gross analyses of total leaf sections have little relevance with respect to concentrations and pressures within conducting sieve tubes. Their so-called polar movement in leaves may well be mass flow along pressure gradients in sieve tubes that are masked by over-all concentration gradients of sucrose in the storage parenchyma.

The effect of the moisture supply upon translocation in *Saccharum officinarum* was studied by Hartt (1967). Low moisture supply decreased the velocity and the percentage rate of translocation. Sucrose not used in growth moved more slowly in the phloem, and was stored in the stalk. Because low moisture curtailed transport to a greater degree than it did photosynthesis, Hartt concluded that the moisture supply has a primary action on translocation. Low moisture retarded the development of a ^{14}C profile in the stem, and a loss in the moisture gradient correlated with a steepened slope of the profile; these results are consistent with a flow mechanism. Hartt interpreted these experiments as pointing to a slow pressure-flow mechanism, particularly at night, but also during the day; superimposed upon this mass transport, Hartt suggested a more rapid process of phototranslocation, independent of sugar gradients and capable of bringing about the accumulation of sucrose at the storage sink.

Although Hartt produced a wealth of information on the physiology of *Saccharum officinarum* in these several researches, there seems to be a misunderstanding, in her discussion sections, of the total function of translocation in the symplast. In the region of a source, there are at least three basic processes that provide osmotically active solutes, and hence, that regulate hydrostatic pressure and activate translocation: (1) the synthesis of sugars, amino acids, and so forth; (2) the mobilization of polysaccharide reserves; and (3) the hydrolysis of proteins to amino acids. All three of these are known to occur in leaves, and one or more may go on in stems, roots, storage organs, and so forth.

In the sink regions, there are again three basic processes that eliminate solutes from the osmotic system: (1) growth involving transfer of the total volume of the sieve-tube content from the phloem to the parenchyma; (2) accumulation that involves conversion of osmotically active assimilates to insoluble forms; and (3) respiration, which goes on constantly, and is especially active during growth. It seems probable that most, if not all, of these processes are affected by light and temperature both, and that translocation may be determined in complex ways not obvious from the mass analysis data provided by Hartt and her co-workers.

All known photoregulated processes—such as growth, flowering, fruiting, and dormancy—may have profound effects upon food distribution via the symplast without affecting mass flow in the phloem in a direct way. Since the sieve tubes of the phloem constitute a complex, ramifying, open, inflated, elastic system, subject to import or export from every living companion cell, phloem parenchyma cell, or border parenchyma cell, it seems almost hopeless to attempt an explanation of translocation reactions from mass analyses such as most of those conducted on *Saccharum officinarum*. Analyses on phloem exudate from insect stylets or cut phloem, while more accurate, are likewise subject to error.

And finally, from the standpoint of the physics of water and sugar molecules, it seems impossible to visualize independent movement of sucrose and water within an open system such as the phloem. For sugar to move through the phloem more rapidly than the water in which it is dissolved [2.5 cm per minute (Hartt and Kortschak, 1963)] would involve the breaking of hydrogen bonds, requiring great amounts of energy; and there is no known mechanism by which energy can be applied to accelerate such movement.

Wardlaw (1968a) wrote an excellent review on the control and pattern of movement of carbohydrates in plants. He attempted to integrate the literature on the movement of assimilates labeled with isotopic tracers, and to assess the relative importance of leaf function and position, of vascular anatomy and organ development, and of sink activity, in controlling food movements under various environmental conditions. He quoted several writers to the effect that low-intensity light reduces not only the photosynthetic rate, but also the proportion of assimilates moved from leaves; and he related these changes to alteration in the products of assimilation, and to a drop in the energy available for transfer within the leaf. He cited a wide range of rates of export of assimilates from the leaves of various plants, and generalized that, initially, export from young leaves is rapid, with a slower continuing transport coming with maturity. The export rate is a complex function of the source–sink relation as determined by factors within the exporting leaf, as well as by external conditions.

Wardlaw stressed the effect of leaf position, and the role of vascular connections, in the pattern of assimilate distribution; supply patterns to growing organs are limited by orthostichy and vascular arrangement, with longitudinal movement predominating over lateral distribution, so far as the velocity of movement is concerned. Bidirectional movement within a given length of stem is often reguired, because the anastomoses of bundles often occur at one or more nodes below the node of leaf insertion, as noted by Bonnemain (1965), Biddulph and Cory (1965), and Crafts (1967b). He discussed various aspects of sink activity, and the role of environmental factors—such as temperature, light, water, and nutrition—in assimilate distribution. He pointed out that the actual process of food transport is probably of minor importance in determining distribution patterns in response to environmental changes. He considered temperature to be more directly related to growth than to either translocation per se or photosynthesis. Light affects translocation through photosynthesis, growth, and development. Water stress exerts direct effects on photosynthesis (Crafts, 1968b) and on growth; its effects upon translocation are indirect. Wardlaw remains uncommitted regarding the mechanism of assimilate movement.

7

Transport of Plant Hormones

Growth Regulators, Natural and Synthetic

Much interest was aroused in the absorption and translocation of growth regulators, as a result of their wide adoption as agents for the thinning and setting of fruits. Donoho et al. (1961) studied the uptake and distribution of NAA, ring-labeled with ^{14}C, in *Malus sylvestris* and *Prunus persica*. Using predisposed environmental treatments involving the control of temperatures and humidities, and applying the tracer in solution to both the upper and lower surfaces of leaves, these workers found that the leaves of *M. sylvestris* grown under cool temperatures (60°F) absorbed more NAA than leaves of trees grown at higher temperatures (70°F), irrespective of relative humidity.

Differences in the measured amounts of cuticle developed by the leaves of *Malus sylvestris* and *Prunus persica* grown under the different environments did not account for absorption differences. Few differences, if any, were found in the weights of the cuticle, although absorption differences were significant. Following the application of labeled NAA, ^{14}C was quickly translocated into the fruits and seeds of both species. High radioactivity was found 4–6 days after the leaves had been dipped into the solution of labeled NAA.

In discussing the mobilization of absorbed organic molecules, Mitchell et al. (1960b) reported that the growth regulator INBA tends to become localized in the terminal bud of *Phaseolus* plants, while 2,4-DI tends to accumulate in the stem and hypocotyl; Table 7.1 gives these data. The authors of this book have noted a similar situation in comparing the distribution patterns of TBA (Mason, 1960) and 2,4-D (Crafts and Yamaguchi, 1964). This comparative difference, however, would seem to be related to the tissue relations of these two molecules, rather than to the mechanism of their transport.

In the use of NAA on *Malus sylvestris* for fruit thinning and for preventing the drop of fruit before harvest, some of the applied compound must enter the plant and move to the sites of action, namely, the developing seeds and the abscission layers. In order to study the movement of NAA in varieties of *M. sylvestris*, Luckwill and Lloyd-Jones (1962) applied ^{14}C-labeled NAA to the spur leaves of 'Miller's Seedling' trees and measured the amounts recovered from the leaf surface, from the leaf tissues, and from the spur and fruits. The radioactivity recovered from the leaf surface amounted to 14.5 percent, all of it as NAA; from the leaf tissues, 15.0 percent was recovered, 12.8 percent of which was NAA; from spur and fruits, 6.0 percent was recovered, of which 1.4 percent was NAA; 64.5 percent of the radioactivity was unaccounted for, and most of this, presumably, had been broken down by ultraviolet light.

Repeating the above experiment on spurs having fruits 3.5 cm in diameter, they found a similar distribution of radioactivity, with 0.2 percent in the seeds, none of which was still in the form of NAA.

Table 7.1. Distribution of radioactivity after application of ^{131}I-labeled INBA and ^{131}I-labeled 2,4-DI each to the primary leaf of a plant of *Phaseolus vulgaris*. (Values are percent of recovered radioactivity.)

Plant part	INBA	2,4-DI
Terminal bud	60	3
First internode	12	37
Hypocotyl	28	56
Roots	0	4

Source: Mitchell, Schneider, and Gauch (1960).

Evidently, like 2,4-D, NAA is strongly retained within living cells, and moves in the vascular system only under the most favorable conditions.

Zweig et al. (1961), using ^{14}C-labeled gibberellin synthesized by a culture of the fungus *Fusarium moniliforme*, studied translocation of this growth regulator in *Zea mays* and *Phaseolus*. Under greenhouse conditions, tracer applied to the midribs of *Phaseolus* leaves showed little translocation; but in plants placed under cellophane bags to create high humidity, tracer applied to the midribs moved throughout the plants. In a time series, plants treated on the cotyledonary node for 8–72 hours had the tracer accumulated in the growing tissues at the stem tips; after 144 and 216 hours, the tracer was more uniformly distributed throughout the roots, stems, and leaves.

The fact that both primary leaves showed light images in autoradiographs, following application to the cotyledonary node, indicated acropetal movement via the xylem; presence of the tracer in the roots, and its accumulation in tip buds and growing leaves, proved simultaneous movement via the phloem. Evidently, gibberellin is like phosphorus, maleic hydrazide, and dalapon, in that it is mobile in both the phloem and the xylem, and may transfer from one tissue system to the other. Translocation of gibberellin was less pronounced in *Zea mays* than in *Phaseolus*, and the authors concluded that, in *Zea mays*, the movement was limited to the phloem.

Kluge et al. (1964) tested the phloem exudate of *Robinia pseudoacacia*, *Tilia cordata*, *Quercus robur*, and *Fagus sylvatica* for gibberellin activity. Huber, Schmidt, and Jahnel (1937) had found auxin activity in the phloem exudate of trees, and Kluge and his co-workers were interested to discover if gibberellins were also involved in the regulation of cambium activity. They obtained the data shown in Table 7.2 when they assayed the activity of the various exudates using test plants of *Zea mays*.

From these results it seems possible that gibberellin activity, combined with auxin activity, may be instrumental in the regulation of cambial activity. Both are synthesized in the leaves of trees, and both are phloem-mobile (Zweig et al., 1961).

Hoad and Bowen (1968) detected gibberellin-like substances in the honeydew of aphids feeding on *Taraxacum*, *Vicia faba*, and *Salix*. This is positive evidence for translocation of those substances in the sieve tubes of the phloem. Hoad and Bowen were unable to determine the site of synthesis of the growth regulators. In *Taraxacum*, there was evidence for their occurrence in mature leaves, but other

Table 7.2. The gibberellin content of the phloem sap of four tree species bioassayed as a function of leaf growth in test plants of *Zea mays*.

Tree species	Phloem sap (ml)	Leaf growth (mm)	Gibberellin content [(μg/ml) × 10^{-4}]
Control		36.6	
Robinia pseudoacacia	6.00	46.6	6
Tilia cordata	2.20	56.2	45
Quercus robur	0.64	41.2	17
Fagus sylvatica	0.96	40.6	11

Source: Kluge et al. (1964).

workers have found evidence for their origin in the roots. In *Salix*, long-day plants produced much more regulator than did short-day plants.

Chin and Lockhart (1965), using *Phaseolus* for their plant material, studied the translocation of applied gibberellin from different application loci. The greatest effect, as shown by stem elongation, resulted from treatment of the stem apex. Treatment on a trifoliate leaf was slightly less effective, and treatment of a primary leaf was appreciably less effective; apparently less gibberellin was moved in the latter case. When leaves were treated after the plants were kept in darkness for extended periods, no response was noted; growth was promoted when the plants were returned to light. The time required for movement of gibberellin from the first trifoliate leaf to the growing stem tip was one hour or less; the maximum growth response was found when the treated leaf remained on the plant for 3 or more hours. In two-branched plants, the untreated branch showed no growth response when gibberellin was applied to the apex of the other branch even where dosage was increased twentyfold; similar results were found when gibberellin was applied to the first trifoliate leaf. Gibberellin moved from a mature leaf to the opposite shoot if this untreated branch had been defoliated. The pattern of gibberellin movement to the opposite shoot was dependent on the position of the treated leaf on the shoot; these results appear to demonstrate that applied gibberellin moves with carbohydrates in the assimilate stream. There is no direct evidence that endogenous gibberellin moves in this way; if it were synthesized in the stem apex, it would have to move out against the stream of inflowing assimilates, or move by a polar mechanism in the parenchyma. Results reported here indicate that applied gibberellin translocates in the phloem.

In some horticultural practices, herbicides are used as growth regulators; 2,4-D, for instance, is used to control fruit size and to prevent fruit drop. For this reason, the effects of such compounds on vines of *Vitis vinifera* have been the subject of considerable research (Leonard and Weaver, 1961; Leonard, Lider, and Glenn, 1966). Further work (Leonard et al., 1967) has explored the results of 2,4-D and picloram treatment on the translocation of ^{14}C-labeled assimilates in the varieties 'Ribier,' 'Tokay,' and 'Thompson Seedless'; the former two were tested in the field, the latter in the greenhouse. In all cases, the herbicide solutions were applied to the same shoots that received a feeding of $^{14}CO_2$ the herbicides being applied when the polyethylene bags used for the administration of $^{14}CO_2$ were opened. Treatments were made at the prebloom, bloom, young fruit, and ripe fruit stages. Samplings were made at the outset (controls), and at 3 days, and 21 days after treatment.

Leonard et al. (1967) found that high concentrations of 2,4-D and picloram interferred with the downward movement of ^{14}C-labeled assimilates in field-grown vines. Interference by picloram was appreciably greater than by 2,4-D. Although basipetal transport was retarded, translocation within the treated shoots continued from the vegetative part to the clusters. Translocation of 2,4-D appeared to follow the same route as ^{14}C-labeled assimilates. Formative effects were absent on untreated grape shoots, although adjacent shoots on the same cordons, treated with 2,4-D or picloram, were killed; however, formative effects were evident on some of the stump sprouts that developed after the vines were cut off. The malformed leaves on the stump sprouts were twelve or more nodes from the base of the shoots, whereas ^{14}C was in the basal leaves. If the basal leaves were beyond the expanding stage at the time the 2,4-D arrived, it would not cause formative effects; such effects occur only on growing leaves. On the other hand, ^{14}C might have been carried into the carbon pool, and utilized in the maturation processes of the leaves.

Rootings of *Vitis vinifera* 'Thompson Seedless' treated with 20,000 ppm 2,4-D or picloram transported less ^{14}C to the roots than did the controls. Treatment with either regulator resulted in a marked increase in the labelling of the stems. This latter effect reflects the well known ability of such compounds to cause the callosing of sieve plates and the inhibition of phloem transport (Crafts and Yamaguchi, 1964). Had the treatments with regulators in the field experiments been made a day or two before the $^{14}CO_2$ feeding, inhibition of translocation would probably have been much more prominent. The experiments of Leonard et al. (1967) indicate that 2,4-D may be safely used in the control of *Convulvulus arvensis* in vineyards, providing the herbicide treatment is made carefully, and at a time when the berries are set and the terminal growth is mostly over.

Eschrich (1968) studied the movement of ^{14}C-labeled IAA in *Vicia faba* plants. He found that this substance moves from the first primary leaf into the stem, and thence to the youngest leaves; old leaves did not incorporate the auxin. Distribution was the same when the tracer was applied to the third leaf. An extract of the stem in 80 percent ethyl alcohol contained labeled IAA, plus two water soluble derivatives, which yielded IAA upon hydrolysis with HCl; one of these derivatives Eschrich took to be indolylacetylaspartic acid. Aphids fed on the stem of a treated plant produced honeydew containing IAA as the only labeled constituent; neither of the above-mentioned derivatives occurred in the honeydew. Aphids fed a diet containing these derivatives produced honeydew containing both substances unchanged. Eschrich conluded that applied IAA moves in the sieve tubes; the other labeled IAA derivatives of the extract are phloem-immobile; they must be formed in parenchyma cells.

Further confirmation of the transport of ^{14}C-labeled IAA in phloem tissue by means of mass flow of the assimilate stream comes from the experiments of Fletcher and Zalik (1965) and Whitehouse and Zalik (1968). These workers treated *Phaseolus coccineus* cotyledons with IAA and its metabolic precursor, tryptophane; both compounds were labeled with ^{14}C, and both were found to translocate upward to the shoot. The possibility of xylem transport was eliminated by an elegant series of experiments involving heat ringing, the uptake and distribution of dyes, and anatomical studies on vascularization. Residue analysis by chromatography showed conclusively that ^{14}C-labeled IAA and ^{14}C-labeled tryptophane move symplastically in a source-to-sink relationship.

Bollag and Galun (1966) studied the translocation and metabolism of ^{14}C-labeled IAA in monoecious *Cucumis sativus* plants. In their studies that were terminated after four hours, approximately 13 percent of the applied IAA moved with the assimilate stream in a source-to-sink relationship, while 81 percent was immobile at the site of application. Longer treatment times (50 hours) did not change the distribution pattern, but did show that 95 percent of the IAA had been metabolized or conjugated.

In contrast to the above studies, Bidwell and Turner (1966), and Bidwell, Levin, and Tamas (1968), found that IAA would translocate from the median trifoliate leaf to the other two trifoliate leaves in *Phaseolus vulgaris*. The absence or presence of IAA in the tissue was evaluated by the increase in uptake of CO_2 produced in untreated leaves with the application of exogenous IAA, and by analysis of ^{14}C-labeled IAA residues. If it is assumed that each of the trifoliate leaves was an equal source of assimilate—exporting equally, and transpiring equally—then the question remains: How did the IAA move from one source leaf to another source leaf against the assimilate stream? The IAA transport velocity was too great for this to have been a diffusion process; on the other hand, normal source–sink relations would imply that the other two leaves should have been bypassed by the basipetally moving assimilate stream. Therefore, IAA must have moved via the xylem to the other two untreated leaves (See Colwell, 1942; Clor et al., 1962, 1963, 1964). Direct phloem transport from one trifoliate leaf to another probably did not take place under the conditions of the experiment.

Additional evidence for the recycling of ^{14}C-labeled IAA was presented by these workers (Bidwell et al., 1966, 1968) in a second experiment where the cut stem of an axial bud was treated. After two hours the ^{14}C-labeled IAA was redistributed evenly between the first and the second sets of trifoliate leaves. Thus, the IAA moved down the phloem into the main stem, leaked into the xylem, and recycled to accumulate in the two developing sets of trifoliate leaves. Crafts (1967b) studied the vascularization and the translocation of ^{14}C-labeled maleic hydrazide in *Glycine max* and found similar distribution patterns involving phloem-to-xylem migration.

Cytokinins

Transport through tissue fragments from agar-block sources to agar-block sinks has been criticized as an unacceptable, unnatural state for the evaluation of the translocation patterns of growth regulators in plants. Pilet (1968) compared *in vitro* assays (transport through stem sections) with *in vivo* assays (translocation in intact seedlings) utilizing *Lens culinaris*. The synthetic cytokinin BAP, labeled with ^{14}C, moved very slowly in an acropetal direction in the *in vivo* studies, and moved not at all in the *in vitro* system. In both assays, ^{14}C-labled IAA gave a predominately basipetal translocation pattern. The *in vivo* basipetal velocity was 5.4 mm per hour, as compared to 6.4 mm per hour for the *in vitro* system. Pilet concluded, therefore, that there is essentially no difference between these two methods for studying translocation.

The cytokinin FAP, or its synthetic physiologically active analog BAP, when sprayed on a mature leaf, will promote the transport of numerous substances from untreated areas of the leaf to this "cytokinin-locus" (Mothes and Engelbrecht, 1961; Nakata and Leopold, 1967; Seth and Wareing, 1964; Morris and Thomas, 1968; Shindy and Weaver, 1967; Kriedemann, 1968a; Quinlan and Weaver, 1969; and Müller and Leopold, 1966a, 1966b). This general type of tissue response, and the translocation associated with it, has been termed "cytokinin-induced transport" by these workers. One of the prominent phenomena associated with cytokinin treatments is that older mature plant tissues, when treated with cytokinin, tend to become physiologically young; thus, cytokinins are ideal model substances for studying regulative principles, or for changing the distribution and exchange of substances within the plant.

Cytokinin-induced transport can be demonstrated with the application of radioactive labeled substrate, after cytokinin treatment, to various plant organs. Mothes, Engelbrecht, and Schütte (1961), and Mothes and Engelbrecht (1961), have found that FAP could induce amino acids to accumulate in cytokinin-treated areas; thus, localized photosynthate sources could be turned into photosynthate sinks. They were able to show—by feeding α-aminoisobutyric acid, a xenobiotic amino acid which is not incorporated into protein—that the process was one of amino acid accumulation, and not one of an increased stimulation of protein synthesis.

Gunning and Barkley (1963) repeated the studies of cytokinin-induced transport done in Mothes' laboratory, and investigated the biochemistry of the cytokinin-induced photosynthate sink. They found that ^{32}P and ^{14}C-labeled glycin were incorporated into the ribosomal fraction, and suggested that the cytokinin has a primary effect on nucleic acid metabolism. Other cytokinin-induced effects were (1) that induced transport was strictly basipetal; (2) that physiologically young tissue was more responsive than old; (3) that respiration was lower; and (4) that the breakdown of chloroplasts was delayed in older tissue. Thus, with these initial physiological observations, it became important to know if the most physiologically active analog was being utilized. Skoog et al. (1967), in an extensive study of the structural activity relationships of cytokinins, evaluated 69 cytokinin analogs by bioassay, using *Nicotiana tabacum* as the test plant. Strong evidence was presented to support the hypothesis that cytokinin activity is limited to the amino-monosubstituted compounds, FAP and BAP being two of the most physiologically active. Hence, an adenine moiety seems to be required, but the exact structure, size, shape, composition, and charge of the substituent groups will also influence biological activity. Skoog and his co-workers postulated that, since cytokinins occur naturally in RNA, a specific role in the control or regulation of amino acid incorporation into protein synthesis seems to be implied.

The translocation of cytokinins has been studied by Lagerstedt and Langston (1967) using ^{14}C-labeled BAP. When the substance was applied to the lamina of detached leaves, it did not move away from the treated area; however, cytokinin-induced transport of photosynthate, and retardation of senescence, were observed in the treated areas. When it was applied over the main vein of *Nicotiana tabacum* leaves, vein loading was observed. The movement of the cytokinin was usually distal to the point of application, and it seemed to occur with the transpiration stream. This observation was further substantiated by treating the distal or proximal halves of detached leaves and holding them in the dark. Leaves treated in the distal half showed no basepetal translocation, while leaves treated in the proximal half showed slight acropetal translocation. In contrast to the above studies with detached leaves, when ^{14}C-labeled BAP was applied to intact leaves that were actively exporting photosynthate, it was observed to move from photosynthetic sources to sinks. Thus, a symplastic translocation pattern was obtained, showing that cytokinin could be transported basipetally out of the source and acropetally to the shoot tip. To verify that it was cytokinin being translocated, and not a radiolabeled metabolite, the cytokinin was extracted from the sinks and shown by bioassay to still have cytokinin activity. These translocation patterns were confirmed by Pilet (1968) with ^{14}C-labeled BAP. Pilet showed that acropetal translocation was very slow in intact plants; and that in detached leaves, the tracer did not move away from the site of application.

Results obtained from analysis and bioassay of xylem sap from root exudates (Itai and Vaadia, 1965; Kende, 1965) suggest that endogenous cytokinins may be synthesized in the root. Reid and Burrows (1968) have shown by extraction and through bioassay of *Glycine max* callus that cytokinins are transported in the xylem sap of *Platanus occidentalis* and *Betula* spp.

The cytokinin analog BTP was compared with BAP in studies conducted by Weaver, van Overbeek, and Pool (1966) on fruit set and development in *Vitis vinifera*. Both cytokinins were effective in inducing food mobilization to the developing fruit, where increasing fruit numbers and size were observed. There was no translocation from the site of application in the case of BAP, while BTP did translocate acropetally from the leaves to the tips of the fruit clusters. This difference in translocation was believed to be due to the increased water solubility of BTP. Additional studies by Shindy and Weaver (1967) with BTP indicated that source–sink relations and the direction of photosynthate transport could be manipulated with BTP. Older leaves on shoots which normally exported basipetally were forced to export acropetally, if the more distal portions of the shoot had been sprayed previously with BTP. Similarly, young leaves that normally supplied photosynthate to the shoot tip could be forced to import more strongly if sprayed with BTP. The chemical composition of the photosynthate which had been mobilized by cytokinin-induced transport was evaluated by Weaver, Shindy, and Kliewer (1969). They showed that BAP treatment of *Vitis vinifera* in the late bloom stage would produce a large accumulation of carbohydrates and a large decrease in the amino acid fraction in the berries.

Kriedemann (1968a) studied the effect of FAP on the accumulation of ^{14}C-labeled photosynthate in *Citrus* fruits. Data from gross autoradiography, fresh weight, and radioactive counting, repeatedly showed that cytokinin-treated fruit was a stronger photosynthate sink than untreated fruit. Similar studies were carried out by Quinlan and Weaver (1969) with BAP. Combinations of cytokinin treatment, leaf darkening and ringing were investigated concomitantly with $^{14}CO_2$ uptake and assimilate distribution in *Vitis vinifera*. With cytokinin application, assimilate distribution could be reversed in fully expanded leaves that were normally exporting photosynthate. Combining cytokinin treatment with exposure to darkness, the strongest sink of all treatments was observed. Analytical separations of the carbohydrate and protein pools accumulating labeled assimilate in these cytokinin-treated areas showed that the carbohydrate level decreased, while the amino acid content increased. This observation is in agreement with the cytokinin mechanism of action proposed by Gunning and Barkley (1963), involving microsomal nucleic acid metabolism, and with the cytokinin-induced amino acid accumulation data of Mothes et al. (1961a, 1961b).

The export of food reserves from cotyledons can be inhibited by the application of cytokinins. Sprent (1968) treated the seeds of *Pisum sativum* with BAP before germination, and noted, after 14–34 days the effect of the substance on morphological and physiological factors. Seedlings showed induction of branching, a reduced net growth of root and shoot, and retardation of cotyledon senescence. The retention of reserve foods from the developing embyro by the cotyledons appeared to be the principal effect of cytokinin treatment of the seeds. Thus, a normally strong food source became a weak provider of food reserves, and the reduced growth rates and branching appears to be the result of this shift in food distribution patterns by the cytokinin-induced mobilization in the cotyledon.

Interaction between growth regulators (IAA and gibberellin), inorganic ions, and cytokinins has been investigated in many laboratories in an intense effort to shift and control the natural or applied hormone balance in plants. Seth and Wareing (1964) found synergism between IAA, gibberellin and cytokinin in source–sink studies with ^{32}P. When used in combinations (for example, IAA with cytokinin, or gibberellin with cytokinin), the uptake of ^{32}P was always greater than the uptake without growth regulator, or with any growth regulator alone. Furthermore, BAP was more active than FAP used in the synergism studies. Similarly, the effect of cytokinin on translocation interactions between ^{14}C-labeled IAA, ^{14}C-labeled sucrose, and $^{32}PO_4^{---}$ was studied by Nakata and Leopold (1967). When these radiotracers were applied to the distal ends of excised primary leaves of *Phaseolus* that were actively exporting assimilates, a typical source—sink pattern was obtained by gross autoradiography. However, if cytokinin was applied to a localized area of the leaf, but more basipetally than the isotope application, there was a strong mobilizing effect toward the cytokinin-treated areas in the case of sucrose (that is, cytokinin areas become photosynthate sinks); this was true to a slight extent with $^{32}PO_4^{---}$, and not at all with IAA.

Morris and Thomas (1968), working with *Pisum sativum* plants decapitated above the third internode, studied cytokinin-induced transport toward and away from the lateral bud. Application of IAA to the decapitated stump prevented growth of the lateral buds and restored the pattern of photosynthate distribution found in the intact plant. FAP applied alone to the decapitated stump enhanced the accumulation of photosynthate in the lateral buds, but FAP with IAA converted the internode into a photosynthate sink.

Cytokinin-induced mobilization has been intensively studied by Müller and Leopold (1966a, 1966b). They used FAP and ^{32}P in detached leaves of *Zea mays*. The transport of the ^{32}P was monitored by scanning the treated leaves in a thin-layer chromatography strip-chart isotope scanner. When the bases of detached senescent leaves were treated with cytokinin, the translocation of ^{32}P from the tips was accelerated, and accumulation of ^{32}P occurred in the cytokinin-treated areas. This was readily correlated with senescence, as monitored by loss of chlorophyll. Nonsenescent leaves of intact seedlings were reported to be unresponsive to cytokinin.

The specificity of cytokinin-induced mobilization and transport of ions was delineated by Müller and Leopold (1966b). Studies with ^{22}Na, ^{86}Rb, ^{36}Cl, ^{131}I, and ^{32}P showed that ^{32}P was the only isotope that moved to a cytokinin-induced sink. This is in agreement with the observations of Gunning and Barkley (1963) that ^{32}P is incorporated into nucleic acids in cytokinin-treated areas of leaves.

The mechanism of cytokinin-induced transport in detached leaves of *Zea mays* was extensively investigated by Müller and Leopold (1966b). When applications of FAP were made to the bases of leaves, and applications of ^{32}P were made to the tips, ^{32}P transport was found to be limited to the axial direction of the leaves, and restricted to the vascular bundles. By use of metabolic inhibitors and steam-killing, the transport could be shown to exist only in the phloem tissue. If two solute sinks exist on the same leaf (for example, a cytokinin-induced sink at the tip of the leaf, and a meristematic tissue or cytokinin-induced sink at the base), they will compete with each other for the same solute (^{32}P applied between the two sinks, for example) and the use of the same vascular system. Thus, the direction and the velocity are determined by which sink has the greater mobilizing force. Furthermore, through radioactive profile analysis and kinetic studies of the transport mechanism, it could be shown that cytokinins accelerate the transport by increasing the velocity, and that the shape of the logarithmic distribution curves fits a mass-flow rather than a diffusionlike transport system.

Milthorpe and Moorby (1969) summarized these findings by concluding that most of the cytokinin-induced mobilization and transport could arise from the stimulation of metabolic sinks near, or at, the point of application; and that treatment with cytokinin may initiate or stimulate many enzymic activities, including the synthesis of protein and nucleic acids, within a few hours. The effect of cytokinin on protein synthesis was investigated by Kuraishi (1968), who found that senescent leaf disks of *Brassica rapa* would incorporate ^{14}C-labeled L-leucine into protein, if also treated with BAP. Kuraishi's results also showed that the protein turnover rate was not increased, but that endogenous protein decomposition was retarded.

In an effort to further understand the mode of action of cytokinin-induced mobilization and transport, Pozsár, Hammady, and Király (1967) studied the incorporation of nucleic acid precursors into nucleic acid and protein fractions, and determined the concentrations of the constituents in *Phaseolus vulgaris* leaves before and after treatment with FAP and BAP. Both compounds stimulated the synthesis of protein and RNA. The BAP was significantly better than

FAP. The extent to which ^{14}C-labeled BAP could be incorporated into several RNA components was investigated by Fox (1966). About 15 percent of the cytokinin was incorporated into the RNA, while the remainder was distributed among several metabolites in the soluble RNA and polynucleotide-free fraction.

Kulaeva, Fedina, and Klyachko (1968), while studying the interrelationships of aging and cytokinin treatment, found that, in excised leaves, there was a disappearance of ^{14}C-labeled adenine from fractions of RNA of high molecular weight, a general decrease in the RNA level, and a net reduction in RNA synthesis. However, when these leaves were treated with BAP, the ^{14}C-labeled adenine was actively incorporated into RNA. They concluded, therefore, that cytokinins have a regulatory role in leaf protein synthesis. The influence of hormones, including cytokinins, on the metabolism of nucleic acid was recently reviewed by Key (1969), who speculated that because cytokinin occurs in certain species of transfer RNA, there is a possibility that cytokinin may be intimately connected to translational regulation of protein synthesis. If this were true, Key further proposed a model involving cytokinin-containing "modulator" species of transfer RNA that would explain both the quantitative and qualitative effects of cytokinins on plant growth and development. However, as he stated, there is no evidence that cytokinin in transfer RNA is associated with the action of cytokinins as growth regulators.

The above experiments with cytokinin-induced transport all show that metabolites move towards regions of high concentration of growth substances, that senescence can be retarded, and that physiologically old tissue may be made to respond as young tissue for short periods of time. It appears, at this time, that cytokinins have no direct effect on phloem sieve tubes or their membranes; whether or not cytokinins have an effect on callose production remains to be answered. Potentially, cytokinin-induced transport may find broad application in the regulation and manipulation of other hormone distribution patterns in phloem transport pathways.

Movement of the Flowering Stimulus

Evans and Wardlaw (1964) studied the translocation of assimilates in relation to the transmission of the flower inducing stimulus in *Lolium temulentum*. They used $^{14}CO_2$ in leaf chambers to label the assimilates, and they prepared the plants by removing all but the sixth leaf, which they then exposed to 8 hours of illumination at

3500 ft-c followed by 16 hours at 20 ft-c. One experiment involved removing some of the plants from this low illumination to total darkness every 2 hours; a second involved removal of treated leaves every 2 hours. Figure 7.1 shows the results of these experiments: the critical photoperiod length was about 16 hours, and within 4 hours of this period, sufficient long-day stimulus moved from the leaf blade to cause flower initiation. Evans and Wardlaw calculated that the minimum rate of movement of the floral stimulus is about 2 cm per hour.

In other experiments, labeled assimilates were translocated to the shoot apex and other parts of the plant from an upper leaf exposed to long-day (inducing) conditions when the lower leaves, exposed to short-day conditions, were variously present or removed; a similar comparison was made of the distribution from an upper leaf exposed to long-day conditions, and that from a lower leaf exposed to short-day conditions. They found that the presence of lower leaves did not reduce the movement of assimilates from the upper leaf to the shoot apex; and that the lower leaf supplied only a small proportion of the

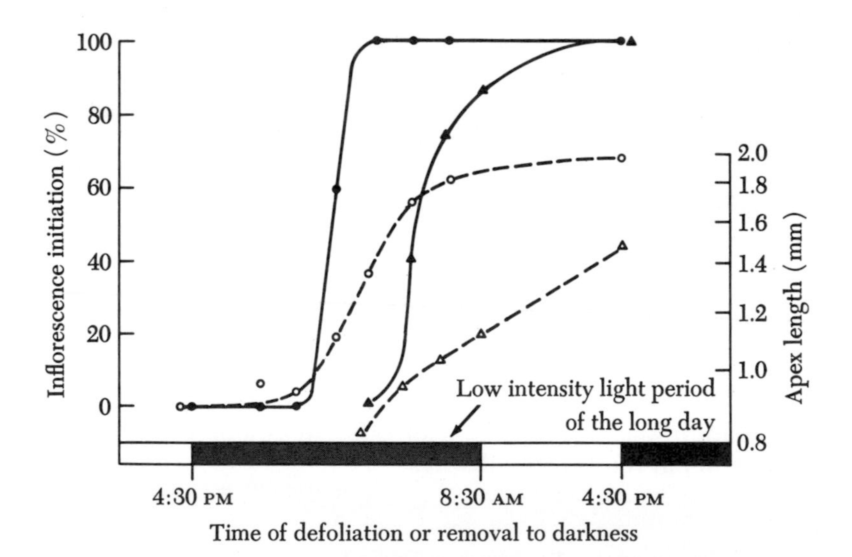

Figure 7.1. The effect of photoperiod length during a single long-day exposure, and the time of defoliation or removal to darkness, upon inflorescence initiation (*solid line*) and apex length (*dashed line*) in *Lolium temulentum* at the time of dissection, three weeks after the long-day exposure. *Circles*, plants removed to darkness at times indicated; *triangles*, sixth leaf blade cut off at times indicated. From Evans and Wardlaw (1964).

assimilates reaching the shoot apex, although it supplied to the roots. Evans and Wardlaw concluded that the inhibitory effect of lower leaves exposed to short-day conditions on floral initiation in *Lolium temulentum* is due neither to their interference with translocation of the long-day stimulus to the shoot apex, nor to their diluting it with assimilates. They suggested that this results, rather, from the production of a mobile inhibitor of initiation in those leaves.

However, Evans and Wardlaw (1966) concluded that while both the floral stimulus and the assimilates may move in the phloem, they can move independently and by different mechanisms. Three kinds of evidence, obtained by using techniques similar to those described above, support the concept that the floral stimulus moves independently of the assimilates. First, the transport velocity of the assimilate and the floral stimulus from the seventh leaf blade of *Lolium temulentum* to the floral primordium was measured simultaneously, and found to be 1.0–2.4 cm per hour for floral stimulus and 77–105 cm per hour for ^{14}C-labeled assimilate. Second, leaves that were only 14–26 percent developed, and not exporting assimilates, were capable of exporting the floral stimulus. Third, inhibitors like diuron could greatly reduce the flowering response without reducing the flow of assimilates to the shoot apex.

King, Evans, and Wardlaw (1968) investigated the translocation of the floral stimulus of *Pharbitis nil* in relation to assimilate movement. They found the average velocity of basipetal movement of the floral stimulus from an apical donor leaf to be 30 cm per hour. Under similar conditions, the translocation of assimilates averaged 36 cm per hour. Working with leaves of varying size, they found that leaf size has a similar effect on the export of floral stimulus and on the export of assimilates. Leaves held in the dark could export floral stimulus at a rate comparable to that of leaves kept in the light, although labeled assimilates apparently did not accompany the stimulus. Because Geiger and Batey (1967) showed that assimilates, labeled progressively earlier in the day with ^{14}C and not translocated immediately, are moved progressively later in the night; King and his coworkers reasoned that their failure to detect the movement of ^{14}C-labeled assimilates, when the floral stimulus was being exported in darkness, did not necessarily mean that assimilates (unlabeled) were not being moved at that time. The much higher velocity of translocation of the floral stimulus in the short-day plant *Pharbitis nil* (30 cm per hour) compared with that in the long-day plant *Lolium temulentum* (1.0–2.4 cm per hour), King and his co-workers suggest, may result from a difference in stimuli in the two plants. It is possible that that in *Pharbitis* is readily sieve-tube mobile, and that that in *Lolium* is able to move only in the parenchyma. Because assimilates move at velocities of 77–105 cm per hour in *Lolium,* the basic mechanisms of assimilate movement in the two plants must be similar.

Weise and Seeley (1964), using greenhouse cultivars of *Chrysanthemum morifolium,* studied the translocation of the floral stimulus. Some of their plants were grown with localized short-day treatments, some with long-day treatments; some 2-branched and grafted plants were grown with one branch exposed to short-day conditions, the other to long-day conditions. Flowering was induced when as few as the upper four mature leaves were exposed to short day conditions. Earlier flowering resulted when one-half or all of a plant was so exposed. With 2-branched plants, a short-day treatment of one branch caused flowering of that branch, and also of the second branch exposed to long-day conditions. Short-day treatment of the stock portion of a grafted plant caused flower bud initiation and development on scions given long-day treatment. Earliness of flower bud appearance, and the number of flower buds on the receptor branch and scion, were stimulated by defoliation of the receptor. These experiments indicate that flowering in *Chrysanthemum* results from synthesis and translocation of a flower-inducing substance that acts as though it moves with the assimilates in plants. Stout (1945) reported similar behavior for the flower-inducing substance in *Beta vulgaris.*

The relations between the transport of ^{14}C-labeled photosynthates and movement of the flowering stimulus in grafted *Silene armeria* plants were studied by de Stigter (1966). He used ^{14}C treatment of leaves to label the photosynthates in his plants, and he determined movement of his tracer by autoradiography. He compared his results with those of Wellensick (1966) for transmission of the flowering stimulus in the same species. He used reciprocal grafts, in which both donor and receptor leaves were treated. He found that transmission from one partner to the other started 7 days from grafting, and that, at 16 days, the union was able to transmit assimilates at high rates. The rates differed widely according to the various combinations and treatments used, being much higher from stock to scion than from scion to stock, regardless of which was donor and receptor. However, appreciable transport from scion to stock could be induced by defoliating the stock, a result easily rationalized in terms of source–sink relations. As shown by de Stigter's figures, the stocks of all plants, except the defoliated ones, had rosettes made up of many mature leaves, a very few of which were sufficient to provide assimilates to the roots; the bulk of the assimilates were moving acropetally into the scions. Only when these rosette leaves were removed was it possible to reverse the flow and obtain basipetal movement.

On the whole, de Stigter's results paralleled those of Wellensick, confirming the idea that the floral stimulus moves in the phloem along with the assimilate stream.

8

Movement of Exogenous Substances

Introduction: Evidence for Associated Movement of Tracers with Assimilates

Ever since Stout and Hoagland (1939) and Colwell (1942) did their pioneering work with radioactive isotopes in studies of translocation in plants, there has been increasing activity in this field of study. Various early reviews (Crafts, 1951, 1961b; Zimmermann, 1960a) covered the literature on this subject up to the first of the 1960's. Hull (1960) tabulated information on herbicides and growth regulators. Schumacher (1967) published an extensive volume on translocation.

Work done during the 1950's showed that labeled tracers, many of which are pesticides, are able to penetrate the cuticle of leaves, to migrate to the vascular tissues, and then to move either apoplastically to the leaf tip or symplastically down the petiole to the stem, and thence to active sinks in shoot tips, storage organs, vascular cambium, and roots. Applied to mature lower leaves, phloem-mobile tracers move to the roots; applied to fully grown upper leaves, they move to the shoot tips; applied to median leaves, they move both basipetally and acropetally, and in doing so, they bypass other mature leaves that are exporting assimilates. Applied to a young importing leaf, they move only acropetally to the leaf tip (Zimmermann, 1960a; Crafts, 1961b). This response is illustrated in Figure 8.1. Inclusion of surfactants, humectants and other adjuvants may increase uptake by the leaves. Adjustment of the *p*H of the applied solution aids penetration of some molecules (Crafts, 1961c). Evidently, such observations indicate that tracers that are able to enter the phloem and move in it are carried along with assimilates from sources to sinks, as in a stream. Those that are xylem-mobile move readily from roots to tops, but seem unable to enter the phloem of leaves, and to accompany assimilates in their normal distribution. Some tracers seem to be both xylem- and phloem-mobile; and some are actively retained in living cells, and so have a restricted distribution (Crafts, 1959a, 1959b, 1961b; see Table 8.1).

Crafts and Currier (1963), in a review on sieve-tube function, enumerated six advantages of using labeled tracers in translocation studies. Such studies show (1) that assimilates in plants follow a source-to-sink distribution pattern, moving out of photosynthesizing leaves to regions of active growth, bypassing mature assimilating leaves in the process; (2) that ^{14}C-labeled tracers may penetrate from the cuticle to the phloem and move readily in the assimilate stream; (3) that tracers may be retained or reabsorbed along the translocation route (which is proof of the distributive function of the phloem system); (4) that, because of the great avidity of certain tissues for compounds such as 2,4-D and IAA, the distribution of these compounds may depend upon the relation between the rate of lateral distribution and the rate of longitudinal translocation; (5) that some compounds migrate from phloem to xylem, and hence may circulate in plants; and (6) that some compounds applied to leaves may translocate to roots and on out into the ambient culture medium (under very favorable conditions, 2,4-D may do this; commonly, maleic hydrazide, MOPA, and TBA show this behaviour).

In a critical discussion of the nature of the protoplasmic connections of the sieve plate, Crafts and Currier emphasized recent evidence by Esau and Cheadle (1961) and Duloy et al. (1961) that these strands, in their natural state, are open tubules that may accommodate a rapid mass-flow of solution. Although their sections were prepared with permanganate and were therefore uncritical, they suggested the possibility that sieve-plate pores are open under natural conditions. This possibility has been confirmed by Cronshaw (1969), Cronshaw and Anderson (1969), Currier and Shih (1968), Johnson (1968), Shih and Currier (1969), and Sjolund (1968).

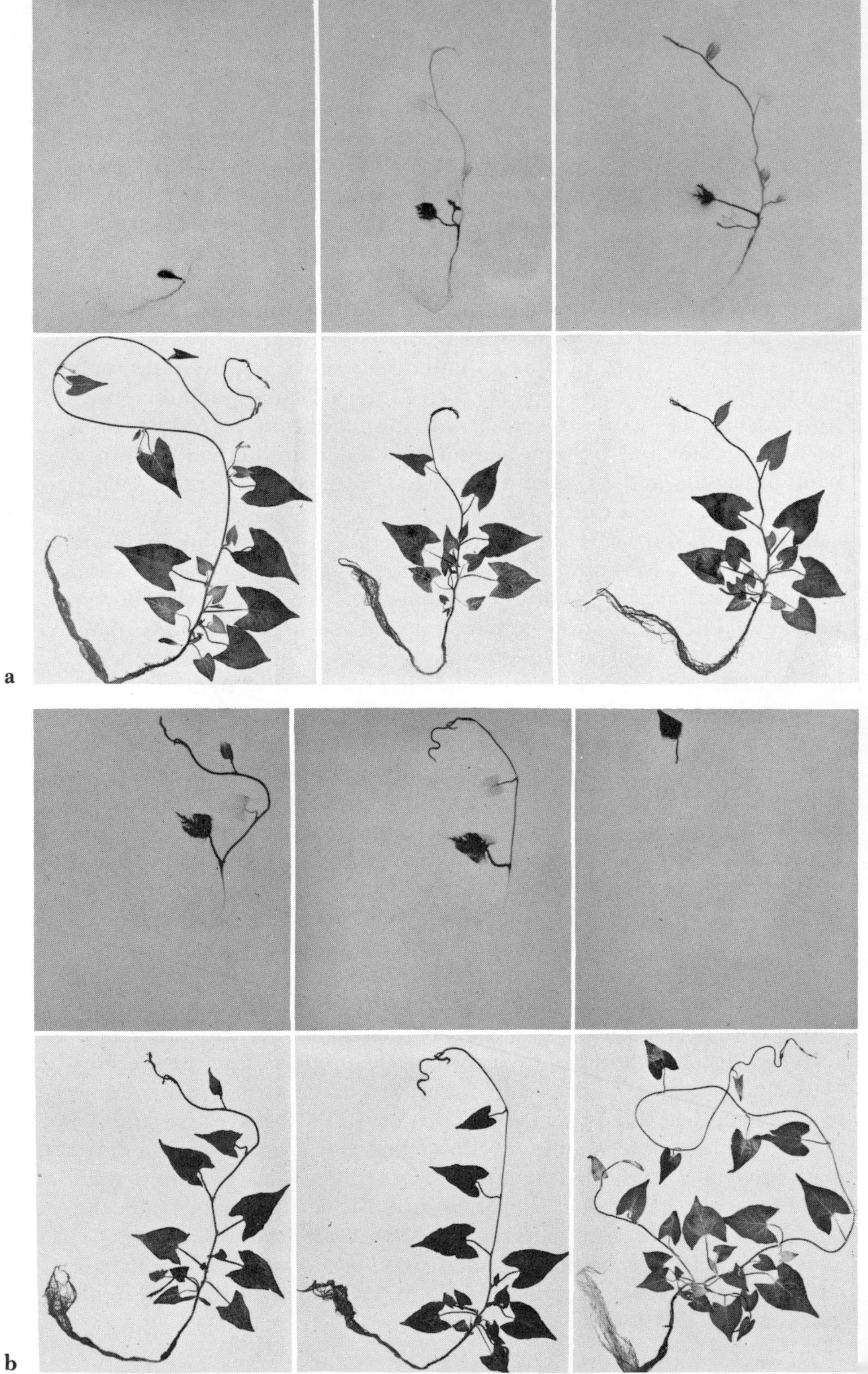
a
b

Thus, for the first time, the phenomenon of phloem exudation and the structure of sieve tubes are reconciled. For many years, botanists have been viewing artifacts resulting from the normal plugging of sieve plates, caused by cutting the tissues in the course of preparing them for microscopic examination. By injecting the killing fluid into the hollow petioles of *Curcubita pepo*, Esau and Cheadle made preparations for electron microscopy that had not been cut before killing.

Crafts and Currier (1963) discussed the structural and functional relations of callose. Although pit callose may be formed in a matter of seconds following wounding, it seems probable that callose is a normal component of functioning sieve elements, and that it may serve to plug these elements as they approach senility. Massive deposits of definitive callose seem to function in this way in old sieve tubes; the function of callose in young sieve elements is still obscure.

Crafts and Yamaguchi (1964) published a book on the use of the autoradiographic method for studying the uptake and distribution of labeled tracers by plants. A few of the high points of this work will be reviewed here; the reader is referred to the original volume for details.

When applied to a leaf of a growing plant, many tracers (including some herbicides) will penetrate the cuticle, and migrate into and along the symplast, accelerated by protoplasmic streaming, until they reach the phloem. In the sieve tubes, they diffuse into the assimilate stream and translocate to distant sinks. Assimilate movement in plants is a transport process involving one or more sources and one to many sinks. Movement from source to sink is a hydrostatic flow along a gradient of turgor pressure, developed osmotically. Metabolic processes in sources and sinks are essential to the functioning of the system. Rapid movement of assimilate and tracer takes place as a mass flow (or pressure flow) through the open sieve tubes of the phloem. Source and sink must both be active for translocation to take place. By manipulation, as by shading or pruning, the assimilate stream may be reversed.

Some solutes (2,4-D, or sugars, for example) move principally in the phloem; others, such as the substituted ureas and uracils, symmetrical triazines, and some surfactants, move freely only in the xylem; some move freely in both phloem and xylem, as do picloram, TBA, maleic hydrazide, and some amino acids. Figure 8.2 shows the

Figure 8.1. Distribution of ^{14}C-labeled 2,4-D in plants of *Polygonum convolvulus* one day after application to: (*a*) cotyledon, leaf 2, and leaf 4; and (*b*) leaf 6, leaf 8, and leaf 10.

distribution of ^{14}C-labeled tracers (namely, 2,4-D, amitrole, maleic hydrazide, and dalapon) one day after application to a primary leaf of *Phaseolus*. Figure 8.3 illustrates the redistribution of these four tracers during a 16-day treatment period.

Crafts (1967a, 1967b) reported on the absorption and translocation of labeled tracers in plants. After reviewing early tracer work, including the study illustrated in Figure 8.9, he compared the various tracers that have been used in translocation studies. Work with labeled tracers, including a number of growth regulators, indicates that these compounds may be grouped into several categories with respect to their distribution patterns in plants. In viewing this classification, however, it should be realized that many factors determine the distribution of a given tracer in a given plant. As pointed out by Crafts and Yamaguchi (1964), the formulation, the locus of application, the growth status of the plant, the dosage, the treatment time, and various other factors, will determine the uptake and distribution of a given chemical. With these points in mind, the following classification is offered (although, undoubtedly, a number of compounds will be shifted from one category to another as information becomes more complete, and many new compounds will be added as new physiologically active substances are synthesized): (1) tracers that enter and move almost solely in the apoplast; (2) those that enter and move almost exclusively in the symplast; (3) those that enter and move freely via both symplast and apoplast; (4) those that move little or none in either vascular system; (5) those that show restricted movement in symplast, in apoplast, or in both; and (6) those that leak from roots into the ambient culture medium. Table 8.1 lists some of the compounds in these categories. For a similar classification of translocated tracers, see Leonard et al. (1966a). Undoubtedly, many other growth regulators, inhibitors, and physiologically active compounds fit into these categories.

It should be apparent that we have considered here only those mechanisms and patterns of solute distribution that are directly related to the vascular systems of organized plants; and that the tracers used have been entirely exogenous materials. While it has been shown that applied auxins (notably IAA and gibberellin) will move in vascular channels once they enter them, polar movement of endogenous regulators through relatively undifferentiated tissues involves entirely different mechanisms, and was not dealt with in the studies mentioned above.

As more and more chemicals of the growth regulator type are used

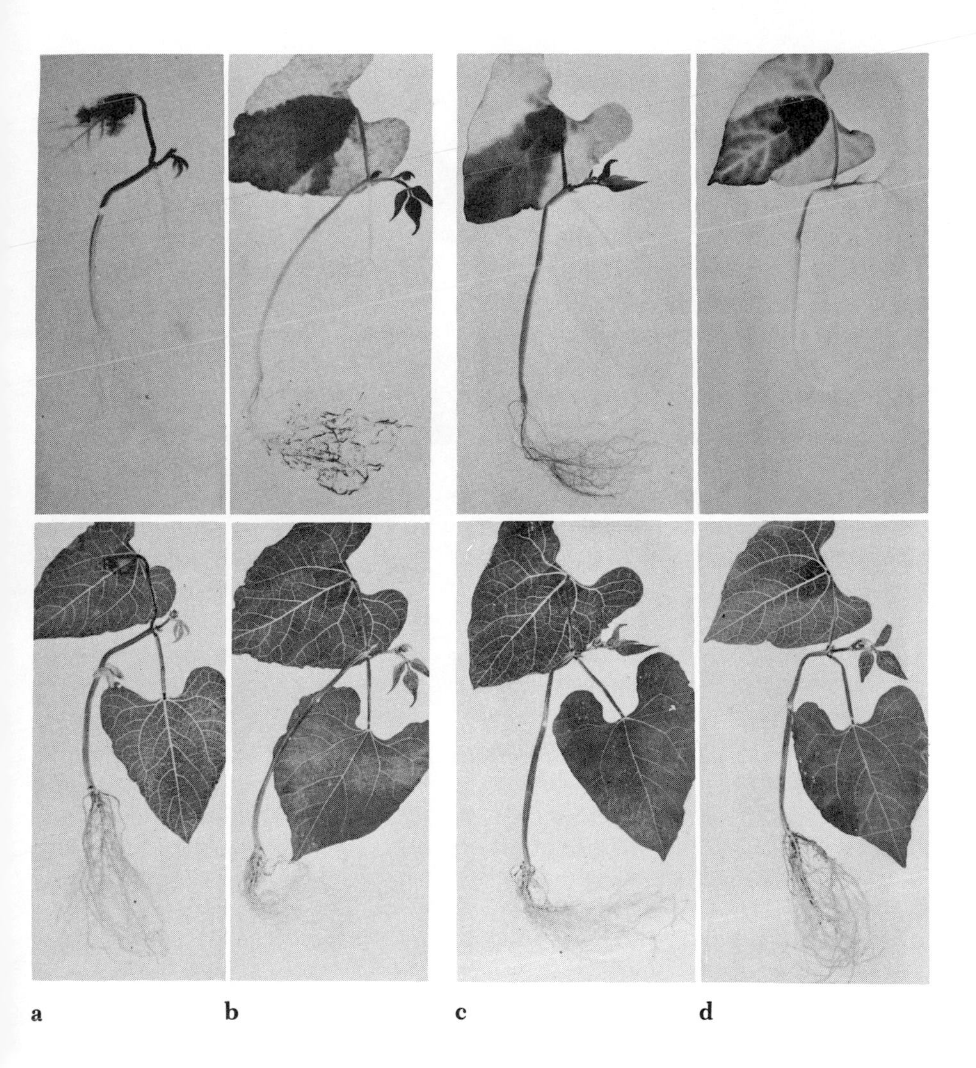

Figure 8.2. Translocation of (*a*) ^{14}C-labeled 2,4-D, (*b*) amitrole, (*c*) maleic hydrazide, and (*d*) dalapon from the primary leaf of *Phaseolus vulgaris* one day after application.

in agriculture, knowledge of their uptake, distribution, and mode of action in plants is of great importance in interpreting the responses to their application, and in understanding their ultimate effects on production. However, because of the national concern for the side effects of toxic chemicals in the ecosystem, man is becoming more and

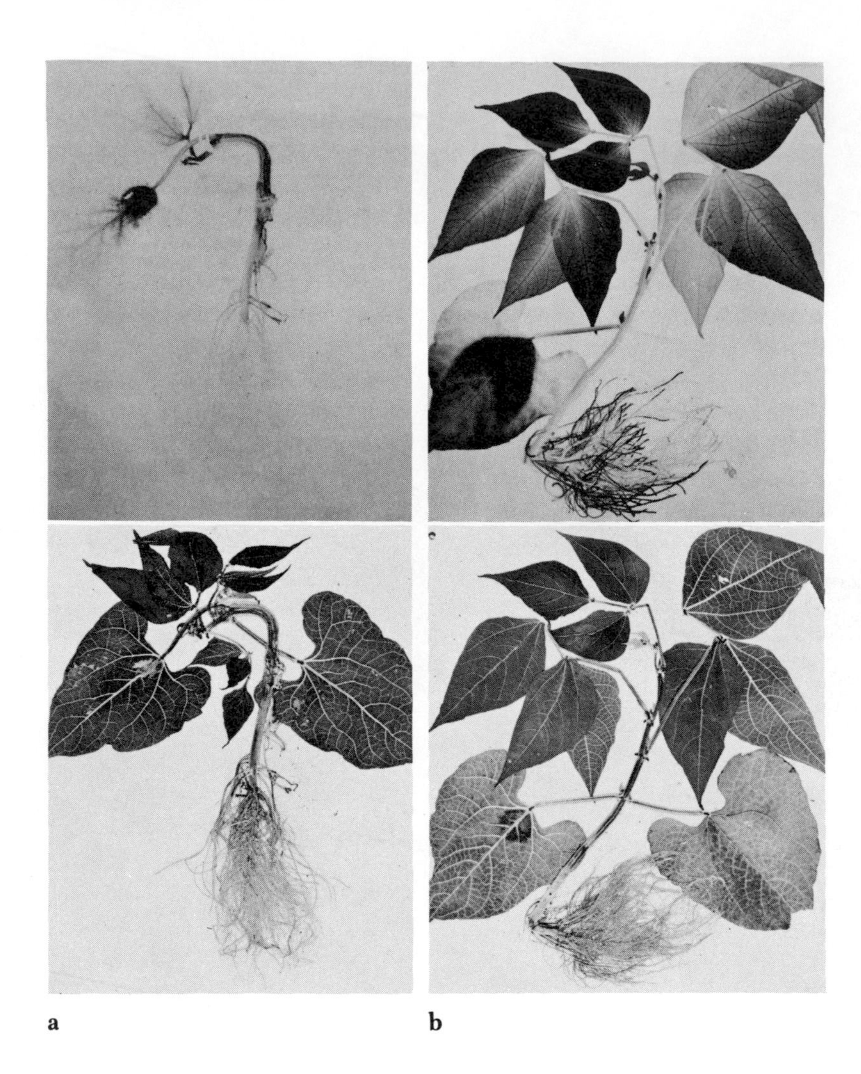

Figure 8.3. Translocation of the same tracers in the same plants as shown in Figure 8.2 during a 16-day treatment period. Formative effects are shown by 2,4-D; the other three tracers show different redistribution patterns.

more judicious in his use of chemicals for pest control and the control of plant growth. The antithesis, however, is that modern man is becoming increasingly dependent upon the use of synthetic chemicals, in the form of drugs, pesticides, food additives, and so forth,

c d

for his well-being. Because the plant is a powerful biochemical storehouse, with the capability of biodegradation or accumulation of the xenobiotic materials, and because translocation processes occur concomitantly, the importance of understanding the translocation mechanisms and extent of transport of any xenobiotic substance becomes paramount. We can no longer afford to be lackadaisical in our concern for knowledge about the total distribution of any foreign compound in our enviroment, including its possible transport and accumulation in plants.

Table 8.1. Mobility of tracers in plants. Compounds having no alternative designation were ^{14}C-labeled. Mobility varies between compounds; it may also vary between plants and between various treatments.

Free mobility			Limited mobility			Little or no mobility
In apoplast	In symplast	In both	In apoplast	In symplast	In both	
Atrazine	Amiben	Amino acids (some)	Barban	2,4-D	Naptalam	DCPA
^{45}Ca	Fenac	Amitrole	Paraquat	2,4-DP	Amino acids (some)	2,4-DB
CIPC	Maleic hydrazide	^{134}Cs	^{210}Po	MCPA	Ammonium thiocyanate	DNBP
$^{36}Cl^-$	Sodium acetate	^{137}Cs		2,4,5-T	$^{77}AsO_4^{---}$	Diquat
Diuron	Sucrose	Dalapon-^{14}C, dalapon-^{36}Cl		TPA	Di-allate	DNOC
Fluometuron		Dicamba			Duraset	Endothall
IPC		^{39}K			EPTC	^{210}Pb
BAP		^{22}Na			^{59}Fe	PCP
Monuron		Picloram			Gibberellin	
Simazine		$^{32}PO_4^{---}$			IAA	
Sodium lauryl sulfate		^{86}Rb			Ioxynil	
^{89}Sr		$^{35}SO_4^{--}$			^{28}Mg	
T-1947		TBA			Propanil	
Tetramine					Sodium benzoate	
Tween 20					^{165}Zn	
Tween 80						

Note: Maleic hydrazide, picloram, TBA, sodium, rubidium, and cesium move from roots into the ambient culture medium. 2,4-D and many other compounds have been found to do this under special circumstances.

Application Methods

The method used to apply a tracer to a plant may largely determine its distribution pattern. Application to the roots by way of the culture medium, or to the roots or stems through cut vascular bundles, results in primary movement to transpiring regions via the xylem. Application of tracers to the foliage in gaseous form (as $^{14}CO_2$ or $^{35}SO_2$, or ^{14}C-labeled EPTC, for instance), or in droplets in liquid form, usually results in acropetal movement via the xylem or basipetal movement via the phloem. The application of a limited volume over a short time through a cut leaf flap may result in phloem movement (Biddulph, 1941), although a larger volume applied over a longer period may result in xylem uptake accompanied by some phloem export. As Colwell (1942) showed, application of a large volume of solution to a leaf may result in uptake and export via the xylem; if the volume applied is small and the area restricted, phloem export is assured.

Trip and Gorham (1968d) studied the translocation of sugars in the vascular tissues of *Glycine max*. They introduced labeled sugars either through the stump of a cut petiole or through a flap of a lateral leaf vein. When the sugars were introduced through a cut petiole, they were transported in the stem via the xylem and were able to pass a steam girdle, contrary to results previously reported. When labeled glucose was introduced via a lateral leaf vein, it was metabolized and translocated as sucrose from the leaf by way of the phloem in a manner resembling that of photosynthate.

To simulate field application of pesticides, Crafts (1956b) applied ^{14}C-labeled 2,4-D to *Convulvulus arvensis* in a formulation in which the tracer was diluted with an emulsifiable acid of commercial origin; the formulation was applied as a spray in the field at a rate equivalent to $\frac{3}{4}$ pound of 2,4-D per acre, including 5 μg of ^{14}C-labeled 2,4-D per plant. This treatment resulted in translocation of the tracer to depths of 16.5 inches in 72 hours. In 8-day, 14-day and 21-day treatments, translocations occurred to a depth of 6 feet, which was the total depth of excavation.

Many factors affect the fate of a chemical applied to the surface of a leaf. Cuticle thickness and permeability, leaf structure, and leaf metabolism are all involved. And pH of the spray solution may alter the penetration process (Crafts and Reiber, 1945).

Crafts (1961c) proposed that there are two routes along which substances may move from the cuticle to the vascular bundles of a leaf.

The first is an aqueous route, characterized by compounds that respond to relative humidity, such as maleic hydrazide, urea, and amitrole. The second is a lipoid route, traversed by compounds having high lipoid solubility. Organic acids and their salts may enter via this route, and adjustment of the *p*H of the applied solution to approximately the isoelectric point favors rapid movement, because the undissociated parent acids are more lipoid-soluble than their salts.

The use of surfactants also may increase uptake. These adjuvants increase the wetting of the cuticle, and in concentrations in the range of 0.5 percent or above, they may alter cuticle permeability.

Bachofen (1962a) studied methods of application in relation to the absorption and transport of ^{32}P in *Phaseolus*. His methods were (1) the application of a droplet to the upper leaf surfaces; (2) the application of a droplet to the lower leaf surface; (3) midrib injection with the flap open toward the leaf base; and (4) midrib injection with the flap open toward the leaf tip. After one hour, the droplet application to either leaf surface gave, by autoradiography, a round spot approximately 2 cm in diameter in the center of the leaf. Injection into the midrib with the flap open toward the leaf base resulted in the total leaf, including the petiole, being infiltrated with ^{32}P; with the flap open toward the tip of the leaf, only the apical half of the leaf was infiltrated.

Table 8.2 shows Bachofen's data on the uptake and distribution of ^{32}P by the four application methods. It proves the superiority of the injection method for getting a tracer into a leaf; it shows also, in the case of *Phaseolus*, the superiority of the lower surface for droplet application. Two points in this work need clarification. First, the treated plants for autoradiography were oven dried, and the spots on the untreated trifoliate leaves are reminiscent of the illustration of an oven-dried leaf in Levi's (1962) paper on an artifact in autoradiography. Second, the leaf injection method must be used with care: If the volume of tracer solution presented to a leaf is in excess of the volume of water transpired during the presentation time, some of the tracer solution may be sucked back into the stem via the xylem and distributed very rapidly to the roots, or to the upper leaves. The petioles of the leaves in many of Bachofen's autoradiographs are black, whereas those of other leaves are unlabeled. This indicates that some of the tracer was sucked back into the xylem. If one is studying normal transport in plants, the droplet application method is less subject to error. (For a detailed discussion of this subject, see Little and Blackman, 1963.)

There has been a great volume of work done on the role of surfactants in the activity of pesticides. To cite a few, Jansen, Gentner, and Shaw (1961) studied the effects of some 63 surfactants on herbicidal activity. These included cationic, anionic, nonionic, ampholytic, and blended compounds. They employed *Zea mays* and *Glycine max* as test plants and 2,4-D, dalapon, DNBP and amitrole as toxicants. Logarithmic increases in concentration of many surfactants, increased herbicidal activity: some surfactants repressed activity, and many were phytotoxic. Surfactant effects varied between the two crop species, and there were no obvious correlations between activity and structure.

Crafts and Yamaguchi (1964) reported tests on ^{35}S-labeled sodium lauryl sulfate. This compound exhibited only apoplastic distribution in *Phaseolus, Gossypium, Hordeum, Glycine,* and *Cucumis.* When extracted from leaves, the ^{35}S was in the form of sulfate.

Smith and Foy (1967) found the activity of paraquat to be enhanced by 8 out of 10 commercial surfactants; significant interactions occurred between paraquat and anionic surfactants. Smith, Foy and Bayer

Table 8.2. Absorption and translocation of ^{32}P applied to leaves of *Phaseolus vulgaris* by different methods.

Method of application	After 2 hr			After 24 hr		
	Amount translocated		Amount absorbed	Amount translocated		Amount absorbed
	As % of amount absorbed	As % of amount supplied	(as % of amount supplied)	As % of amount absorbed	As % of amount supplied	(as % of amount supplied)
Droplet to upper surface	3.97	0.33	8.3	29.5	2.5	8.5
Droplet to lower surface	1.24	0.17	13.7	20.5	3.0	14.6
Midrib injection toward base	3.80	1.58	41.6	30.4	30.2	99.3
Midrib injection toward tip	0.65	0.28	43.1	9.0	8.8	97.8

Source: After Bachofen (1962a).

(1967) found four biodegradable surfactants to provide excellent wetting of the foliage of *Zea mays,* and to be equal or superior to commonly used commercial formulations; all were slightly toxic when applied alone to Z. *mays.*

Movement of Inorganic Tracers

Arisz (1964) studied the translocation of ^{36}Cl in excised *Vallisneria* leaves. He concluded that ^{36}Cl was actively absorbed into the symplasm of these leaves and could move from the point of uptake to the sink at rates of 4 cm per hour. However, when the chloride content was measured by chemical analysis and compared to the isotopic concentration, it could be shown that the ^{36}Cl had rapidly reached equilibrium with the unlabeled chloride and, that once absorbed, the ^{36}Cl did not leak out of the symplasm.

The role of boron in sugar translocation has received much attention in recent years. One explanation of this role is that boron forms a negatively charged complex with sugar; this complex is thought to be more readily moved across cell membranes than is sugar alone. A contrasting idea is that boron is essential to the metabolic activity of meristematic regions, and hence that its presence is responsible for the development of normal sinks.

Dyar and Webb (1961) studied the relationship between boron and auxin in ^{14}C translocation in *Phaseolus vulgaris.* Using NAA as their auxin, and applying $^{14}CO_2$ to one leaf of *P. vulgaris* 'Black Valentine,' they found that both terminal bud treatment and root treatment enhanced the movement of sugars in boron-deficient plants. Roots of boron-deficient plants were stunted; NAA not only increased linear growth of roots to near normal, it stimulated lateral root formation, hence adding materially to the number of meristematically active sinks. Dyar and Webb concluded that boron-deficient plants, prior to phloem necrosis, possess a functional translocation system that fails to function because substrates are not being utilized; and that boron is not necessary to sugar translocation per se, but that it plays an essential role in the biosynthesis of auxins, which, in turn, stimulate growth, thereby activating sinks for translocation.

McNairn and Currier (1965) studied the relation of boron to callose, and found that five ppm of boric acid could inhibit cell-to-cell translocation by boron-induced callose in the pit fields of *Phaseolus vulgaris* leaves; however the induction of callose in sieve plates occurred only after the marked formation of pit callose in mesophyll cells. They concluded that, although boron is translocated almost exclusively in the apoplastic continuum with accumulation at sites of water loss, there must be sufficient boron leakage into the symplast, after the pit necrosis due to boron, to effect sieve plate callose formation. These observations bring up an interesting question. Is it possible that boron, at low concentrations, forms complexes with callose precursors and renders them ineffective, thus stimulating sugar translocation (Mitchell, Schneider, and Gauch, 1960); and that, at higher concentrations, it stimulates callose synthesis? Boron would then play a regulatory role in callose biosynthesis.

Hewitt (1963) studied the effect of boron sprays on the translocation and distribution of endogenous carbohydrates in *Malus sylvestvis* 'York Imperial' as measured by ^{14}C-labeled sucrose. Trees were sprayed with a 369 ppm boric acid solution during full bloom, or immediately after, for 3 successive years, in an attempt to elucidate the function of boron sprays in relieving bitter pit in the fruits. However, regardless of the source, time, or frequency of the application, the treatments had no measurable effects on the distribution of assimilates. Different leaf–fruit ratios likewise had no effect, and there was no relation between boron content and ^{14}C activity. Girdling was ineffective. Bitter pit control by boron sprays must have some other mechanism than the enhancement of assimilate movement.

Many workers have shown that calcium, applied to leaves, is relatively immobile in living plants (Biddulph et al., 1959). Because Kessler and Moscicki (1958) had reported basipetal movement of foliarly applied calcium following treatment with TIBA, Taylor et al. (1961) made foliar applications of ^{45}Ca to plants of *Lycopersicon esculentum* in an attempt to elucidate this problem. Applying 48 and 96 μc of ^{45}Ca, and allowing 24 and 48 hour migration times, they sampled fresh and oven dried plants and counted the activity in a gas flow counter. They were unable to confirm the results of Kessler and Moscicki, finding, if anything, reduction in counts in the TIBA-treated plant samples. Likewise, treatment with IAA had no effect on calcium movement.

Dehydration of intact ^{45}Ca-treated plants in a forced draft oven resulted in an over-all increase in radioactivity of a hundredfold, compared with plants that had been cut into separate fractions before drying. This effect was observed regardless of treatment with TIBA or IAA. Possibly, the explanations of Biddulph et al. (1959) and Pallas and Crafts (1957), to the effect that an antifact of xylem movement could be involved, may be correct.

Ziegler (1963a) studied the movement of ^{45}Ca in the fruit of *Cucurbita maxima*. He injected the tracer in solution into the basal end of growing fruits and analyzed fruits and peduncles after 24 hours. He found that most of the tracer had moved basipetally, some being present in the pendunele at a distance of 14 cm from the point of injection. Only an insignificant trace was found 12 cm toward the apex of the fruit. Ziegler interpreted these findings as indicating a back-flow via the xylem, as postulated by Münch (1930).

Xylem injection, girdling, and ^{47}Ca (an isotope emitting high-energy gamma rays) were employed by Thomas (1967, 1968) for the study of calcium transport in *Cornus*. Transfer from xylem to phloem in the tagged trees was demonstrated by presence of ^{47}Ca in the bark above the stem segment from which all bark had been removed. In a second experiment, using ^{45}Ca, vein loading by ^{45}Ca was observed, and quantitative data showed that stable calcium and ^{45}Ca were evenly distributed. Additional quantitative data, with respect to concentrations of stable calcium, showed that the intervein areas of the leaf contained 78.5 percent of the total calcium in the leaf, but that on a concentration per unit area basis, the veins had a higher concentration. Thus, autoradiographs showed a darkening over the veins.

Calcium resembles the urea and triazine herbicides in being only xylem-mobile. Phosphorus and zinc are phloem-mobile; ^{32}P circulates freely in plants. Figure 8.4 illustrates distribution of calcium, phosphorus, and zinc, following application to the leaves of *Phaseolus*.

Although calcium has been shown repeatedly to be phloem-immobile (Biddulph, Cory, and Biddulph, 1959; Levi, 1968a; Fischer, 1967), its immobility is not due to lack of uptake by the symplast: Thomas (1967) and Eschrich, Eschrich, and Currier (1964) have shown ^{47}Ca and ^{45}Ca, respectively, to be present in phloem tissue. The latter workers studied the uptake of ^{45}Ca by *Cucurbita maxima* roots, and found distribution throughout the plant. Histoautoradiographs clearly showed that ^{45}Ca is taken up by newly deposited callose, where it forms insoluble complexes. They concluded that callose precursors give up organic phosphorus during callose formation, which combines with ^{45}Ca to form an insoluble salt at the site of synthesis.

Compared with other ions reaching the leaf, phosphorus is particularly mobile, being retranslocated extensively (Fischer, 1967). The degree of its retranslocation appears to be determined by the rate of its movement into and out of the phloem (Greenway and Gunn, 1966). Phosphorus translocated to young leaves originated with old leaves, while rapidly expanding mature leaves obtain their phosphorus directly from the roots. Further, Greenway and Gunn (1966) show evidence to support the concept that the export or retranslocation of

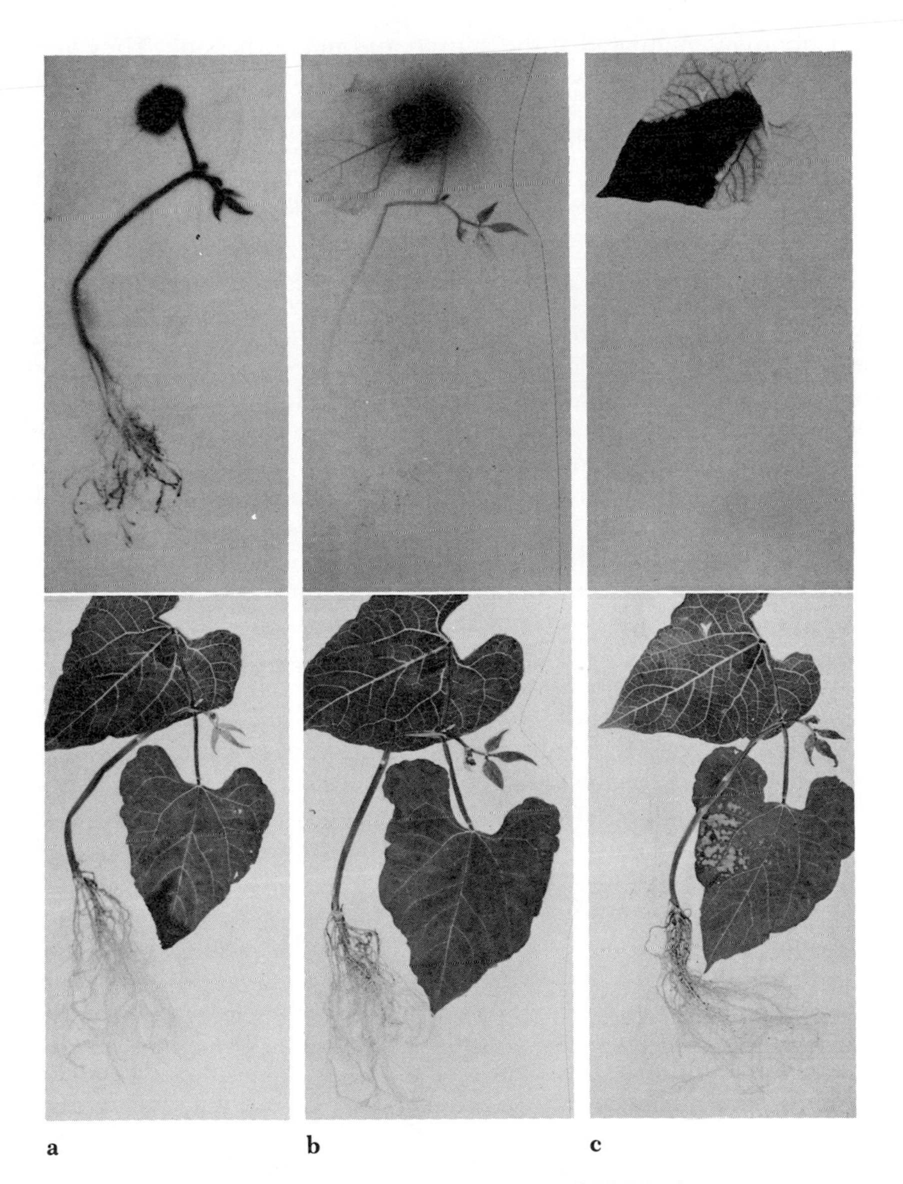

Figure 8.4. Translocation of (*a*) ^{32}P, (*b*) ^{65}Zn, and (*c*) ^{45}Ca during one day following application to the primary leaf of *Phaseolus vulgaris* plants. Symplastic distribution is shown by ^{32}P and ^{65}Zn; ^{45}Ca shows apoplastic movement only.

phosphorus is regulated at the source and not at the sink. They visualize competition between the photosynthetic parenchyma and phloem cells at the source as the dominant factor. Thus, when high concentrations of phosphorus are supplied to the source, retranslocation is rate-limited, because the sites transporting phosphorus into the phloem are saturated. If low concentrations of phosphorus are supplied, release from individual cells becomes the rate-limiting factor.

Bieleski (1966b) evaluated an active accumulation mechanism in excised phloem tissue for its role in phloem transport. He found that both $^{32}PO_4^{---}$ and ^{14}C-labeled sucrose are accumulated in phloem tissues (sieve tubes, phloem fibers, companion cells, and phloem parenchyma) against concentration ratios of the order of 10^3 or higher. The rate of accumulation of $^{32}PO_4^{---}$, $^{35}SO_4^{--}$, and ^{14}C-labeled sucrose was 4–35 times higher for phloem tissues than for comparable parenchyma tissue. The results strongly suggest that the first stage in the movement of a metabolite from the leaf into another part of a plant is the active accumulation of that metabolite into the veins. Of course, the experimental results do not separate companion cells (which are thought to be very specialized centers of high metabolic activity) from the sieve tubes, or from closely associated phloem parenchyma cells. Further, these results lend support to the notion of vein loading as viewed by gross autoradiography, but do not prove that the sieve tubes themselves are loaded. Another limitation to this type of study is that excised tissue that does not have an active source-sink relationship cannot be compared with the tissues of an intact plant, where the accumulation of solutes probably does not occur, since transport away from the source is continually in process.

Spanner and Prebble (1962) made an attempt to accurately measure the movement of cesium down the petiole of *Nymphoides peltatum*. Applying ^{137}Cs to the bared phloem near the base of the lamina, and counting at various intervals along the petiole, they obtained data for movement of the tracer over a total period of 44 hours. These data are recorded in Figure 8.5. In an analysis of these data, the writers plotted the logarithm of activity in the petiole against the distance along the petiole from the point of application, and produced a series of curves for different time periods after tracer application. The slopes of these curves are an index of velocity: if straight, they indicate a constant velocity; if convex upward, they indicate decreasing velocity with distance; if concave upward, they indicate increasing velocity with distance. Plotting their data (Figure 8.6), Spanner and Prebble found that the curves were concave upward; they went to great lengths to explain this phenomenon mathematically. What they seem not to have appreciated is the probability that their tracer, in addition to riding the assimilate stream down the petiole, also migrated to the xylem at the point of application, translocated into the lamina, diffused from xylem to phloem, and was exported thence down the petiole, adding steadily to the concentration of the stream, and thereby to its apparent velocity. Since the shortest time period in Figure 8.6 is 2 hours, there would have been ample time for this to have taken place. In discussing Spanner and Prebble's results, Moorby et al. (1963) stated:

> The differences between the results obtained with carbon and cesium may be due simply to the different natures of the plant material and the translocate. However, the ^{137}Cs measurements were made over a period of many hours, and in this time movement in the reverse direction may have influenced the results.

Spanner and Prebble decided that they could not make a satisfactory estimate of the linear velocity of movement from their data. They

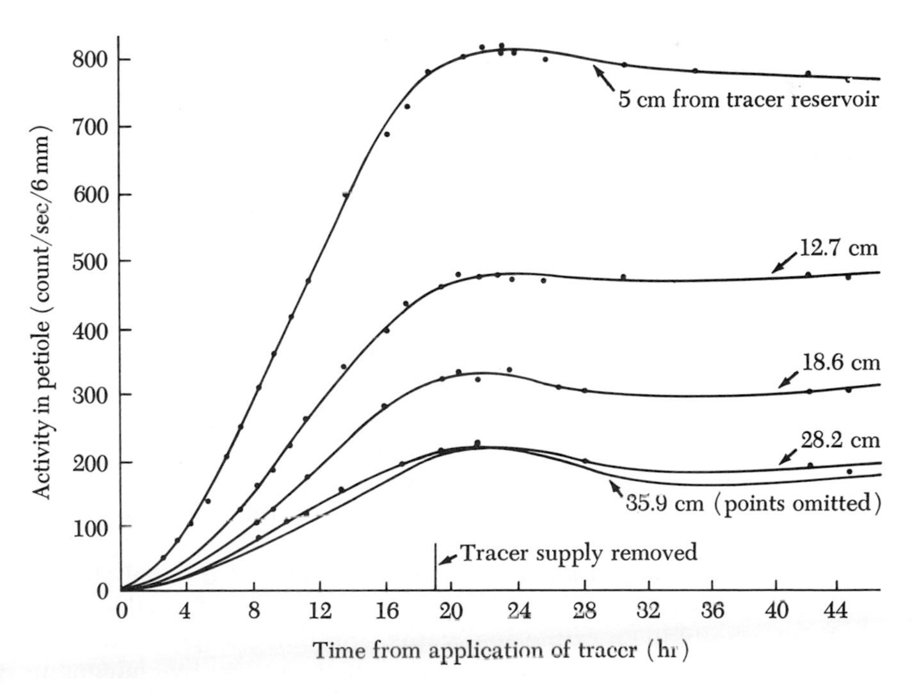

Figure 8.5. Movement of ^{137}Cs down the petiole of *Nymphoides peltatum*, recorded at 5 points over a period of 44 hours. The counter monitored a petiole length of about 6 mm. From Spanner and Prebble (1962).

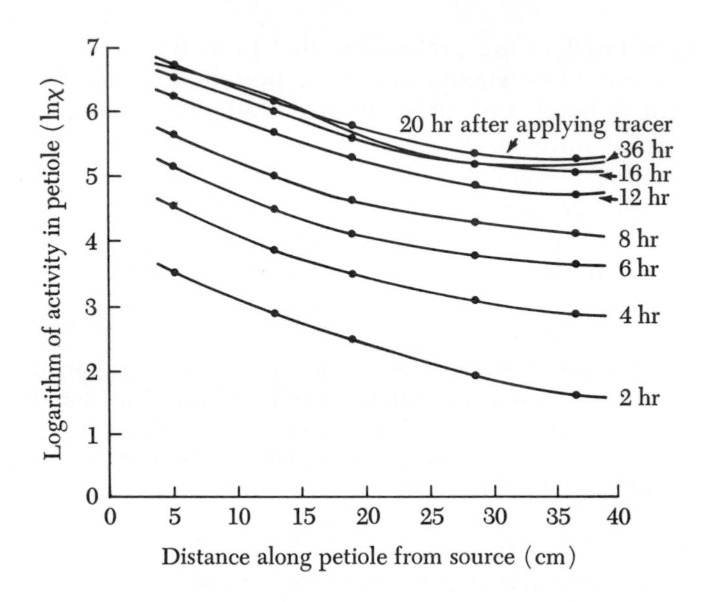

Figure 8.6. Semilogarithmic plot of the fall-off in ^{137}Cs activity with distance along the petiole of *Nymphoides peltatum*. From Spanner and Prebble (1962).

indicated the difficulty in making such measurements, and suggested that most published estimates are too low.

Moorby (1964) followed the foliar absorption and translocation of ^{137}Cs in *Pisum* seedlings. Cesium is known to be phloem-mobile (Fischer, 1967), and its uptake and distribution by plants is important in the problem of radioactive fallout. Moorby found that ^{137}Cs entered leaves through the cuticle primarily, and that its uptake can be reduced by darkness and by inhibitors. Since he had previously shown that the translocation of sugars is not light sensitive, Moorby proposed that the effects of light on translocation were through photosynthesis. This notion is borne out by the fact that prometryne applied to the cesium-treated leaf reduced translocation in the light as much as it did in darkness; the major effect of darkness, therefore, is to stop photosynthesis.

Taken as a whole, Moorby's results indicate that a metabolically controlled process is involved in the transfer of cesium from the mesophyll into the phloem of the leaf veins, and that changes in the distribution of cesium were also governed in some way by changes in the carbohydrate metabolism of the source leaf. All of these observations and conclusions are compatible with the concept of metabolic uptake by the symplast and transport by a pressure-flow mechanism.

Iron is generally considered to be immobile in plants; once it has been incorporated in a leaf, it will not be redistributed to younger tissues. Certain plot trials, however, in which iron sulfate sprays were applied to chlorotic plants of *Sorghum vulgare* growing on alkaline soils in California, have indicated that there is some movement; some of the new leaves of the chlorotic test plants became green. To further explore this phenomenon, Brown et al. (1965) made spot applications of ^{59}Fe to the leaves of young plants of *Sorghum*, *Gossypium*, and *Phaseolus*. Treatment of *Sorghum* was for 4 days, and of *Gossypium*, 5 days; root treatment of *Phaseolus* plants lasted for 48 hours. Autoradiography of leaf-treated *Sorghum* plants showed both apoplastic and symplastic translocation; ^{59}Fe concentration in root tips and young leaves was high, and in older leaves it was low. Image intensity was maintained in root tips and the terminal shoot tips during the 4-day treatment time, indicating continued movement. There was light, uniform labeling of mature leaves, indicating migration from phloem to xylem in the way that maleic hydrazide and dalapon move.

Similar treatment on the cotyledon of *Gossypium* gave a faint image of the young expanding leaf and shoot tip, and showed a little basipetal movement in the hypocotyl. Root treatment gave a strong image of the whole plant. Concentrations of ^{59}Fe in various leaves of root-treated *Phaseolus* plants showed the following values; leaf 1, 560 count/min; leaf 2, 485 count/min; leaf 3, 335 count/min; leaf 4, 510, count/min; leaf 5, 1035 count/min; and leaf 6, 1450 count/min. The increasing concentrations from leaf 3 to leaf 6 indicate that there was redistribution of the tracer from old to young leaves, which is good evidence for phloem transport. Brown and his co-workers concluded that ^{59}Fe is reasonably mobile in young plants, and that it shows both xylem and phloem mobility.

The transport and distribution of ^{59}Fe was studied in *Solanum tuberosum* by Wurster and Smith (1966), who combined grafting and radiotracer techniques. When shoots from plants that had been cultured on ^{59}Fe were grafted onto nonradioactive rootstocks, phloem transport of the tracer downward to young leaves and axillary shoots below the graft union occurred within a few days. Furthermore, a considerable amount of the iron was translocated to the tubers.

Eddings and Brown (1967) studied the absorption and translocation of ^{59}Fe by two varieties of *Phaseolus vulgaris* (red kidney bean and small white bean) and by *Sorghum vulgare* and *Lycopersicon esculentum*. They found that ^{59}Fe from $^{59}FeCl_3$ was translocated most actively by *L. esculentum,* less actively by *S. vulgare,* and least by small white bean. Stomata were found to be important in iron absorption; surfactants markedly increased absorption.

Magnesium transport in plants was investigated by Steucek and Koontz (1970), who found ^{28}Mg to be more mobile than calcium in the phloem, but less mobile than potassium, rubidium, and cesium. By using a 25 μc dose of ^{28}Mg, they proved that magnesium is phloem-mobile in *Phaseolus vulgaris.* Separation of the xylem and phloem by a plastic film confirmed this phenomenon.

The absorption and translocation of foliar and root applications of ^{35}S in plants of *Brassica napus* and *Phaseolus vulgaris* was studied by Panak and Szafranck (1967). When ^{35}S was applied to the foliage, it exhibited limited symplastic transport to the roots and young leaves; and in older leaves, some vein loading was noticable. When the period of absorption was increased from 6 to 48 hours, the relative concentration of ^{35}S increased in all the plant parts. Root applications through the hydroponic culture medium resulted in uniform labeling of the whole plant, with slightly higher concentrations in the younger leaves. The apices and margins of the leaves showed the highest concentrations of ^{35}S when it was applied to the roots. These observations are indicative and characteristic of an apoplastic type of transport.

Tso and Fisenne (1968) supplied radioactive lead and polonium to plants of *Nicotiana tabacum,* by soil, stem, and leaf-surface applications, in order to study their patterns of translocation and distribution. Foliar applications of ^{210}Po resulted in limited phloem transport from young leaves, while ^{210}Pb applied in the same way did not move at all. When ^{210}Po or ^{210}Pb was applied to mature leaves, there was no transport away from the treated area. Root applications were typically apoplastic, with ^{210}Pb accumulating in younger, upper leaves and ^{210}Po accumulating in lower, older leaves. In general, these two heavy inorganic elements are almost immobile in plants.

Creger and Allan (1969) found that—although strontium is readily absorbed by the roots of growing plants of *Phaseolus vulgaris,* and is accumulated in the tops—when ^{89}Sr was applied to *P. vulgaris* seed coats, it moved into the cotyledons of the germinating seed and thence to the shoot axis and into leaves 1 and 2, where it continued to accumulate for up to 18 days. In plants of *Gossypium hirsutum,* ^{89}Sr absorbed for a period of 2 days moved into stems and leaves, and then retranslocated from old to young leaves. This indicates that strontium may be phloem-mobile, at least in some plants.

Redistribution and interrelationships of these inorganic compounds was evaluated and summarized in the reports of Fischer (1967) and Levi (1968a, 1968b). These publications also serve as an excellent summary to this section. Fisher (1967), reporting on studies made with radioisotopes, pointed out that, although hydrogen, phosphorus, and potassium are readily redistributed in plants via the phloem, the mobility of calcium in the phloem is extremely slow. Rubidium and cesium behave like potassium in being readily moved; strontium and barium are like calcium. Lithium is more like the alkaline earths than the alkali metals in its phloem mobility. There appears to be a competition between potassium, rubidium, and cesium in the process of absorption, as between calcium, strontium, barium, and lithium. Fischer concluded that the lack of phloem mobility of the latter four elements is caused primarily by a failure of absorption. Sodium and magnesium, which may be absorbed into the roots by a specific carrier system, occupy a position between the above two groups, with respect to phloem transport.

Levi (1968a) studied the distribution of ^{22}Na, ^{43}K, ^{86}Rb, ^{134}Cs, ^{32}P, ^{35}S, ^{45}Ca, ^{65}Zn, and ^{85}Sr in *Phaseolus* plants following uptake by leaves and roots. Autoradiographic and counting methods were used. The studies revealed specific relations between the leaves and roots of *Phaseolus,* and two distinct patterns of distribution in the leaves following uptake by the roots. In the leaf-to-root distribution experiments, Levi found that ^{32}P, ^{43}K, ^{86}Rb, ^{134}Cs, and ^{22}Na moved from the point of application and were extensively redistributed. By contrast, ^{45}Ca, ^{85}Sr, ^{89}Sr, ^{35}S, and ^{65}Zn did not move from the region of application under Levi's experimental conditions. These results differ from those of Crafts and Yamaguchi (1964) and Pereira et al. (1963), who found zinc to be appreciably mobile from both leaf and root applications. When root-to-leaf distribution was studied, Levi found that ^{32}P, ^{35}S, ^{36}Cl, ^{45}Ca, ^{65}Zn, ^{86}Rb, ^{89}Sr, and ^{134}Cs all entered the roots, moved upward in the transpiration stream, and labeled the leaves. The patterns of distribution varied, however. Autoradiographs indicated that ^{36}Cl, ^{45}Ca, ^{65}Zn, and ^{89}Sr appeared first in all of the main veins of the leaf, progressing thence to smaller veins, until all the leaf network was labeled; only later did the isotopes label the mesophyll. By contrast, ^{32}P, ^{35}S, ^{86}Rb, and ^{134}Cs moved out of the midribs and directly across areas of the mesophyll without any notable accumulation in the veins. Distribution was seldom uniform; but it seemed to follow a set pattern, occurring into definite areas, of which six per leaf could be described. These are the tip, medium, and basal areas, for each side of a primary leaf. After a 6-hour experimental period, labeling of the primary leaves indicated that the treatment of roots directly below one primary leaf resulted in the labeling of that leaf with no label in the opposite primary leaf. Exposing roots at right angles to the primary leaves resulted in labeling the half-leaves on the side of the treated roots. Distribution into one or more of the six areas noted above followed the treatment of one or more groups of roots, or one or more single roots of a group. Treatment of roots deficient in phosphorus, sulfur, or calcium resulted in much slower movement of ^{32}P, ^{35}S, or ^{45}Ca from the roots to the leaves. Although the experiments just described indicate that chlorine, calcium, zinc, and strontium move most rapidly in the apoplast and are absorbed by living cells more slowly than are phosphorus, sulfur, rubidium and cesium, these effects are probably only relative; this, because the first listed elements finally arrive inside living cells, as indicated by the essential nature of zinc for cell metabolism, the occurrence of calcium oxalate crystals in living cells, and the presence of chlorine and strontium in cell sap.

In further experiments, Levi explored the fate of the elements that were accumulated in specific areas of the leaves. He used ^{32}P and ^{45}Ca in these experiments, the absorbing roots being removed after the initial uptake period of 6 hours. Samples were taken at the time roots were removed, and after periods of 7 and 14 days. Table 8.3 indicates the redistribution patterns. Both isotopes translocated readily to the trifoliate leaves that expanded during the 7- and 14-day redistribution periods; ^{32}P moved into the untreated parts of the root systems, but ^{45}Ca did not.

Levi pointed out that there is a relation between the mobility following leaf absorption and the distribution following root uptake. The three phloem-mobile elements phosphorus, rubidium, and cesium had, after root uptake, an interveinal distribution pattern. Calcium, strontium, and zinc, on the other hand, accumulated mainly in the veins following root uptake. These latter elements could move laterally in the xylem much more easily than phosphorus. They could move very rapidly in the transpiration stream, but their chemical binding by living cells (possibly on the cell wall-cytoplast interface) gives them an apparent immobility. (Eschrich, Eschrich, and Currier, 1964).

Sodium and sulfur did not fit the above pattern. Sodium, very mobile after leaf treatment, was found in the veins after root treatment; sulfur, not mobile when applied to leaves, showed interveinal distribution after root uptake. When roots of older *Phaseolus* plants were supplied ^{32}P or ^{35}S, the isotopes labeled only the main veins of the mature primary leaves, but accumulated in mesophyll of the growing trifoliate leaves. This emphasizes a point that is come upon many times: namely, that young leaves may serve as active sinks for assimilate import; but that mature leaves do not readily change from source to sink, even when shaded.

Levi pointed out that the redistribution of calcium from old to new growth may take place via the xylem, whereas phosphorus moves both acropetally and basipetally (in other words, it shows typical symplastic movement). Redistribution patterns found in these studies suggest that the circulation of nutrients in plants is restricted to discrete rings, with little cross transfer between these closed systems. To visualize the simultaneous presence of a number of conducting units that allow circulation from regions of leaves to parts of root systems, and vice versa, and that are independent from each other, yet slightly connected, may be a more satisfactory way of picturing the distribution of nutrients in plants. In another paper, Levi (1968b) showed the relative losses of foliarly applied ^{22}Na, ^{134}Cs, and ^{86}Rb from the roots of *Phaseolus* plants. Levi concludes that young *Phaseolus* plants may normally excrete certain elements into the culture medium, and that this may account for the unmeasured loss of isotopes from plants under certain experimental conditions.

Transport of Chelating Agents

Chelates of various minerals play a unique role in plant nutrition. Many of the problems of mineral uptake that are due to solubility relationships, interference from other metals, and variations in soil *p*H and soil type, can be resolved through ligand formation. Research

Table 8.3. Translocation of ^{32}P and ^{45}Ca in *Phaseolus vulgaris* plants following a 6-hour uptake period. (Values are averages of 5 plants.)

Isotope	Days after removal of treated roots	Translocation (% of absorbed activity)					
		Primary leaves	Trifoliate leaves	Lateral growth	Old stem	New stem	Roots
^{32}P	0	73.6	10.2		12.4		3.8
	7	37.6	35.0		14.9		12.5
	14	22.2	28.4	13.8	5.9		29.7
^{45}Ca	0	41.0	29.8		29.2		0.0
	3	66.3	29.1		4.1	0.5	0.0
	7	65.5	30.8		3.6	0.1	0.0

Source: Levi (1968a).

since 1955 on the translocation of chelates in plants has dealt primarily with EDTA ligand–metal complexes, and has centered on whether or not chelates are translocated intact. Chelation in the absorption and translocation of mineral nutrients was reviewed by Stewart (1963).

De Kock (1955) studied the translocation of ^{59}Fe chelated with EDTA by means of gross autoradiography. Root applications of the chelated tracer resulted in vein-loading, with a gradient into the interveinal areas; but with time, the ^{59}Fe decreased disproportionately faster in the veinal than in the interveinal areas. He concluded that retranslocation via the phloem accounted for the loss of ^{59}Fe from the tissue.

Other studies with iron chelates, using ^{14}C- and ^{15}N-labeled EDTA as the chelating agents, confirmed the chelate uptake by the root and its translocation to the top of the shoot (Jeffrys, Hale, and Wallace, 1961). Jeffrys and his co-workers also used the ligand EHPG: they used nonradioactive iron chelated with ^{14}C-labeled EHPG, and ^{59}Fe chelated with unlabled EHPG, to establish iron–ligand translocation ratios. They found the uptake and distribution to be dependent on the *p*H of the nutrient solution. When the *p*H was 4.0, more of the ^{14}C-labeled EHPG was taken up; and when the *p*H was 8.5, more of the ^{59}Fe was absorbed into the apoplast.

Further confirmation that plants separate chelates into metals and ligands came from the work of Tiffin, Brown, and Krauss (1960), and Tiffin and Brown (1961). They supplied plants of *Zinnia* sp., *Helianthus annuus*, and *Glycine max* with iron chelates of EDTA, EHPG, and DTPA in nutrient solutions. The chelates were variously metal-labeled (with ^{59}Fe) and ligand-labeled (with ^{14}C). Data from radiometric and spectrophotometric analyses of plant exudates and the nutrient solutions indicated that plants deficient in iron took up a large amount of ^{59}Fe and very little of the ^{14}C-labeled ligand, and that plants adequately supplied with iron took up less ^{59}Fe but about the same amount of ^{14}C-labeled ligand. The exudate of *Helianthus annuus* had a concentration of ^{59}Fe that was 8 times higher than that of the nutrient solution, while the chelating capacity of the nutrient solution was increased sevenfold.

Hale and Wallace (1961), using two double-labeled chelates (the ^{59}Fe chelates of ^{14}C-labeled EDTA and ^{14}C-labeled DTPA), studied the uptake and distribution of these chelates by seedlings of *Persea americana*, *Citrus aurantifolia*, and *C. sinensis*. They found ready movement of the chelates past a girdle into the fruit and leaves, following application to the roots; the movement, therefore, was via the xylem. Branch injection resulted in movement into the fruits and leaves. Studies conducted to ascertain the extent of retranslocation from primary leaves of *Phaseolus vulgaris* into new growth were negative; thus, it appears that there is little or no phloem transport of these chelating agents.

Although it is recognized that naturally occurring chelates play a vital role in the biochemistry of plants, there is relatively little information about the form in which the natural chelates are translocated (Stewart, 1963). All amino acids are capable of chelating metal ions to some degree. Histidine and cysteine are especially reactive; they bind metal ions very strongly, and could play an important role in metal transport in phloem tissue, and in plant nutrition in general.

Foreign Compounds of Natural Origin

The chemical composition of phloem exudate was reviewed in Chapter 5, and the forms in which assimilates move were discussed in Chapter 6. In general, the various chemical constituents that have been found in the assimilate stream indicate that a broad spectrum of inorganic and organic compounds can be translocated in the phloem. Carbohydrates, sugar alcohols, sugar phosphates, organic phosphates, organic acids, amino acids, nucleic acids, enzymes, and growth regulators have all been shown to be present in phloem sap; although we know they are present in the sap, we have only limited knowledge of the velocity and quantity of their long-distance transport through phloem tissues. Limited, too, is our knowledge of the effect a naturally occurring endogenous compound (an amino acid, for example) will have on phloem transport when it is applied to the system at concentrations higher than those which occur in nature.

Many compounds that occur endogenously in the phloem sap are incapable of being assimilated into the symplast from external applications because of physicochemical factors, such as polarity, a lack of cuticular penetration, an incapacity for diffusion across membranes, or the absorption of the compound to various tissue and cellular components. However, if an exogenously applied organic compound can overcome these barriers, then phloem transport should occur concomitantly with the mass flow of the assimilate stream. Since the solvent in the mass transport of materials is water, studies using isotopic water and solute should reveal confirmation of mass flow.

Simultaneous feeding of ^3H_2O and uniformly ^{14}C-labeled fructose to the basal portions of cut petioles of *Glycine max* resulted in normal movement of the labeled fructose, but no movement of HTO (Gage and Aronoff, 1960a). Photosynthate labeled with ^3H from vapor-feeding of leaves also moved normally, whereas HTO did not move. Similar results were obtained using seedlings of *Cucumis sativus.* Gage and Aronoff proposed that the solute was moving without a corresponding movement the solvent-water. On the basis of quantitative comparisons, Gage and Aronoff concluded that ^{14}C, ^{36}Cl and ^3H-labeled photosynthate may be transported by the same type of mechanism, whether the plants were fed the tracers from solution through cut petioles, or from gaseous applications to the leaves. This conclusion throws much doubt upon the validity of their work, because uptake and movement into a plant through a cut petiole must take place via the xylem, whereas export from a leaf treated with $^{14}CO_2$ or tritium must occur through the phloem. The plant fed through the petiole has no source, and the xylem sinks are transpiring organs; the gas-treated plants had normal source–sink relations, with movement taking place via the assimilate stream. Furthermore, when the petioles of *Cucumis* or *Glycine* plants are cut, phloem exudate flows over the cut surfaces, and is rapidly sucked into the cut xylem conduits; these may be blocked to varying degrees, rendering the results of such tests difficult to analyze.

In a subsequent paper, Gage and Aronoff (1960b) described autoradiographic tests on *Glycine max, Phaseolus vulgaris, Cucurbita maxima,* and *Cucumis sativus,* in which ^3H_2O vapor was applied to the leaves. The paper is illustrated with histoautoradiographs of *Cucumis sativus,* showing localized radioactivity over the young sieve tubes in transverse section, and a more general distribution of radioactivity over the longitudinal sections. As with all such work, interpretation of the results is difficult because of the difficulty of determining whether the localization of activity represents mobile tracer or accumulated tracer. Because of the nature of the procedures involved, most of the mobile molecules are probably lost, and the localized intensity represents tracer retained in the section by incorporation or by accumulation in storage cells.

When the third trifoliate leaf of a plant of *Glycine max* that has had the remainder of the leaves removed is allowed to take up ^3H_2O as a vapor in the dark for periods of 30–40 minutes (until vapor equilibrium is achieved), followed by a period of 20–30 minutes of light and CO_2 uptake, a basipetal gradient of HTO is generated in the petiole (Choi and Aronoff, 1966). When the HTO incorporated into photosynthate was separated from the free HTO, the gradients remained much the same; the incorporated tritium, however, was found to be 3–4 times more plentiful than the free HTO. Approximately half of the total radioactivity administered moved acropetally with the transpiration stream, and one-tenth moved basipetally. When the leaf was darkened for the entire treatment, no movement occurred. If the roots were removed, the quantity of free HTO generating the gradient and with steam-killing of the phloem of the petiole, basipetal transport almost completely stopped. All of these data support the concept of mass flow, with the solute and solvent moving in the same assimilate stream and generating parallel gradients.

In contrast to the above work, Plaut and Reinhold (1967) used ^3H_2O as the solvent for ^{14}C-labeled sucrose, which they applied to the foliage of young plants of *Phaseolus vulgaris* in order to study the effect of water stress on the simultaneous transport of the two isotopes. Gross autoradiography, chemical analysis, radiometric counting, and steam girdling showed that (1) ^{14}C-labeled assimilate moved directly from the treated area into the phloem of the midvein; (2) under conditions of water stress, a tenfold decrease in the amount of ^{14}C-labeled assimilates translocated away from the treated area; (3) HTO and ^{14}C-labeled assimilate both were transported; and (4) there was a threefold increase in translocation of HTO away from the treated area. However, similar data for transport from the primary leaf to the roots, a good measure of long distance phloem transport, showed that translocation of ^{14}C-labeled assimilate was reduced by approximately one-half as a result of stress, while HTO increased threefold. Steam-girdling and stress almost totally inhibited the transport of photosynthate, although they had little effect on HTO transport. Plaut and Reinhold concluded that assimilate movement in the leaf is by phloem transport, and that one or more steps in the HTO transport process are stimulated by water stress.

Neeracher (1966) used ^3H_2O, *Zea mays*, and histoautoradiography to localize the conducting tissue involved in HTO movement. Foliar applications of ^3H_2O, made by dipping the tips of *Z. mays* leaves in ^3H_2O solutions, indicated in short term experiments that only phloem and the sclerenchyma were labelled at the front of activity. By studying the distribution patterns of the water insoluble fractions, they could show that the sclerenchyma was not the channel for translocation. HTO incorporated into water-insoluble fractions accumulated in active metabolic sinks. Tritiated assimilates bypassed leaves older than the treated leaf and accumulated in young leaves; but in the leaf sheath, tritiated assimilates moved upwards and downwards. It was shown that simultaneous translocation could occur in opposite directions from this meristemactic tissue. During the course of these investigations, the velocity of translocation was repeatedly measured and found to be 60 cm per hour.

Similar studies were conducted by Trip and Gorham (1967), who made some improvements in the histoautoradiographic technique to minimize movement of water-soluble substances and to obtain high resolution. Histoautoradiographs were made from the petioles of *Cucurbita maxima* plants that had been fed glucose-6-^3H and $^{14}CO_2$; they showed patches of silver grains located in the phloem regions, and in the sieve tubes in particular. Trip and Gorham concluded that longitudinal translocation occurs in the sieve tubes, rather than in other phloem cells. It is interesting to note that the freezing technique that they used did not produce phloem plugging, and that their longitudinal sections show open sieve-plate pores. Schmitz and Willenbrink (1968) repeated these studies with D-glucose-6-^3H and found the activity localized in the phloem tissue, and in the slime plugs in particular. They used the "dry-technique" for preparing the histoautoradiographs, and their results suggest phloem injury. However, they concluded that the sieve tubes are the conducting channels for the long-distance transport of the assimilates.

Sorbitol, the primary assimilate of *Malus* spp., was ^{14}C-labeled and applied to the underside of actively exporting *Malus* leaves by Williams, Martin, and Stahly (1967). The translocation source-to-sink relations of the labeled sorbitol were compared to those of ^{14}C-labeled sucrose. Chemical fractionation, isotopic counting, and gross autoradiography confirmed that ^{14}C-labeled sorbitol was translocated with the photosynthate to the newly developing fruit, and that the rate of translocation was faster than for ^{14}C-labeled sucrose. Chemical fractionation showed that about half of the ^{14}C-labeled sorbitol was translocated, arriving at the sink as sorbitol, but that the remainder served primarily as substrate for the synthesis of other sugars and small amounts of amino acids and organic acids.

Lactose labeled with ^{14}C was shown by Husain and Spanner (1966) to be phloem-mobile, after it was applied to the tips of young leaves of *Lolium*. Chopowick and Forward (1969) found that if ^{14}C-labeled L-alanine was applied to the leaf surface of *Helianthus annuus*, the bulk of the radioactivity exported to sinks from the treated leaf was in the form of ^{14}C-labeled sucrose.

Eschrich and Hartmann (1969) studied the translocation and biochemical behavior of D-phenylalanine and L-phenylalanine applied to the first primary leaf of *Vicia faba*. They found that DL-phenylalanine moved into the sieve tubes and appeared in the honeydew of aphids feeding on the third primary leaf. Chemical fractionation of the leaves showed that DL-phenylalanine and *N*-malonyl-D-phenylalanine were the primary components of the residual radioactivity. D-phenylalanine was taken up quickly by the sieve tubes and quickly transported to parenchyma tissue, where it was converted to the phloem-immobile *N*-malonyl-D-phenylalanine. By contrast, L-phenylalanine was completely incorporated into protein in young leaves, translocated in large amounts in older leaves and did not become incorporated into phloem-immobile compounds, as did D-phenylalanine.

Crafts and Yamaguchi (1964) reported experiments on the translocation of ^{14}C from eight ^{14}C-labeled amino acids. Table 8.4 gives the label locations, specific activities, and exposure times of these tracers. In these studies, Crafts and Yamaguchi found that some amino acids are more mobile in the phloem than others, and that some excel in xylem mobility. In general, greater quantities of amino acids were transported when applied to leaves than when applied to the roots. Some of them (DL-valine and tryptophan, for example) act like 2,4-D, in that they accumulate in the roots and fail to ascend to the tops. Others, such as L-valine, seem to move about as freely in the transpiration stream as they do in the assimilate stream; they resemble picloram and TBA in their translocation properties. Figures 8.7 and 8.8 illustrate these experiments.

Yamaguchi and Islam (1967)—using the same eight ^{14}C-labeled amino acids and the herbicides 2,4-D, amitrole, and monuron—made

Table 8.4. Properties of ^{14}C-labeled amino acids used in translocation studies.

Acid	Label location	Specific activity (mc/mmole)	Exposure time (days)
DL-Arginine	guanido-	2.2	6.4
Glycine	2-	3.0	4.7
L-Histidine	uniform	9.3	1.5
DL-Lysine	1-	8.95	1.5
DL-Phenylalanine	3-	1.3	11.0
DL-Tryptophan	3-	6.64	2.1
L-Valine	1-	5.73	2.5
DL-Valine	1-	6.05	2.3

Source: Crafts and Yamaguchi (1964).

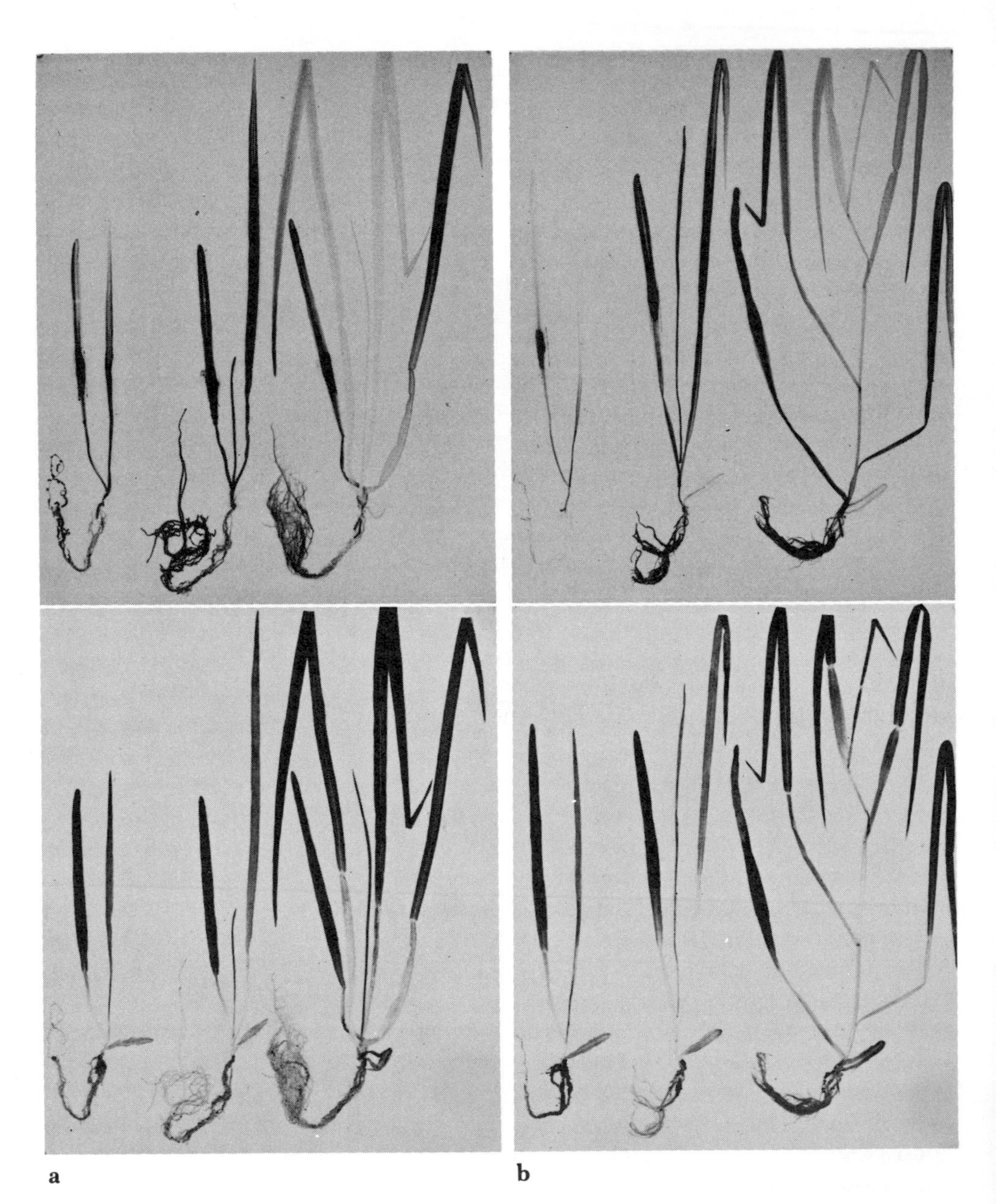

Figure 8.7. Distribution patterns of ^{14}C-labeled tryptophan applied to the leaves of *Hordeum vulgare,* 1, 4, and 14 days after treatment: (*a*) *H. vulgare* 'Atlas'; (*b*) *H. vulgare* 'Atsel.'

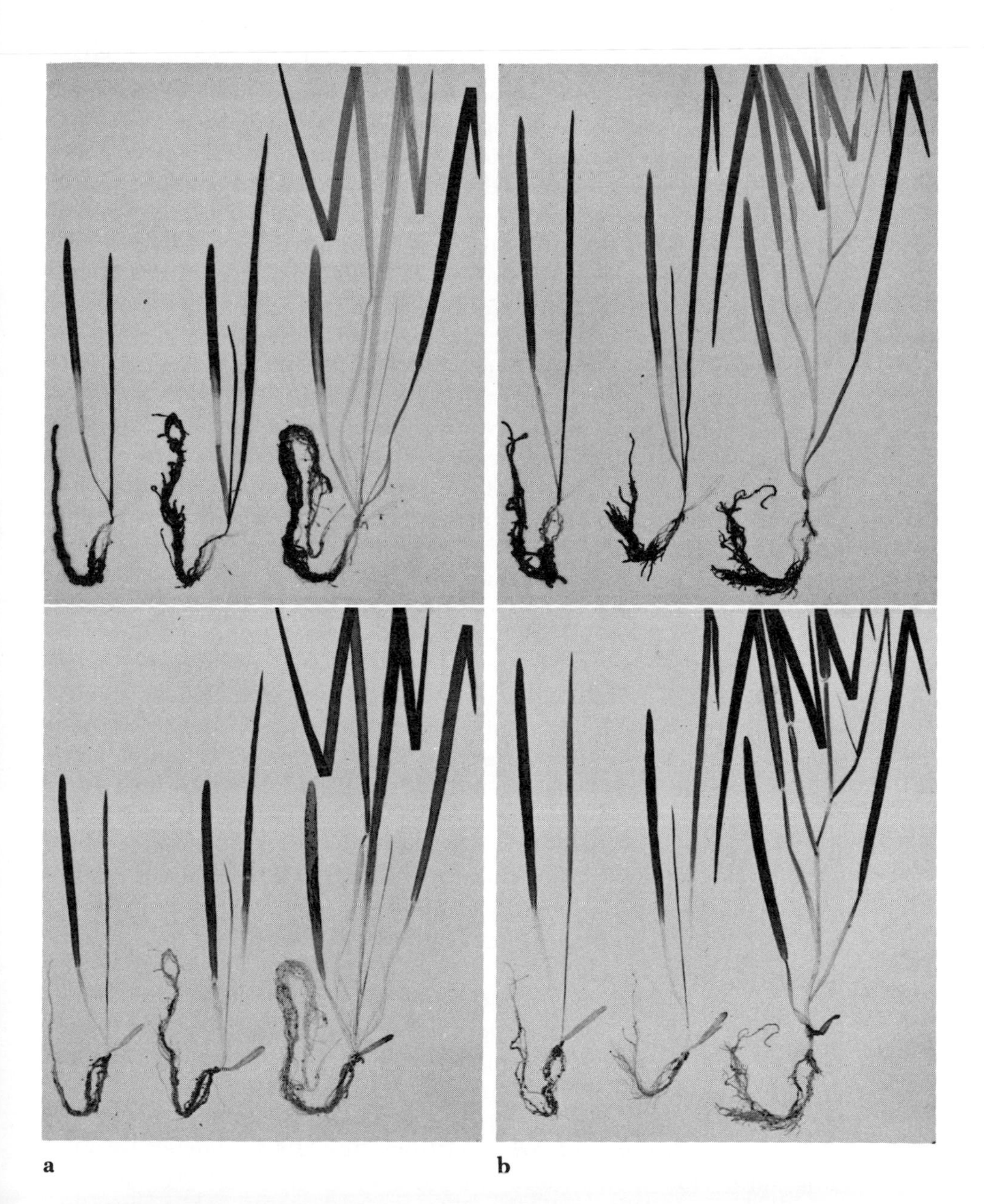

Figure 8.8. Distribution patterns of ^{14}C-labeled L-valine applied to the roots of *Hordeum vulgare*, 1, 4, and 14 days after treatment: (*a*) *H. vulgare* 'Atlas'; (*b*) *H. vulgare* 'Atsel.'

studies of translocation in the *Hordeum vulgare* varieties 'Atlas' and 'Atsel.' The labeled compounds were applied to leaves of some plants and to the roots of others, and were allowed to act for 1, 4, or 14 days. All of the amino acids applied to the leaves moved fairly uniformly; L-valine applied to the roots moved about as much as the same substance applied to the leaves; DL-lysine and DL-tryptophan moved weakly after being applied to the roots; in *H. vulgare* 'Atlas,' glycine moved equally well from roots and leaves; and in *H. vulgare* 'Atsel,' glycine moved more strongly from the leaves than from the roots.

Extraction and chromatography of labeled L-valine proved that this compound moved for up to 77 hours, but very little thereafter. All other amino acids tested showed the greatest accumulation in bud tissues and root tips one day after application; translocation was probably slowed after that time. Compared with the amino acids, 2,4-D showed an even more limited symplastic distribution; transport from a treated leaf possibly lasted less than 11 hours. Amitrole, on the other hand, translocated and retranslocated to the extent that each new leaf emerging from the main axis was heavily labeled, even for up to 14 days; the label intensity also continued to increase in the roots. The translocation of monuron was entirely apoplastic. The distribution of labeled amino acids resembled that of labeled sugar from ^{14}C-labeled urea: both exhibited the most intense accumulation of solute in the shoot apex and root tips after one day, with the largest accumulation in the second leaf, and in the tip of the third. The distribution of amino acids and the distribution of sugar were similar in the time series; this was due not only to the similar characteristics of mobility of the solutes, but also to their being readily metabolized and incorporated into the tissues, and to their being depleted in the assimilate stream between 4 and 14 days after application. Amitrole, by comparison, was much less readily metabolized or incorporated. Movement in the treated leaf from the spot of treatment was both acropetal (apoplastic) and basipetal (symplastic). Complete and uniform labeling of leaves and roots, even in the 14-day test, indicates that this compound continues to move for 4 days or more after application.

Working with *Beta vulgaris* seedlings, Joy (1962) showed by autoradiography that when glutamine, glutamic acid, and glycine were added to the culture solution, they moved into the leaves within 1 to 3 days, with the greatest amount being concentrated in the youngest leaf. Because steam-ringing the petiole of a leaf prevented this import, Joy concluded that the movement from root to leaf took place via the phloem. When the nonprotein amino acid α-aminoisobutyric acid was used, movement took place via the xylem, as steam ringing had no inhibitory effect. Joy assumed that the selective action in these plants took place in the roots. From more recent work (Crafts and Yamaguchi, 1964; Yamaguchi and Islam, 1967) it seems more probable that both kinds of amino acid moved from roots to leaves within a few hours of treatment. The protein amino acids were apparently transferred from the xylem to the phloem and exported to young expanding leaves via the assimilate stream. The α-aminoisobutyric acid, on the other hand—being, like the urea herbicides, unable to enter and move in the symplast—moved into the leaves via the transpiration stream, and so was unaffected by ringing. It is difficult to visualize a mechanism by which an amino acid could move from root to leaf in the phloem of a *Beta vulgaris* seedling against the assimilate stream, which is moving sucrose in the opposite direction.

Chopowick and Forward (1970) found that when uniformly ^{14}C-labeled L-alanine was applied to the surface of mature *Helianthus annuus* leaves, it was absorbed and metabolized very quickly in the light. After a 15-minute absorption period, export began and ^{14}C was distributed within the plant much as it is when $^{14}CO_2$ is assimilated by similar leaves. The bulk of the ^{14}C exported from the absorbing region was in the form of ^{14}C-labeled sucrose. There were only minute amounts in alanine and other amino acids; these, including glutamic and aspartic acids, were present and heavily labeled in the regions of absorption. Within 30 minutes of the time of initial application, ^{14}C was found in all parts of the plant except in the mature leaves; the labeled alanine continued to enter the treated leaf for the entire 4-hour absorption time. The selective export of sucrose in these experiments indicates a mechanism whereby this sugar is moved preferentially into the phloem within the leaf. Chopowick and Forward suggest that this critical step may well be entry into the sieve tube.

Phloem translocation of amino acids and protein hydrolysate from older leaves to sinks has been demonstrated by Joy and Anticliff (1966). Glutamate, glycine, and protein hydrolysate—all labeled with ^{14}C—were applied to the upper surface of actively exporting leaves of *Beta vulgaris*. After 24 hours 5–12 percent of the absorbed material could be found in the photosynthate sinks; if steam-girdled, translocation stopped. Interestingly, the nonprotein amino-acids (α-aminoisobutyric acid and D-alanine) were absorbed, but little was translocated. Joy and Anticliff correlated their data with studies on xylem transport from root uptake of the same compounds. A complete mixture of amino acids was taken up by the roots and translocated acropetally to all parts of the plant. Sarcosine, betaine, β-alanine, and 2-aminoisobutyric acid were more generally distributed. Brennan, Pate, and Wallace (1964), and Dezsi, Barkoczi, and Palfi (1967) have shown similar results for xylem transport. Bolli (1967), on the other hand, found that 2-aminoisobutyric acid was not transported in the xylem apoplastically. Bolli also observed, by means of histoautoradiography, that a high concentration of the amino acid isotopic activity could be observed in the phloem tissue of the shoot when the uptake of amino acids was via the roots.

The phloem transport and distribution of oleanolic acid, a triterpenoid, was investigated by Kasprzyk, Wojceichowski, and Czerniakowska (1968). It was proven that the rate and the direction of transport of oleanolic acid in *Calendula officinalis* plants depend on the stage of development of the vegetation and the position of the leaf on the plant. Upper leaves transported oleanolic acid primarily toward the younger leaves and to the inflorescence, and lower leaves transported the acid to the roots. Proof was given that the root, although unable to synthesize the acid, is able to accumulate it from the shoot. This steroidlike molecule, a five-membered ring with one carboxyl group and one hydroxyl group, was the first compound studied that indicated that lipids can be transported in the phloem.

Other indications that the phloem may transport lipids came from the research of Bykhovskii, Sokolov, Fomina, and Kabanov (1968), who studied the translocation of resin and metabolites in the phloem tissue of the main shoots of *Pinus* sp. The transport of ^{14}C-labeled assimilate and ^{14}C-labeled oleoresin was studied under conditions of tapping and in the absence of tapping. The downward movement of ^{14}C-labeled assimilate was observed to be 8–16 cm per hour, while the oleoresin and monoterpenes have the lower translocation rate of 5 cm per day. Tapping the trunk increased the transport of ^{14}C-labeled assimilate.

Yoshida (1967), in his studies on nicotine synthesis in *Nicotiana tabacum* plants, showed that ^{14}C-labeled L-glutamic acid, ^{14}C-labeled L-arginine, or $^{14}CO_2$ could be administered to the lower leaves of the plants and translocated to the roots, where their ^{14}C was incorporated into nicotine. After 24 hours, most of the ^{14}C-labeled nicotine was translocated to the shoot. When the shoots were topped or placed in darkness, the incorporation of the substrates into nicotine was enhanced; but when the shoots were sprayed with maleic hydrazide, their nicotine content decreased. The data gave no indication that the nicotine could be retranslocated in the phloem.

In studies on the movement of solanaceous alkaloids from the tops of *Datura stramonium* and *Atropa belladonna* to the roots of *Lycopersicon esculentum* through bridge-graft unions, Wilson (1959) found by analysis that *D. stramonium* tops transmitted traces of alkaloid in 5½ weeks, 1.88 mg in 11 weeks, and 2.30 mg in 19 weeks. From *A. belladonna* tops, 0.85 mg of alkaloid had moved to the *L. esculentum* roots in 3 weeks, 4.47 mg had moved in 9½ weeks, 11.35 mg in 12½ weeks, and 23.71 mg in 17½ weeks. Wilson concluded from anatomical examination of the graft unions that vascular continuity had been established within 3 weeks; in older unions, considerable secondary phloem and xylem had been produced. Wilson concluded that the downward-directed stream was moving in the phloem, further evidence for the movement of endogenous tracers with food in the assimilate stream.

Virus Movement

The translocation of viruses in plants has been studied for over 40 years; C. W. Bennett had concluded by 1927 that virus movement was correlated with food movement. Subsequent work (Bennett, 1927, 1934, 1935, 1937, 1940a, 1940b, 1944, 1956; Bennett and Esau, 1936; Esau, 1938, 1941, 1963; Esau, Cronshaw, and Hoefert, 1966, 1967; Esau and Cronshaw, 1967a, 1967b; Cronshaw and Esau, 1967) has emphasized the relations of virus movement to phloem ontogeny, to plant tissues in their role as viral hosts, and to the movement of food in plants.

In order to understand the distribution and effects of viruses in plants, it is important to realize that viruses differ in their tissue relations; some viruses are virulent in parenchyma and vascular tissues, just as others are localized in the phloem. It is important, too, to bear in mind that plants vary in food conduction rates, and in their responses to experimental manipulation and virus infection. Tobacco mosaic virus moves slowly in parenchyma, but upon entering sieve tubes it is transported at rates at which foods are known to move. Curly top virus moves rapidly when it is inoculated directly into functioning sieve tubes, and a general inoculation of a field of *Beta vulgaris* plants of a susceptible variety can be disastrous.

When a plant of *Beta vulgaris* is defoliated, or shaded, new leaves developing from a crown act as sinks for the movement of food and viruses. Leaves of *Nicotiana tabacum, Phaseolus vulgaris,* and many other plants will, when shaded, senesce, mobilize and export carbohydrate reserves and amino acids derived from proteins, and then abscise. Thus, in contrast to *B. vulgaris* such leaves act temporarily as sources for food movement.

Symplastic movement of viruses, tracers, and pesticide molecules occurs in all plants. Accelerated in cell lumina by protoplasmic streaming, such movement may attain values in the range of 0–5 cm per hour.

One further generalization that has been thoroughly substantiated by recent work with tracer molecules is that exogenous materials, if they can enter the phloem, move rapidly from source to sink of food movement. The mass rate of movement (weight delivered per unit of time) depends upon the rate of entry into the symplast, the rate of migration into the sieve tubes, the velocity of phloem transport, the amounts removed from the phloem by surrounding parenchyma cells during transport, and the activity of the sink. Velocities of movement of up to 150 cm per hour in the petioles of *Beta vulgaris* were early recorded by Bennett (1934); movement was slower in *Nicotiana tabacum.*

Bennett has published several excellent reviews on virus movement in plants (Bennett, 1940a, 1940b, 1956). In his 1956 review, Bennett pointed out that, whereas nearly all viruses are able to enter the parenchyma and invade various plant parts, some viruses have difficulty in entering the phloem; and that those that reach their highest concentrations in the parenchyma may not be able to increase in the phloem. Tobacco mosaic virus has been shown to be able to pass through long sections of *Lycopersicon esculentum* stems without producing infection in certain sections. Evidently, it is carried along in the assimilate stream in varying concentrations, depending upon its rate of entry into the sieve tubes, and undergoes little or no multiplication en route (Bennett, 1956).

Viruses that produce leaf curls and general yellowing without mottling cause injury to the vascular system; symptoms usually appear in the phloem parenchyma, and advance just behind the region of sieve-tube maturation. Curly top virus occurs in relatively high concentrations in the naturally occurring phloem exudate from petioles and blades of infected *Beta vulgaris* leaves. It also is high in the phloem exudate from the roots of infected plants. Curly top fails to pass phloem rings that sever phloem continuity on stems of *Nicotiana tabacum.* The exact locus of multiplication of this type of virus has not been determined (Bennett, 1956).

Schneider and Worley (1958) found that southern bean mosaic virus inoculated into a trifoliate leaf of *Phaseolus vulgaris* 'Black Valentine' passed down into the hypocotyl and migrated laterally into a plant of another variety of *P. vulgaris* (pinto bean) that had been approach-grafted to it. From there it entered the xylem of the second plant, moved upward in the transpiration stream past a steam ring, multiplied, and caused symptoms in the upper leaves. This indicates that the virus can migrate from phloem to xylem. Thus, as reported later (Schneider and Worley, 1959a) southern bean mosaic virus will translocate into and out of tracheary elements, a property that tobacco mosaic virus does not have. Caldwell (1934) showed that tobacco mosaic virus introduced into xylem of *Lycopersicon esculentum* would move upward into leaves, but that symptoms were not produced unless the tissues were macerated so the the virus particles were able to escape from the tracheids.

Schneider and Worley (1959b) reported that virus detected beyond steamed regions was present as a result of transport through the steamed regions, not as a result of chance introduction, or of multiplication in cells of the steamed region. Since there was some downward movement of the virus from the infected host in the approach-grafted *Phaseolus vulgaris* plants when there was no steaming, Schneider and Worley suggested that southern bean mosaic virus moves rapidly in the phloem when properly introduced. Because the molecular weight of this virus is estimated at 6.6 million and the diameter of the particle at 300 Å, it seems that either it or its nucleic acid portion can pass through xylem walls, phloem walls, and living membranes. Once inside a cell, the movement could be symplastic.

In a following paper, Schneider and Worley (1959c) reported that southern bean mosaic virus, introduced via the dead steamed portion of a stem, infected leaves far above the point of introduction; and that multiplication at the point of inoculation is therefore not necessary. Infection via the stem resulted very rapidly in symptoms several feet above the point of inoculation; if inoculation was directly to the leaves, necrotic lesions appeared in 3 days at the earliest. Evidently, upward movement is rapid, and the movement to the symplast is unimpaired by the xylem cell walls.

Concerning the movement of virus from a point of inoculation to other tissues, Bennett (1956) cited evidence for the movement of tobacco mosaic virus from one epidermal layer to another in 24 hours; Worley and Schneider (1963) cited similar information for southern bean mosaic virus. Other viruses may take 2–5 days or more to reach the phloem and move out of an inoculated leaf. Because some viruses show a correlation of rate of movement with the number of primary infections per host plant, it seems evident that the expression of symptoms may not be a satisfactory criterion for determining the rate of virus movement. Furthermore, when two or more viruses infect a plant, the presence of one may hasten the export of the other (Schneider, 1965); and when several viruses are inoculated simultaneously, some may move more rapidly than others. Plant age, the locus of inoculation, and the food transport function of the plant are further complicating factors in determining the rate, direction, and velocity of virus translocation.

Worley (1965b, 1968) described the role of phloem fibers in the translocation of southern bean mosaic virus in *Phaseolus vulgaris.* Using symptomatology, a partial girdling technique, infectivity assay, fluorescense microscopy, and fluorescent antibodies he showed vertical transport of the virus at velocities of up to 12 mm per day after the first appearance of necrosis. Horizontal movement across the cortex parenchyma occurred at the rate of 0.5 mm in 2–3 days; after it arrived in the fibers, the virus moved 8 mm in 24 hours and 16 mm in 48 hours. This was a 30-fold increase in the velocity of movement; such increases varied from 10- to 100-fold in different plants.

The most rapid velocities of virus transport in phloem are those recorded by Bennett for the movement of curly top virus (Bennett, 1934, 1937, 1943, 1944, 1960). Using his triple-crowned plants of *Beta vulgaris* Bennett (1960) inoculated three viruses; all three traveled at comparably rapid velocities into defoliated crowns. Under the same environmental conditions, movement into undefoliated crowns varied: tobacco mosaic virus produced symptoms in 34 days, beet yellows virus required 57 days, and curly top virus had not produced symptoms after 140 days. Curly top virus is strictly limited to the phloem, and so would have to move counter to the assimilate stream in order to enter a leafy crown. Darkening of the crowns of *B. vulgaris* plants has the same effect as defoliation, because it induces production of new leaves. That many viruses move out of the roots of plants very slowly, if at all, is another expression of this phenomenon. Because roots constitute a constant sink for assimilates, outward movement via the phloem should occur only in spring from roots having stored reserves; there should be little or no such movement in herbaceous annual plants. Those viruses that produce systemic infections from root inoculation may move via the xylem, as does the virus of Pierce's disease.

Mitchell, Schneider, and Gauch (1960), in a review of particle movement in plants, cited the work of C. W. Bennett (1956). They pointed out that viruses may move from cell to cell via plasmodesmata, and that such movement in parenchyma is relatively slow; and that viruses may move rapidly through sieve tubes—they cited Bennett's value of 60 inches (150 cm) per hour for the movement of curly top virus in petioles of *Beta vulgaris*. The third pathway of virus movement is the xylem, and viruses that can live and move via the transpiration stream (those of Pierce's disease and southern bean mosaic, for example) may move at a velocity of 12 inches (30.5 cm) per minute.

Mitchell and his co-workers went on to point out that virus distribution in plants is influenced by the reactions between the virus and its host. Some viruses (curly top virus, for example) are limited to the phloem; others (the virus of Pierce's disease, for example) are limited to the xylem; and yet others may infect most of the living cells of a plant, including those of the phloem and xylem (the mosaic viruses are of this kind). Restriction of a virus to a certain tissue does not necessarily indicate that the plant is unable to translocate the virus particles to other locations within the plant; it may simply indicate that the virus cannot infect, live in, or multiply in, the uninvaded tissues. Insect vectors play a role in the transmission of viruses in that they feed on different tissues.

In studies on inoculation methods in relation to the spread of viruses, superficial treatment of plants with southern bean mosaic virus resulted in its superficial local lesions; deep inoculation of the virus into stem xylem resulted in rapid translocation to the tops of plants. Young leaves are rapidly injured as a result of such inoculation, but old leaves do not show lesions unless they are wounded to release the virus into living cells. Viruses have been very useful as translocation tracers, as Bennett (1956) has indicated; but the above example shows that care must be taken in the interpretation of experiments in which they are used.

In studies on the movement of beet yellows virus, Bennett (1960) caged viruliferous green peach aphids (*Myzus persicae*) on rapidly growing leaves of *Beta vulgaris;* and by periodic leaf removal, checked on the time required for the infection to be carried a distance of 10 cm. After 10 minutes, no virus had moved from an inoculated leaf; after 20 minutes, virus had moved out of 6 of 67 leaves. The percentage of leaves from which the virus had moved 10 cm increased with time up to 2 hours; after that it varied but did not increase farther. When the virus was mechanically inoculated it took more than 48 hours to move a distance of 10 cm; in the majority of the plants it required over 72 hours to move that far. Inoculation of this type seldom gets into the phloem, and the movement time is extended by the time required for the virus to move through the parenchyma and into the phloem. Using *Chenopodium capitatum* plants, export of the virus required 40 minutes from the time of aphid inoculation;

when mechanical inoculation was used, there was no infection at a distance of 10 cm from the locus of inoculation until the fourth day. Using his split-beet technique, Bennett obtained strong evidence for movement of beet yellows virus with assimilates in the phloem.

Worley and Schneider (1963) and Schneider (1964) observed movement of southern bean mosaic virus from an inoculated leaf of *Phaseolus vulgaris* into the opposite leaf. This kind of movement has not been observed for radioactive tracers, except where there was injury to the treated leaf (Crafts, 1956a), or when the plant was placed in a supersaturated atmosphere (Clor, Crafts, and Yamaguchi, 1962, 1963, 1964). In the above case, the presence of virus in the opposite leaf was not detected until the fifth or sixth day after inoculation; it then appeared to be rather evenly distributed throughout the mesophyll tissue, which would seem to indicate movement into the leaf via the xylem. Schneider (1964) considered that the virus may have moved into the opposite leaf via the phloem. If it did, it would seem that the route was along the phloem parenchyma rather than the sieve tubes: a distinct possibility, because at least 6 days were required for the symptoms to appear. An alternative possibility would be export from the inoculated leaf via the phloem, migration of the infective agent from phloem to xylem, and movement into the opposite leaf in the transpiration stream. Esau, Cronshaw, and Hoefert (1967) suggested the possibility that viruses may destroy the plasma membrane of a sieve tube and pass directly through the wall as complete or incomplete particles.

Virus workers are giving increased attention to the mobile forms of viruses. Gierer and Schramm (1956) and Fraenkel-Conrat (1956) have shown that the nucleic acid of tobacco mosaic virus, free of its protein, may cause infection. Although the plasmodesmata have been considered to be too small to conduct viruses, Esau, Cronshaw, and Hoefert (1967) showed beet yellows virus particles in both sieve pores and plasmodesmata. Livingston (1964), studying plasmodesmata in living and dead tissues fixed in various ways, has found them to be structurally similar in angiosperms and gymnosperms; plasmodesmata in both groups averaged 0.2 μ in diameter.

In his review on the introduction, translocation, and distribution of viruses, Schneider (1965) covered the topic of the mobile forms of viruses. Some workers have suggested that tobacco mosaic virus moves between cells as the nucleic acid free of its protein. The enzyme ribonuclease prevents the formation of lesions when it is introduced into a cell that serves as a pathway between the inoculated leaf and the receptor leaf: thus, the enzyme hinders infection: either the virus is inactivated, or it does not move in a ribonuclease-sensitive form. The use of strains of defective tobacco mosaic virus has given evidence that this virus can move from cell to cell as nucleic acid, but that it does not translocate in phloem in that form. Possibly, the acid is not compatible with the alkaline *p*H common to phloem sap. In the views provided by Esau, Cronshaw, and Hoefert (1967), the particles in the sieve tubes and those in the plastids of chlorenchyma cells appear to be similar.

When *Cuscuta* was grown on *Nicotiana* infected with tobacco mosaic virus, the virus passed into the *Cuscuta* plants, but only in small amounts; apparently the whole nucleoprotein particles moved from the host to the parasite (Schneider, 1965). Some workers have questioned that viruses multiply in the sieve tubes, for if they did so multiply, phloem exudates of infected plants would contain the mobile forms of the viruses. Vectors, presumably, may transmit a form of virus that is immediately translocated; this seems to be true of streak and curly top viruses, which move rapidly soon after their introduction.

Absence of inclusion bodies in the mature sieve tubes of *Beta vulgaris* plants infected with beet yellows virus (Esau, 1960) and in the sieve tubes of *Nicotiana tabacum* infected with tobacco mosaic virus (Esau, 1941) suggests that these viruses do not multiply in these enucleate cells. Worley (1965a) reported evidence for the presence of the antigen of southern bean mosaic virus in the sieve tubes of systemically infected plants of *Phaseolus vulgaris* 'Black Valentine'; his test, however, did not distinguish between multiplied virus and translocated virus.

Worley (1965b) found that southern bean mosaic virus moves vertically in the stems of *Phaseolus vulgaris* from superficial inoculation in the phloem fibers. The velocity of movement was 8 mm per day, a movement that could well occur in the symplast of young fibers by diffusion accelerated by protoplasmic streaming.

Kluge (1967) found that he could infect *Cucumis sativus* plants with Cucumis Virus 2A by inoculating the phloem exudate of infected plants into healthy plants of the same species. The infectious factor could not be removed by dialysis or treatment with phosphodiesterase. Kluge thought it improbable that the virus is transported in the sieve tubes as units of low molecular weight, or as unprotected RNA. Rod shaped particles (345 μ by 23 μ) were identified in the infectious sieve-tube sap by electron microscopy. The same particles could be found in sap extracted from infected leaves, but not in sap from healthy plants. Kluge concluded that Cucumis Virus 2A is transported in the sieve tubes as complete particles. He shows electron microscopic pictures of the particles in expressed leaf sap and in sieve-tube sap.

Esau, Cronshaw, and Hoefert (1967) made an electron microscope study of the relations of beet yellows virus to the phloem and to the movement of substances in the sieve tubes. In the minor veins of bundle ends in the leaves of *Beta vulgaris*, beet yellows virus particles were found both in parenchyma cells and in mature sieve elements. In parenchyma, the particles were not found in vacuoles; in sieve tubes having no vacuoles, the particles were found scattered throughout the cell. In dense aggregations, the particles tended to assume an orderly arrangement in both parenchyma cells and in sieve elements. Most of the sieve elements containing virus particles had mitochondria, plastids, endoplasmic reticulum, and a plasmalemma, as in normal, mature sieve elements. Some sieve elements showed signs of degeneration. Virus particles were present in the pores of sieve plates, in the plasmodesmata connecting the sieve elements with parenchyma cells, and in the plasmodesmata between parenchyma cells. The distribution of virus particles in the phloem of *Beta vulgaris* is compatible with the concept that plant viruses may move through the phloem in the sieve tubes, and that the movement is a passive transport by mass flow. The observations of Esau and her co-workers also indicate that the beet yellows virus moves from cell to cell in the sieve tube in the form of complete particles, and that this movement may occur through sieve-plate pores and through plasmodesmata (Esau, 1968).

Esau, Cronshaw, and Hoefert discussed in some detail the mechanism of virus movement in sieve tubes, particularly with respect to virus multiplication. Since mature sieve tubes lack nuclei and have a paucity of ribosomes, it is difficult to visualize virus replication in mature sieve elements. The virus might enter immature sieve elements and multiply therein, and so be in a position to translocate as the elements mature. The amounts of virus observed in sieve elements by Esau and her co-workers was variable. The observation that uninfected regions occur in stems during the initial postinoculation spread indicates that the virus moves as complete particles, and that those particles are transported on the assimilate stream. Esau and her co-workers suggested that they are distributed in a source-to-sink pattern of assimilate movement.

It has commonly been assumed that viruses move from cell to cell in parenchyma tissues via the plasmodesmata. Esau, Cronshaw, and Hoefert provided electron micrographs of virus particles in sieve-plate pores and in plasmodesmata. By scaling their micrographs it appears that the virus particles are around 60–100 Å in diameter, and the plasmodesmata 500–1000 Å in diameter; thus, it is obvious that the particles can move through these connections. Esau and her co-workers, considering the driving force responsible for cell-to-cell movement, pointed out the possibility that this movement may also follow the source-to-sink pattern. The cytoplasmic core of the plasmodesma, they suggested, may be modified by the virus infection, or the particles may mechanically push it aside. From the density of particles shown in some of their illustrations, it seems that the turgor pressure within the sieve tubes may be sufficient to push the particles through into the adjacent parenchyma cells. Movement through sieve-plate pores presents no problem; the latter appeared to be 0.25 μ or more in diameter (Esau, 1968, figure 73).

Esau and Cronshaw (1967a), following their work on beet yellows virus, studied the relation of tobacco mosaic virus to host cells using *Nicotiana tabacum* plants systemically infected with the virus. Although most of their observations involved identification of crystalline inclusions, X-bodies, and virus particles in a variety of cells, they included descriptions of the contents of both young and mature sieve tubes. In one of the mature sieve tubes, they found virus particles in an ordered array and surrounded by degenerated material forming an ovoid body that appeared to be attached to the parietal cytoplasm by membranous fragments (their figure 10). In the lumen of the cell there was a reticulum, which, because of thin sectioning, appeared to be composed of rods of various length. Although they labeled this filamentous material a virus, it differs in arrangement and appearance from the virus rods in the ovoid body within the sieve tube, and from the ordered virus particles in a neighboring parenchyma cell. It resembles the plasmatic reticulum of Behnke and Dörr (1967). Esau and Cronshaw considered the presence of virus in sieve elements to be in harmony with the idea that rapid long-distance transport of tobacco mosaic virus takes place in sieve tubes.

In a companion paper, Esau and Cronshaw (1967b) described the protein component (slime) of the sieve-tube protoplast of *Nicotiana tabacum* that they designated as P-protein. This consists of tubular elements that resemble one of the tubular components that they term "X-component" that occurs in cells of *N. tabacum* infected with tobacco mosaic virus. The P-protein and the X-component can be distinguished, in that the latter occurs in straight tubules that have a more orderly arrangement, that are somewhat larger in diameter, and that (at least in some preparations) are some what less dense to electrons (compare their figures 10 and 11). Apparently, the X-body tubules may occur singly, in filaments of three tubules, or as longitudinally arranged aggregates. They make no mention of the possible presence of the permanent plasmatic filament system that occurs in the mature sieve tubes of many species. They state that the P-protein disperses in the sieve element, and that the tubular elements are broken down into smaller units characterized by regularly spaced cross-striations. They suggest that the X-component is probably the noninfectious viral protein that occurs, along with complete, infections virus particles, in the sap of plants infected with tobacco mosaic virus.

In a third paper, Cronshaw and Esau (1967) described ontogenetic studies of the sieve-element protoplasts of *Nicotiana tabacum.* The P-protein, they found, originates in small groups of tubules in the cytoplasm. These subsequently enlarge to form the compact masses of tubules, with an average diameter of 231 ± 2.5 Å, known as P-protein (or slime) bodies. Later the P-protein bodies disaggregate, and the tubules become dispersed throughout the cell; this occurs at the same time as the breakdown of the tonoplast and the nucleus. Cronshaw and Esau stated that the P-protein tubules become reorganized into smaller elements with an average diameter of 149 ± 4.5 Å. They designated the 231 Å tubules as P1-protein, the 149 Å fibrils as P2-protein. These smaller fibrillar components appear to be striated, and it is doubtful that they are tubular. The P2-protein fibrils fill the pores of mature sieve plates, and fray out into the lumina of the sieve-tube elements on the two sides of a sieve plate (Figure 2.8, page 35), and they evidently form a continuum throughout the sieve tube. Cronshaw and Esau stated that they are denatured within the pores into electron-opaque material; the plates are very obviously plugged.

In view of the fact that the virus particles of beet yellows are able to pass through the sieve-plate pores, Cronshaw and Esau considered the possibility that their observations implied that the fibrils in the sieve plate had moved in on the assimilate stream, and become fixed there by their preparation procedures. This would indicate that the P2-protein was being transported through the sieve tubes, and that it had been retained in a denatured state as a result of cutting. The symmetrical distribution on each side of the sieve plate, however, rules against this mechanism. They discarded a third possibility, that the filling of the sieve pores with P2-protein is a normal condition, on the basis of the presence of virus particles in the pores. However, if this P2-protein can be considered as the living filamentous reticulum that occupies only a small portion of the pore lumen and becomes denatured by hydration and swelling, the views of Cronshaw and Esau would simply represent filament-plugged sieve plates. In their normal state, the pores should be pervious to solutions, virus particles, and tracers. The heavy callose surrounding the tubule contents in Cronshaw and Esau's figures 10 and 13 would seem to imply the development of callose to such an extent that the diameters of the pores had been appreciably decreased, possibly by the very rapid response described by Currier (1957) and Eschrich (1965).

Esau (1967b) wrote an excellent review of the relation of host-plant anatomy to virus infection, specializing in the recent work with electron microscopy, wherein she discussed localized and systemic infections, growth disturbances and necroses induced by mosaic viruses, and the relations of viruses to meristems and reproductive tissues. Of particular interest is her discussion of phloem degeneration caused by yellows diseases. Citing results with curly top and beet yellows, she explained that the first degenerative changes are localized around the first mature sieve elements in the young roots and shoots. Initial symptoms include increase in chromaticity followed either by immediate degeneration and necrosis or by extensive hyperplasia and necrosis. Hyperplastic tissue in aster yellows and curly top infections shows an abnormal abundance of sieve elements having the usual cytologic characteristics but deviating in form, size, and arrangement; companion cells are sparse. Such degenerate tissue may have such large accumulations of callose that the cells become obliterated. In virus infections not associated with tissue proliferation the phloem conduits may die and collapse, a phenomenon which Esau terms "necrotic obliteration." The two sorts of effects viruses have on phloem – necrotic obliteration and hyperplastic growth – may not be mutually exclusive; in two viral tree diseases – tristeza of citrus and pear decline – the initial necrosis below the bud union is followed by stimulated meristematic activity resulting in replacement phloem having abnormal structure.

Discussing the routes of movement of viruses in plants, Esau and her co-workers pointed to the recent observation of virus particles in sieve plate pores and in plasmodesmata between parenchyma cells (Esau, Cronshaw, and Hoefert, 1967). They noted the high concentration of virus particles in some sieve elements, and suggested that transport follows a source-to-sink pattern, the particles accompanying assimilates and water by a mass-flow mechanism. They suggested further that cell-to-cell movement in parenchyma tissues may be regulated by the same forces; that is, that the intercellular movement of viruses parallels the movement of normal cell contents. This brings to prominence the question of the effect of the cytoplasmic cores of the plasmodesmata. Although the plasmodesmata are wide enough

(2000 Å) to accommodate a group of particles (each around 100 Å), the exact method by which the resistance to movement is overcome is of some importance. Esau stated that in the study on beet yellows (Esau, Cronshaw, and Hoefert, 1967) the cells containing the virus showed no plasmodesmatal core. Could the plasmodesmata, like the sieve-plate connections, be tubular in nature? Or, could protoplasmic streaming proceed from cell to cell through these narrow cytoplasmic bridges? These questions have yet to be answered?

In her recent book, Esau (1968) stated:

> The movement between the parenchyma cells and the sieve elements is, undoubtedly, a two-way exchange. Virus can be assumed to be carried by the water associated with the photosynthate delivery from the parenchyma cell to the conduit. Parenchyma cells also withdraw sugars from the sieve elements and the movement in that direction would provide the mechanism for carrying the virus from the conduit to the parenchyma cell. If this cell is still free of virus, a new infection would be established. Since the phloem is spread as a continuous system through the entire plant, and since the food moves toward sinks in various parts of the plant, particularly the growing parts, the virus transported with the food can effectively invade the entire plant and continue to invade the new growth increments.

Esau provided numerous illustrations to substantiate her interpretations of virus transport.

The specific properties of the beet yellows virus have been determined to such a degree that there is little doubt as to its relation to the yellows disease (Esau, Cronshaw, and Hoefert, 1967). Doi et al. (1967) in Japan described structures resembling mycoplasma bodies in the phloem elements of plants naturally infected with mulberry dwarf, potato witches'-broom, Japanese aster yellows, and paulownia witches'-broom. Similar organisms have since been identified from plants infected with American aster yellows, corn stunt, white leaf disease of sugar cane, yellow dwarf of rice, Crimean yellows from the USSR, stalbur from Rumania, and para-stalbur and clover dwarf from Czechoslovakia. All of these diseases are borne by leaf hoppers. In the past, these diseases were arbitrarily classified as virus diseases on the basis of their transmission by leaf hoppers and the type of disease syndrome exhibited. The results of observations with the electron microscope and experiments with chemotherapy now suggest that they may be caused by mycoplasmalike organisms. Such bodies have been identified in a leaf hopper infected with corn stunt. Because these bodies, as viewed in electron micrographs, range up to 0.5 μ in diameter, it is questionable that they would be phloem-mobile in many plants. Filamentous forms as small as 0.1 μ in diameter and up to 1.0 μ in length have been found both in leaf hoppers and in corn phloem cells. So-called "elementarylike bodies," ranging from 80 to 90 mμ are also found in the electron micrographs (Maramorosch et al., 1968). These could very well be infantile or alternate forms of the organism; they should be readily phloem-mobile. Even these have diameters about ten time larger than beet yellows virus particles.

Movement of Dyes and Fluorescent Tracers

Schumacher (1933) introduced the use of fluorescein (fluorescein-K in water, 0.1 percent w/v) as a fluorescent phloem-mobile dye. Because of the uniqueness of this fluorescent tracer it is worthy of review. Schumacher found that fluorescein was absorbed into sieve tubes and parenchyma cells. At the subcellular level, accumulation occurred in the nucleus, protoplasmic threads (endoplasmic reticulum), starch grains, and chloroplasts. It was not absorbed into the cell walls or the vacuoles. Cells with high metabolic activity (guard cells and companion cells, for example) were capable of accumulating higher concentrations than the surrounding parenchyma cells. Classical translocation characteristics were observed: (1) a concentration gradient away from the point of application was quickly generated in the sieve tubes; (2) source–sink relations were characteristic, with the tracer moving towards the strongest assimilate sink; (3) nonliving tissue was nonconducting; (4) when applied to leaves in the upper part of the plant, fluorescein moved down the petiole and then up to the growing point; (5) transport was influenced by temperature, with the highest rates occurring at about 30°C; and (6) rates under optimum conditions were about 30 cm per hour.

Further studies on fluorescein and some of its analogs were conducted by Bauer (1949), who confirmed the above results of Schumacher, and concluded in a later publication (Bauer, 1953) that the driving force for the transport of fluorescein originated in the leaf blade in which assimilate was being synthesized. Some fluorescein analogs showing phloem activity were rhodamine G (*m*-diethylamino analog of fluorescein) and rhodamine B (isothiocyanate of rhodamine G). The latter compound readily forms fluorescent conjugates with proteins. Eosine Y, the tetrabromo analog of fluorescein, has been shown by Schumacher (1933) to be toxic to phloem sieve cells. This is probably because of the extensive halogenation, which renders it more lipid-soluble. Similar results might be expected of erythrosin (the tetraiodo analog of fluorescein) and mercurochrome (the hydroxymercuridibromo analog of fluorescein). Other dyes and fluorescent indicators found to be phloem mobile by Bauer (1949) were arthanitin and berberine bisulfate (a water-soluble, fluorescent, natural product).

Numerous investigators (Willenbrink, 1957b, 1966a; Eschrich, Currier, Yamaguchi, and McNairn, 1965; and Eschrich, 1967) have confirmed the earlier observations of Schumacher and have concluded that there is no difference in the translocation behavior of fluorescein and that of ^{14}C-labeled assimilate in phloem tissue. Eschrich (1967) combined this unique tracer with the equally unique aphid stylet technique to study bidirectional flow in *Vicia faba.* If aphids were forced to feed on an internode between a lower leaf that had been treated with fluorescein and an upper leaf treated with ^{14}C-labeled urea or ^{14}C-labeled sodium bicarbonate, the simultaneous transport of the two tracers could be investigated. Using this technique, Eschrich found that 42 percent of the aphids produced double-labeled honeydew, and that the remainder produced honeydew labeled with either ^{14}C or fluorescein alone. The conclusions from many detailed trials were that double labelling could come about by bidirectional translocation in a single sieve tube, or by unidirectional movement involving a looped feedback system at the tissue level. Crafts (1967b) studied similar bidirectional transport in *Glycine max* seedlings and concluded that there was no evidence of bidirectional movement within a single file of sieve-tube elements.

Translocation of Growth Retardants

The group of chemically unrelated compounds known as growth retardants and represented by compounds like CCC, phosphon, AMO 1,618, and Alar, all have some phytophysiological effects in common. The response to growth retardants varies markedly in effectiveness, depending upon the plant species. Common group characteristics include stunting or dwarfing of plant growth, shortening of internodes, alteration of oxidase patterns, manifestations of antiauxin activity, inhibition of gibberellin biosynthesis, and inhibition of the incorporation of mevalonic acid into endogenous substrates and cholesterol. Knowledge related to their translocation patterns and cotransport with assimilate could help explain the large number of physiological and biochemical observations that have been made with respect to growth retardants, but only a limited number of studies have been conducted. Some information, however, can be obtained by inference from other physiological observations.

Edgerton and Greenhalgh (1967) applied ^{14}C-labeled Alar to the young fruit and foliage of *Malus sylvestris* 'McIntosh' and determined the distribution in the fruit at frequent intervals during the growing season. Within 24 hours, 50 percent of the applied dose of labeled Alar was metabolized, and the remainder was concentrated in the flesh and seeds. After five weeks, no residue could be detected on the surface of the fruit; levels in the flesh were diluted by growth, and no further accumulation in the seeds took place. The distribution in the treated branches was measured during the dormant season. Alar labeled with ^{14}C was observed to accumulate in flower buds, vegetative buds, cluster bases, and 1-year-old bark and xylem. Some translocation from treated branches to untreated branches was observed. These data show that Alar is translocated in the phloem and, most likely, with the assimilate stream.

Because of the possible horticultural usefulness of Alar, Sachs et al. (1967) studied its penetration, translocation, and metabolism under greenhouse conditions. Penetration proved to be extremely variable: within a single species, treated under comparable conditions, uptake varied as much as tenfold in successive greenhouse experiments. Relative humidity was one of the major factors affecting penetration; at 95–100 percent RH, 65 percent of the Alar applied penetrated in 24 hours; at 50–60 percent RH, penetration was only about 2 percent. Under common summertime greenhouse conditions (RH around 70 percent), 30–35 percent uptake occurred. In seedlings, basipetal translocation exceeded acropetal translocation by a factor of five in 72-hour trials. This probably reflected the activities of source and sink. Hydrolysis of Alar, when it occurred, took place on the leaf surface; growth inhibition induced by Alar was not correlated with hydrolysis of the molecule.

Similar studies with ^{14}C-labeled Alar were conducted by Moore (1968). Applications of the tracer to cut stem bases of three monocotyledons (*Sasa* sp., *Convallaria majalis* and *Zantedeschia* sp.) resulted in its uptake and translocation throughout the plant. Foliar applications to single leaves of 3 monocots and 5 dicots showed uptake and translocation to other shoots and to the roots. Moore concluded that ^{14}C-labeled Alar is readily mobile in the phloem and xylem. Residue analysis showed no metabolic breakdown of Alar.

Differences in the growth retardant activity were due to species selectivity at the metabolic level, not to degradation of Alar or to lack of uptake and translocation.

Only a limited amount of information is available on the translocation of AMO 1,618. Hanawa (1968) peeled the embryo out of the seed coat of *Sesamum indicum*, removed one of the cotyledons, and treated the embryo with AMO 1,618. After 48 hours, the first leaf primordium developed into a double leaf on the decotylated side of the shoot apex. Under similar conditions, 2,4-D induced gamophylls that resembled the double leaf. Hanawa concluded that AMO 1,618 induces the formation of double leaves by causing structural changes in the apical meristem and leaf primordia, whereas 2,4-D induces gamophylly by stimulating the interprimordial region in the peripheral zone of the shoot apex. CCC, IAA, NAA, gibberellin, and maleic hydrazide had no effect on producing the double leaf or gamophyll. These data seem to indicate that AMO 1,618 is translocatable with the assimilate stream moving from the region of food storage to the developing shoot and is thus phloem-mobile.

Further indications that AMO 1,618 is phloem-mobile come from the research of Macchia (1967), who treated the foliage of *Pisum sativum* seedlings with AMO 1,618 at concentrations of 10^{-4} to 10^{-5} M. Expansion growth was limited, apical dominance was decreased, and stem diameter and the volume of parenchyma cells were increased. He concluded that the diameter of the stem was increased by stimulation of the mitotic activity of the cambium. Of special importance was the observation that AMO 1,618 decreased the size of sieve tubes. In treated plants, slime bodies were present, the nucleus of the sieve tube degenerated later than in the control plants, and slime plugs appeared near the sieve plates. The mitotic activity of the apical meristems did not change. Additional physiological phenomena were delayed lignification, inhibition of primary starch synthesis, and the stimulation of secondary starch accumulation in parenchyma cells.

The translocation of CCC was investigated with ^{14}C-labeled CCC by Birecka (1967). Approximately half of the tracer applied to roots of *Triticum* plants was translocated to the tops and was detected in all parts of the plant. When applied to the fifth lamina from the top, it influenced elongation of the culm of the treated shoot and the tillers; when applied to the flag leaf, it retarded the growth of the uppermost internode only. CCC was not metabolized in *Triticum*, but it did increase the choline in all tissues studied. Blinn (1967), using ^{14}C-

labeled CCC, confirmed Birecka's results and concluded that the tracer was metabolically inert. The unchanged compound was the only radiolabeled material found as residue in wheat foliage, roots, and grain. In wheat plants, it was absorbed slowly from foliar applications, and small amounts were translocated to the roots. Birecka's observations of xylem transport of CCC from root applications could not be confirmed by Adedipe, Ormrod, and Maurer (1968) who found soil applications to the roots of *Pisum sativum* to be ineffective in modifying plant growth. However, foliar applications at the five-node stage were very effective in increasing plant height, fresh weight, internode length, and dry weight. The pigment content of the plastids of young leaves was reduced more by foliar application than by soil application.

Humphries (1968) studied the growth retardation effects of CCC on the leaves of *Beta vulgaris* and concluded that CCC diverted the assimilates to the roots. He observed the same effect on *Hordeum vulgare, Avena sativa, Secale cereale,* and *Triticum aestivum*. CCC has been shown by Macchia (1967) to reduce the size of vessels and sieve tubes in *Pisum sativum*, an effect noted above for AMO 1,618. This sieve-tube response is undoubtedly related to assimilate distribution, but the exact mechanism is not understood at this time. Hundreds of papers in the literature, however, report a growth retarding effect on cereals, where the practical application of CCC is to prevent lodging. All of these reports seem to be related to photosynthate source–sink distribution patterns; and conclusions from the above work with ^{14}C-labeled CCC seem to indicate that it is phloem mobile, metabolically stable, and effective in reducing the size of sieve tubes. Thus the effect on sieve-tube differentiation may account for the growth retardation.

Information on translocation patterns of phosphon is not available. Morphactin-type growth retardants (CF1 and FL, for example) have recently been reviewed by Ziegler (1970). The long-distance translocation of the ^{14}C-labeled butyl ester of CF1 was shown by autoradiography and aphid-feeding studies to be predominantly in the phloem. In *Vicia faba*, translocation of CF1 was from the primary leaves to the apical meristem, which is a region of CF1 accumulation.

Abscisic acid is a newly discovered plant growth hormone that has many plant growth retardant characteristics. Addicott and Lyon (1969) have recently reviewed the physiology of abscisic acid. The first definitive characterization of abscisic acid translocation comes from the work of Milborrow (1968), who used ^{14}C-labeled (±)-abscisic acid to study the metabolism and transport in *Platanus occidentalis* and *Phaseolus vulgaris*. He concluded that three times more abscisic acid was transported basipetally than acropetally. He found three radioactive metabolites which appeared to be nontransportable in these tissues.

Further indirect evidence for abscisic acid being transported with the assimilate stream comes from the work of Evans (1966) who applied abscisic acid to the leaves of *Lolium temulentum* and obtained a significant reduction in the flowering response. This suggests that abscisic acid either interferes with the synthesis of the floral stimulus at the source, or inhibits the response to the floral stimulus at the apex. This could be interpreted as mass flow of abscisic acid, floral stimulus, and assimilate from the source leaf to the apex. Hoad (1967), Bowen and Hoad (1968), and Lenton, Bowen, and Saunders (1968) have shown that abscisic acid is present in phloem exudate by collecting honeydew from feeding aphids on *Salix viminalis* and analyzing it by gas liquid chromatography and spectropolarimetry. The concentration found was approximately 10 μg/100 ml of phloem sap. Similar studies on xylem sap obtained by water displacement show abscisic acid to be present at concentrations of 1–5 μg/100 ml of xylem sap.

In deciduous trees, the ability of abscisic acid to transport from mature leaves to apical buds and induce dormancy indicates that it is transported in the phloem with photosynthate (see Addicott and Lyon, 1969). Because abscisic acid is easily metabolized by plant tissues, the active molecule for phloem transport may be the abscisyl glucoside. Confirmation of this and more detailed characterization of the translocation patterns for abscisic acid will have to await additional radiotracer studies.

Other antimetabolites that cause defoliation have been investigated by Kraft and Bokarev (1967). These workers applied ^{35}S-labeled diethylxanthogen trisulfide to the foliage of *Gossypium hirsutum* plant and followed the translocation by gross autoradiography. The defoliant was not translocated in the plant, was not extracted from the plant by chloroform, and tended to persist in the young leaves. Information related to the translocation of other defoliants is lacking.

The (+)-limonene derivatives of quaternary ammonium salts have been evaluated for growth retardant activity and translocation in plants by Newhall and Pieringer (1967). Application to the stems of *Citrus paradisi* and *Phaseolus vulgaris* resulted in downward movement with the photosynthate; foliar applications, however, were not effective. Similar results were obtained by Edgington and Dimond (1964) for a series of closely related analogs; they found that these quaternary ammonium derivatives are strongly absorbed to the tissue and are cation-dependent for mobility. Since the work of Edgington and Dimond is more closely related to fungicides, it will be discussed in greater detail in the following section.

Movement of Chemotherapeutants and Fungicides

One way to control disease in plants is through chemotherapy, wherein compounds are introduced into the plant for direct or indirect control of pathogen. Dimond (1965) divided chemotherapeutants into two categories: (1) systemic fungicides or systemic toxicants which are directly toxic to the corpus of the pathogen, or are converted within the plant to toxic molecules: and (2) compounds that alter the processes of pathogenesis of either the host or the pathogen. The ultimate usefulness of the chemotherapeutant depends upon whether it is suitably mobile, nonphytotoxic, reasonably stable to the *in vivo* metabolism of the plant, and functional in combating disease.

Early work on the uptake and movement of chemotherapeutants was reviewed by Dimond and Horsfall (1959) and Crowdy and his coworkers (Crowdy, Green, Grove, McCloskey, and Morrison, 1959; and Crowdy, Grove, and McCloskey, 1959). More recent reviews on the translocation of antibiotics include those of Goodman (1962) and Dimond (1965). The reviews of Summers (1968), on synthetic organic fungicides, and of Sijpesteijn and van der Kerk (1965), on the fate of fungicides in plants, provide insight into the importance of phloem mobility for this class of compounds. The majority of the chemotherapeutants and fungicides are water-soluble materials that are applied to roots and translocated in the transpiration stream. Some of these materials are phloem-mobile.

Dimond and Horsfall (1959) reported that streptomycin sprayed on primary leaves of *Phaseolus vulgaris* showed antibacterial action as far away as the fourth trifoliate leaf, and that this action persisted for 11 days. Streptothricin and pleocidin were translocated both basipetally and acropetally in *P. vulgaris* and *Nicotiana tabacum*. Immersion of leaves in solutions of antibiotics resulted in enhanced uptake and translocation, but some of this movement was via the xylem. Apparently, the more mobile of the antibiotics are either acidic or neutral, basic and amphoteric compounds giving anomalous results. This is true also for the movement of dyes in the xylem.

Gray (1958) studied the translocation of a number of antibiotics in several crop plants in a search for systemic plant protection. He found that streptothricin and pleocidin, applied to intermediate leaves of *Phaseolus vulgaris* and *Nicotiana tabacum* plants at 1 percent concentration, were readily translocated in high amounts—downward to older leaves, and upward to young leaves. When the primary leaves of *P. vulgaris* plants were sprayed with these compounds, young untreated leaves became immune to common bacterial blight. Under like conditions, streptomycin, dihydrostreptomycin, neomycin, oxamycin, bacitracin, and actinomycin failed to move out of sprayed leaves. When primary leaves of *P. vulgaris* plants were immersed in a solution of 1000 ppm streptothricin, the antibiotic was transported downward to the roots and upward to younger leaves. Several of the antibiotics that failed to move out of sprayed leaves were translocated upward and downward when applied by immersion. Streptothricin and pleocidin applied by immersion passed steam rings on petioles and stems indicating xylem movement as a result of this application method. Streptothricin was absorbed by roots and intact stems of plants and translocated to the tops. Movement from primary leaves of *P. vulgaris* to young untreated leaves when application was by spraying indicated that streptothricin and pleocidin may be phloem-mobile; all of the other experiments described by Gray seemed to imply xylem transport; apparently the above two antibiotics are both phloem- and xylem-mobile; some of the others are xylem-mobile only.

In a comprehensive review covering absorption and translocation of growth regulators and pesticides, Mitchell, Smale, and Metcalf (1960) described the plant anatomy involved in these processes. They discussed the various aspects of penetration, and migration across cuticle, epidermis, and cortical tissues, and they cited many references to data on translocation via phloem and xylem. They take up in turn the absorption and translocation of plant regulators, of therapeutants, and of insecticides. In all of these cases there is convincing evidence cited for some compounds being readily moved via the transpiration stream in the xylem, for others being distinctly phloem-mobile, for some that move freely in both of these tissue systems, and some that apparently are not readily moved because they cannot penetrate to the vascular channels. They cite the interesting example of one plant regulator applied to stems of young *Phaseolus vulgaris* plants that moved in sufficient amounts into their seeds to alter internodal elongation in their offspring. These second-generation plants translocated to their seeds the regulator they had received, and thus affected the third generation. Foy (1961b) described a similar experience with dalapon in *Triticum aestivum.*

In the case of therapeutants, Mitchell, Smale, and Metcalf (1960) reported that some of these materials are readily translocated from roots to leaves, and that others fail to move out of the roots. Downward movement from leaf or stem application was not detectable following light application; when applied at 2000 ppm or more, downward movement was found. Most leaf applications apparently result in movement through the leaf but little or no export from the leaf. In this connection, Phelps and Kuntz (1965) studied the movement of the antibiotics cycloheximide and oligomycin in *Quercus ellipsoidalis.* Foliar applications resulted in basipetal transport into the stem; most of the original dose, however, was present in the leaves. Transport in the xylem tissue was very efficient as observed from root, cut branch, or trunk injections.

Maier (1960) studied the absorption, translocation and retention of streptomycin in *Humulus lupulus* plants using a bioassay to determine the antibiotic, and a standard curve of log streptomycin concentration against mean bacterial inhibition to determine streptomycin values. Rapid absorption followed by upward translocation was observed when streptomycin in lanolin paste was applied to the first node of stems. Maximum concentrations were attained in tips, 74 cm above the point of application, sooner than in intervening leaves. This distribution indicates phloem movement, at least in the stem above young leaves that normally import assimilates. Movement from the first node into mature leaves must have taken place via the xylem. Rates of movement from point of treatment to growing tip were 4.0 and 4.6 cm per minute for streptomycin absorbed from nitrate and sulfate preparations. These rates must include xylem movement from node 1 to mature assimilate-exporting leaves, transfer from xylem to phloem, and then, movement via the phloem to importing leaves and shoot meristem. The fact that root samples failed to provide measurable amounts of streptomycin confirm the postulation of xylem movement from node 1.

Davey and Papavizas (1961) found that streptomycin applied to *Coleus blumei* is absorbed and transported basipetally into the roots within 24 hours. When plants growing in soil were treated with doses of 1000 μg and 3600 μg of the antibiotic and the rhizosphere tested 4 and 8 days later, the gram-negative rhizosphere bacteria had dropped 28 percent and 22 percent, respectively; this effect largely disappeared by the 12th day. Using ^{14}C-labeled streptomycin, Davey and Papavizas found that the tracer moved downward about 6 inches in 6 hours, and into the root system in 24 hours. This proves that the bactericidal effect noted above resulted from translocation of the antibiotic.

Apoplastic systemic fungicides are desirable in combating wilt diseases (Dimond, 1965). Thus, for internal fungistasis in the xylem, griesofulvin, chloroamphenicol, streptomycin, and the sulfonamides (Irgafen, thiazamide, and DDS, for example)—which move readily in the xylem (Brian, 1966)—are applicable. Streptomycin applied foliarly appears to be phloem-immobile. Brian (1966) reviewed the evidence for phloem immobility and offered the following as support: (1) ^{14}C-labeled streptomycin does penetrate the cuticle, but moves only acropetally in treated leaves; and (2) if the lamina is dipped in solutions containing the antibiotic, no movement into the petiole occurs. Streptothricin, a streptomycin-like cationic antibiotic, if applied to the primary leaf of *Phaseolus vulgaris,* could move basipetally to the main stem and then acropetally to the shoot apex. There was some evidence of retranslocation based on steam-girdling experiments.

The phloem mobility of the herbicide 2,4-D is well established (*see* Herbicide Distribution, p. 231). The translocation of some fungicide analogs has been evaluated by Zygmunt (1968) who finds 2-nitrovinylphenoxyacetic acid and sulfamoylphenoxyacetic acid to be phloem-mobile and effective systemic compounds under some conditions. This is an interesting cross-correlation of structure–activity relations between herbicide and fungicide activity.

The translocation of three systemic fungicides for prevention of cytospora canker disease in *Prunus persica* was investigated by Williams and Helton (1967). Cycloheximide acetate, cycloheximide thiosemicarbazone, and 8-quinolinol phosphate were sprayed on the foliage of three-year-old *P. persica* trees 14 days after being inoculated with the fungus *Cytospora cincta.* All three compounds gave effective control and were observed to translocate from foliage to roots.

Systemic characteristics of eight nitrophenol analogs were evaluated by El-Zayat et al. (1968) in studies against the fungus *Erysiphe polygoni* on *Phaseolus vulgaris* plants. The nitrophenols were found to be translocated from a primary leaf to younger trifoliate leaves higher on the plant. If applied to the roots, there was some effect; but fungicidal dosages were phytotoxic. The most effective derivative of the eight analogs tested was 2-chloro-4(1′,1′,3′,3′-tetramethylbutyl)-6-nitrophenol.

Indirect evidence for phloem transport of the methyl ester of D-1991 was presented by Ogawa, Manji, and Bose (1968). Blossoms of *Prunus avium, P. persica, P. persica nectarina* and *P. armeniaca* were sprayed with D-1991 for control of brown rot blossom blight.

Fruit removed from the trees and inocculated with the fungus *Sclerotinia fructicola* showed effective control from D-1991 applications. This indicated that the D-1991 translocated from sprayed leaves to the fruit in sufficient concentration to be effective fungicidally.

Antifungal antibiotics of the polyoxin type have been investigated for control of sheath blight of *Oryza sativa* plants (Sasaki et al., 1968). Crude polyoxin fractions were applied foliarly and observed to penetrate the tissue of *O. sativa* and *Phaseolus vulgaris* plants. Although some symplastic transport was observed, the bulk of the transport was apoplastic in the transpiration stream.

In summary: Although thousands of fungicides have been synthesized and tested for biological activity, the availability of effective symplastic systemic fungicides remains limited. Those fungicides currently characterized as being systemic when subjected to critical testing for mobility in plants usually are only apoplastic systemic fungicides. With the availability of radiolabelled fungicides and the use of translocation testing as described by Crafts (1967b), more definative results with respect to phloem mobility should become available.

Nematocides and Rodenticides

Because many compounds applied to plant foliage are known to enter the tissues, migrate to the phloem, and translocate into roots; and because a few, following such movement, actually pass out of roots into the soil (see Table 8.1); it has been considered possible that systemic nematocides may be found that will be effective when applied to the foliage as sprays. Peacock (1966) used TCPP in experiments on nematode control on *Lycopersicon esculentum* plants. This compound is apparently phloem-mobile, and, used in sprays at 5000–10,000 ppm, it has proved able to protect tomato plants from subsequent experimental infection with root-knot nematodes (*Meloidogyne* sp.). Attempts at control of nematodes already in the plant or at protection from naturally arising multiple infections have not given an acceptable degree of control. Experiments under partially controlled conditions proved that light intensity between 500 and 1800 ft-c had no effect on the degree of protection. TCPP proved more effective on plants growing in a humid atmosphere and in moisture deficient soil than in dry air and moist soil. Sucrose as an adjuvant had a slight enhancing effect; spraying both surfaces of *L. esculentum* leaves was roughly twice as effective as spraying either upper or lower surfaces alone. Although certain organophosphorus compounds have also given indications of systemic nematocidal activity, none has proved effective in commercial practice to date. Nevertheless, the fact that such compounds exist may be taken as an indication of the promise that lies in the future for discovery of useful systemic nematocides that may be effective when applied foliarly.

Rohde and Jenkins (1958) reported that an extract from roots of *Asparagus officinalis* resistant to a nematode, when applied to foliage of tomato, was translocated into the roots where it brought about a nonreversible paralysis of the nematodes.

In relation to root-knot nematodes and a study of the metabolic degradation of Nellite nematocide, Meikle (1968) presented evidence for the translocation of this compound in *Cucumis sativus* plants. Applications of phenyl-1-^{14}C-Nellite or dimethyl-^{14}C-Nellite to the nutrient solution resulted in apoplastic transport primarily to the margins of the older leaves (areas of high transpirational loss); some ^{14}C appeared in the new growth.

Radwan (1966, 1967) studied the translocation of the powerful rodenticide tetramine in *Pseudotsuga menziesii, Rubus* sp. and *Dactylis glomerata.* Using ^{14}C-labeled tetramine, he found that apoplastic transport could take place from soil or root applications; foliar applications, however, showed the labeled tetramine to be completely immobile. Retranslocation did not occur, and he concluded that tetramine did not behave as a systemic rodenticide.

Translocation of Insecticides

The mode of action and translocation of systemic insecticides in plants was reviewed in 1957 by S. H. Bennet. He defined systemic insecticides as those which are absorbed and translocated to other parts of the plant, thus rendering untreated areas insecticidal. A classification based on residual behavior has been proposed by Ripper (1967), as follows: (1) stable systemic insecticides, which are not metabolized by the plant; (2) endolytic systemic insecticides, in which the toxic compound is present in its original form when taken up by the insect; and (3) endometatoxic systemic insecticides, which are transformed in the plant from nontoxic compounds into toxic compounds, so as to be insecticidal when taken up by the insect. The role of metatoxic reactions in the translocation of systemic insecticides has recently been reviewed by Metcalf (1967). In both the systemic and endometatoxic action, the plant plays a very active role in producing the final sink for the insecticide and in rendering it insecticidal. Ultimately, the plant must degrade the insecticide and eliminate it from the enviroment.

For purposes of discussing the relationship of systemic insecticides to translocation, a classification restricted to transport is more useful, and the following is proposed: (1) general systemics, those compounds that move freely in both the xylem and phloem; (2) symplastic systemic insecticides, those that are phloem mobile; and (3) apoplastic systemic insecticides, those that are transported freely in the xylem. Table 8.5, from Mitchell, Smale, and Metcalf (1960), presents data indicating the relative mobility of some six systemic insecticides in *Citrus limon.* Table 8.6 shows the differing mobilities of Systox and amiton compounds in *Gossypium hirsutum.* Apparently, the polarity of these compounds is of overriding importance in their penetration of cuticle and stem tissues; nonpolar molecules seem the most able to move through such tissues and translocate in vascular bundles. Thus, the Systox thiolisomer is much more mobile than is its methylsulfonium salt, and amiton base is more mobile than its oxalate; the salts, in these cases, are polar, and the ions seem to have difficulty in moving into the transporting tissues. Both of these ions

Table 8.5. Absorption and translocation to terminal leaves of systemic insecticides applied to roots or stem of *Citrus limon.* (Application in equivalent amounts of 20 μliter topically or 100 mg via water culture. Values are radioactivity in leaves in ppm.)

Insecticide	Applied to	Time			
		1 day	2–3 days	1 week	2 weeks
Schradan	Roots	9	41	49	79
	Stem	153	375	1085	888
Dimefox	Roots	30	129	38	257
	Stem	47		32	81
Systox thiono	Roots	30	39	235	80
	Stem	8	38	182	284
Systox thiol	Roots				
	Stem	67	708	1680	
Amiton (base)	Roots	8	12	24	248
	Stem	198	737	1257	3120
Amiton oxalate	Roots	1	1	21	202
	Stem	6	49	120	176

Source: Mitchell, Smale, and Metcalf (1960).

Table 8.6. Translocation of ^{32}P-labeled systemic insecticides into leaves of *Gossypium hirsutum* from stem applications of equivalent amounts. (Values are radioactivity in leaves in ppm.)

	Days after treatment			
	1	2	4	7
Systox thiol isomer	520	690	1283	765
Systox thiol isomer sulfoxide	33	40	225	313
Systox thiol isomer methyl sulfonium salt	34	156	118	54
Amiton thiono isomer base	25	340	740	856
Amiton thiol isomer base	162	532	512	920
Amiton thiol isomer oxalate	15	28	30	90

Source: Mitchell, Smale, and Metcalf (1960).

are positively charged (cations) and hence would be attracted to and held by the negatively charged cuticle. Phosdrin, after absorption through the roots of *Pisum sativum,* is redistributed away from roots, stem, and mature leaves to the growing tip and young leaves. This distribution represents primary movement from roots to transpiring leaves via the transpiration stream, followed by secondary movement into meristematic regions of the shoot tip via the phloem. Apparently, this insecticide moves freely in both vascular tissue systems. Such systemic distribution is very important in the control of aphids and other insects that feed primarily on the phloem or on young growing stems and leaves.

From this review by Mitchell and his co-workers, it is apparent that many systemic insecticides are effective when applied to bark of the main stem or trunk. Again, these materials seem able to penetrate the bark and phloem, to cross cambium, to move up in the transpiration stream to the mature foliage, and from there, to move into young tissues by redistribution via the phloem. Systox, dimefox, shradan, Systox thiolisomer, amiton, disulfoton, Thimet, phosphamidon, Pyrazoxon, and isolan are some of these compounds.

In studies on the penetration and translocation of dimethoate, an organophosphorus insecticide, de Pietri-Tonelli (1965) found this compound to readily penetrate the cuticle, epidermis, and chlorenchyma of leaves, but to move only apoplastically in the tissues. Long-distance transport took place in the xylem from trunk or root applications; basipetal movement via the phloem was slow and involved only small amounts. In prolonged tests (59 days), dimethoate concentration was higher in the bark than in the wood of twigs, branches, and trunks of *Citrus limon* trees; this de Pietri-Tonelli attributes to radial transfer from the xylem.

Contrasting with de Pietri-Tonelli's results with dimethoate, Randall (1962) evaluated the systemic effectiveness of dimethoate, menazon, and phosphamidon from foliar applications of aqueous solutions to the apex and branch tips of spruce. Translocation was bioassayed with the spruce budworm *Choristoneura fumiferana*. The results showed that all three compounds could be translocated basipetally, and that phosphamidon was more effective than the other two insecticides.

These results can best be explained from the results of Lucier and Menzer (1968), who studied the metabolism of dimethoate in relation to its mode of entry into the plant. Degradation occurred very slowly from foliar applications, and very rapidly from stem injections; therefore, the longer the insecticide could be kept away from plant hydrolytic enzymes the more residual its action. When the insecticide did come in contact with the enzymic system of the plant, several metabolites were formed. Undoubtedly, these metabolites have different rates of translocation.

The systemic capabilities of Systox were reviewed by Metcalf (1966). He noted that ^{32}P-labeled Systox applied to the lower surface of *Vicia faba* leaves was translocated to the remainder of the plant. The labeled Systox could be transported from one primary leaf to another at levels up to 1 percent of the applied dose. When applied to lower leaves, movement could be detected upward to new trifoliate leaves of *Phaseolus vulgaris* plants. In *Solanum tuberosum*, limited amounts of Systox were transported from the foliage to the tubers. All of these observations are in general agreement, and they characterize this organophosphate insecticide as a phloem-mobile compound. Another compound which appears to move with the photosynthate from foliage to roots, or upward to more distal shoots is isolan, which was studied in woody ornamental trees by Treece and Matthysse (1959).

S. H. Bennet (1957) reviewed the literature associated with the transport of schradan. From 1–4 percent of a foliar application is transported per day away from the site of application with the assimilate stream. Shading and ringing experiments were effective in halting translocation. The presence of schradan in nectar secretions after application to leaves further indicates phloem mobility. These results were generally confirmed by Rediske and Johnson (1965) in their studies on *Picea sitchensis* and *Abies grandis.*

DDT is not translocated in plants, according to the studies of Eden and Arthur (1965) and Ware (1968). These investigators used ^{14}C-labeled DDT on *Glycine max* and on *Medicago sativa,* respectively. Some chlorinated hydrocarbons can be translocated in plants. Weisgerber et al. (1968) have shown that ^{14}C-labeled endrin can translocate from the foliage of *Brassica oleracea* to the roots. Residue studies show, however, that the metabolite endrin-ketone is present, as well as another water-soluble metabolite. The translocated ^{14}C is probably in a water soluble metabolite.

Johnson and Zingg (1967) have found that four systemic organophosphate insecticides will translocate from foliage into cones of *Pseudotsuga menziesii.* Solutions of the compounds Bidrin, Azodrin, methyl demeton, and dimethoate were brushed onto the foliage and cones of *Pseudotsuga menziesii* in late May. There was no cross-movement from one side to the other in the cones; fair movement from base to tip, however, did take place. Phytotoxicity increased with increasing concentration of the applied dose; 8 percent solutions caused heavy injury. At concentrations of 4 and 8 percent, movement occurred from foliage to cone, and vice versa. Determinations were by bioassay using cone midges (*Contarinia oregonensis*).

The importance of endometatoxic reactions in the activation of systemic insecticides was recently reviewed by Metcalf (1966). He explained that Systox, Thimet, disufloton, and dimethoate can be rapidly activated to more toxic compounds by the hydrolytic enzyme system of a plant. Furthermore, the products of these reactions are more water soluble, more rapidly hydrolyzed, and possibly more translocatable. Two primary reactions are responsible: the oxidation (or desulfuration) of P=S to P=O; and the oxidation of sulfide to sulfoxide, and eventually on to sulfone. Detailed studies by Metcalf's research group have shown that Temik is an excellent example of a new carbamate-type insecticide that undergoes endometatoxic activation to a more powerful *in vivo* metabolite, the sulfoxide Temik analog.

Metcalf summarized these metabolic transformations as "delay factors" and pointed out how vital they are in designing effective systemic insecticides. Thus, through the alkylthio group, lipid solubility is attained for cuticular penetration and membrane transport; but after endometatoxic activation, water solubility is characteristic of the molecule. With these final characteristics, faster, more specific translocation can take place and the compound can be more rapidly degraded. But, while this metabolic action is going on and delaying the metabolism of the compound, active toxic action is being carried out against the target insect. Whether reactions of this nature will transform apoplastic systemic insecticides into more effective symplastic systemic insecticides will be the subject of considerable research in this coming decade.

While the many studies on virus movement in plants are very gratifying in that they seem to provide strong evidence that rapid transport in the phloem is by mass flow with the assimilate stream, the rather few tests on antibiotics, fungicides, and insecticides provide little information of use in the characterization of phloem transport. However, the fact that an increasing number of compounds give evidence of being phloem mobile and biocidal at the same time provides hope for the discovery of effective systemic agents for control of plant pests and diseases. When one considers the array of compounds in Table 8.1 that move about in plants without being immediately toxic, it seems certain that continued search is justified. The characteristics of the readily phloem-mobile molecules may well provide a template to match in the search for such chemicals.

Herbicide Distribution

The uptake and distribution of herbicides has been the subject of valuable studies, and has provided answers to problems in translocation physiology as well as helping in the effective use of weed killers. Early studies utilized the growth regulating symptoms of 2,4-D for tracing movement (Day 1950, 1952; Crafts, 1956a). With the synthesis of ^{14}C, this valuable isotope was soon incorporated into the 2,4-D molecule, and an ideal tracer was available for study (Crafts, 1956a). After development of satisfactory methods of autoradiography (Yamaguchi and Crafts, 1958; Crafts and Yamaguchi, 1960, 1964), many experiments were carried out to elucidate the mechanisms of herbicide absorption and translocation. Because autoradiography could be carried out using intact plants subject to various experimental conditions, critical results have been obtained in quantity.

Pallas (1960) studied the effects of temperature and humidity on foliar absorption of ^{14}C-labeled 2,4-D and ^{14}C-labeled benzoic acid by *Phaseolus vulgaris*. Using autoradiographs as a qualitative estimate, and counting as a quantitative measure of absorption and translocation, Pallas found increases both in uptake and in transport with increases in temperature between 20° and 30°C. At humidities varying from 34 to 48 percent, less of the tracers was absorbed and translocated than at humidities in the 70–74 percent range. Movement of 2,4-D in the leaf followed the assimilate stream and was confined to the vascular bundles. In leaves treated with labeled benzoic acid, there was similar export from the treated areas of the leaf; but in addition, this tracer exhibited apoplastic movement in the treated leaf, indicating a somewhat slower pickup into the symplast. By extraction and chromatography, it was determined that the bulk of the radioactivity was being translocated as free 2,4-D and benzoic acid (or dissociable salts thereof) within the 8-hour treatment time.

Foy (1961a, 1961b) determined the absorption, translocation, and metabolism of ^{36}Cl- and ^{14}C-labeled dalapon in *Gossypium hirsutum* and *Sorghum vulgare* by autoradiography, counting, and chromatography. The tracers were applied to leaves and roots of the plants. Foy found that dalapon readily enters and moves through leaves and roots of *G. hirsutum* and *S. vulgare;* cuticular sorption was very rapid. Acute injury at the point of application in foliage reduced or prevented systemic distribution; in the absence of such injury, absorption and movement continued for 2 weeks or more. Dalapon absorbed through roots, severed veins, or severed petioles was readily translocated via the xylem in the transpiration stream. Absorbed through leaf surfaces, dalapon moved readily via the phloem in the assimilate stream. Transfer from phloem to xylem also occurred, and dalapon was found in the culture solution of *G. hirsutum* plants that had been treated on the foliage. In both the tolerant *G. hirsutum* and the susceptible *S. vulgare,* retranslocation from initial accumulation sites occurred in response to shifts in sink activity (growth and maturation of successive leaves). Despite its ready mobility, some dalapon was retained along the transport routes; restriction was notable in the intercalary meristem regions of young sorghum leaves. Neither penetrability nor metabolic inactivation plays a major role in determining selectivity of dalapon; translocation likewise did not appear to be involved. Foy concluded that the key to selectivity must reside in the protoplasm of the two species.

In his second paper, Foy (1961b) reported that dalapon was absorbed, translocated, redistributed and accumulated principally as the intact molecule, and that it remained essentially unmetabolized for long periods, particularly in dormant or quiescent tissues; after 10 weeks, 85–90 percent of an applied dose was recoverable as dalapon from fruits of *Gossypium hirsutum.* Dalapon stimulus was traced into the third generation in *Triticum aestivum* following preplant application of the compound at a rate of 4 pounds per acre. Slow metabolic decomposition resulted in release of ^{36}Cl from ^{36}Cl-labeled dalapon and the incorporation of ^{14}C into other compounds. Possibly the mode of action may involve inhibition of pantothenic acid synthesis and disturbance of coenzyme-A and pyruvate metabolism.

Crafts (1959a, 1959b) reported on the relative mobility of a number of tracers. Using ^{14}C-labelling on organic molecules, he found that ^{14}C-labeled 2,4-D is absorbed by leaves and transported in the phloem along with assimilates. When assimilates are moving rapidly, 2, 4-D distribution is extensive; when they are moving slowly, 2,4-D movement is restricted. Applied to *Hordeum vulgare* plants, the following mobility series may be demonstrated: 2,4-D $<$ IAA $<$ amitrole $<$ maleic hydrazide $<$ dalapon. Maleic hydrazide and dalapon were found to move from the phloem into the xylem, and so, to lightly label the mature leaves of *H. vulgare*. Such leakage was positively correlated with mobility.

The herbicide 2,4-D is an agricultural chemical of great importance. Zweep (1961) applied 2,4-D, methylene-labeled with ^{14}C, to winter and spring *Hordeum vulgare* plants at the 1-, 2-, 3-, 4-, and 5-leaf stages and noted the distribution of the tracer after 24 hours. Autoradiographs of plants treated in the 1-leaf stage had very little radioactivity in the roots, a situation that Zweep attributed to the fact that the seed dominates in the nutrition of the root at that stage of growth. Good movement to roots was observed in plants in the 2-leaf and 3-leaf stages; no differences were noted between spring and winter varieties of *H. vulgare*. At the 4- and 5-leaf stages much less movement of tracer to roots was found; increasing the dosage had little effect; treatment of tillers did not lead to tracer movement into roots. Zweep concluded that there is evidence for a block at the base of *H. vulgare* leaves that prevents 2,4-D from moving to other parts of the plant. Evidence for such a situation, as reported by Forde (1965), will be discussed in Chapter 9.

Greenham (1962) used orthoarsenic acid and 2,4-D in translocation studies on *Lygodesmia juncea,* an economic pest in Australia. Varying conditions of soil oxygenation proved to have no effect on translocation of the arsenic acid. Exposure to darkness for 4 days prior to treatment resulted in enhanced depth of kill of *L. juncea* roots by both orthoarsenic acid and 2,4-D, a result that Greenham attributed to the sustained high carbohydrate content of the leaves. A 0.25 M orthophosphate solution at *p*H 3.0 increased the translocation of radioactive 2,4-D compared with citrate and glycine solutions of equal molarity. Possibly, phosphate metabolism influences translocation through activation of source or sink. In time experiments lasting 7 days, radioactivity in roots increased rapidly with time;

counts leveled off logarithmically between the third and seventh days. There was a gradient of decreasing activity in young plants from the petioles to the roots with no focal point of concentration. Radioactivity in roots decreased through a series of increasing dosage of radioactive 2,4-D, as shown in Table 8.7. Injury to leaf tissue or phloem or both was regarded by Greenham as a major factor limiting translocation in *L. juncea*.

Little and Blackman (1963) studied the movement of 2,4-D, 2,4,5-T, and IAA in *Phaseolus vulgaris* using labeled forms of the compounds and employing autoradiography, counting, and chromatography techniques. They used both the cut-flap method of midrib injection and application to the cotyledonary node immediately following excision of a cotyledon as rapid methods for getting the tracers into plants. The uptake time, in both instances, was around 5 minutes. Using the cut-flap method, they applied only 10 μliter per treatment; and by heat-ringing, they proved that the artifact of xylem stream reversal was not involved. Export from the leaf was via the phloem. They found the velocity of downward transport to be 10–12 cm per hour for 2,4-D and 2,4,5-T, and 20–24 cm per hour for IAA. This confirms the finding of Crafts (1959b) that IAA is more mobile than 2,4-D. Because both of these compounds are retained along the route of export by absorption out of the flowing assimilate stream, these relatively low velocities probably reflect relative retention during both penetration and export, more than they do the actual velocities of assimilate flow. This is suggested by Little and Blackman's studies on distribution for times up to 27 hours, in which they

Table 8.7. Influence of the concentration of 2,4-D on its absorption and translocation in young *Chondrilla juncea*. (Formulation included ^{14}C-labeled 2,4-D, 0.1 M citrate, and 0.5 percent Tween 20 at *p*H 4.0.)

Concentration of 2,4-D	Radioactivity in roots [$\log_{10}$ (avg count/root)/10 min]
3.4×10^{-3} M	4.272
6.7×10^{-3} M	4.198
10.0×10^{-3} M	4.090
Least significant differences:	0.154 for $P = 0.005$
	0.209 for $P = 0.01$

Source: Greenham (1962).

found that penetration into the leaf and export from it were greatest when the application was through a midrib flap, less when the application was to the lower leaf surface, and least when the application was made to the upper surface. Accumulation in the main axis was highest in the first internode and upper hypocotyl. After 27 hours, no tracer had appeared in the opposite primary leaf, indicating no transfer from the phloem to the xylem for the three growth regulators being studied.

Penetration, translocation, and metabolism of 2,4-D and 2,4,5-T in two cucurbits—the wild *Sicyos angulatus* and the cultivated *Cucumis sativus*—was studied by Slife et al. (1962). Both herbicides were labeled in the carboxyl position with ^{14}C, and autoradiography, counting, and chromatography were employed to obtain data on the responses being studied. Typical distribution of 2,4-D was achieved in 24 hours, and there was little change in 4 or 8 days. The pattern of 2,4,5-T translocation continued to change during an 8-day period, indicating a higher mobility for this compound. Metabolic studies indicated that considerably more 2,4-D than 2,4,5-T was absorbed by these plants, but that 75 percent of the absorbed 2,4-D was metabolized within 24 hours. The absorbed 2,4,5-T underwent only slight metabolism after 8 days. Evolution of $^{14}CO_2$ was about 10 times greater from 2,4-D–treated plants than from 2,4,5-T–treated plants. The greater phytotoxicity of 2,4,5-T on these plants seems to be related to the inability of the plants to detoxify this compound as rapidly as 2,4-D. In the field, 2,4,5-T is known to control *S. angulatus* at a much lower dosage than is required of 2,4-D. Both compounds are effective against the cultivated *C. sativus*.

Bidirectional movement of tracers in *Glycine max* seedlings was reported by Crafts (1967b). Using autoradiography as a method, and four ^{14}C-labeled herbicides, Crafts showed that movement of 2,4-D is largely restricted to the phloem; that monuron moves in the xylem and cell walls; that amitrole moves in the phloem, xylem, and cell walls; and that maleic hydrazide moves like amitrole, and may, in addition, migrate from phloem to xylem, and hence, circulate like phosphorus in the plant (Figure 8.9).

Applied to one cotyledon of a seedling of *Glycine max*, 2,4-D moves acropetally through the stem to the expanding trifoliate leaf and shoot tip in moderate concentration, and through the hypocotyl to the roots in higher concentration. Applied to the epicotyl, 2,4-D moves both acropetally and basipetally in strong concentration; in both cases, both primary leaves are completely bypassed (Figure 8.9a). Applied

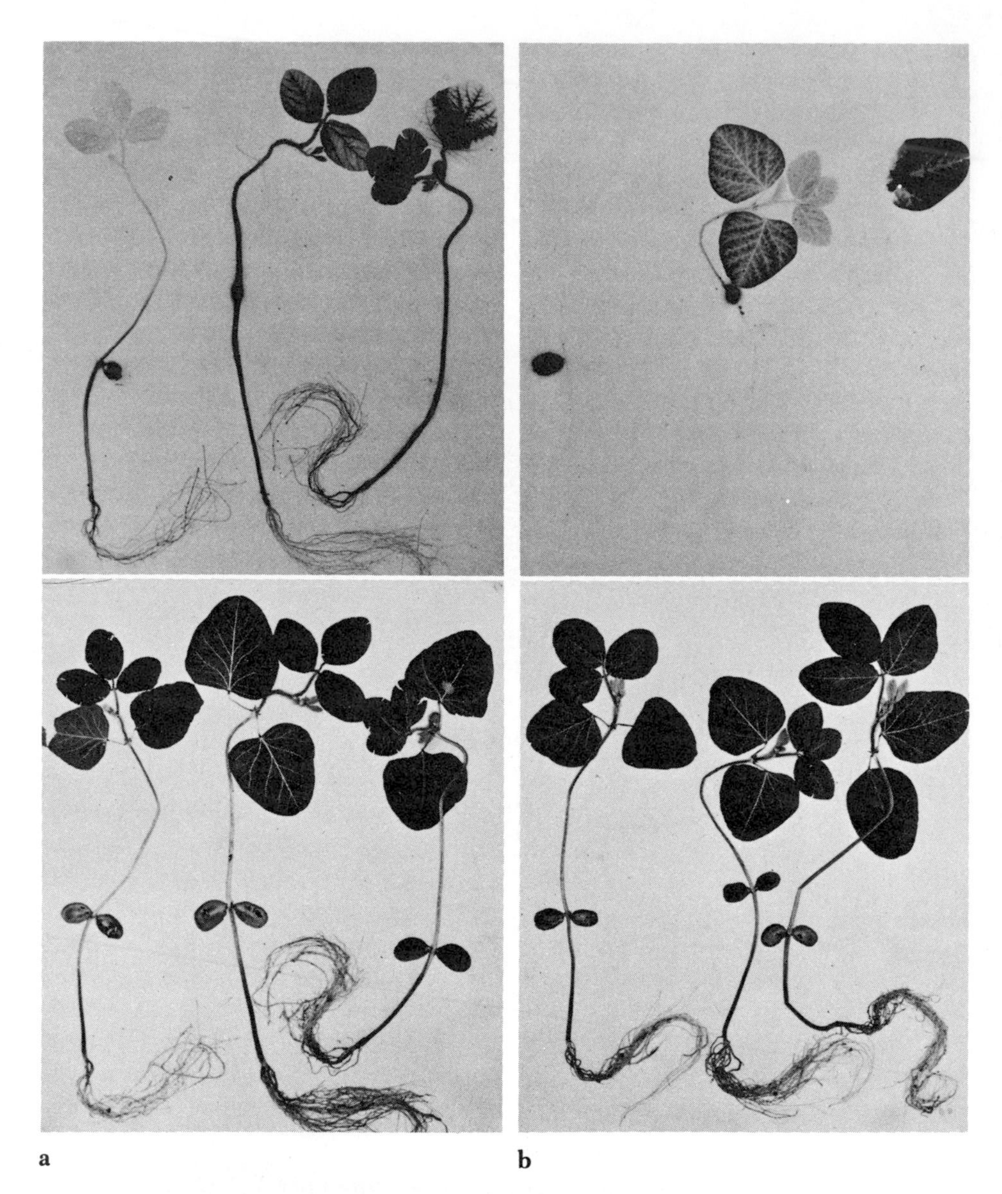

Figure 8.9. Autoradiographs (*above*) and photographs (*below*) of treated plants of *Glycine max* showing distribution of (*a*) 2,4-D, (*b*) monuron, (*c*) amitrole, and (*d*) maleic hydrazide, applied to cotyledon (*left*), epicotyl (*center*), and primary leaf (*right*). Treatment time was 2 days. From Crafts (1967a).

c d

to one primary leaf, 2,4-D moves both to the trifoliate leaf and shoot apex and to the roots; the opposite primary leaf is bypassed. Such movement is completely symplastic.

Monuron applied to a cotyledon moves only acropetally; it fails to move out of the cotyledon. Applied to the epicotyl, it apparently seeks out the xylem and moves acropetally, labeling both primary leaves strongly, and labeling the trifoliate leaf and shoot apex less strongly; there is no basipetal movement below the point of application. Likewise, applied to one primary leaf, it makes the typical apoplastic wedge; none is exported from the leaf (Figure 8.9b).

Applied to a cotyledon, amitrole translocates both acropetally to the trifoliate leaf and the bud and basipetally to the roots; both primary leaves are bypassed. Applied to the epicotyl, amitrole assumes this same distribution pattern, but, in addition, both primary leaves are labeled: the one on the treated side strongly, the opposite one weakly. Applied to a primary leaf, amitrole moves acropetally to form an apoplastic wedge; it also moves acropetally in the stem, labeling an expanding trifoliate leaf with moderate intensity, and a younger trifoliate very strongly; roots are labeled; the opposite primary leaf is bypassed (Figure 8.9c).

When maleic hydrazide is applied to cotyledon, epicotyl, or primary leaf, both cotyledons, both primary leaves, and all trifoliate leaves are labeled. The treated organs all have high-intensity labeling, the trifoliate leaves and shoot apices are strongly labeled, and the untreated cotyledons and primary leaves are weakly labeled. From these results, it seems evident that 2,4-D is distributed via the symplast exclusively, and that monuron is distributed exclusively via the apoplast; amitrole, from cotyledon or primary leaf treatments, shows apoplastic wedges, but distribution to other parts is symplastic. Applied to the epicotyl, principal movement is via the symplast, but some tracer is able to enter and move in the xylem. Maleic hydrazide is distributed principally via the symplast, but a small amount migrates to the xylem and moves in the transpiration stream. Figure 8.9d shows these results.

Yamaguchi (1965), in an analysis of 2,4-D transport, has described a number of factors that determine the distribution of this herbicide in plants. Using plants of *Phaseolus vulgaris, Gossypium hirsutum, Hordeum vulgare,* and *Glycine max,* and high activity ^{14}C-labeled 2,4-D (12.3 mc/mmole), Yamaguchi treated by both leaf and root applications. He found that DNP, anoxia, and starvation reduced the absorption and retention of 2,4-D by roots and leaves. Phloem translocation was likewise reduced, and apoplastic mobility was increased. A *p*H effect on root absorption resulting in a greater uptake at low *p*H values was apparently composed of two processes: (1) a loose physical binding, and (2) a tight metabolic absorption. The 2,4-D held by loose binding could be largely leached out into tap water by two 1-hour soakings; by this method, 10 times as much 2,4-D was released at pH 3 as at pH 11. The 2,4-D metabolically absorbed was also 10 times as great at pH 3 as at pH 11. Over the pH range of from 3 to 11 the metabolic fraction was 2–4 times as great as the loosely bound fraction. The symplastic movement of 2,4-D to root tips of *Gossypium hirsutum* and *Phaseolus vulgaris* plants is well illustrated in figures 8, 9, and 10 of Yamaguchi's paper.

The pattern of absorption, translocation, and ultimate distribution of ^{35}S-labeled EPTC was investigated by Yamaguchi (1961). This compound, an alkyl thiolcarbamate, is a popular herbicide used by the soil-incorporation method. Because it is volatile, and thus subject to absorption by leaves as well as by roots, it is important to understand its behavior in plants. Yamaguchi applied the tracer to leaves in a small closed chamber sealed to the leaf with lanolin; to grass leaves, he applied the compound in 55 percent alcohol solution containing 5 percent glucose and 1.18 percent labeled EPTC. Application to roots was in a solution in water with Tween 20 as cosolvent.

EPTC labeled with ^{35}S proved to be mobile via both phloem and xylem of *Datura stramonium, Convulvulus arvensis, Phaseolus vulgaris,* and *Zea mays*. In *C. arvensis,* there was apparently some migration from phloem to xylem, with attendant light labeling of mature, untreated leaves; this was not true of the other species treated. There was slight apoplastic (acropetal) movement in treated leaves, and strong symplastic movement. From root application, ^{35}S-labeled EPTC moved into roots, ascended in the transpiration stream, and labeled all foliage strongly. In addition to this general distribution, there was a concentration of the label in the developing leaves, buds, and root tips, indicating secondary transport via the phloem.

Nalewaja (1968) investigated the uptake and distribution of ^{14}C-labeled di-allate by roots and coleoptiles of *Avena fatua,* a susceptible plant, and *Triticum aestivum, Hordeum vulgare,* and *Linum usitatissimum,* resistant plants. The pattern of ^{14}C uptake and movement was similar in *A. fatua, H. vulgare,* and *T. aestivum;* thus differential uptake and distribution did not account for the selectivity of this herbicide. Translocation of di-allate was via both symplast and apoplast; exposure of roots of seedlings to the herbicide resulted in its accumulation in the root tips of all four species; application to the middle of the shoot or to the base of the mesocotyl resulted in movement both acropetally and basipetally.

Smith and Davies (1965) studied the absorption and translocation of dalapon, amitrole, and paraquat by *Paspalum distichum*, a serious pest in irrigation ditches. Dalapon was labeled with ^{36}Cl, and amitrole and paraquat with ^{14}C. The plants each had two stolons growing over water, and one stolon of each plant received treatments of one chemical on four different loci. Autoradiographs were used to determine distribution of the chemicals. Both dalapon and amitrole accumulated in the apices of the treated stolons and in some of the young shoots growing as tillers from the base of the plant. The ^{14}C from paraquat showed xylem translocation predominantly, with no apical accumulation and little basipetal translocation. These results show that young basal shoots are sinks for assimilate movement, but that there is little transport between mature stolons. Evidently, a very thorough spray job will be necessary if herbicide treatment of this pest is to be successful.

The translocation of the herbicide dicamba has been studied in *Cirsium arvense* by Chang and Vanden Borne (1968) and in *Cyperus rotundus* by Magalhaes, Ashton and Foy (1968). In *Cirsium arvense,* the results were in harmony with a source-to-sink system of translocation in the phloem; the tracer moved from a mature leaf throughout the symplast, and even out into the culture medium. Accumulation in young leaves followed uptake through the leaves or roots, indicating transfer from xylem to phloem in the leaves. Leaves retained a substantial portion of the absorbed dicamba; after 54 days, 63.1 percent of the recovered radioactivity in a treated leaf was still in the form of unaltered dicamba. The remaining 36.9 percent was in the form of an unidentified metabolite; less change occurred in other plant parts. During a 54 day period, one-fifth of the dicamba applied was recovered as radioactive CO_2. In a dosage series, Chang and Vanden Born found that as dosage increased the transport system became saturated; with higher dosage the system became partially inactivated.

In *Cyperus rotundus,* dicamba was slowly translocated, following application to leaves or roots. Applied to leaves, dicamba moved both acropetally and basipetally and became widely distributed throughout the foliage; it accumulated in meristems. Though excreted into the culture medium by roots, dicamba was barely detectable in underground organs; it also passed through rhizomes and tubers into daughter plants. Root-applied dicamba was distributed throughout the plants; it was low in concentration in tubers but high in leaf tips. Translocation was enhanced under low light, and distribution was most thorough in plants that were actively growing. After flowering, export from a treated leaf decreased. Dicamba remained unchanged in *Cyperus rotundus* plants during the 10-day duration of the experiments; and the label excreted from roots was shown by chromatography to be dicamba. The writers concluded that the initial transport of foliarly applied dicamba is via the phloem, but that the tracer migrated from the phloem to the xylem at the connection of the stem and basal tuber. This probably accounts for the accumulation often found in stem tips, since these import via both the phloem and the xylem. Magalhaes and his co-workers suggested that the lack of dicamba metabolism in *Cyperus rotundus* may account for its high susceptibility to this herbicide.

In 1961, J. F. Pereira, a graduate student in the Department of Botany of the University of California at Davis made a detailed study of tracer uptake and movement in *Coffea arabica*. Results of autoradiographic tests using ^{14}C-labeled 2,4-D, maleic hydrazide, amitrole, monuron, dalapon, and urea, as well as $^{45}CaCl_2$, $^{65}ZnCl_2$, and $NaH_2{}^{32}PO_4$, were reported (Pereira et al., 1963). As a part of this study, Pereira did some work using ^{77}As. This isotope is short-lived and difficult to work with. However, some results not reported by Pereira and his co-workers indicate that $^{77}AsO_4{}^{---}$ moves like $^{32}PO_4{}^{---}$ in plants. More recently, Holt et al. (1967) studied translocation of AMA in *Cyperus rotundus*. By visual symptoms of toxicity and by arsenic analyses, Holt and his co-workers determined that arsenic was translocated from shoots laterally into tubers that were separated by at least four tubers from the treated shoot. The tuber on the far end of the chain from the treated shoot tended to have a higher arsenic content than the ones in between, and the amount of translocated arsenic tended to be higher in those tubers that were actively growing. There was no apparent relationship between arsenic content and the ability of tubers to produce new shoots. Initial regrowth following treatment was often in the form of multiple shoots. Death of tubers following repeated treatment was considered to be due to depletion of food reserves or bud supply, rather than to arsenic accumulation. AMA was absorbed and translocated from *Cyperus rotundus* leaves through the basal bulb into the tuber chain. Since viable tubers often contained more arsenic than some killed tubers, Holt and his co-workers concluded that food depletion associated with resprouting, interruption of oxidative phosphorylation, and exhaustion of bud supply were all involved in the death of tubers. Lower arsenic levels in single tubers that had received several applications may have resulted from translocation of arsenic to newly developing tubers that were not analyzed. Chlorosis on shoots from untreated tubers also indicated the transport of arsenic into food sinks. These results would fit well the concept of methylarsonate moving like phosphate from sources to food sinks.

Further studies carried out by Duble, Holt and McBee (1968) using ^{14}C-labeled DSMA confirmed the mobility of the organic arsenical in *Cyperus rotundus*. DSMA apparently moves in both apoplast and symplast; movement from old leaves was primarily basipetal; from younger leaves, movement was acropetal into shoots, a pattern indicating phloem transport. Accumulation of DSMA and AMA took place in active growth sites where rapid utilization of assimilates occurs. There was no evidence for redistribution of these compounds.

Translocation and metabolism of dicamba in *Sorghum halepensis* and *Phaseolus vulgaris* were studied by Hull and Weisenberg (1967). Dicamba has been shown to be highly toxic to the rhizomes of *S. halepensis*. To relate this toxicity to the general responses to dicamba, a comparative study using *S. halepensis* and *P. vulgaris* was made in which translocation and metabolism were measured. Rapid symplastic transport of ^{14}C-labeled dicamba was observed when the herbicide was applied to a mature leaf of young *S. halepensis*, or to one primary leaf of *P. vulgaris*. Evidently, there was migration of the tracer from phloem to xylem, because all of the leaves above the site of application had light uniform labeling. Basipetal transport to the roots of either species, or to a rhizome serving as a propagule for *S. halepensis*, was restricted. Some accumulation took place in shoot meristems, particularly in *P. vulgaris*. Movement from roots to shoots was sluggish; the roots held little of the label. The total recoverable ^{14}C did not vary appreciably during 5 days; apparently there was little decarboxylation of the herbicide. Although over 50 percent of the ^{14}C remained as the unaltered herbicide after 5 days, two metabolites of dicamba were found in *S. halepensis*. On acid hydrolysis, dicamba was released from one metabolite; in *P. vulgaris*, only dicamba was recovered 5 days after treatment.

Maestri (1967), in studies on the physiological effects on plants of the herbicide endothall, used the ^{14}C-labeled molecule to detect translocation in *Phaseolus vulgaris* and *Cucumis sativus*. He found this compound practically immobile when applied to leaves or roots of *P. vulgaris*, and only somewhat mobile in similar tests with *C. sativus*. Pretreatment with endothall at 10 mg per liter stunted *P. vulgaris* plants and greatly reduced translocation of labeled amitrole applied to a leaf; it also inhibited absorption and translocation of labeled monuron applied to the roots. When applied through the cut stems of endothall-treated plants, monuron moved normally; evidently, endothall applied to the roots causes injury that prevents the normal uptake and transport of monuron. Further experiments showed that a spray treatment of the leaves of *P. vulgaris* with endothall prevents vein loading and translocation of ^{14}C-labeled assimilate. Evidently, endothall brings about rapid local contact injury to plant tissues and prevents the normal functioning of sources and sinks.

Thomas and Seaman (1968) found that ^{14}C-labeled endothall is readily absorbed by the foliage of *Potamogeton nodosus* and translocated to the shoot tips, developing secondary plants, and roots. Application to mature leaves, stem, and winter buds all gave the same patterns of symplastic movement, with mature untreated leaves being bypassed. Application of labeled endothall, atrazine, and diuron to roots resulted in no movement into the tops of these submerged aquatic plants. Thomas and Seaman concluded that there was little or no movement of the transpiration stream in the xylem. The fact that labeled endothall failed to move from roots of *P. nodosus* they attributed to its toxicity to root tissues. They proposed that it not only injures roots, but that it is strongly adsorbed so that it does not enter the xylem. This may account for its strong herbicidal action on weeds germinating in the soil.

Leonard and Weaver (1961) used ^{14}C-labeled 2,4-D and ^{14}C-labeled amitrole in studies on absorption and translocation by leaves of *Vitis vinifera* 'Tokay.' This variety is highly sensitive to 2,4-D, and much of the early work on 2,4-D drift centered on the injury to vines growing in the Lodi area in California. This injury resulted from airplane application of 2,4-D to grain fields west of Lodi in the delta of the Sacramento River. Leonard and Weaver found that both 2,4-D and amitrole were readily absorbed by the leaves of *V. vinifera* 'Tokay': amitrole absorption took place mainly in three days, whereas 2,4-D absorption continued for three weeks or more. During translocation both 2,4-D and amitrole showed a source-to-sink pattern of distribution; sinks included cambium, growing axillary and apical buds, tendrils, clusters, and immature leaves. The ^{14}C from treatments with labeled 2,4-D moved into flower clusters, accumulating in anthers, pedicels, and young berries. Labeled amitrole concentrated in small berries.

When applications were made to leaves after the berries were about half grown, 2,4-D concentrated in and near the skin of the berries; amitole was more generally distributed, although very little was found in the seeds. The 2,4-D in fruits was recoverable 38 days after treatment; the quantity present was less than one-third that from comparable treatments made during flowering. Amitrole in its original form was not recoverable, except when the treated leaf was covered with a polyethylene bag—a treatment shown by Clor et al. (1962) to greatly enhance uptake.

Translocation was only acropetal during the prebloom stage. Upward translocation occurred from fully expanded leaves below a cluster and from less mature leaves above a cluster. In one of Leonard and Weaver's experiments, translocation became partially reversed, continuing toward the apex, but also occurring basipetally into the clusters and the stem below them. Leaves one-half or less the width of mature leaves imported the tracers; the blades were more uniformly labeled with amitrole than with 2,4-D. Labeling with amitrole reached a maximum 3 days after treatment; with 2,4-D, the intensity of the label increased up to 23 days after treatment. These results relate back to the translocation times described above. As leaves enlarged and began to export, the entire vein appeared to remain labeled when amitrole was the tracer, but only the outer portions of the veins (cortex) appeared to remain labeled when 2,4-D was used. Labeling from both tracers became greatly faded by 112 days after treatment.

The original free form of 2,4-D was present in the shoots of *Vitis vinifera* 'Tokay' for at least three weeks; one-third the total radioactivity of clusters was recoverable as 2,4-D 112 days after treatment. The label from amitrole was present in a narrow ring of xylem about two-thirds in from the cambium; this confirms the experience of Yamaguchi and Crafts (1959). Amitrole appears to remain for only a short time in the shoots or fruit before it is complexed or broken down. Amitrole, as such, was not recoverable from clusters 3 days or more after treatment.

Pallas (1963) studied the absorption and translocation of 2,4-D and 2,4,5-T in *Acer rubrum, Liriodendron tulipifera, Liquidambar styraciflua,* and *Fraxinus americana.* Both unlabeled and ^{14}C-labeled compounds were used as sprays on the plants, which were being grown in cans. *A. rubrum* is known to be resistant to the phenoxy herbicides; *Liriodendron tulipifera* is susceptible to 2,4,5-T. Results proved that 2,4,5-T was not absorbed or translocated as effectively as 2,4-D. Autoradiograms showed a good distribution of ^{14}C from both 2,4-D and 2,4,5-T in *Liquidambar styraciflua, Liriodendron tulipifera,* and *A. rubrum,* and poor movement of 2,4,5-T in *F. americana,* probably due to poor absorption. Absorption and translocation continued throughout the 16-day experimental period. The resistance of *A. rubrum, Liriodendron tulipifera,* and *Liquidambar styraciflua* to 2,4-D, as shown in spray tests with cold 2,4-D, was not a result of poor absorption or translocation. In *A. rubrum,* a major fraction of ^{14}C activity was in an unknown compound having an R_f of 0.07. Chromatographic analysis proved that the major ^{14}C activity in the plants treated with 2,4,5-T was in the original unmetabolized compound; this inability of plants to inactivate 2,4,5-T may account for its superiority over 2,4-D as a herbicide against woody species.

Leonard, Lider, and Glenn (1966) used a number of labeled herbicides in studies on absorption and translocation by plants of *Vitis vinifera* 'Thompson Seedless.' The compounds used were 2,4-D,

2,4,5-T, dicamba, amitrole, paraquat, monuron, diuron, atrazine, and simazine, all labeled with ^{14}C. Neither 2,4-D nor paraquat, and only traces of 2,4,5-T, translocated from roots to shoots. Dicamba, amitrole, diuron, monuron, simazine, and atrazine moved readily from roots to shoots. The substituted ureas and *s*-triazines were mobile only in the xylem or cell walls (apoplast), although dicamba and amitrole were mobile in apoplast and symplast. Both 2,4-D and 2,4,5-T moved almost entirely in the symplast. Paraquat moved slightly, if at all, in either apoplast or symplast (compare Table 8.1).

The stability of the herbicides in plants varied, as judged by the detection of label in ethanol-insoluble products. Dicamba was considerably more stable than 2,4,5-T, which, in turn, was more stable than 2,4-D. Diuron was more stable than either atrazine or simazine. Amitrole was largely converted into ethanol-insoluble products in 30 days. Paraquat was not extractable with ethanol. These results in *Vitis vinifera* confirm many of the observations of Crafts and Yamaguchi (1964) and Crafts (1967a) on other plants.

Translocation studies on *Acer rubrum* and *Fraxinus americana* growing in 1-gallon containers, by Leonard, Bayer, and Glenn (1966), have been carried on using ^{14}C-labeled amitrole and ^{14}C-labeled 2,4,5-T as tracers. Applications were made to leaves and stems; and some trees, having been sprayed with these two herbicides, were fed $^{14}CO_2$ to find the effects of herbicide treatment on normal assimilate movement. Leonard and his co-workers found that amitrole applied to either leaves or stems was absorbed and transported throughout the plants. By contrast, 2,4,5-T was not exported from leaves and did not enter and move in the phloem after application to stems. When $^{14}CO_2$ was fed to leaves, both amitrole and 2,4,5-T caused significant transport of ^{14}C-assimilates into the veins of fully expanded leaves from the leaf treated with $^{14}CO_2$. Translocation to roots was reduced if the leaves treated with $^{14}CO_2$ had been sprayed previously with either herbicide. When the treated leaf had not been sprayed, transport of labeled assimilates to roots was only moderately affected by amitrole or 2,4,5-T applied to the remainder of the shoot. *A. rubrum* and *F. americana* are two species that are resistant to treatment with 2,4,5-T in the field, and the results of Leonard and his co-workers indicate why. *A. rubrum* and *A. circinatum* have responded most strikingly to spray treatments with the heavy ester of 2,4,5-T at the time of budbreak in the spring. At this time, there are no expanded leaves to act as traps for the herbicide; response upon the part of the plants involves very limited translocation, and fairly thorough coverage is necessary to achieve successful control.

Morton (1966) determined the distribution of 2,4,5-T in *Prosopis juliflora* seedlings using the ^{14}C-labeled tracer. Translocation was primarily basipetal from the point of application when the temperature was 70°F. At 85°F, translocation was both acropetal and basipetal; and at 100°F the tracer moved only into the upper stem. Table 8.8 presents the results. Analyses of leaves, stems and roots of the seedlings at greenhouse temperature (72–93°F), at 70°F, and at 100°F proved that soluble sugars were highest in all fractions in the plants at the 70°F treatment temperature; sugars were highest in roots at all temperatures. High concentrations of sugars apparently reflected high transport conditions.

Norris and Freed published a series of papers on the absorption, translocation and metabolism of a number of chlorophenoxy herbicides by *Acer macrophyllum,* a major pest of managed softwood forests in the Pacific Northwest (1966a, 1966b, 1966c). The variables considered in their studies were formulation, degree of chlorination, and structure of the side chain. Using 5-year-old seedlings, they applied the herbicide solutions to leaves and, after a 72-hour treatment period, washed the treated leaves with alcohol to determine the unabsorbed fraction. They then sampled the treated leaves, new growth, older stems, and roots. Table 8.9 shows the results of counting these samples. It is evident why ester formulations of these materials have been most successful in controlling woody plants: they are absorbed to a much greater extent than other formulations. The ^{14}C from these

Table 8.8. Relation of 72-hour exposures to various temperatures to the distribution of 2,4,5-T* in *Prosopis juliflora* seedlings.

Tissue fraction	Tissue radioactivity [(count/min)/0.1 g fresh wt]		
	At 70°F	At 85°F	At 100°F
Growing tip	0	198	0
Upper leaves	0	22	0
Upper stems	0	76	134
Lower leaves	0	0	0
Lower stems	116	46	0
Roots	65	48	0
Amount absorbed (μg)†	1.68	1.70	2.48

Source: Morton (1966).
*Carboxyl-labeled with ^{14}C.
†5 μg tracer supplied.

materials also reaches the roots in appreciable quantities. The first of Norris and Freed's experiments tested three formulations of 2,4-D and 2,4,5-T; their second experiment determined absorption and translocation of ester formulations of 2,4-D, 2,4,5-T, and their propionic acid analogues (Table 8.10). Absorption of all was high, but the 2,4,5-T and 2,4-DP accumulated to a higher degree in roots. Because of the small quantities reaching the roots, Norris and Freed consider translocation to be a critical factor in the control of *A. macrophyllum.*

Because the above techniques measured only ^{14}C, Norris and Freed determined the unmetabolized herbicides in *Acer macrophyllum* leaves 72 hours after treatment (1966b). Table 8.11 shows the results for the acetic acid and propionic acid analogues tested. The propionic acid compounds proved to be appreciably lower in stability. Extending their studies, they treated the leaves of seedlings—then they sampled treated leaves, new growth, stems, and roots, and determined, by chromatography, the unmetabolized molecules. Table 8.12 presents these data. The greatest differences appear in roots, 2,4,5-T being much the stabler compound in these. This gives a clue to the superior performance of 2,4,5-T in the field; evidently, the combined effects of translocation and stability make for appreciably more of the active chemical reaching the roots.

In their third paper, Norris and Freed (1966c) report on the absorption, translocation, and metabolism of the butyric acid analogue of 2,4-D. Compared with the other molecules reported in Table 8.14, 2,4-DB was only 12.3 percent absorbed; however, with only 77.5 percent retained in the treated leaves, new growth had 4.5 percent, stem 11.8 percent, roots 6.2 percent, and the herbicide in roots totaled 14.7 μg, over twice the amount of any of the other compounds. Because the effectiveness of 2,4-DB results from its oxidation to 2,4-D, Norris and Freed treated plants with 2,4-DB, allowed a 3-day reaction time; then they extracted treated leaves, stems, and roots, and determined, by gas chromatography, the amount converted to 2,4-D. Table 8.13 presents these data. It is evident that roots of *Acer macrophyllum* are able to convert a large portion of 2,4-DB to 2,4-D; the leaves and stems convert much less. Norris and Freed concluded that the low conversion by foliage makes for more efficient transport to roots, because 2,4-D is known to inhibit phloem transport. These results of basic physiological and biochemical research on herbicide performance are essential to an adequate understanding of the use of these chemicals in the field. In the control of woody plants and perennial herbaceous weeds it is the active herbicide in the roots that determines the results.

Eliasson (1965) studied the interference of the transpiration stream

Table 8.9. Distribution of ^{14}C-labeled acid, amine, and ester formulations of 2,4-D and 2,4,5-T in *Acer macrophyllum* 72 hours after treatment.

Treatment	Number of plants	Absorption (%)	Translocation (% of absorbed activity)				Herbicide in roots (μg a.e.)
			Treated leaves	New Growth	Stem	Roots	
2,4-D	2	2.9	77.0	10.0	8.1	4.9	2.8
2,4-D amine*	3	1.7	68.8	17.4	9.7	4.0	1.4
2,4-D ester†	4	20.8	95.4	2.6	1.4	0.5	2.3
2,4,5-T	2	2.3	71.3	21.4	6.5	0.7	0.3
2,4,5-T amine*	3	0.7	62.0	17.6	7.5	12.9	2.0
2,4,5-T ester†	3	16.5	95.9	2.1	0.6	1.3	4.3

Source: Norris and Freed (1966a).

* Triethanolamine salt.

† 2-Ethylhexyl ester.

Table 8.10. Distribution of ^{14}C-labeled esters of 2,4-D, 2,4,5-T, and their propionic acid analogues in *Acer macrophyllum* 72 hours after treatment.

Treatment	Number of plants	Absorption (%)	Translocation (% of absorbed activity)				Herbicide in roots (μg a.e.)
			Treated leaves	New Growth	Stem	Roots	
2,4-D ester*	4	20.8	95.4	2.6	1.4	0.5	2.3
2,4,5-T ester*	3	16.5	95.9	2.1	0.6	1.3	4.3
2-4-DP ester*	2	27.4	94.9	2.8	1.1	1.2	6.2
Silvex ester*	4	22.1	96.1	3.0	0.6	0.3	1.1

Source: Norris and Freed (1966a).

* 2-Ethylhexyl ester.

Table 8.11. Recovery of unmetabolized ^{14}C-labeled herbicides from the foliage of *Acer macrophyllum* 72 hours after treatment.

Treatment	Number of leaves	Recovered herbicide (% of total recovered activity)
2,4-D	3	95.0
2,4,5-T	4	93.3
2,4-DP	4	87.2
Silvex	4	84.1

Source: Norris and Freed (1966b).

Table 8.12. Distribution of unmetabolized ^{14}C-labeled herbicides recovered from the foliage of *Acer macrophyllum* seedlings 7 days after treatment.

Treatment	Number of plants	Recovered herbicide (% of total recovered activity)			
		Treated leaves	New growth	Stem	Roots
2,4-D	2	94.1	82.5	62.2	44.7
2,4,5-T	2	93.7	96.3	67.1	85.4

Source: Norris and Freed (1966b).

Table 8.13. The distribution of 2,4-DB and its oxidation product 2,4-D in *Acer macrophyllum* seedlings 72 hours after foliar treatment with the 2-ethylhexyl ester of 2,4-DB.

Plant part	Herbicide recovered (μg)		Conversion to 2,4-D (%)
	2,4-DB	2,4-D	
Treated leaves	992	178	16.8
Stems	128	6	5.0
Roots	63	136	70.9

Source: Norris and Freed (1966c).

Table 8.14. Distribution of ^{14}C-labeled esters of five chlorophenoxy herbicides in *Acer macrophyllum* seedlings 72 hours after treatment.

Treatment	Number of plants	Absorption (%)	Translocation (% of absorbed activity)				Herbicide in roots (μg a.e.)
			Treated leaves	New growth	Stem	Roots	
2,4-D ester*	4	20.8	95.4	2.6	1.4	0.6	2.3
2,4,5-T ester*	3	16.5	95.9	2.1	0.6	1.3	4.2
2,4-DP ester*	2	27.4	94.9	2.8	1.1	1.2	6.2
Silvex ester*	4	22.1	96.1	3.0	0.6	0.3	1.1
2,4-DB ester*	3	12.3	77.5	4.5	11.8	6.2	14.7

Source: Norris and Freed (1966c).

* 2-Ethylhexyl ester.

with the basipetal translocation of foliarly applied 2,4-D and 2,4,5-T in *Populus tremulus*. Treatment was to the lower leaves of one shoot of forked plants. The dosage was 0.5 percent 2,4-D or 2,4,5-T in a lanolin paste containing 50 percent water; on each plant, approximately 100 mg of paste was spread over the centers of the upper leaf surface. Steam rings were used to limit translocation to the xylem. Increases in the length of shoots, changes in epinastic curvature, and physiological concentrations of regulator (as determined by bioassay) were used to measure results. Under his conditions, Eliasson found translocation taking place most rapidly into the tops of treated branches. Steam ringing did not influence this acropetal distribution.

When the treated branch was excised above the treated leaves, a more rapid and larger transport into the untreated branch was found. Ringing tests proved that the basipetal translocation took place via the phloem, and that acropetal translocation within the untreated shoot occurred in the xylem. Only minute quantities of 2,4-D could be extracted from roots 3 days after treatment. Eliasson concluded that, although basipetal translocation of these tracers takes place in phloem, under the conditions of his experiment the 2,4-D and 2,4,5-T were easily transferred from phloem to xylem, where they move in the transpiration stream. This may be an important factor limiting the movement of these herbicides into roots.

Eliasson discussed his results in comparison with those of Crafts (1961b). The difference lies in the much higher dosage used by Eliasson. In large doses, the herbicides cause injury that renders the tissues permeable and releases the toxicants from the phloem into the xylem. Crafts and Yamaguchi (1964) cited an example of this, in which the concentration of 2,4-D around the roots of *Phaseolus vulgaris* plants was greatly increased; under those conditions, the toxicant was released from the ground parenchyma into the xylem and carried to the tops in lethal amounts.

Badiei et al. (1966) found that translocation of 2,4,5-T in *Quercus marilandica* in Oklahoma was influenced by the time of treatment, soil moisture stress, and by the sugar status of the plants. Translocation was high in June, and it declined with passing time and maturation of foliage. The free sugar level in roots was low in May and early June rising to a maximum in August and declining in September. Figure 8.10 shows these trends. Apparently, the depleted root sinks were responsible for active transport in early summer. Soil moisture, too, had a profound effect on translocation. In Table 8.15, the absorption and translocation of ^{14}C-labeled 2,4,5-T are reported for three conditions of soil moisture. As found by Pallas (1960) and Wardlaw

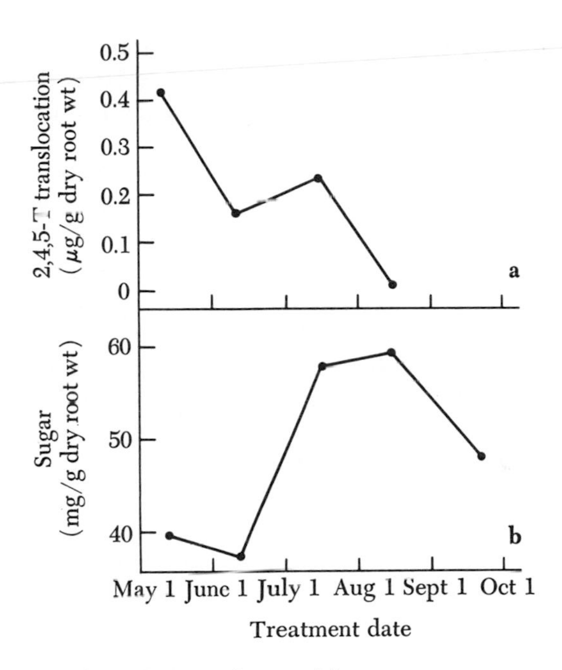

Figure 8.10. Relation of the concentration of ^{14}C-labeled 2,4,5-T in the roots of *Quercus marilandica* to the time of application (*a*); and the free-sugar content of *Q. maritandica* roots at the time of 2,4,5-T treatment (*b*). From Badiei et al. (1966).

Table 8.15. Effect of soil moisture level on absorption and translocation of ^{14}C-labeled 2,4,5-T by *Quercus marilandica* seedlings. (Values are averages of eight determinations.)

Soil moisture (%)	Amount absorbed (as % of amount supplied)	Amount translocated (as % of amount absorbed)	
		From treated area	From treated leaf
2.8	50.1	16.0	13.2*
7.5	54.1	22.4	
16.0	57.2	37.1	33.2

Source: Badiei et al. (1966).

*Significantly different at the 5% level of probability.

(1967), moisture stress inhibited translocation; in Badiei's experiments, translocation was inhibited to about 50 percent of normal. Dark treatment inhibited 2,4,5-T transport, and sucrose treatment enhanced export from treated leaves.

Picloram is an important new herbicide that is absorbed and translocated following either spray application to foliage or treatment through the soil. Bovey et al. (1967) made studies on the use of picloram against the woody plant *Acacia farnesiana* in Texas. They employed 8- to 10-month-old plants growing in 10-inch plastic pots; treatments included foliage application, soil application, and foliage and soil applications together. Picloram effects were evaluated visually, and concentrations in tissues were measured by gas chromatography. Soil application proved to be more effective than foliar spray, and defoliation reduced the effectiveness of spray treatments; a period of 24 hours was required to obtain maximum effects from foliar application. Concentrations of picloram in roots were similar from soil and foliage treatments. One great advantage of picloram over the chlorophenoxy herbicides is the fact that entry via roots is so effective. Thus, in regions of summer rainfall, spray application may be followed by washing of the herbicide into the soil, where its uptake will be continued by the roots; such enhanced uptake may bring about the death of plants by dosages that would not kill by foliar uptake alone.

Autoradiography and steam girdling were used by Wells, Hurtt and Reid (1969) to study modes of distribution of ^{14}C-labeled picloram in *Phaseolus vulgaris* 'Black Valentine.' Steam girdling 2 cm above the primary leaf node did not prevent movement of picloram into the trifoliate leaves. Girdling 2 cm below the primary leaf node likewise did not prevent it; when girdles were placed both 2 cm above and 2 cm below the primary leaf node, transport was greatly reduced. These results emphasize the point, noted in Table 8.1, that picloram is mobile in both the phloem and the xylem. These studies indicate further that this herbicide may migrate from phloem to xylem: in this respect, it resembles maleic hydrazide.

The extreme mobility of picloram probably explains, at least partly, its great effectiveness against perennial plants; not only does it move readily from foliage to roots, it may migrate on out into the rhizosphere surrounding the roots. In one situation, however, this proved detrimental: When used to control certain woody species in forests, it injured not only the target species but the crop trees as well.

Davis et al. (1968) recorded the effects of paraquat and 2,4,5-T on the absorption and distribution of picloram in *Prosopis juliflora,*

Acacia farnesiana, Ilex vomitoria, and *Phaseolus vulgaris.* Paraquat reduced transport of picloram in *Prosopis juliflora, A. farnesiana,* and *Phaseolus vulgaris.* Paraquat enhanced the uptake of picloram by *I. vomitoria,* but did not affect transport. The uptake and transport of picloram by *Prosopis juliflora* increased in the presence of 2,4,5-T; the uptake and transport of 2,4,5-T decreased in the presence of picloram.

Increasing ratios of 2,4,5-T to picloram up to 16:1 continued to increase the uptake and transport of picloram; the inverse was true when ratios of picloram to 2,4,5-T were increased. These herbicide interactions are complex and difficult to interpret. The most apparent effects are possibly related to the effects of picloram and 2,4,5-T as growth regulators and toxicants and the effect of paraquat as a contact toxicant, the actions being upon sources and sinks rather than upon the translocation process.

Lund-Høie and Bayer (1968) studied absorption, translocation, and metabolism of amitrole in *Pinus ponderosa* and *Abies concolor* using 3-year-old seedlings. Both ^{14}C-labeled and unlabeled amitrole were used, and the studies involved autoradiography and chromatography. Amitrole proved to be the toxic compound in both species, and it was mobile via symplast and apoplast; some circulation of the compound was also observed. Amitrole migrated slowly from phloem to xylem, and during this lateral movement, part of the amitrole was metabolized. The detoxification rate seemed to be dependent upon the rate of migration, and it was positively correlated with the intensity of cambial activity. Translocation to roots was greatest when the growth of shoots was at a standstill. Nelson (1964) found a similar result for assimilate movement in *Pinus strobus,* indicating that movement of amitrole is associated with transport of assimilates. The presence of ^{14}C-labeled amitrole in old needles, following application to needles and to exposed phloem, indicates redistribution via the xylem.

Herbicides are being used in increasing quantities to defoliate forest trees in the tropics; some of this is to allow penetration of light to the forest floor, and some is to reduce obscuration in military operations. Herbicides are also used to kill forest trees preparatory to clearing for agricultural use of the land. The chlorophenoxy compounds have been outstanding in these applications, and 2,4,5-T has been the most prominent.

In order to obtain more knowledge of the performance of 2,4,5-T in its use to kill tropical trees, Sundaram (1965) treated trees of susceptible and resistant species by painting the trunks with a labeled solution of a commercial 2,4,5-T formulation in diesel oil. Treatments

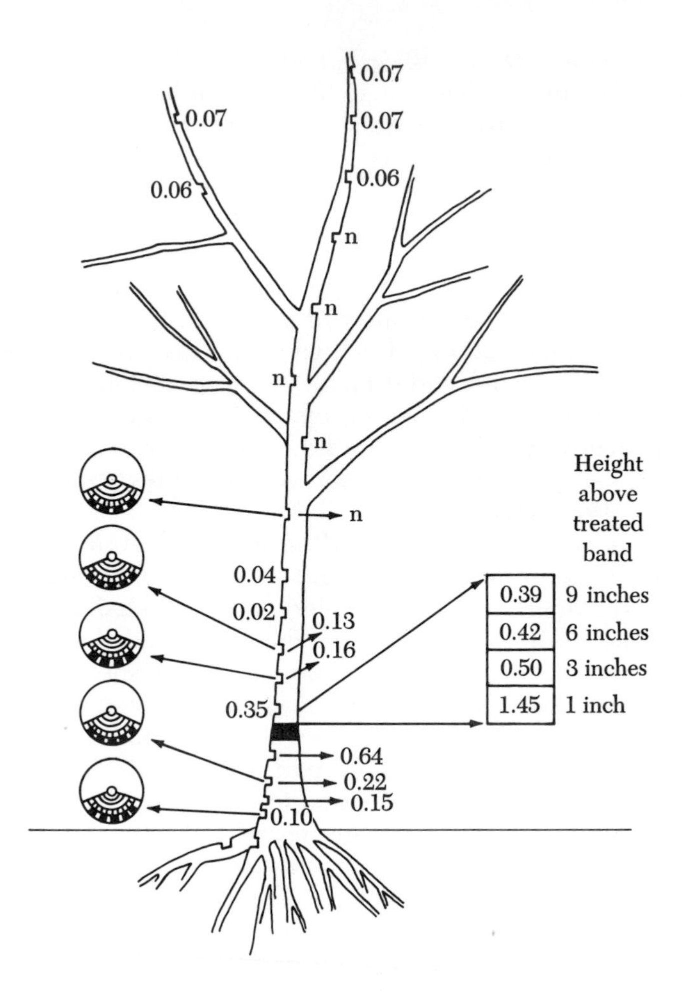

Figure 8.11. Diagram showing the distribution of ^{14}C-labeled 2,4,5-T in four susceptible trees (*Celtis milbraedi*). Cross sections of trunk record the radial distribution of radioactivity in trees treated in the wet season. The outer rings indicate bark; the others indicate 1-cm increments. *Broad radial bars,* high activity; *thin bars,* negligible activity. From Sundaram (1965).

were made during the dry season and repeated in the wet season. The trees were felled and sampled after the susceptible trees had died. Figure 8.11 shows the distribution of radioactivity in four susceptible trees, and Figure 8.12 shows the same for five resistant trees. The pattern of distribution shows little upward movement but marked downward movement in the susceptible trees. By contrast, the toler-

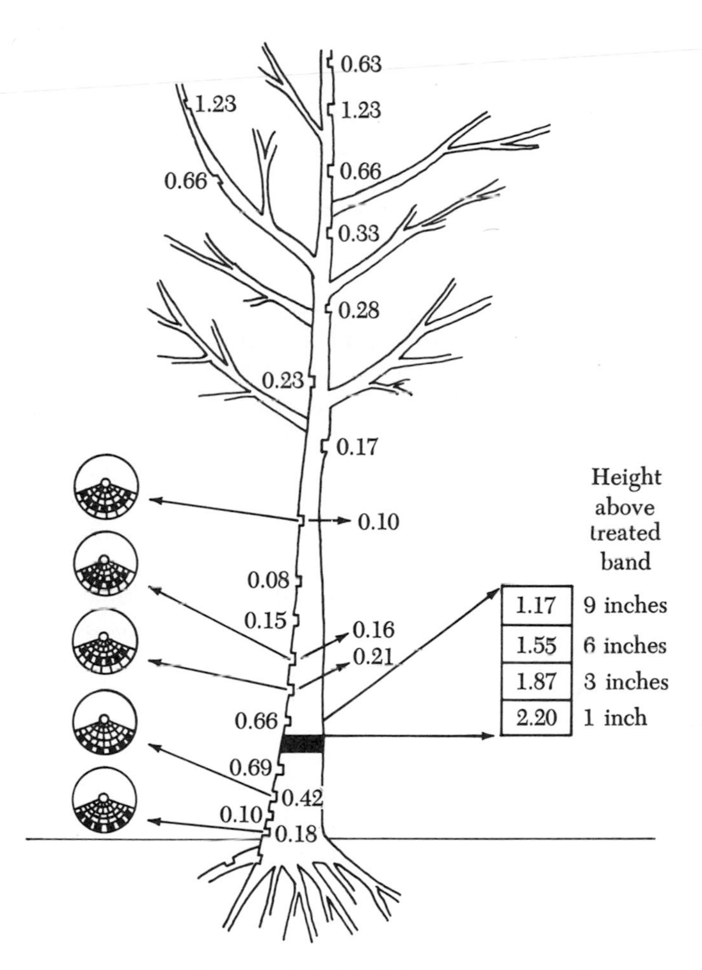

Figure 8.12. Distribution of ^{14}C-labeled 2,4,5-T in five tolerant trees (*Xylopia quintasii* and *Ricinodendron heudelotti*). Values are means of species and seasons. Cross sections of trunk record the radial distribution of activity in trees treated in the wet season. Symbols are the same as in Figure 8.11. From Sundaram (1965).

ant trees moved much more herbicide acropetally and about the same amount downward. Mobility in the phloem was less than in the xylem, and the killing of the sensitive species was due to retention of the chemical in the phloem, resulting in effective girdling of the trees. Sundaram concluded that failure of 2,4,5-T to kill the tolerant trees was due to the extensive upward movement in the apoplastic xylem tissues via the transpiration stream. Radial movement into the heartwood and pith was also greater in the resistant species.

Redistribution Patterns

It has long been known that some mineral nutrients redistribute in plants, moving from maturing leaves to young growing leaves and to apical meristems. For example, a young plant of *Lycopersicon esculentum,* fed liberally on phosphate, may subsequently grow to maturity in a phosphorus deficient soil. Calcium, on the other hand, is known to become fixed following its initial movement into leaves, and unless this element is constantly available in the culture medium, young leaves become deficient and the plants fail.

Kursanov reported (1956a) that when $^{14}CO_2$ was supplied to leaves, labeled sugars synthesized in the leaves moved down into roots penetrating their thinnest branches. Figure 6.2 (page 111) shows this phenomenon in *Zebrina pendula.* Figure 6.3 (pages 112, 113) shows that even root hairs absorb the labeled sugar. Kursanov reported that the descending assimilate stream in his experiments attained a velocity of 70 cm per hour.

In the roots, the sugars undergo glycolytic transformation resulting in the formation of pyruvic acid; this acid, by means of an enzyme, incorporates CO_2 derived from the soil to form oxalacetic acid. This, in turn, is reduced and transformed into malic acid, part of which is then converted into citric, ketoglutaric, and other acids.

Such organic acids formed in roots move up in the transpiration stream at velocities up to 180 cm per hour and label mature leaves that fail to import the original sugars via the phloem. Such labeling is evident in Figure 6.2 in contrast to the plant shown in the frontispiece, which was labeled with ^{14}C-labeled amitrole. There is no enzyme in roots capable of converting amitrole into a xylem-mobile compound.

The carbon captured from the soil is thus moved into the leaves along with the ^{14}C that was originally fed to the plants. The acids from the roots, upon reaching the leaves, are reduced in the process of photosynthesis to form carbohydrates, proteins, and so forth. Some of the sugars are, in turn, translocated back into the roots, and the cycle is repeated. Thus, according to Kursanov, carbon may circulate

---

Figure 8.13. Distribution patterns of ^{32}P applied to (*a*) leaves and (*b*) roots of *Phaseolus vulgaris* at dosages of (*left to right*) 0.2, 0.4, and 0.8 μc per treatment. Leaf treatment results in symplastic movement; root treatment leads to apoplastic distribution via the transpiration stream, which may be followed by migration to phloem in the leaf and symplastic translocation to the shoot tip. Translocation time was 4 days.

a
b

in plants, and such redistribution conserves carbon once it enters the plant from either the atmosphere or the soil.

Biddulph and Markle (1944) established the fact that ^{32}P may circulate in *Gossypium hirsutum*. They obtained evidence for ready transfer of phosphorus from phloem to xylem during its migration in the assimilate stream; they found that up to 40 percent of the mobile phosphate moved in the acropetal direction.

O. Biddulph et al. (1958) determined the distribution patterns of ^{32}P, ^{35}S, and ^{45}Ca in *Phaseolus vulgaris* when the tracers were applied to the roots for one hour and samples were autoradiographed at periods of 0, 6, 12, 24, 48, and 96 hours after the removal of the tracers. Calcium, being relatively immobile in the phloem, was used as a basis of comparison for the distribution of sulfur and phosphorus. The autographs indicated that ^{32}P was very mobile, and that it continued to move from roots to foliage throughout the 96-hour period. In addition, it built to a maximum in the primary leaves in 6 hours; thereafter, it decreased in concentration as it was exported via the phloem to younger importing leaves. Even the first and second trifoliate leaves retained less ^{32}P than they originally acquired. Since it is well authenticated that ^{32}P moves from leaves to roots via the phloem (O. Biddulph, 1941), it is obvious that this essential element may circulate in plants.

In the primary leaves of *Phaseolus vulgaris*, ^{35}S accumulated to a high concentration during the 1-hour absorption period; concentration was also high in the young trifoliates. The concentration decreased during the first 12 hours to a mere trace, which finally disappeared by 96 hours. Meanwhile, the young trifoliates absorbed the tracer for up to 48 hours; in 96 hours, the ^{35}S was diluted by growth, and there was only moderate redistribution into the newly developing trifoliates. The dosage of ^{35}S was 10 μc/liter/plant; that of ^{32}P and ^{45}Ca was 88 μc/liter/plant. Evidently, the supply of ^{35}S became limiting with time, and Biddulph and his co-workers suggest that it is captured metabolically and, presumably, incorporated into leaf proteins.

The isotope ^{45}Ca moved from roots to leaves more slowly than ^{32}P and ^{33}S; by 6 hours, it was high in concentration in all foliage. The

---

Figure 8.14. Distribution patterns of ^{14}C-labeled amitrole in *Phaseolus vulgaris* during a 4-day translocation period. Leaf application (*a*) results in both apoplastic distribution (apoplastic wedge) and symplastic movement to sinks in roots and shoot tips. Root application (*b*) shows typical apoplastic distribution, followed by symplastic redistribution. Dosages were (*left to right*) 0.05, 0.2, and 0.8 μc.

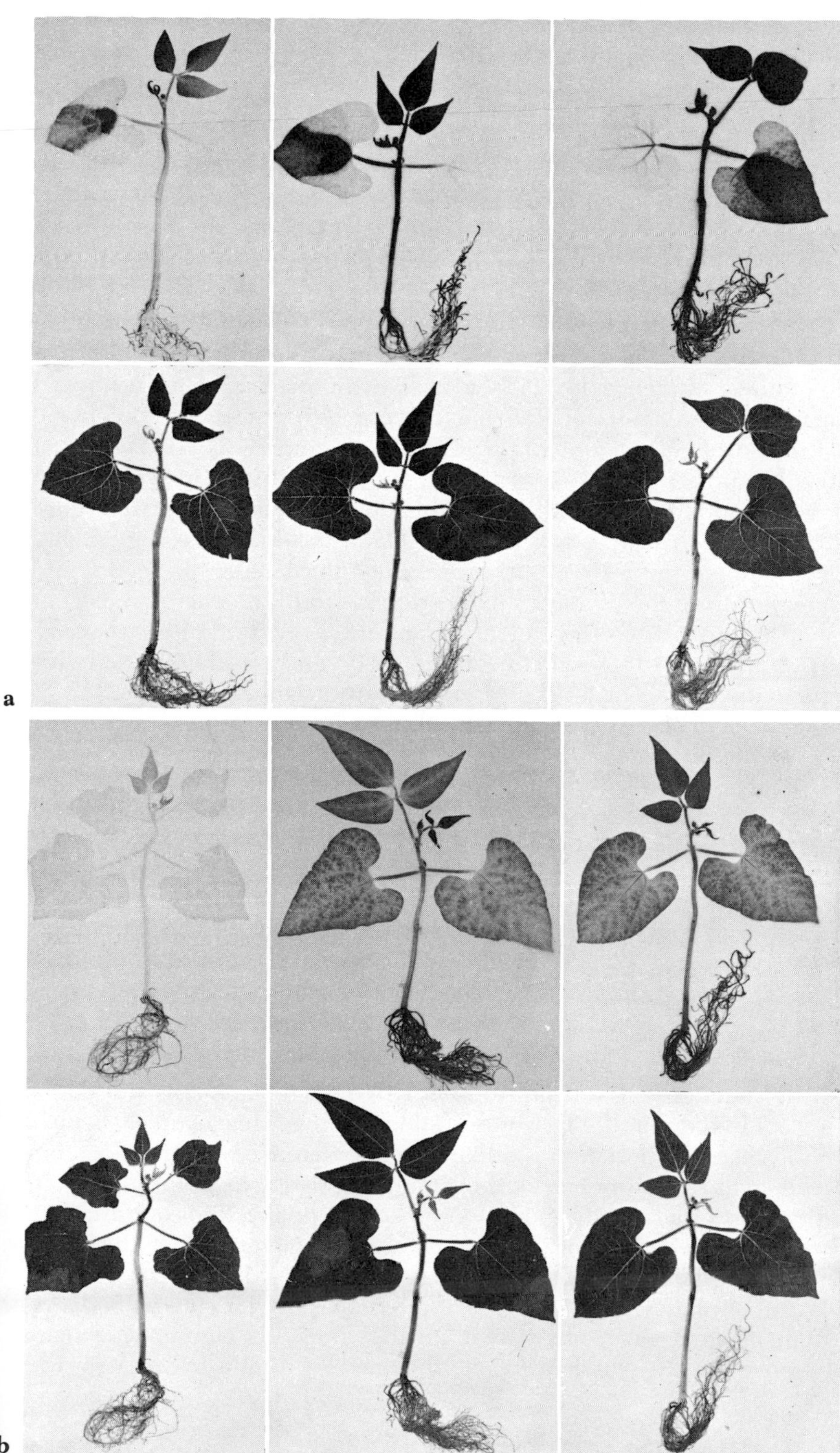
a
b

calcium that arrived in leaves evidently stayed there, and that which was absorbed by roots continued to move upward so that trifoliates one and two became highly labeled during the first 24 hours. Thereafter, concentrations in these leaves were diluted by growth; and as the root supply became exhausted, newly developing leaves were not labeled. These results confirm the fact that calcium is extremely low in phloem mobility.

Crafts and Yamaguchi (1964) gave several illustrations of redistribution patterns of tracers in plants. As mentioned, Figure 6.2 (page 111) shows evidence for carbon redistribution following its initial transport to the roots of *Zebrina*. Figure 6.5 (page 119) shows ^{32}P distribution in *Hordeum vulgare* after 1-, 4-, and 16-day translocation times. Comparing the 1-day with the 16-day picture, it is evident that ^{32}P continued to move from the application spot for many days. Figure 8.13 shows a dosage series with ^{32}P. Comparison of Figures 8.2 and 8.3 illustrates the redistribution of amitrole, maleic hydrazide, and dalapon during the 16-day translocation period. The compounds continued to move from leaf to leaf during the growth of *Phaseolus vulgaris* plants; 2,4-D becomes bound very early and fails to move appreciably. Figure 8.14 shows amitrole redistribution.

Figure 8.15 shows the redistribution of ^{65}Zn in *Phaseolus vulgaris* and *Hordeum vulgare* in 16 days. This isotope, though slow to enter the plant, continues to move from leaf to leaf, and so to concentrate in the young developing leaflets. Figure 8.8 shows redistribution of L-valine during 4- and 14-day translocation times. The failure to label the very young leaves probably represents exhaustion of the ^{14}C supply, rather than failure of movement in the assimilate stream.

Evans and Wardlaw (1964), studying the translocation of assimilates as related to movement of the floral stimulus in *Lolium temulentum*, found that the presence of lower leaves did not reduce the movement of assimilates from the uppermost expanded leaf to the shoot apex. The lower leaf supplied only a small portion of the assimilates reaching the shoot apex, but it supplied much to the roots. Evans and Wardlaw concluded that the inhibitory effect of lower leaves in short days on inflorescense initiation is due not to interference with translocation of the long-day stimulus to the shoot apex, not to dilution effects, but rather to their production of a transmissible inhibitor of initiation. They estimated that the minimum rate of movement of the floral stimulus is about 2 cm per hour.

Wardlaw and Porter (1967) extended the studies on the redistribution of stem sugars in wheat during grain development. Sugars labeled with ^{14}C, which were synthesized by single leaves near the

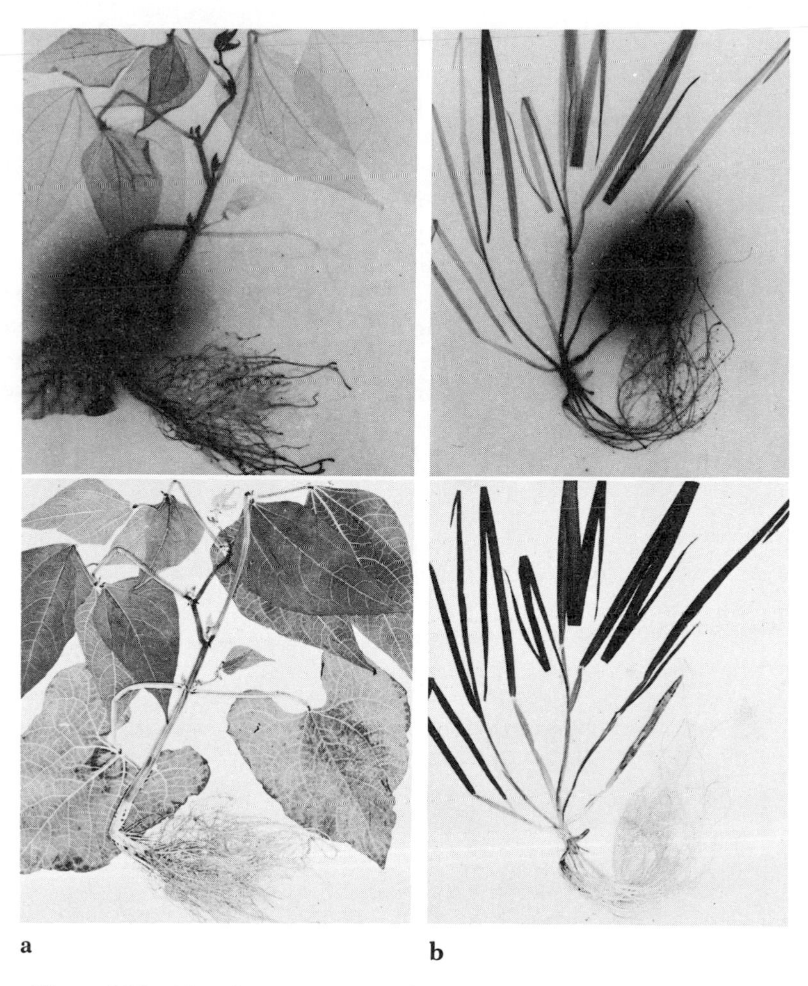

Figure 8.15. Distribution patterns of ^{65}Zn in (*a*) *Phaseolus vulgaris* and (*b*) *Hordeum vulgare*, with translocation allowed to go on for 16 days.

time of anthesis, were redistributed, in part, to other plant organs and, in part, converted into polysaccharides within the stem. Only a small portion of stem sugar was lost in respiration. Upward transport occurred almost entirely out of the top two internodes; contribution to the final weight of the ripe ear was at the most only 5–10 percent. Downward movement of sugars from the third and lower internodes was to newly formed tillers; there was little or no redistribution into the roots.

Hull (1969) found clipping ineffective as a means of altering the distribution patterns of ^{14}C-labeled assimilates and ^{14}C-labeled dalapon in *Sorghum halepensis*. As the plants matured to the flowering stage, the translocation of assimilates into rhizomes increased; this was closely related to the degree of growth of secondary rhizomes. Labeled dalapon followed a translocation pattern similar to that of assimilates. The production of flowers and early seed appears to have a low priority for assimilates compared to the growth of rhizomes. Marked decline in retention of assimilates by leaf and stem tissue during flowering suggests that these tissues may store photosynthates until new rhizome tissue has been produced. At flowering, secondary rhizomes provide active sinks for food storage. The rapid increase in growth of secondary rhizomes between the boot and flowering stages may result from the utilization of reserves stored in culm and leaf tissues.

III
Translocation Mechanism

9

Mechanisms Proposed to Explain Transport Phenomena

Introduction

Diffusion was probably the first mechanism proposed to explain solute movement in the phloem. It was mentioned in the early literature, but was soon discarded when the real nature of the inverse-square relation was given serious consideration. By the turn of the century, protoplasmic streaming as suggested by de Vries (1885), and pressure flow, as evidenced by phloem exudation (Hartig, 1858; Nägeli, 1861), were the predominant concepts. At one time, Dixon (1923) proposed that solutes move both upward and downward in the xylem, and Curtis (1925) suggested that all solute movement in plants is confined to the phloem.

Münch (1927) first presented his pressure-flow hypothesis in 1926; he elaborated it considerably in 1930. According to Münch, assimilates synthesized in leaves are transported to regions of utilization via the sieve tubes of the phloem along a gradient of turgor pressure developed osmotically (Figure 3.1, page 48). By this mechanism, water from the xylem moves into the sieve tubes in the leaves, establishing a high turgor pressure. In the regions of food utilization and storage, this water is pressed out as the assimilates are consumed or stored, returning to the xylem; around 5 percent of the water in the plant would be circulated in this way.

The classical work of Mason, Maskell, and Phillis, as described in their review of 1937 (Mason and Phillis, 1937) clarified the situation, reestablishing the role of the xylem in the transport of mineral elements absorbed by roots, and clearly showing that assimilates follow a source-to-sink pattern of distribution, with rapid longitudinal movement taking place in the phloem. In contrast to Münch, they proposed that assimilates in the phloem move by "accelerated diffusion" at rates " . . . about 40,000 times as rapid as would be expected from physical diffusion in water . . ." and that ". . . transport occurs through stationary cytoplasm." They suggest that ". . . the mechanism activating diffusion consists of some special organization of the cytoplasm, maintained by metabolic energy, whereby the resistance to solute movement is so reduced that materials diffuse in the sieve-tube at rates comparable with those in a gas." They found sugars, nitrogen, phosphorus, potassium, magnesium, and chlorine to be phloem-mobile; calcium was immobile in the phloem.

The concept of assimilate movement along surfaces was suggested as early as 1917 by Mangham and revived in 1934 by Clements. Honert (1932) proposed that a rapid displacement of matter, resulting from surface-active forces, may take place along the surface of water; he suggested that the interfacial surface between water and another liquid might constitute such a channel of transport. He demonstrated this mechanism in an apparatus containing a layer of water upon which a layer of ether was superposed; when potassium oleate solution containing an excess of potassium hydroxide was placed in a chamber in contact with the water–ether system, an indicator (chlorphenol red) dissolved in the water changed color in a moving pattern indicating a rapid spread of the potassium oleate along the water–ether interface as a result of the altered interfacial tension. He suggested that the interface boundary between protoplasm and vacuole was the probable site of movement. He recognized that protoplasmic streaming might enter the picture, but thought this might be the consequence instead of the cause of transport; the primary cause of transport he attributed to the concentration difference of the substance being transported.

Czapek (1897), as a result of experiments involving ringing, heat, chloroform, and plasmolyzing agents, concluded that assimilate movement takes place through the sieve tubes, but that the living protoplasm in some way accelerates transport by a mechanism resembling secretion. Schumacher (1933) applied fluorescein to leaves and found that it moved in the sieve tubes. He visualized a mechanism dependent upon the metabolic activity of the living cells. Kursanov (1961) viewed transport in the phloem in a similar way: it is highly energy-dependent and closely tied to phloem metabolism.

From this brief description it is evident that at least 5 different theories have been elaborated to account for assimilate transport in the sieve tubes. While these various theories have been propounded and advocated by plant physiologists through the years, there are, in reality, two basic concepts involved. These are: (1) pressure flow or mass flow, a system in which solutes and the solvent water move in solution from source to sink, as in a stream—this has been termed the "assimilate stream" by Huber, Schmidt, and Jahnel (1937); and (2) a mechanism involving independent movement of different solutes and movement independent of the solvent water—in such a mechanism, each solute would move along its own independent gradient and, presumably, the water in lumina of the sieve elements would be static (Crafts, 1938, 1951, 1961b).

The pressure-flow mechanism is based upon the phenomenon of phloem exudation as evidenced by the outflow of sap from cut phloem (Crafts, 1932, 1936, 1939a, 1961b; Huber and Rouschal, 1938; Huber, Schmidt, and Jahnel, 1937; Tammes, 1933, 1951, 1958; Tammes and Die, 1964, 1966; Ziegler, 1956; Zimmerman, 1958a, 1958b; and others). Phloem exudation is also evident in the many sap collections from severed aphid stylets (Kennedy and Mittler, 1953; Mittler, 1953, 1957, 1958; Peel, 1963, 1965, 1966, 1967; Hill, 1960, 1962, 1963; Eschrich, 1963b, 1967; Weatherley et al., 1959; Peel and Weatherley, 1959, 1963) and by tests on aphid honeydew (Mittler, 1959, 1962).

These observations are further supported by analyses of phloem sap, which indicate that it carries the nutrients required for growth. They are strengthened by the source–sink pattern of food, virus, and tracer distribution found in literally hundred of experiments. They are substantiated by the recent work of Hammel (1968), in which the turgor pressure of sieve tubes and its gradient in *Quercus* have been measured directly by means of a special needle and manometer; Hammel found turgor pressures ranging from 7.4 to 20.4 atmospheres, and measurements taken at 6.3 meters and 1.5 meters gave quite consistent evidence for larger values at the upper level. And they are backed up by the fact that a simple model, consisting of a differential osmometer (Figure 3.1, page 48), can be set up that demonstrates the nature of the process.

The alternative hypothesis is based upon experiments designed to show independent movement of different solutes in the phloem (Mason and Phillis, 1936; Mason et al., 1936; Gustafson and Darken, 1937; Palmquist, 1938; Biddulph and Cory, 1960; Eschrich, 1967). It is further backed by tests using chilling, anoxia, anaesthetics, inhibitors, and toxic substances, all of them designed to prove that phloem transport is dependent upon metabolic activity of the sieve tubes. Further, it has been called upon to reconcile the fact that the pores in the sieve plates of plants appear to be blocked by protoplasm or slime, and hence, to be unable to accommodate a mass flow having a velocity of transport in the range of 50–100 cm per hour. Mason and Maskell (1928b) recognized the apparent futility of these proposals, yet they could not accept the mass-flow mechanism in the face of the evidence for plugged sieve pores.

Canny and Markus (1960) studied respiration of carefully prepared isolated strips of *Vitis vinifera* phloem to find out if a clue to the mechanism of translocation might be found in phloem metabolism. Using phloem slices 2.0 by 0.5 cm by 100 μ thick, cut in 0.25 M sucrose buffer solution, they measured respiration by infrared gas analysis, or the Warburg manometer, and obtained a value of 220–230 μliter per gram fresh weight per hour—values consistent with others found in the literature, but appreciably lower than those reported by Russian workers. They found evidence for ready synthesis of sucrose from glucose and fructose, obtaining an increase in sucrose of 0.2 mg per gram fresh weight per hour. They also recorded a permeability to sucrose of 2 mg per gram fresh weight per hour for their slices in 10 percent sucrose buffer. Attempts to induce translocation in their phloem slices by presenting ^{14}C-labeled sucrose to one end of a slice failed.

Although Canny and Markus concluded that measurements such as those described above and those of Kursanov (1956b), Willenbrink (1957a), and Ziegler (1958) may never elucidate the translocation process, they seem at a loss to understand why. If normal translocation requires an active source and an active sink, as has been indicated by innumerable tests with tracers, it is apparent that there can be no transport in isolated strips of phloem. Also, as indicated by Eschrich (1956, 1965) and Currier (1957), callosing of sieve plates may occur within seconds of injury to the phloem; both cut ends of the sieve tubes were undoubtedly plugged by slime or filaments, and eventually by callose. Under these conditions, about all that was achieved in this work was a measure of respiration of injured cells.

Further, if one accepts pressure flow as a logical mechanism, then the driving force of transport is built up in the sieve tubes of the protophloem of the fine veins of leaves; and the respiration of the conducting sieve tubes of the petiole, stem, or roots represents only the energy utilized in maintenance of the relatively inert elements. The respiration actually measured in such experiments is largely that of the phloem parenchyma, and so should not differ greatly from that of other parenchyma tissue. Canny (1960a) and Canny and Markus (1960) found this to be true.

Using excised shoots of *Vitis vinifera* set up in the laboratory with their cut ends in water, Canny (1960a) exposed leaves to $^{14}CO_2$ and measured the evolution of $^{14}CO_2$ from petiole and stem sections in order to obtain a measure of ^{14}C-labeled sucrose during translocation. He found that soon after exposure of a leaf to $^{14}CO_2$, the evolution of $^{14}CO_2$ by the petiole starts and gradually rises until a steady state is established 7–10 hours after the start of the exposure; this lasts for about 3 days. During this steady state, about 5 percent of the ^{14}C-labeled sucrose present in the phloem is lost in 7 hours. This loss corresponds to about 200 μliters CO_2 per gram fresh weight per hour, somewhat lower than the output of CO_2 for the total petiole. Using values obtained from the literature for phloem transport, Canny estimated that the loss of sucrose in transit is between 0.3 and 5.0 percent of that passing through the petiole. Commenting on Zimmerman's finding of a lowering of total sugar concentration with distance down the trunk of *Fraxinus*, Canny—differing with Zimmerman's interpretation of this as evidence for normal distributional loss—concluded that it results from respiration; and hence, that the lowering concentration would be a consequence, rather than a reflection, of the cause of the translocation process. Canny concluded that the energy expended in the petiole of *Vitis vinifera* would be four times that calculated as necessary to drive a streaming mechanism. Since Canny used excised shoots from which the root sinks were removed, translocation in his experiments was not normal. Listing the five theories presented above, Canny pointed out that these theories have been modified, and that most plant physiologists choose some form of mass flow or accelerated streaming.

Having narrowed the possibilities to two, Canny discussed the various examples of translocation velocity to be found in the literature, and stressed the advantages of using mass transfer (grams per square centimeter per hour) over any attempt to measure the linear velocity of flow as such. Averaging a number of mass-transfer rates obtained from the literature (Canny's table 1), he arrived at a figure of 3.6 g dry weight per square centimeter per hour, which, expressed as grams per square centimeter of sieve tube lumen per hour, gives a mean of 20 for stems. In these terms, the value of Weatherley et al. (1959) of 1.0 μliter of 10 percent sucrose per hour for sieve tubes of diameter of 23 μ turns out to be 24 g sucrose per square centimeter of sieve tube per hour.

Canny accepted the proposition that transport takes place via the sieve tubes, but pointed out the difficulty of mass flow across sieve plates if the pores are full of cytoplasm. Canny gives a valuable analysis of the interpretation of tracer studies, pointing out that the sensitivity of the detecting device, the amount of tracer supplied, and the length of time of exposure all enter into the accuracy of determination of rates. Probably, Moorby et al. (1963) have come nearest to meeting Canny's criticisms; their value of 60 cm per hour should be well within the acceptable limits.

Canny recalculated the value of the energy requirement for translocation of Mason, Maskell, and Phillis (1936) using corrections suggested by Palmquist (1938) and Spanner (1963), and arrived at a value of 26 cal per milliliter of sieve tube per day, equivalent to 0.004 g sucrose per milliliter of sieve tube per day. Referring to his previous (Canny, 1960a) value of 0.05, he argued that the rough agreement between this energy requirement and his value for the breakdown of sucrose in *Vitis vinifera* petioles suggests that this is the range of energy expenditure in which we should be looking for the requirements for translocation. Because translocation was not normal in the petioles that Canny studied—for the respiration measured related more to sieve-tube maintenance and parenchyma metabolism than to the driving force of transport—and because protoplasmic streaming is too slow by one or two orders of magnitude to account for assimilate movement, we cannot but question the significance of Canny's suggestions.

In studies on polar movement of sugars into bundle ends in the leaves of *Beta vulgaris*, Brovchenko (1965) found that the thin conducting bundles contain 3–4 times more sugar than the adjacent assimilatory tissues, indicating that absorption into the sieve tubes of protophloem in bundle ends is an active process. Whereas mesophyll tissue contained mainly hexose sugars (and fine veins were also high in glucose and fructose), bundles of the second and first order of branching, midvein bundles, and petiole bundles were high in sucrose. Brovchenko concluded that the synthesis of sucrose takes place along the whole conducting system. Sucrose labeled with ^{14}C was hydrolyzed when brought into contact with mesophyll and thin leaf-bundles. Parenchyma cells of mesophyll absorbed glucose, fructose, and sucrose at a much lower rate than did conducting tissues, in spite of the fact that they discharge their internal sugars into the external solution much more readily. Brovchenko concluded that one factor controlling translocation from assimilatory to conductive tissue may be the higher capacity of the conducting bundles for absorbing sugar.

The difficulty in interpreting results like those above is that mass analyses of tissues, no matter how accurate, cannot give the concentrations of sugars in the cytoplasm of the cells involved in transport; values often reflect the amounts in vacuoles more than they do the concentrations in the cytoplasm. Sugars are synthesized in chloroplasts and moved symplastically to sieve tubes. Just as sugar concentrations in the cell walls are maintained at extremely low values by the semipermeable plasmolemma, those in vacuoles may be many times lower than in the cytoplasm outside the tonoplast. Thus, migration of sucrose from chloroplasts to cytoplasm to sieve tubes may take place down a gradient that is masked by vacuolar concentrations, in analyses such as those reported by Brovchenko. One thing his results do indicate is that vein tissues have a high capacity for storing sugars, a point also brought out in Hartt's work. The fact that sucrose was hydrolyzed when brought into contact with tissue sections or vein bundles has little significance, for the cutting of such tissues releases enzymes normally contained in the cytoplasm, and hence, artifacts may be produced.

The time course of translocation of ^{14}C from blades of *Saccharum officinarum* was studied by Hartt and Kortschak (1967), whose methods were the analysis of tissues and radioactive counting of successive punched samples from a single blade. The mass-rate of translocation, in terms of percentage of original counts, was highest immediately after $^{14}CO_2$ treatment. Morning-made photosynthate moved in higher proportion during the morning than during the afternoon in 90-minute periods in light. Afternoon-made photosynthate translocated as well as, or better than, morning-made photosynthate for the initial hour in the light. The translocation rate was lower by night than by day for several successive days.

The ^{14}C-labeled sucrose translocated by day was made primarily by photosynthesis; the sucrose translocated by night came from the conversion of organic acids, organic phosphates, and insoluble residue. The radioactive constituent of the residue that was converted to sucrose was tentatively identified as a xylan, with constituents of glucose, xylose, and glucuronic acid; most of the ^{14}C was in the glucose fraction. Hartt and Kortschak postulated different mechanisms for translocating sucrose by day and by night: pressure flow for nocturnal transport, and phototranslocation by day.

They gave no hint as to a mechanism by which sucrose molecules may be accelerated to move independent of the water in the sieve tubes. By examining the effects of light upon leaf temperature, upon protoplasmic streaming in chlorenchyma cells, and upon those processes responsible for migration of sucrose molecules from chloroplasts to sieve tubes, it seems possible that a satisfactory answer to the problem of rapid translocation by day might be found. It seems physically impossible for sucrose molecules to move through static water at velocities up to 360 cm per hour.

The effects of auxins upon assimilate movement in *Glycine max* have been investigated by Hew, Nelson, and Krotkov (1965, 1967). Their method involved removal of the apical meristem of the plant, applying IAA or gibberellin, allowing the plant to stand for 30 minutes, feeding $^{14}CO_2$ to one of the primary leaves, allowing an additional 30 minutes, and then sampling the various parts of the plant. The addition of growth substances stimulated translocation in 3 ways: it increased the total amounts of ^{14}C translocated; it increased the rate of translocation of ^{14}C-labeled sucrose; and it affected the distribution pattern throughout the plant. Because use of ^{14}C-labeled IAA has shown that the action of IAA is on longitudinal translocation in the stem, Hew and co-workers concluded that their experiments did not concern transfer of assimilate from mesophyll to phloem.

De la Fuente and Leopold (1965) conducted experiments that clearly distinguished between the translocation of sucrose and ^{32}P in the phloem and the polar transport of auxins in parenchyma cells. They measured the movement of auxins, sucrose, and ^{32}P through petiole sections from *Phaseolus vulgaris,* using agar blocks as donors and pieces of filter paper as receptors. Translocation of the auxins IAA and NAA was inhibited by metabolic inhibitors, whereas translocation of sucrose and ^{32}P was accelerated. Accumulation of solutes in the tissue was suppressed by the inhibitors, probably accounting for the increased transport of sucrose and ^{32}P. With increased age of leaves, polar transport declined, while translocation of sucrose and ^{32}P increased up to about 24 days. De la Fuente and Leopold concluded that the contrasting characteristics of the two systems indicate that they are mutually exclusive; sucrose and ^{32}P seem unable to move in the polar system, and IAA and NAA are prevented from moving in the passive system. Because, in petiole sections, the sieve tubes would be plugged at both ends, preventing normal rapid phloem transport, it is difficult to interpret experiments of this type. Possibly, sucrose and ^{32}P moved in the phloem parenchyma; auxins normally exhibit polar transport in undifferentiated ground parenchyma. Crafts

(1959a, 1959b) found that ^{14}C-labeled IAA applied to leaves of normal plants of *Hordeum vulgare* was translocated into the roots.

Because much current research attempts to support or refute the various theories of translocation, it seems desirable to consider each theory in detail. The following sections are devoted to critical analyses of data presented in support of these theories.

Protoplasmic Streaming

Although De Vries suggested protoplasmic streaming as a mechanism to account for the rapid transport of foods in plants in 1885, it was not until Curtis became involved in ringing experiments that this mechanism again gained notice. Curtis (1935), in ringing experiments on *Ligustrum*, found that leaves on unringed stems gained 115.9 percent in total nitrogen, whereas leaves on ringed stems gained only 22.7 percent. From this type of information, Curtis reasoned that the phloem—which he had removed by ringing—was the principal conduit for mineral transport. Since carbohydrates were known to be descending the same stems, Curtis proposed that solutes were ascending and descending simultaneously through the phloem, and that protoplasmic streaming was a logical mechanism to account for these processes.

What Curtis failed to account for was the 22.7 percent nitrogen that *did* ascend the stems of the ringed plants. The stems were devoid of phloem; the nitrogen had to have gone up through the xylem. It is now evident that ringing injured the newly formed, most pervious xylem elements and starved the roots of sugar, thereby lowering xylem transport to 22.7 percent of normal. Mason and Maskell (1928a) found that ringing reduced water flow in the xylem within 13 hours; this effect increased with time. This fundamental misinterpretation by Curtis led to a whole series of experiments attempting to support the hypothesis of protoplasmic streaming.

The hypothesis of protoplasmic streaming has been favored by Biddulph (1941), Swanson and Böhning (1951), and others; its most recent advocates, however, are Thaine and Canny. Thaine (1961) announced his transcellular protoplasmic streaming theory of translocation in 1961. According to his description, there are transcellular tubular strands in sieve tubes. These strands, he maintained, traverse the sieve-tube lumina and pass through the sieve pores, providing a structural basis for cytoplasmic communication throughout the vascular system. Within the strands, Thaine detected microscopic particles, which he found to occur in close linear order about 3 μ apart; the particles measured 0.5 μ in diameter, and the strands were 1.5–2.0 μ thick. The particles, he observed, would pass along the lumina of the sieve-tube elements and through the sieve plates within the transcellular strands; they were seen to move in opposite directions in single sieve tubes, sometimes within adjacent strands. The velocity of movement was 3–5 cm per hour. Although this is from one to two orders of magnitude below measured translocation velocities for assimilates, Thaine proposed that other constituents of the sieve tubes may move faster, particularly in stems with long internodes. Thaine claimed to have observed transcellular particle movement through as many as 10 sieve-tube elements in the linear file of a single sieve tube. He proposed that these particles might be involved in long-distance movement of mobile materials from a leaf to the stem, the apex, or the root; such particles could move out of a leaf loaded with mobile materials and then move back to the leaf after discharging their contents. This would presuppose a circulatory system.

In a second paper, Thaine (1962) elaborated on his first announcement. He described two types of linear structures that he had seen, by phase microscopy, in plant cytoplasm: fine threads 0.1–1.0 μ in diameter seen in hair cells, where they formed endoplasmic systems along streaming pathways in parietal cytoplasm and in transvacuolar strands; and small plastids and mitochondrialike particles, which moved along the fine threads. Similar fine threads, small plastids, and mitochondrialike particles occurred in phloem exudate and transcellular strands. The transcellular strands, 1–7 μ in diameter, were seen in sieve-tube elements, phloem parenchyma, border parenchyma, and cortical cells. The movement of small plastids across end walls in border-parenchyma cells, the appearance of the same structures within strands in phloem cells, and the occurrence of small plastids in successive drops of phloem exudate, were taken as evidence for their participation in translocation. Particle movement, Thaine felt, occurs through transcellular strands in association with fine threads, motivated by a transcellular form of protoplasmic streaming.

Thaine (1962) diagrammed a *Cucurbita* sieve tube (his figure 1) having six coarse strands traversing the elements shown and passing directly through the sieve plates. Crafts (1932) saw and illustrated strands in *Cucurbita* sieve tubes, but he saw nothing comparable with those postulated by Thaine. Furthermore, Crafts described the cessation of protoplasmic streaming in maturing sieve tubes. The strands that stream in young sieve tubes slow down and become much less prominent as the elements mature; they cannot be identified in the large, old, functioning elements that are the conduits for phloem exudate; filamentous reticulum in such cells is below the resolving power of the light microscope.

Canny (1962b) considered the mechanism of translocation in detail. Of prime importance is the fact that Canny accepted and promoted the observations and concepts of Thaine (1961, 1962), including the cinefilm demonstration purporting to show streaming of transcellular strands in *Primula.* Many of Canny's conclusions must be viewed in light of that fact.

In a short note on the velocity and energy requirement of translocation, Spanner (1962) corrected Canny's (1961) calculation of the energy requirement for protoplasmic streaming as a transport mechanism, and showed that a streaming mechanism based on cyclosis must be ruled out on energy considerations alone. He criticized Canny's (1960b) calculation of velocity on the basis of Canny's assumption that there is little leakage of labeled assimilate to surrounding tissues. Peel and Weatherley (1962) made the same criticism. Certainly, as a distribution system, the function of the sieve tubes is to deliver assimilates to cells along the translocation route, so these criticisms seem valid; and if they are valid, they explain the very low velocity calculated by Canny. Spanner further questioned Canny's view that the ". . . slope constant of the logarithmic profile . . ." may be ". . . independent of the plant or substance. . . ." This same question was raised with respect to exogenous tracers in the consideration of Canny's paper.

Spanner (1963) analyzed the idea that the translocation profile of a tracer must show an exponential fall-off in concentration with distance. He questioned the method used by Vernon and Aronoff (1952) to estimate transport velocity, and adopted Horwitz's (1958) suggestion of an irreversible leakage to explain the concave curve he found when he plotted log of activity in petiole against distance along petiole for movement of ^{137}Cs in *Nymphoides peltatum.* This paper by Spanner illustrates again the futility of attempting a mathematical analysis of phloem transport. First, he neglected the fact that all materials that move in the assimilate stream enter the sieve tubes by a process analogous to diffusion. Thus the profile of an entering tracer must be logarithmic at the start. Second, he took as an arbitrary assumption the irreversible leakage assumed by Horwitz. Because the phloem is a distribution system, any solute—and particularly an exogenous tracer like ^{137}Cs—will move out of the sieve tubes throughout their length, the amount lost being controlled by the activity of the surrounding parenchyma cells as sinks for that particular solute.

Third, Spanner failed to consider that the velocity of transport may vary, depending upon activity of all sources and sinks. Thus, light at the source, temperature throughout, and growth in the sinks are all involved in transport velocity. Water potential in the xylem (Wardlaw, 1967), aeration of leaves and roots, and such phenomena as diurnal fluctuation in root pressure may also determine velocity of flow. Finally, in his *Nymphoides* experiments (as will be explained elsewhere), Spanner failed to consider the possibility that ^{137}Cs may have penetrated into the xylem, moved in the transpiration stream to the lamina, and returned via the phloem, resulting in a buildup in activity along the petiole. This would have produced the concave upward curve shown in Spanner's figure 3 (see Figure 8.5, page 185).

Having accepted the transcellular strand mechanism of Thaine (1962) to explain solute translocation in phloem, Canny enlisted the aid of a mathematician to test and elaborate the theory (Canny and Phillips, 1963). To one familiar with the real problems of assimilate transport, this paper is nothing short of an exercise in futility. Again, it might be well to enumerate the incorrect assumptions upon which Thaine's theory is based.

(1) Thaine assumed that transcellular streaming occurs in functioning sieve tubes. Using petiole material from *Primula obconica*, Esau et al. (1963) examined the phloem in fresh sections. They stated: "No streaming occurs in the mature sieve elements, and their starch grains remain stationary." This is in agreement with many previous observations (Crafts, 1932).

(2) Thaine saw and recorded streaming in sectioned material. From the analysis on sieve-tube plugging (Chapter 5) it should be evident that translocation was not going on in the sections Thaine used. Thus, it is apparent that streaming, as Thaine saw it, does not depend upon translocation.

(3) Swanson and Geiger (1967) performed experiments in which petioles were cooled to a temperature that stops streaming; yet, after a pause, translocation resumed. Thus, translocation as recorded by Swanson and Geiger does not depend upon streaming.

(4) The velocity of streaming, as recorded in the literature, of 2–6 cm per hour is too slow, by an order of magnitude or more, to account for known transport velocities. Canny and Phillips, in referring to this discrepancy (between 3–5 cm per hour and 50 cm per hour or more) say, "It is likely that this velocity is maintained by a sol-gel interface forming an active surface on the inner boundary of the strands." This leads to the next point.

(5) Even if the strands were streaming at 50 cm per hour, they would still be entirely inadequate, because—from accurate measurements—the cytoplasm in mature functioning sieve tubes occupies only around 5 percent of the lumen volume (see Thaine, 1964b), and only one-half would be streaming in the right direction. Sucrose, on the other hand, may be present in a solution of up to 25 percent concentration. Thus, the moving sugar may occupy more space than the whole cytoplasm and, certainly, infinitely more space than the sol-gel interface on the inner boundary of the strands.

(6) Finally, to quote Esau et al. (1963):

> The transcellular strands described by Thaine (1961, 1962) and interpreted by him as providing the structural basis for a transcellular cytoplasmic streaming are merely lines caused by diffraction of light from walls out of focus. These lines are as clearly visible in dead cells as in living.

For the reasons given above, it seems that the mathematical treatment produced by Canny and Phillips is of little value. This is particularly true when it gives rise to such statements as these:

> Transport is analogous to diffusive transport in that the steepness of the concentration gradient is proportional to the rate of transfer. . . . Transfer is found to be rapid over a short path and slower over a longer path; there is a maximum distance beyond which a given rate of transfer cannot take place. . . . If the time over which the translocation of radioactive solute has proceeded is small compared with strand diffusion time, a wave-like profile of radioactivity with distance is produced, and if large, the profile is of a diffusive kind.

When one considers that the sieve-tube system is made up of living semipermeable elements with an osmotic pump at its head, and intimately connected throughout its length with active parenchyma cells by means of protoplasmic connections, it is difficult to see how so much significance can be attributed to radioactivity profiles of moving tracers. Along the total system, sugars are being imported or exported at varying rates, depending upon supply and demand; sugars are supplied by photosynthetic tissues and utilized in actively growing regions. Some tracers, such as ^{14}C-labeled 2,4-D, are accumulated by living cells along the translocation route, whereas ^{14}C-labeled amitrole moves freely and piles up at the ends of the channels in shoot and root tips. Some tracers (sugars, for example) are incorporated, whereas others are redistributed from leaf to leaf up a stem; phosphorus is both incorporated and redistributed. The bypassing of mature leaves by all phloem-limited tracers, the basipetal transport from lower leaves and acropetal transport from upper, and the various distribution patterns of a number of tracers (depending primarily upon the ease with which various tissues export or import them), all seem to preclude the possibility of diffusion, or diffusionlike processes, entering into the rapid longitudinal movement of assimilates through the phloem.

Thaine (1964a) described exudation from the cut ends of fine veins of leaves of *Glycine max*. He claimed that the exudate was fluid endoplasm, with which two types of cytoplasmic particles were associated: (1) small particles with a diameter of 0.5 μ; and (2) plastids, 1.0 by 3.0 μ. These latter were green in color, and stained blue-black with iodine. The cytoplasm was exuded only from cells of the phloem, but further identification was not made. From these observations, Thaine concluded that sieve elements contain normal amounts of cytoplasm, and that his observations are consistent with the transcellular interpretation of protoplast structure in sieve-tube elements. From his illustrations, it appears that the so-called sieve-tube cytoplasm extruded from the cut surface of leaf sections is actually cytoplasm from cut phloem parenchyma. Cut sieve tubes of *Cucurbita*, which exude a coagulable sap, would produce many times the volume shown in Thaine's figures. And plastids from sieve tubes characteristically stain wine red with iodine; those from phloem parenchyma stain blue-black.

Thaine (1964b) elaborated his transcellular strand theory of translocation and attempted to reconcile results of different kinds of translocation experiments with the protoplasmic streaming mechanism. He considered reconciliation possible because he considered transcellular streaming to be primarily a fluid-flow phenomenon, deriving its motive force (like other forms of protoplasmic streaming) from metabolic energy. Thaine described transcellular strands as made up of the following: (1) a boundary membrane, which separates the strand from the vacuolar fluid; (2) longitudinally oriented threadlike structures, which are possibly endoplasmic reticulum; (3) microscopic and submicroscopic particles, including mitochondria; and (4) a matrix of fluid endoplasm. While he found no evidence to pinpoint the exact location of the transport mechanism within the cytoplasm, he pointed out the possible role of microscopically visible particles, the endoplasmic reticulum and the fluid endoplasm. In this mechanism, concentrations or pressure gradients develop outside the actual pathway of long-distance transport; these are regarded as initiating or regulating translocation, but not as providing its driving force. Thus, the solutes of the vacuolar fluid would be relatively immobile in the longitudinal direction, whereas the contents of the transcellular strands would move rapidly, not being subject to structural or physiological barriers. Thaine viewed such movement as "activated" mass flow; he regarded phloem exudate as leakage from damaged sieve-tube elements. Thaine considered superficial the discrepancy between observed rates of streaming (3–5 cm per hour) and measured translocation velocities (50–300 cm per hour), in that submicroscopic particles and fluid strand contents may move faster than microscopic particles. Curtis (1935) made this same suggestion, but no one has been able to substantiate it.

Mitchell and Worley (1964) observed rotational streaming of cytoplasm in phloem fibers of *Phaseolus vulgaris* at velocities of 2.5–6.5 cm per hour with an average velocity of 3.6 cm per hour. Worley (1965b) showed that southern bean mosaic virus moved vertically in *P. vulgaris* stems in these fibers. Virus movement in fiber bundles was about 8 mm per day, which was 30 times as fast as movement through cortical parenchyma. Worley (1966) further showed that droplets of castor oil injected into individual phloem fibers moved along with indigenous particles on rotational cytoplasmic currents.

Thaine (1964b) estimated that 5 percent of the cross-sectional area of the sieve-tube lumen is occupied by the parietal cytoplasm, and points out the limiting nature of this small conducting channel for circulation streaming. He estimates that transcellular strands occupy 25 percent of the lumen section, and hence, that they provide a conduit five times as large. And in addition, Thaine would have had less specialized tissue (presumably phloem parenchyma) involved in translocation in an effort to bring his mechanism within the limits of mass transfer. He even cited the old work of Mason (1926) and more recent studies of Brouwer (1953) indicating interruption of sieve tubes in *Dioscorea* as evidence that sieve tubes may not constitute the total structural mechanism for long-distance transport. This question has recently been cleared by Behnke (1965a, 1965b), who has shown that the glomeruli of these stems are simply modified sieve tubes.

Esau, Engleman, and Bisalputra (1963) controvert Thaine's transcellular streaming hypothesis. Through careful studies on petiole cells of *Primula obconica* they found that streaming is limited to phloem parenchyma; no streaming occurred in mature sieve elements. The transcellular strands they claimed to be artifacts, mere lines caused by the diffraction of light from walls out of focus. They found these lines as clearly visible in dead as in living cells.

Parker (1964a) reported seeing strands in mature sieve tubes in the bark of *Carpinus caroliniana, Acer rubrum* and *Fagus grandiflora.* These strands appeared as two membranes representing opposite sides of tubules with pores in their centers. In tangential sections from *A. rubrum,* the tubules appeared to penetrate the sieve plates at points where pores could be seen. In view of Thaine's (1961) concepts, Parker considered the strands that he saw to be of some significance to assimilate movement.

In a second paper, Parker (1964b) described strands in the petioles of eight broad-leaved tree species and in the bark of *Pinus strobus* and *Liriodendron tulipifera.* In *Tilia* and *Fraxinus,* Parker saw particles about the size of mitochondria moving along pathways that apparently crossed sieve plates. The particles appeared to move independently of one another, sometimes colliding, sometimes separating, but often following the same paths in the same direction for several minutes. They could also be seen moving in opposite directions—apparently in one sieve-tube element, but along different pathways.

Continuing his studies, Parker (1965) stained the phloem of several tree species with a staining mixture containing bromphenol blue and mercuric chloride that stains amino-acid residues blue. His illustrations show protoplasmic strings stretched from sieve plates that are reminiscent of illustrations from *Cucurbita* (Crafts, 1932, figures 27 and 28); *Solanum tuberosa* (Crafts, 1933, figures 13, 16, and 17); certain tree species (Crafts, 1939b, figure 5; Evert and Murmanis, 1965, figure 15).

There is no doubt that such strands exist in the sieve tubes of many species. Evert claimed that they consist of slime. Possibly, they are remnants that do not break down. More probably, they are parts of the reticulum of plasmatic filaments described by Behnke and Dörr (1967), or the P2-protein of Cronshaw and Esau (1967). That they can in any way provide the driving force for solution flowing through the sieve elements at velocities up to 300 cm per hour seems impossible. It is also difficult to reconcile their universal presence with the empty appearance of the mature, functioning sieve tubes of many species. For example, in Figure 2.4 (page 26), from Esau (1941), the inner reticulum of strands in the mature sieve tubes must be of minor consequence. In comparing figure 1 of Tamulevich and Evert (1966) with their figures 20–25, one is impressed with the paucity of contents of the mature sieve tubes; their figures 26 and 27 showing slime do not include the long, straight strands shown by Parker.

Lawton and Biddulph (1964) produced some excellent illustrations of *Dioscorea* sieve plates, along with histoautoradiographs showing accumulations of silver grains that correspond with the sieve fields of the sieve plates. The plants had been fed $^{14}CO_2$, and the massed silver grains were interpreted as aggregates of radioactive translocates. In their interpretations, Lawton and Biddulph concluded that their results did not support pressure flow, because, in pressure flow, radioactive build-up should cover the whole sieve plate, and not just the sieve fields. They also questioned whether their evidence supported the electroosmosis theory of Spanner (1958). They felt that the distribution of radioactivity shown would correspond to the strands of cytoplasm carrying radioactive translocate and passing through the sieve plate at the sieve fields, as suggested by Thaine's (1962) transcellular streaming theory. However, they admitted that they found no evidence for sieve tube cytoplasm arranged longitudinally in strands along the sieve tube. A simpler and more logical interpretation would be that, in spite of their precautions to prevent artifacts, their method of cutting and quick-freezing allowed time for callose to be synthesized on the sieve plates, and that the accumulations of silver grams corresponded to callose cylinders around the sieve-plate pores. Labeling, in this case, would have been ^{14}C that by incorporation had been built into the callose cylinders; and these, owing to their increased mass, were constricting the pores. Lawton (1966) has shown callose deposits on sieve plates of *Dioscorea* spp. by fluorescence in ultraviolet light and by polarized light. Northcote and Wooding (1966) produced convincing illustrations of callose formation from ^{3}H-labeled glucose on sieve plates of *Acer pseudoplatanus*. They postulated that, in their materials, labeled callose was deposited in response to wounding of the tissue.

In continued studies, Thaine (1965) made careful observations on protoplasmic streaming and particle movement in *Tradescantia* hair cells, *Primula* leaf hair cells, and mesophyll cells of *Phlox* and *Glycine*. Phloem exudate from stems of *Cucurbita pepo* was fixed and viewed under the microscope. Cells of the fine veins of leaflets of *Glycine* were also studied. Thaine extended his transcellular streaming theory to account for observed patterns of particle movement and the activity of streaming cytoplasm, which he interpreted in relation to the development and transformation of the endoplasmic reticulum. He presented evidence for a dynamic association between particles and endoplasmic reticulum in circulation streaming in hair cells; this structural association was also described in mesophyll cells and in cells along the pathways of transported materials. Evidence was claimed for cell-to-cell movement of particles in sieve tubes along parallel tubules of endoplasmic reticulum. Thaine disagreed with the concept that movement in protoplasmic streaming can be explained as a phenomenon of mass flow (compare Canny, 1962b).

Problems of sieve-tube cytology and physiology were attacked by Thaine, Probine, and Dyer (1967). Following the previous reasoning of Thaine, these workers attempted to substantiate the transcellular streaming mechanism of phloem transport. They controverted the claims of Esau, Engleman, and Bisalputra (1963) to the effect that the streaming shown in Thaine's cinefilm did not take place in sieve tubes. They did not agree that transcellular strands are artifacts resulting from diffraction effects; and they gave convincing evidence for the internal fibrillar structure in mature sieve tubes described by Crafts in 1932. They distinguished between slime, which appears to be amorphous under the light microscope, and strands, which have a linear arrangement. They showed that sieve tube exudate of *Cucurbita* may acquire a stranded appearance when it becomes set in a fluid matrix. This is very obvious when the exudate is collected in 50 percent alcohol (Crafts, 1961b, figure 6.10).

Thus, it seems that—although protoplasmic streaming is an intriguing phenomenon, and one that must be intimately involved in the symplastic movement of assimilates from mesophyll to phloem, of tracers from epidermis across mesophyll to phloem, and of all solutes from phloem to their ultimate loci in various sinks in the plant body—to attribute rapid longitudinal translocation in mature sieve tubes to streaming protoplasm is futile. Streaming is too slow, the cytoplasm in mature sieve tubes occupies too little space, and the many distribution patterns of labeled tracers in plants are just not right for a mechanism involving independent movement of different molecules, and of solutes and water.

Metabolic Movement

It should be perfectly obvious to anyone familiar with the physiology of plants that translocation is dependent upon metabolic energy. The problem is to pinpoint the locus at which the energy is applied and to explain the mechanism involved. Protoplasmic streaming is a vivid manifestation of cell metabolism, in which energy is utilized in viscosity changes and interfacial activity, with the result that the cytoplasm moves in various ways within the cell lumen. Such movement certainly accelerates the diffusional movement of solutes from chlorenchyma to phloem and from phloem to receiving cells along the symplast in parenchyma tissues. That it enhances rapid movement of the assimilate stream in mature sieve tubes is doubtful, for reasons already explained.

Any mechanism responsible for accelerating diffusion in cells would necessarily depend upon metabolic energy. The difficulty here is that no mechanism has been suggested. It seems highly improbable that such a mechanism actually exists for accelerating diffusion to the extent required to account for rapid, long-distance transport in mature sieve tubes, because there is no known physical, structural organization upon which it could be based. Likewise, energy would be needed to establish and maintain the gradients essential to effect movement along surfaces. The problem lies in finding surfaces with the required properties and in the proper loci to effect translocation along sieve tubes. As explained previously, if the sugar in a sieve tube occupies more space than the cytoplasm, it seems difficult to visualize surface movement as being adequate or realistic.

In the pressure-flow mechanism, the metabolism of green cells produces assimilates having osmotic activity. Metabolism motivates the migration of molecules from chlorenchyma to phloem; a low basic metabolism maintains the sieve tubes in their living, functioning state. Metabolism is essential in the export of assimilates to their ultimate sinks in meristems, ground parenchyma, storage tissues, and so forth. Where the pressure flow concept differs from all of the other theories is in the fact that it does not call upon metabolism to move the assimilates from cell to cell along the functioning sieve tubes. The osmotic pump at the source and the active absorption processes in the various sinks provide the driving force for pressure flow.

Finally, metabolic movement (as this term is used in the literature) implies that energy is applied over the whole length of the transport channels, including the sieve tubes; and that each species of moving molecule or ion moves along its own independent gradient. An example is Spanner's (1958) theory, which pictures a cycling of K^+ through the sieve pores and back via the companion cells so that each sieve plate acts as an electroosmotic pump. Fensom (1957, 1958, 1959) proposed a metabolic mechanism involving bioelectric potentials that bring about electroosmotic flow across barriers in the phloem (sieve plates) and, hence, provide energy to drive translocation.

Kursanov (1961) reviewed the transport of organic materials in plants. He questioned the mass- or pressure-flow theory and emphasized the role of metabolism in food distribution. He gave data on the sucrose content of phloem conduits, on the fact that sucrose is the predominant transport material in reciprocal grafts of *Helianthus annus* and *H. tuberosus*, and on the amino acid and organic acid content of leaf veins of *Rheum rhaponticum.* He quoted translocation velocities of 40–120 cm per hour from the Russian literature; and he stressed the high respiration intensity of fibrovascular bundles, as compared with mesophyll and cortex parenchyma cells of a number of plants.

Table 9.1 presents data that Kursanov quoted concerning the enzyme content of vascular bundles of *Beta vulgaris.* Kursanov listed AMP, ADP, ATP, UMP, and UDPG as the acid-soluble nucleotides of those tissues. Citing data on the effect of cooling on transport, Kursanov stressed the role of metabolism in food movement, but failed to describe a physical mechanism capable of carrying on long-distance transport. Furthermore, most of the data on the content of sugars, acids, and enzymes present in the conducting system were derived from analyses of complete vascular bundles, of which sieve tubes make up only 10–20 percent. Most of the remaining cells were phloem parenchyma cells, which are known to be high in the substances reported. As for respiration rates on isolated vascular bundles, they should have been high as a result of injury from the isolation process. In addition to producing much injury, isolation results in rapid phloem plugging; hence, no transport was in progress in the analyzed tissues.

Kursanov cited work of Belikov, who, using $^{14}CO_2$ on *Glycine max,* found that low-level leaves send their products mainly to the roots; but as the fruits develop, ever increasing numbers of leaves pass their

Table 9.1. Relative activity of enzymes of glycolysis and oxidation in vascular tissues of *Beta vulgaris.*

Enzyme	Relative activity
Aldolase	+++
Apyrase	++
Cytochrome oxidase	++++
Acid phosphatase	++
Phosphorylase	+
α-Galactosidase	+
Hexokinase	+++
Phosphohexose isomerase	+
Invertase	+
Succinic dehydrogenase	++

Source: After Kursanov (1961).

assimilates to them; and at length, even those at the lower levels follow suit. Young leaves that have not completed their growth are able to photosynthesize, but they do not yield their assimilates to other parts of the plant; they continue for a long time to import assimilates from the nearest mature leaves. Exchange of products between mature leaves at different levels virtually ceases; Kursanov found it impossible to bring about a flow of assimilates to a mature leaf even by prolonged darkening. That this source-to-sink pattern of assimilate distribution can be readily rationalized in terms of pressure flow, but that it makes no sense for a metabolic transport mechanism, seems to have escaped Kursanov completely.

In a later paper, Kursanov (1963) pictured a mechanism of metabolic transport involving the reactions DPN $\longrightarrow$ DPNH and ATP $\longrightarrow$ ADP. He explained that

> if the path between the sieve cells is obstructed by protoplasmic bridges (and this at present seems most probable), the role of the living contents of sieve cells as a factor controlling the composition of substances undergoing transport and their rate of movement must be much more marked. In this case, each sieve plate zone must function like a glandular cell, absorbing substances from the vacuolar solution, fixing them with appropriate carriers or particles (Thaine, 1961), transporting these complexes through the connecting strands of the protoplasm via sieve plates and, finally, again releasing them into the vacuolar solution of the next sieve cell.

After reviewing Thaine's (1961, 1962) ideas, Kursanov remarked:

> Let it be noted that the fibrous structure of the protoplasm of the sieve cells, which is their characteristic peculiarity (Kollmann, 1960a, b) is very reminiscent of the construction of a conductor extended in the direction of the flow of transport. . . . The range of possibilities here extends from the free flow of solutions through a series of sieve cells to the continuous metabolic transport of substances in contact with the protoplasm. . . . However, no matter what the solution of the problem of the mechanism of transport in the phloem may turn out to be, metabolism of the conducting cells remains an essential aspect of this process.

These words of Kursanov characterize the thinking of the advocates of metabolic movement in the phloem. The physical bases for such a mechanism are the plugged condition of sieve plates, and the claimed demonstration of independent movement.

Fensom (1957) proposed that bioelectric potentials of plants may be involved in transport, both in the xylem and in the phloem. Using the Helmholtz equations of electroosmosis, he calculated values for sap-stream velocities. Although the direct use of these equations did not yield a satisfactory answer, by assuming that observed potentials in plants are metabolically maintained, and using the Poiseuille equation for flow in capillaries, he found a streaming potential for xylem of 1.0 mv. If the mechanism involves barriers across which electroosmosis occurs, then a streaming velocity of 36 cm per hour was calculated.

Fensom (1958) conducted experiments in which he measured xylem exudation. He proposed, from electropotential determinations, that overall differences in potential would arise from the combined potentials of xylem and phloem, being the two main components of a binary system manifesting opposed polarity with a certain amount of feedback linking them to each other. He recognized the possibility that the potentials he measured might have been the result, rather than the cause, of the observed flow; but he argued that, because the changes in potential preceded the changes in flow by 20–100 seconds, the potentials provided the driving force for flow. Fensom did not point out that the xylem and its contents are relatively acid (*p*H 5.5) as compared with phloem (exudate *p*H 7.0); that the phloem became plugged within minutes of the excision of his plant parts; and that respiration at cut surfaces was probably very high, due to wound healing processes. His assumption of barriers in the xylem and phloem does not coincide with our present understanding of the anatomy of these two tissue systems. The energy involved in the differences in potential that he recorded, even if maintained by metabolism, seems small compared with the energy available for bringing about transpiration pull in the xylem and pressure flow in the phloem. Both of these forces derive their energy directly from the sun, and are practically inexhaustible under normal field conditions.

In a third paper, Fensom (1959) suggested that the hydronium ion is the most important ion involved in diffusion and subsequent ion exchange; and that regions of hydronium ion production and consumption may bring about circulation and movement of water. Since Fensom did not consider these processes in relation to assimilate movement, further discussion of his paper is irrelevant here.

Duloy and Mercer (1961) studied respiratory metabolism in vascular and ground tissues of *Apium graveolens, Cucurbita pepo, Vitis vinifera,* and *Verbascum virgatum.* They found that the respiratory quotients for both vascular and parenchyma tissues were close to unity, suggesting that the primary substrate for respiration was soluble carbohydrate. By using various substrates in respiration studies, they concluded that both tissues have respiration resembling the tricarboxylic acid cycle. By using inhibitors, they concluded that, in both tissues, electron transfer is mediated by the cytochrome oxidase system, to which is linked the succinic dehydrogenase system.

From these studies, Duloy and Mercer decided that the respiratory pathway in phloem is identical with that in parenchyma. Although the pathways of respiration in phloem and in ground tissue are alike, the respiratory activity of phloem is higher than in other tissues. However, when respiration rates are compared on the basis of protein nitrogen, the differences are small. Duloy and Mercer suggested that the high rate found in phloem is a result of having more cytoplasm per unit volume of tissue. Thus there seems, from these results, to be no unique type of respiration in phloem. This might be expected on the basis of the low activity of mature sieve tubes. The high cytoplasmic content of phloem parenchyma and companion cells may well account for the high respiratory level of phloem. These cells are actively engaged in moving assimilates into sieve tubes in source areas and removing them in sink regions; both are active processes requiring metabolic energy.

Ullrich (1961) performed many experiments on the relation of oxygen supply and aerobic respiration to translocation in the phloem. He found that the transport of fluorescein from the lamina through the petiole of *Pelargonium zonale* was not delayed by depriving the exposed central petiole bundle or the lamina of the leaf of oxygen. Ullrich concluded that oxygen per se takes no part in transport through sieve tubes, a fact that he correlates with the absence of intercellular spaces in vascular bundles and the deep embedding of the latter within the tissues enclosing them.

Using isolated tissue of *Heracleum sphondylium*, Ullrich found an intensive anaerobic fermentation, lasting for many hours. This was stimulated by 0.1 M sucrose, and markedly inhibited by 0.01 M hydrocyanic acid. Because he found normal respiration in the presence of oxygen, Ullrich assumed that the phloem of this plant has a complete respiratory system. Hydrocyanic acid gas reliably and reversibly inhibited transport when applied locally to exposed vascular bundles of *Pelargonium* petioles; a dose ten times stronger was required to cause inhibition by application to the lamina of the leaf. Because transport did not require oxygen, yet was sensitive to cyanide, Ullrich used the benzidine-hydrogen peroxide method to localize the peroxidase system. A high activity was detected in the sieve tubes, companion cells, and phloem parenchyma cells of *Passiflora coerulea* and *Morus alba*. This reaction was sensitive to cyanide, suggesting a peroxide-peroxidase system as part of the cyanide-sensitive respiration; such a reaction would have a low oxygen requirement, provided peroxides were present.

Ullrich considered the energy metabolism of phloem to be peculiar in its low oxygen requirement, despite the fact that it seems to be a complete system. Because of the cyanide sensitivity, he considered heavy-metal enzymes essential, and suggested that anaerobic respiration and respiration aided by the peroxidase system may constitute the energy supply. Energy-rich compounds or bound oxygen carried in the phloem may serve in such a mechanism. Although Ullrich does not stress it, it seems evident that O_2 from photosynthesis of leaves, or green chlorenchyma cells of petioles or stems, may constitute one source of oxygen. Because Ullrich had no means for distinguishing between sieve tubes and parenchyma cells in his determinations, his results, applied to all phloem cells, give no clues to special transport mechanisms directly related to respiration.

Braun and Sauter (1964) showed that the nucleate cells of phloem glomeruli of the Dioscoreaceae are rich in cytoplasmic constituents and are high in acid phosphatases. The sieve-tube elements, companion cells, sieve plates, and sieve fields are all high in acid phosphatases. Braun and Sauter proposed that an active transport of assimilates, involving phosphorylating and dephosphorylating processes, takes place within the glomeruli. Thus, the sieve-tube elements themselves—especially the sieve plates—may be implicated in the active transport. Just how they could serve as pumps, however, was not stated.

The relation of metabolism to translocation in the phloem is obvious. The exact role of metabolism in this regard, however, is problematical. We know that it is involved in the synthesis of assimilates, and their migration from mesophyll to border parenchyma, and that it is probably involved in their transfer into conducting sieve tubes. Metabolism is involved also in the movement of solutes, including assimilates, from sieve tubes into companion cells and phloem parenchyma in sinks that occur along the entire route of translocation. Thus, metabolism is responsible for the osmotically active solutes that provide the driving force for pressure flow in the sieve-tube system; and, as a factor in growth, accumulation, and storage, it is an agent in the transfer of solutes from sieve tubes to neighboring parenchyma cells in the sink regions. Metabolism is obviously involved in the transformation of these solutes into protoplasm, starch, cell walls, and the multitude of end products, including CO_2, that mark the final disposition of translocates in plants.

It is much less obvious how metabolism might bring about movement in the phloem of an assimilate, such as sucrose, independent of its associated solvent water. Sucrose is known to form hydrogen bonds with water: normally, the sucrose molecule in solution carries six molecules of water of hydration, and this complex must be surrounded by a cloud of water molecules less firmly held. For sucrose to move more rapidly or more slowly than water in the phloem, large amounts of energy would be needed to break the hydrogen bonds and to overcome the frictional resistance.

As previously mentioned, most of the work reported in this chapter was done prior to the demonstrations by Esau and Cheadle (1961), Duloy et al. (1961), and more recently by many others, that the sieve plates of mature sieve tubes are open, so that mass flow across them may take place. The interesting thing is that many translocation physiologists writing since 1961 have overlooked or ignored this profound discovery, and have continued to seek ways in which metabolism might serve in assimilate movement along sieve tubes, other than by maintaining those elements in their open conducting state, as explained in the opening paragraphs of this chapter. No plant physiologist conversant with modern literature in this field can fail to appreciate the tremendous strides that have been made in the past two decades in our understanding of the processes of metabolism. And no one can conceive that a function so demanding of energy as translocation would not require metabolic support. On the other hand, it seems futile to attempt to explain a hydraulic process, such as long-distance movement of water and assimilates at velocities up to 300 cm per hour, on the basis of metabolism without at least attempting to visualize a physical mechanism. It is one thing to postulate a metabolic pump, and an entirely different thing to describe it and explain its functioning. By contrast, the well known action of an osmotic pump (such as that shown in Figure 3.1, page 48), the recognition of the ubiquitous occurrence of phloem exudation, and the unique patterns of tracer distribution found in translocation research would all seem to point to a mass-flow mechanism.

Activated Diffusion

The activated diffusion theory of assimilate transport was proposed by Mason and Maskell (1928a). They stated, for example, that "The presence of a sugar gradient in the direction of sugar movement is obviously in accord with the view that some process analogous to diffusion is a factor in transport." Concerning mass movement, they remarked that

> The most substantial objection to the bark, or rather the phloem, being the channel of transport has been that a mass movement of solution at rates as high as those reported . . . is impossible. . . . It is indeed evident that a mass movement of solution at these rates is very improbable either through the parenchyma cells of the bark or through the colloid-filled sieve tubes.

Mason and Phillis (1937) spoke of "accelerated diffusion," and while they thought in terms of a diffusionlike process, they had no explanation for the extremely rapid rate of translocation shown in their experiments. Although they mentioned protoplasmic streaming, they were aware of its obvious shortcomings with respect to velocity. Having rejected mass flow because of "colloid-filled sieve tubes," they chose accelerated diffusion—an even less logical mechanism, in terms of physical principles. That the diffusion gradient that they found in *Gossypium hirsutum* might be converted by osmosis to a turgor gradient sufficient to overcome resistance to longitudinal movement of solution in sieve tubes apparently did not occur to them. The artifact of sieve-plate plugging was a handicap for these workers, as it has proved to be for so many others.

Activated diffusion was mentioned by other physiologists involved in translocation studies, but Canny (1962b) has been the most dedicated proponent of this mechanism in recent years. He visualized a mechanism that combined activated diffusion, protoplasmic streaming (Thaine, 1961, 1962), and mass flow. In Canny's defense of this theory, the rationalizations he proposed did not meet the criticisms that were made. Activated diffusion seems entirely inadequate for explaining rapid longitudinal movement of assimilates and other solutes in the phloem of active plants.

Penot (1965) introduced a concept of attraction by sinks to explain a positive action of receiver sites on phloem translocation. Attraction, Penot stated, may be (1) suppressed in buds, roots, stems, and storage organs; (2) dependent upon the ratio of receiver to donor; (3) restricted by reducing donor leaf surface, or by varying the size of the receiver; and (4) increased by activating starch hydrolysis. He observed that, by local application of FAP, attraction by the treated organ may also increase; and he reported that FAP stimulates phloem translocation of sugars and inorganic nutrients, and heals sublethal effects of thermic shock and respiration inhibitors, which tend to decrease attraction by the treated organ. FAP also causes a local retention of phosphorus and starch in leaves.

Penot distinguished between chlorine, which exhibits free diffusion, and sugars, phosphorus, and rubidium, which display oriented movement. The movement of chlorine toward leaves Penot terms "free diffusion," and movement of phosphorus, rubidium and sugar he terms "oriented diffusion." Defoliation, Penot found, stimulated emptying of sugars and phosphorus from the remaining leaves, but it slowed the export of chlorine. Thus, Penot cited Mothes's mention of specificity of attraction in connection with movement of sugar, phosphorus, and rubidium. Penot concluded that molecular attraction implies translocation through diffusion rather than through mass flow.

Hartt and Kortschak (1963) stated that

> At least two forces are involved in translocation, a push and a pull. . . . The push from the leaf is due to light. . . . The pull comes from below. . . . Supplying a sink, for example growing lalas or suckers, increases the rate of translocation.

Thus, Penot and Hartt and Kortschak imply that a sink, removed by appreciable distance from the source, has some means by which it may pull or attract solutes moving in the phloem; yet no physical mechanism comparable with transpiration pull in the xylem has been described.

A much more logical explanation of the results of such experiments would seem to be that movement takes place from source to sink along a positive pressure gradient, with accumulation or utilization of solutes lowering turgor in the sinks, thus activating flow. "Attraction" or "pull" seem to imply the existence of a property—such as magnetism—possessed by molecules and capable of acting over a considerable distance. No known physical mechanism can exert force in this way over meters of space, as, for example, between the foliage and the roots of tall trees. Turgor gradients of sufficient magnitude can very well account for such movement.

Penot's specificity of molecular attraction probably relates to the possible migration of some elements from phloem to xylem, and thus, to reverse flow, which does not fit a simple source–sink pattern. As explained in previous chapters, maleic hydrazide and dalapon do this; presumably, phosphorus may do it, too, under some conditions; sodium, potassium, rubidium, and chlorine may also, given the proper circumstances.

Canny and Askham (1967) cautioned researchers using tracers

about the significance of tracer distribution. Convinced that the translocation system in plants (phloem) is a diffusion-analogue system, they analyzed the distribution pattern expected when a tracer is introduced between the source and the sink. They showed that the movement would not follow the source–sink polarity of the established movement, but that it would diffuse in all directions, and that it might, therefore, enter an exporting leaf. They felt that the faint images of untreated mature leaves found in the autoradiographs of many tracer studies supported their diffusion-analogue idea. They described an experiment with aphids feeding on *Vicia faba* plants, in which, after supplying $^{14}CO_2$ to a mature leaf, they claimed to have found labeled aphids on other mature, exporting leaves. This, they felt, supported the idea that, although the normal movement of assimilate was out of the mature leaves, ^{14}C-labeled sucrose diffused out of the treated leaf, along the stem, and into untreated, mature leaves, where it was accumulated by feeding aphids.

In their introductory statement, Canny and Askham proposed that the differences in distribution between labeled and unlabeled translocate (that is, between ^{14}C-labeled and unlabeled sucrose) cannot be ignored. They proceeded, then, to illustrate their point by picturing a model in which labeled tracer is introduced into a system having an established gradient, and showing that, by diffusion, the tracer might produce a pattern related to its own diffusion gradient, and not to the established gradient. This is inherent in the processes of diffusion, but it has little relevance for translocation in plants at velocities of 100 cm per hour or more.

In the experiments described by Crafts (1967b), in which the ^{14}C-labeled compounds 2,4-D, monuron, amitrole, and maleic hydrazide were applied to the epicotyls of young plants of *Glycine max* (Chapter 8), four distinct distribution patterns were found, depending on properties of the molecules. These patterns have been demonstrated in *Zebrina pendula, Brassica napus, Hordeum vulgare,* and the tuber tissue of *Solanum tuberosum* (Crafts, 1961b). The above patterns do not follow the predictions of Canny and Askham in any single case; they show source-to-sink movement via phloem, via xylem, and via both xylem and phloem (that is, xylem movement superimposed on phloem movement, in the case of maleic hydrazide). Never was there free independent diffusion. Mature leaves were bypassed.

Canny and Askham went back to Mason and Maskell's original work (1928a, 1928b) and argued that the kinetics of translocation were described by the steady-state form of Fick's law. This follows from the fact that the process of movement into the phloem is diffusional;

it does not prove that subsequent movement is diffusional, just as the profile of tracer-movement does not substantiate a diffusional mechanism. The point that Canny and Askham seemed not to appreciate is that, once in the phloem, the sugar gradient is transformed osmotically to a turgor gradient, and that this turgor produces the driving force for rapid translocation through functioning sieve tubes.

Colwell proved in 1942 that ^{32}P applied to a leaf of *Curcurbita* shows up soon in the phloem exudate; Clor (1959) established this beyond doubt, and innumerable tests have since shown that ^{14}C-labeled sucrose moving out of a mature leaf is picked up by aphids feeding on the stem and excreted as honeydew; the same tracer is present in exudate from stylets (Weatherley et al., 1959).

Images of mature leaves have been obtained occasionally in tracer studies. Leonard found that many coniferous trees contain compounds that react with x-ray film to form pseudoautographs (Leonard and Hull, 1965); and a number of tracers, including maleic hydrazide and dalapon, migrate from the phloem to the xylem, where they move to leaves in the transpiration stream and produce light images in all transpiring organs (Crafts and Yamaguchi, 1964; Crafts, 1967b). As for ^{14}C-labeled assimilate, such effects are not seen in short-term experiments; but in trials running 24 hours or more, images resulting from movement from roots of metabolites that move in the transpiration stream may be found (Figure 6.2, page 111).

There are a few spots in the autoradiographs in Canny and Askham's paper that appear to correspond in location to aphids, but there are no traces of the vascular bundles leading to them. Possibly, self-absorption by the leaf tissues eliminated these; but we wonder at the high concentration of ^{14}C in the aphids with no indication of honeydew near their posterior parts. Did they produce no honeydew during the 6-hour feeding periods? We would expect to see a cloud of fine dots near each aphid. The aphid does not concentrate sugars in its body, but tends to excrete most of them while retaining mainly the amino acids (Mittler, 1953). That the aphid should obtain any labeled sugar within the 6-hour period, considering the large supply of unlabeled sugar in its immediate vicinity, is difficult to explain. Although selective and preferential uptake have been accurately recognized, this can scarcely be taken as an occurance of one of those two phenomena.

From Canny and Askham's plates 1 and 2, it is evident that labeled sugar moved to the shoot apex and to the roots during the feeding period. The only rational explanation of their results would seem to involve movement of labeled aphids from other labeled tissues or, possibly, the import of labeled metabolites into the mature leaves via the xylem, and their concentration by preferential uptake from phloem bundles that were screened out of the autographs by self-absorption. What Mason and Maskell (1928a, 1928b), Canny (1962b) and Canny and Askham (1967) have all failed to do is to provide some acceptable explanation for the discrepancy between normal diffusion of sugar in water and the postulated "activated" type of diffusion. In the absence of such explanation, all of the results cited in their papers can be interpreted more logically in terms of a pressure-flow mechanism.

Surface Migration

During the early years of the present century, there was tremendous interest in the application of the concepts of physical chemistry to biological systems. A great number of studies were made on the permeability of cells to ions, on the accumulation of solutes by cells, and on equilibria between cells and their culture media. One important phenomenon that intrigued biologists was adsorption—the holding of ions or molecules on surfaces in higher concentration than that of the bathing media. Adsorption of various solutes, including dyes, became the subject of laboratory exercises in physiology classes; and the Gibbs-Thompson rule, stating that substances that lower the surface energy of a system tend to accumulate at the surface, was tested and elaborated.

In 1917, Mangham proposed that adsorption might provide a mechanism that could account for assimilate transport in plants. Describing what essentially amounts to the symplast concept as related to the protoplasmic continuum in living plants, he visualized sugar movement as the result of a wave of disturbance and readjustment of equilibrium propogated along the cytoplasmic continuum of sieve tubes. By means of a very graphic description, he explained how the waves of disturbance would be sent out from all points at which sugar is transformed, and would spread in all directions, subject to structural limitations. He went on to describe a hydraulic system, comparing the parenchymatous cells with their vacuoles to small tanks, the sieve tubes to the main supply line, and the plasmodesmata to small-bored pipes leading away from the supplying tanks. He described the mature sieve plates as having numerous fine perforations, each lined with a protoplasmic tubule through which the fluid contents of adjacent segments are continuous.

Having described the structure of the plant in this way, and appreciating the presence of concentration gradients along the phloem, it is difficult to understand why Mangham chose surface translocation over a pressure-flow mechanism. He visualized fluctuations in concentration as constantly occurring throughout the living tissues, especially during periods of vigorous growth and storage. He considered that such fluctuation would be propogated along the inner surface of the sieve-tube cytoplasm, and that sieve plates would offer little resistance to such movement. He described sieve tubes as richly provided with connecting threads on their lateral walls adjoining the companion cells, and the walls of the latter freely connected with the phloem parenchyma. Thus, solutes could move throughout the plant "... *without ever passing through the plasmatic membrane or outer surface of protoplasm in contact with the cell wall.*" [Mangham's italics.]

Mangham was apparently unaware of the demonstrations of phloem exudate by Hartig (1858), Nageli (1861), Zacharias (1884), and Kraus (1885). He had little knowledge of phloem anatomy and the limitations it placed on the transport mechanism in terms of available conduit. Further, he did not have the advantage, as later physiologists did, of the knowledge provided by the elegant studies of Mason and Maskell (1928a, 1928b).

Honert (1932) made the next contribution to the surface migration concept. He visualized a mechanism involving a continuous interface in the phloem along which assimilates might move. He explained that when a drop of liquid insoluble in water, the molecules of which contain an electropolar group, is placed on a water surface, it will tend to spread rapidly over this surface; this spreading takes place in the direction from lower to higher surface tension, and will not stop until the tensions are equalized. Honert proposed that the interfacial surface between water and another liquid might constitute a channel of transport, and he constructed a model to demonstrate its action. This was made up of a tube 95 cm in length connecting two small bulbs that were open on top. He half-filled this apparatus with slightly acidulated water containing chlorphenol red and added a layer of pure ethyl ether saturated with water. To one bulb he added 25 ml of potassium oleate containing excess potassium hydroxide. When this was done, the indicator changed from orange-red to purple, and the color change spread along the water–ether interface. Honert measured the velocity of spread in seconds. Although the velocity was high (2.5 cm per second) for the first 25 cm, it was much lower for the full 95 cm (0.3 cm per second). From this observation, it seems that Honert's mechanism obeys an inverse power function, and it seems unlikely that it could possibly account for assimilate movement in tall trees.

In an analysis of translocation in plants, Clements (1934) proposed that although assimilates move in water, they move independent of it. He argued that if mass movement were taking place, the gradients shown by Mason and Maskell would be washed away more rapidly than they could be formed; hence, one would expect a uniform distribution of solutes. How one familiar with the formation and utilization of assimilates by plants could arrive at such a conclusion is hard to understand.

Continuing his analysis, Clements—recognizing that sugars must move down a steep gradient—proposed that each assimilate moves independently, that they move as though they were gases moving in air, that there is an impelling force operative in the phloem, and that the highly colloidal condition of the sieve-tube protoplasm is a necessary part of the mechanism. He asked, then, if surface forces kept active by respiration of living cells might not be at play.

Citing statements by Chen (1951) and Biddulph, Biddulph, and Cory (1958) to the effect that ions applied to leaves, upon reaching the stem, may move in an upward or downward direction; and quoting Swanson and Whitney (1953) to the effect that two or more ions, applied to a leaf, may move in different directions simultaneously, or in the same direction but at different rates; Mitchell, Schneider, and Gauch (1960) proposed that rapid and even bidirectional movement of ions in phloem is by rapid movement along interfaces. They mentioned rates of observed movement of 20 cm per second for migration of ions along surfaces, and they pointed out that surface migration would take care of the problem of simultaneous bidirectional movement in a single sieve tube. They noted, however, that ^{32}P and ^{60}Co were found to move out of leaves only when sugar was being translocated, and they posed ". . . the possibility that there is more to the mechanism of translocation in the phloem than simple movement of ions and other substances over interfaces."

Among contemporary translocation physiologists, Canny (1962b) proposed that assimilate movement in the phloem may consist of a combination of activated diffusion, protoplasmic streaming, and mass flow. Presumably, adsorption on surfaces would be involved in the streaming mechanism. Nelson (1962) suggested that interfacial movement may be responsible for high-velocity transport of solutes. Canny and Phillips (1963), discussing movement of particles observed microscopically to travel at 3–5 cm per hour, considered that, in intact plants, rates may be considerably higher. Admitting that velocities up to 50 cm per hour would be required to account for measured rates of dry weight transfer, they said: "It is likely that this velocity is maintained by a sol–gel interface forming an active surface on the inner boundary of the strands."

Thaine (1965), in a paper on surface associations between particles and the endoplasmic reticulum in protoplasmic streaming, cited evidence for the idea that particles may move from cell to cell in sieve tubes along surfaces of parallel tubules of the endoplasmic reticulum. Thaine reported an association between particles and linear structures in fresh, unstained exudate from the phloem of fine veins of *Glycine max:* he noted a similar structural association between single files of particles and linear structural elements in *Cucurbita pepo* phloem exudate. Thaine proposed that the mechanism of particle movement depends on physicochemical interactions between the external surfaces of mobile particles and the surfaces of the endoplasmic reticulum. Thaine observed particles moving faster than the changes of form and location characteristic of endoplasmic reticulum, proving that particle movement does not depend on a mass flow of fluid pushed along by the cortical gel, or on adhesion of particles to the endoplasmic reticulum. Separate movement of particles and endoplasmic reticulum indicates that a motive force is generated by surface interactions, and that these interactions, in turn, influence the endoplasmic reticulum. Motive force for particle movement and development and the transformation of endoplasmic reticulum depends, in Thaine's view, on the activation of molecular structures at the membrane interface by electrons derived from high energy intermediates of metabolism. The development of endoplasmic reticulum through immobile regions appeared to him to be related to the orientation and frequency of particle movement. When the endoplasmic reticulum system was not developed, particles moved along irregular pathways, apparently at a sol–gel interface. In immobile parts of the cytoplasm, particles apparently moved on the endoplasm side of a flattened double-membrane envelope, which lines the cell at the external boundary of the fluid endoplasm.

Although this work by Thaine treats mainly with hair cells, and is thus not relevant to assimilate movement in sieve tubes, it may prove to be of great significance in the mechanism of solute movement in palisade parenchyma, spongy parenchyma, and border parenchyma of leaves, and in companion cells and phloem parenchyma throughout the plant. Protoplasmic streaming is an obvious stirring mechanism by which molecules in the cytoplasm of a cell may be rapidly equilibrated. Although it is difficult to see how surface forces might accelerate movement to velocities higher than those exhibited by organelles in the cytoplasm, they certainly must come into play in the general distribution process.

Assuming that the open nature of sieve-plate pores might be an artifact of potassium permangante fixation, and stating that ". . . the pores of the sieve plate are occupied with a finely fibrous, electron opaque material . . ." and that ". . . the plugged nature of the sieve plate appears a severe obstacle to bulk flow of phloem materials . . ." Bouck and Cronshaw (1965) proposed that ". . . this plugged region is the site of rapid transport through some active process. . . ." They went on to propose a two-phase system,

> that is, an active pumping through the sieve plate and perhaps a cytoplasm-mediated transport through the sieve cell. . . . The construction and location of the sieve tube reticulum seems ideally suited for surface transport.

It is interesting to note that, although they gave much evidence for the presence of the filamentous reticulum, which they would have derived from the endoplasmic reticulum and from vesiculation and breakdown of the nuclear envelope, they made no mention of the tremendous loss of plasmatic materials shown so graphically in Figure 2.4 (page 26). They cited no literature on phloem exudation or tracer movement.

Evert and Murmanis (1965) suggested the possibility that ". . . strands that run from sieve element to sieve element through the sieve-plate pores play some role in translocation." In contrast to Bouck and Cronshaw (1965), they said that ". . . most strands observed in *Tilia* possessed no structures which could be identified as endoplasmic reticulum." Proposing that enzymes might be associated with slime and, consequently, with strands, they cited Esau (1948) as implicating slime strands with callose removal from sieve plates of *Vitis*, and Braun and Sauter (1964) as reporting strong acid-phosphatase activity in sieve elements of the Dioscoreaceae. They said:

> The possibility that enzymes capable of mediating phosphorylating and dephosphorylating processes are associated with strands that run from sieve element to sieve element and possibly from companion cell to sieve element, leads us to speculate that important energy-releasing reactions involved in phloem transport take place within the sieve-element protoplast and that the movement of assimilates occurs in association with, or along the surfaces of strands. We visualize many streams of assimilates flowing through a more or less stationary fluid medium.

Considering the number and variety of tracers reported to be mobile in the phloem (in Chapters 7 and 8), and that the phloem-mobility of these is closely associated with the movement of assimilates (reported in Chapter 6), it seems highly unlikely that a metabolically mediated movement along surfaces of strands in sieve tubes could be of significance in the overall transport mechanism in the phloem. Certainly, all of these materials—including sugars, amino acids, and a host of herbicides and other pesticides—cannot move along respiration-activated pathways at similar velocities and in measurable quantities, and assume comparable distribution patterns, including the bypassing of mature leaves. And so—ignoring the fact that the known rates of protoplasmic streaming are from one to two orders of magnitude too slow, and that the total cytoplasmic sheath in the sieve tube occupies only 5 percent or less of the lumen area, whereas sugars may occupy 20 percent or more—it still seems improbable that translocation, intimately associated with sieve-tube cytoplasm, can meet the requirements laid down by known translocation rates.

Mass Flow

After Hartig's 1858 description of sieve tubes and his work on phloem exudation, many plant physiologists thought of assimilate movement in terms of mass flow. But with the introduction of refined killing, embedding, microtoming, and staining techniques, most microscopic work on phloem anatomy involved these preparatory procedures; and the attendant artifact of phloem plugging gave rise to the widespread view that sieve plates were really not sievelike, but that their pores were full of slime or protoplasm. This opinion was held by Dixon in 1922; and it has been a common opinion that has continued to be expressed in recent years by many (for example: Kursanov, 1963; Hepton et al., 1955; and Weatherley et al., 1959).

With the work of Mason and Maskell (1928a, 1928b), it became obvious that food movement followed a source–sink pattern, and that if the sieve tubes were open, then a turgor gradient developed osmotically might provide the driving force. Münch proposed such a mechanism in 1927, and with the publication of his book *Die Stoffbewegungen in der Pflanze* in 1930, the "Münch mechanism" became well known and widely accepted by plant physiologists. Though confused about the actual channels of conduction in the phloem, Crafts (1931, 1932, 1933, 1948, 1961b) accepted the mass-flow mechanism as providing the driving force for phloem translocation. Bennett (1937,

1940a, 1940b) recognized the relation of food movement to the rapid distribution of viruses in plants, and visualized a flow mechanism to explain it.

With the introduction of growth regulators as herbicides, their movement with foods in plants was soon recognized (Mitchell and Brown, 1946; Linder et al., 1949; Day, 1952). Anderson (1957) found a similar relation in the distribution of amitrole. Likewise, the correlation of the movement of radioactive tracers with the distribution of food was early recognized (Colwell, 1942), and within two decades, scores of papers on tracer movement mentioned this relationship. In fact, this correlation is apparent in almost every paper reviewed in Chapters 6–8.

In spite of this wide acceptance of the mass-flow mechanism many plant physiologists chose other concepts. The greatest obstacle continued to be the opinion that the sieve plates were blocked.

Fischer showed in 1885 that, by killing a cucurbit plant by placing it intact in hot water, callose plugging could be avoided. And Kennedy and Mittler (1953) proved that phloem exudation through excised aphid mouthparts would go on for hours, delivering phloem sap at rates commensurate with normal flow of the assimilate stream. Yet it was not until Esau and her students conceived the idea of killing without cutting—by injecting the killing fluid into the hollow stem or petiole of a cucurbit plant—that it could be reliably shown that sieve plates are normally open; that is, that they are indeed sieve-like, as described by Hartig. With the demonstration that the sieve plates of functioning sieve tubes are open to accommodate mass flow; and with a clear understanding of the three distinct mechanisms of sieve-plate plugging; it seems, at last, that the serious objections to the pressure flow mechanism have been eliminated.

Ziegler and Mittler (1959) used aphids to sample the sieve-tube sap of the umbellifer *Heracleum mantegazzianum* and the conifer *Picea abies*. Sucrose was the only sugar detected in their sap samples, which were obtained from severed stylets. This sugar occurred in the umbellifer at a concentration of 24 percent, and in the conifer at 10 percent. From *H. mantegazzianum*, sap equivalent to the contents of 5500 sieve-tube elements exuded through severed stylets; from *P. abies*, a volume was obtained equal to the contents of 50 sieve cells having dimensions of 28 $\mu \times 2750$ μ.

Ziegler and Vieweg (1961), using a delicate thermoelectric method on isolated phloem strands of *Heracleum mantegazzianum*, were able to obtain a reliable measure of the velocity of flow of the assimilate stream. Their measured temperature values lay between 35 and 70 cm per hour, rates that correspond well with many quoted in the literature. Because the flow was basipetal along the phloem, it was considered to correspond to a mass movement of solution in the phloem.

In their paper on the physiology of the sieve tube, Weatherley, Peel, and Hill (1959) presented some of the most convincing evidence for mass flow that has ever appeared in the literature. Using the aphid-stylet method of Kennedy and Mittler (1953), they collected exudate in measured quantities under a variety of experimental conditions, and revealed many of the significant properties of the phloem system as a transport mechanism. Assuming that the aphid stylets puncture a single sieve tube, a fact since well confirmed by Zimmerman (1961) and Evert et al. (1968), they sought evidence for the relation of sieve-tube exudation to normal translocation. Their experiment is illustrated in Figure 9.1. From the results shown it is evident that incisions alongside the stylets paralleling the long axis of the stem had no serious effect on exudation rate, a finding that follows from phloem anatomy. When a transverse incision was made 6.5 cm vertically above the stylets, the exudation rate soon slowed down; a cut 2 cm above the stylets caused complete stoppage.

Normally, the sieve-tube sap contained around 10 percent sucrose and 0.5 percent amino acids. When the plants were held in the dark for several days, the leaves turned yellow, as they normally do in autumn; the sugar contents of the exudate soon fell to a mere trace, whereas the concentration of amino acids rose to as high as 5 percent. Mittler found in 1958 that the concentration of amino acids in naturally senescing leaves followed a similar course. Thus, Weatherley and his co-workers pointed out that the stylets appeared to tap a normal channel for the transport of food from the leaves.

Studying next the permeability relations of sieve tubes, Weatherley and his co-workers determined the permeability constant of the sieve-tube system; this turned out to be 254 μ per atmosphere per minute. The usual constant for permeability of plant cells to water is around 0.2 μ per atmosphere per minute. Calculating similar constants for sugar, assuming a sugar concentration difference of 10 percent between sieve tubes and the surrounding cells, they arrived at a figure of 1.3×10^{-2} g mole/μ^2/mole concentration gradient/min. The usual figure for plant cells is between 10^{-17} and 10^{-20}. Thus, for sugar and water to arrive in sieve tubes by normal diffusion, permeabilities would have to be several orders of magnitude greater than normal.

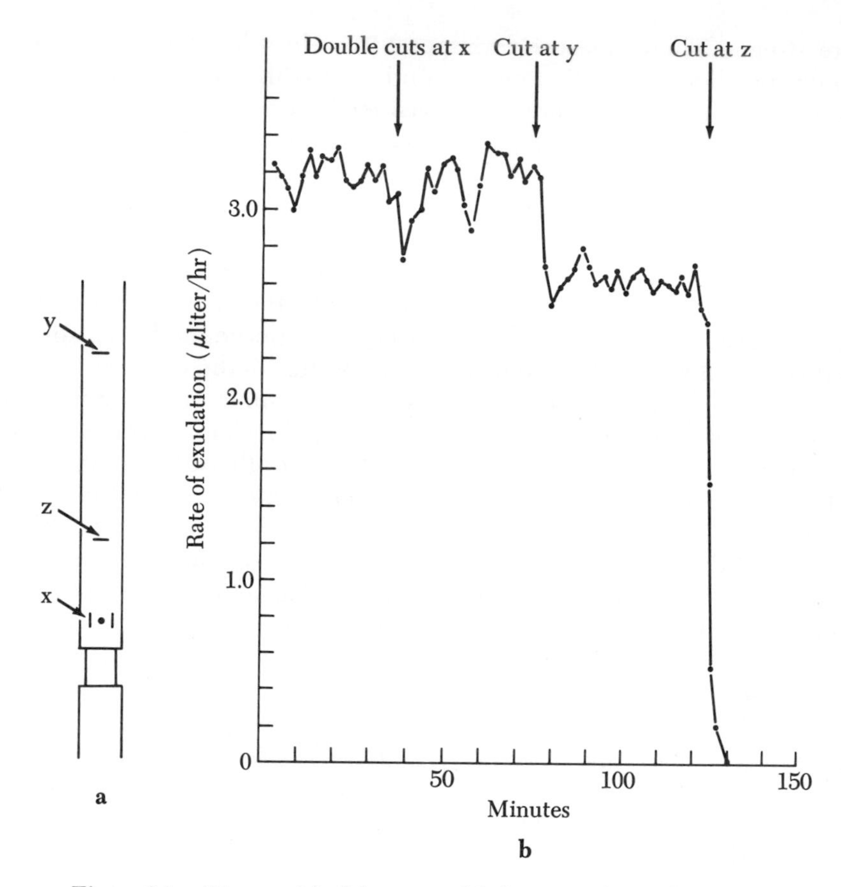

Figure 9.1. Diagram (*a*) of the stem of *Salix viminalis* used by Weatherley et al. (1959) for experiments on phloem exudation through aphid stylets. *Locus X*, parallel cuts beside the stylets; *locus Y*, horizontal cut 6.5 cm above the stylets; *locus Z*, horizontal cut 2.0 cm above stylets. The graph (*b*) shows the effects of the various incisions upon rate of exudation. From Weatherley et al. (1959).

Sugars, quite evidently (and possibly water), enter the sieve tubes via the symplast, and permeability barriers are thereby avoided. And within phloem, a special rapid longitudinal transport is indicated.

Experiments with whole plants proved that exudation rates determined by measuring stylet exudation were similar to those determined by other means. An average rate at the stylet site proved to be 100 cm per hour, a rate that coincides with many of those found with tracers. Some rates much higher were found in these experiments, too, as has been true also in studies of tracer movement (Webb and Gorham, 1964; Hatch and Glasiou, 1964). Wilting was found to cause a marked decrease in exudation rate with an accompanying increase in the concentration of the exudate; when the water supply to the roots was normal, exudation was remarkably constant in rate.

In girdling experiments, unless the girdles were close to the stylets, Weatherley and his co-workers found that exudation was not stopped. This indicated a rapid plugging of the sieve plates at the cut ends and a switchover in source of supply from the leaves to the storage cells of the stem. Using isolated stem segments, they found that there was no polarity of movement (compare Crafts, 1936). They ascertained that a minimal length of about 16 cm of stem, involving 800–1000 sieve elements, was required to produce full exudation rates. When the water potential on the inside of the bark was lowered by running sucrose or mannitol solutions through the xylem, the rate of exudation from stylets was drastically lowered, while the concentration of sucrose in the exudate rose, approaching a value of 50 percent after 1½ days; but exudation did not stop. Figure 9.2 shows the results of the experiment in which sucrose was used as the plasmolyticum; Figure 9.3 shows the results of the one in which mannitol was used. In the latter experiment, when the mannitol solution was replaced by water, stylet exudation returned slowly to normal. In order to obtain more rapid responses to changes in water potential, a bark-strip technique was used; the results are shown in Figure 9.4. Obviously, the changes took place much more rapidly, but the results were essentially the same. These experiments indicate that the sieve-tube system of the phloem acts like a ramifying, turgid, osmotic system, with sugar being introduced in leaves, or—in their absence—along the phloem system by storage parenchyma cells.

Weatherley and his co-workers concurred with Mittler (1958) that stylet exudate is a genuine sample of the assimilate stream, coming from a single sieve-tube element. Leaf mesophyll cells, although normally serving as the source, play no unique role; storage parenchyma anywhere along the sieve tube may act as a source or sink for mobile assimilates, according to circumstances. Although this conclusion applies to woody stems such as those of *Salix* spp., it may not apply to other types of plant materials: this will be discussed at some length later in this chapter.

Weatherley and his co-workers clearly distinguished between the processes of exchange of water and solutes between sieve tubes and surrounding cells, and rapid longitudinal transport along the sieve

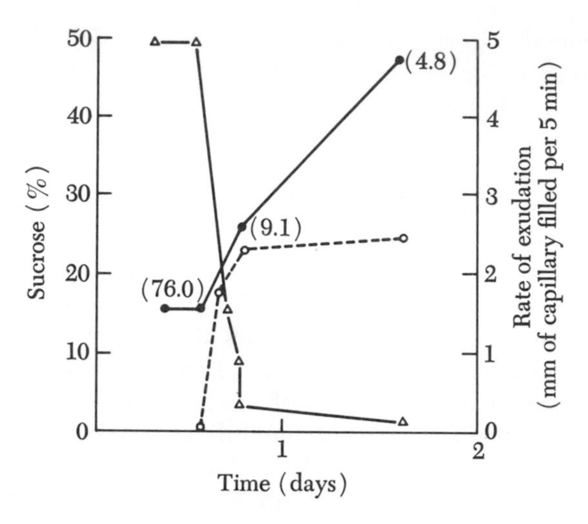

Figure 9.2. The effect of passing a 25 percent sucrose solution through the xylem vessels of *Salix viminalis* upon the rate of exudation. *Dashed line with open circles,* concentration of sucrose in xylem effluent; *solid line with triangles,* rate of exudation from stylets; *solid line with solid circles,* concentration of sucrose in exudate. Sucrose secretion values are given in brackets. From Weatherley et al. (1959).

tubes. For import into sieve tubes and export from them, an active transfer mechanism was invoked, with water following potential gradients passively by diffusion; the surface area of 800 or more elements is adequate for uptake. With a discrepancy of several orders of magnitude between diffusion constants of sucrose, active transfer was considered. However, in view of the symplast concept, this latter type of movement may not be necessary; sugar potential in plastids is such that starch is commonly formed early in the day. Possibly, sucrose may also move passively from chloroplast to sieve tube down a concentration gradient along continuous cytoplasm. There, streaming in mesophyll cells may serve to accelerate movement.

Because their paper was published before the demonstration that sieve plates are normally open, Weatherley and his co-workers had difficulty reconciling their results with a mechanism of mass-flow through blocked sieve plates. They found the turgor pressure of the sieve tubes to be in the neighborhood of 10 atm, which they considered adequate to produce the flows that they measured. They ex-

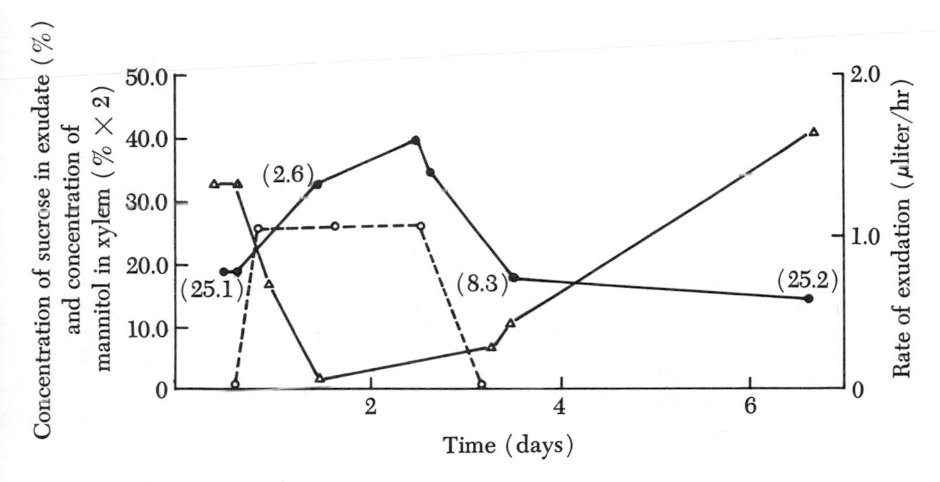

Figure 9.3. The effect of passing a 15 percent mannitol solution through the xylem vessels of *Salix viminalis* upon concentration of sucrose in the exudate and the rate of exudation. *Dashed line with open circles,* concentration of mannitol in xylem effluent; *solid line with triangles,* rate of exudation from stylets; *solid line with solid circles,* concentration of sucrose in exudate. Sucrose secretion values are given in brackets. From Weatherley et al. (1959).

plained movement into and out of sieve tubes as taking place along gradients of sucrose potential. The motive power for pressure flow they took to be the secretion of solutes into the sieve tubes and their withdrawal from them. They took their experiment illustrated in Figure 9.2 to be an indication of the high sucrose potential that exists in cells surrounding the sieve tubes, because secretion into the sieve tube continued when its contents were as much as 50 percent sucrose. Their view was further borne out by the fact that stylet exudation will continue from whole plants after the leaves have wilted from water deficiency, and by the fact that phloem exudation from cut phloem will take place from wilted leaves (Crafts, 1936).

Duloy et al. (1961) gave an excellent analysis of the effects of metabolic inhibitors on translocation. They pointed out that such inhibitors are seldom completely effective; that they may enter the xylem, pass to the leaves, and affect assimilation at the source; and that they may also move with assimilates into sinks, and inhibit absorption from the sieve tubes. They suggested, further, that inhibitors may affect the semipermeability of the parietal cytoplasm of the sieve tubes; that they might increase the viscosity of sieve-tube sap; and that they might change the colloidal state of slime.

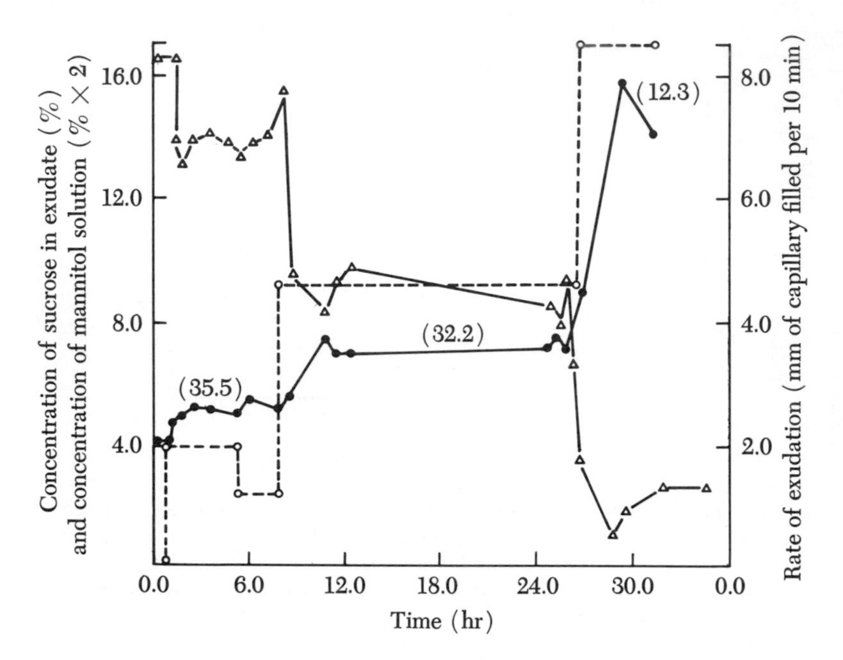

Figure 9.4. The effect of irrigating the inner surface of the bark of *Salix viminalis* with mannitol solutions. *Dashed line with open circles,* concentration of mannitol solution; *solid line with triangles,* rate of exudation from stylets; *solid line with solid circles,* concentration of sucrose in exudate. Sucrose secretion values are given in brackets. From Weatherley et al. (1959).

Concluding that translocation in mature sieve tubes is by mass flow, Duloy et al. speculated as to the role of companion cells and phloem parenchyma. They considered companion cells to be involved in lateral movement of assimilates and in maintenance of the structure of the metabolically inert sieve tubes. Parenchyma, they suggested, may be concerned with temporary storage of solutes. In view of their common content of starch this seems reasonable; the storage period may be brief, as in the second internode of *Triticum aestivum* (Wardlaw and Porter, 1967), or prolonged, as in deciduous trees during dormancy.

Wiersum (1967) presented an excellent set of calculations on the influx of water and calcium, and the efflux of phosphorus, potassium, and dry matter constituents (mainly sugar), from *Solanum tuberosum* seed pieces. Assuming that the tuber was importing water and calcium via xylem and exporting organic material and phloem-mobile minerals, he calculated the water balance of the seed piece and estimated the influx and efflux of substances shown in Table 9.2.

Table 9.2. Derived (*in italics*) and estimated values for the amounts of influx and efflux of substances in sprouting tubers of *Solanum tuberosum.*

Weeks from planting	Water balance of the tuber (ml)			Efflux of P_2O_5 (mg)	Efflux of K_2O (mg)	Increase of CaO (mg)	Concentration of CaO in influx (mg/ml)
	Net H_2O increase	Total influx	Phloem efflux				
3	*15.5*	37.5	22.0	*25.0*	*100*	*21.8*	0.58
4	*18.7*	60.9	42.2	*48.0*	*179*	*35.4*	0.58
5	*19.9*	78.0	59.0	*66.9*	*231*	*48.9*	0.63
6	*19.1*	85.1	66.0	*75.8*	*266*	*54.9*	0.64
7	*22.0*	94.2	72.2	*81.9*	*279*	*57.8*	0.61
8	*15.4*	91.1	75.7	*85.9*	*297*	*60.3*	0.66

Source: Wiersum (1967).

In the course of 8 weeks, he calculated that the efflux of water was 75.7 ml, the amount of dry matter 1415 g, the amount of potassium (as K_2O) 300 mg, and the amount of phosphorus (as P_2O_5) 86 mg. Thus, the total calculated efflux was approximately 14.9 g, and the concentration of the efflux was 19.7 percent. Wiersum concluded that the results obtained in this rather crude calculation support the validity of the mass-flow hypothesis. They bring to mind two observations reported on growth of fruits that takes place by absorption of the assimilate stream: fruits of *Pisum sativum* have been noted to grow to maturity with little uptake of calcium, and fruits of *Citrus sinensis* in orchards treated with monuron contain no residues of this herbicide, because calcium and monuron are not phloem-mobile, as shown in Table 8.1.

Hammel (1968) provided direct evidence for high turgor pressure in the phloem of *Quercus borealis,* the plant upon which Huber et al. (1937) performed their classical work. Using a special needle to pierce the bark and to allow for exudation into a manometer, Hammel measured the turgor pressure of phloem exudate directly. He measured turgor pressure at 6.3 m and 1.5 m above the ground; 61 measurements at the upper level averaged 15.13 atm, and 62 at the lower level averaged 13.85 atm. Hammel stressed the elastic nature of the phloem conduits.

Trip and Gorham (1968c) made studies on the comparative movements of ^{14}C-labeled sugars, ^{3}H-labeled photosynthates, and HTO in *Curcurbita.* In their first experiment, they applied $^{14}CO_2$ to a leaf and painted HTO on the under surface; exposure time was 30 minutes. They found that ^{14}C was distributed fairly uniformly throughout the leaf and petiole; HTO was concentrated in the treated area with a small amount in the midrib and a trace in the rest of the leaf blade; none reached the petiole. These results are natural: ^{14}C-labeled photosynthate was being provided by the leaves for normal symplastic distribution, while transpiration water from the roots was moving apoplastically in the leaf and evaporating into the atmosphere. HTO applied to the leaf surface would have to diffuse against this current in order to enter the phloem of leaf veins. While $^{14}CO_2$ is actively absorbed into the symplast, there is no evidence for active water movement into, or between, plant cells.

In further experiments in which HTO or HTO and ^{14}C-labeled glucose were introduced through flaps, according to the method of Biddulph and Cory (1957), the tritiated water moved through the leaf and petiole along a nearly flat gradient; ^{14}C-labeled glucose was similarly distributed. The constant ratios of ^{14}C to ^{3}H that were observed suggested to Trip and Gorham a concurrent movement of sugar and water (mass flow); and they concluded that such solution flow cannot be excluded as a translocation mechanism. Thus, the excellent work of Trip and Gorham on translocation was brought into line with evidence from phloem exudation and tracer movement in support of the mass-flow mechanism.

In three excellent papers on the kinetics of the translocation of ^{14}C in *Glycine max*, Fisher (1970a, 1970b, 1970c) found that the translocation profile moved down the stem with little or no change in shape. With pulse-labeling, sucrose made up 90–95 percent of the activity in the stem up to 2 hours; less than 3 percent of the activity was in an insoluble form. Two-thirds of the petiolar sucrose was in the assimilate stream, and one-third was in a stationary pool; sucrose was slowly accumulated from the translocation stream. Thus, loss from the stream was slow, and Fisher concluded that the kinetics of ^{14}C translocation must be determined primarily by factors operating within the leaf, rather than in the stem. Fisher mentioned a special paraveinal parenchyma tissue in the leaves of *G. max* that has similarities to the transfer cells described by Gunning, Pate, and Briarty (1968).

Since there was no spreading of the profile, the results are considered to be inconsistent with an active diffusion mechanism. Using pulse-labeling, Fisher found the specific activity of leaf sucrose to attain a maximum within 5 minutes, whereas that of exported sucrose did not reach a maximum until at least 20 minutes had elapsed. Fisher proposed that the source of sucrose in the leaf consists of two compartments; he proposed that his paraveinal parenchyma might constitute one compartment, and the mesophyll the other.

The ^{14}C kinetics of two steroids resembled those of sucrose. Fisher suggested that these compounds might act as membrane carriers of sucrose between the translocation stream and surrounding cells. Because the need for such carriers has not been demonstrated, the authors of this book consider it probable that the steroids found were degradation products from sieve tube organelles (slime) that were riding along on the assimilate stream (compare Biddulph and Cory, 1965).

Visualizing various mathematical models, Fisher studied the effects of leaf size, leaf shape, and translocation velocity on the rate of tracer export from a leaf. He found that the dimension of these parameters approximated the time required for the tracer to reach the petiole from the tip of the leaf. Designating this dimension as the "kinetic size," he found it to have little effect in *Glycine max;* with plants having larger leaves, it would be reasonable to anticipate larger effects.

Fisher described source-pool kinetics in *Glycine max* as a two-compartment system; kinetics in the petiole were approximated by a similar model, represented by the translocation stream and tissues outside the stream. A combination of these models, he found, resembles the translocation kinetics of *G. max*. Comparing his models with others in the literature, Fisher found his to be in substantial agreement with those of Evans, Ebert, and Moorby (1963) but to be inconsistent with the model of Canny and Phillips (1963) based on movement of transcellular strands.

In a consideration of pressure drop in sieve tubes, Fisher assumed that the Hagen-Poiseuille equation is valid. He apparently did not appreciate the weaknesses of this equation when applied to flow through sieve plates of the protophloem of leaves. (The modifications proposed on page 400 should throw new light on this process.) Thus, Fisher's questioning of the adequacy of the mass-flow mechanism may not hold up under critical analysis.

Fisher's argument for the reversibility of phloem transport would seem logical in view of the point, made so often in this book, that the phloem is a distribution system. The quick reversibility proved by the experiments of Weatherley et al. (1969) and Peel and Weatherley (1962) when they girdled stems, provides evidence for this conclusion. These kinetic studies of Fisher would seem to fit well the requirements for mass flow of the assimilate stream through an open system of phloem conduits (sieve tubes).

Evidence in favor of the mass-flow mechanism could be marshalled from many of the experiments reported in Chapters 6 and 8 on movement of assimilates and tracers. Because the chapters following this one will consider the mass-flow mechanism in detail from many aspects, it will not be dealt with further here. It suffices to point out at the end of this discussion on mechanism that the mass-flow mechanism can now be reconciled with phloem anatomy (open sieve plates and a living, continuous, stationary, filamentous reticulum in functioning sieve tubes), with the various aspects of phloem exudation, and with the great variety of distribution patterns of labeled tracers in plants. With regard to the alternative hypotheses of protoplasmic streaming, metabolic movement, activated diffusion, and surface migration, there are many unexplained and apparently unexplainable inconsistencies. A few of these are: a lack of evidence for protoplasmic streaming in mature sieve tubes; the lack of any plausible mechanism to account for the application of the energy of metabolism to long distance transport of sugars, amino acids, and so forth, through the phloem conduits at velocities of 100–300 cm per hour; the lack of any concept of how the physical process of diffusion could be speeded up to the required values; and the lack of any explanation of how the process of adsorption to surfaces could possibly account for rapid transport, when its fundamental action is to hold molecules to surfaces, not to propel them along surfaces. Coupled with these weaknesses are the spatial relations of sieve tubes that indicate that assimilates moving in sieve tubes occupy several times as much space as the parietal layer of cytoplasm. And finally, these alternative hypotheses have no way of accounting for the correlated flow of assimilates and water in a stream, as occurs in phloem exudation. Although the objection may be made that phloem exudation is not a normal process, it is, nonetheless, a real, measurable, physical phenomenon that seems to be quantitatively related to assimilate transport. As previously stated (Crafts, 1961b, page 151), when we can adequately explain phloem exudation we will be well along on the way of explaining assimilate translocation. The research reported in this chapter offers many insights into the processes of phloem exudation. If we can accept them at face value, they provide convincing evidence for a rational mechanism of phloem function.

The Source

Without being facetious, it seems that the best place to start an analysis of the functions of the pressure-flow mechanism of assimilate movement in plants is at the source. The simple concept of this mechanism illustrated in Figure 3.1 would seem to imply that the leaves of a plant constitute the source, and that the main stem, root system, meristems, and fruits make up the sinks. The work of Weatherley and his associates (1959) surely dispels this simple picture and proves that the whole phloem makes up an integrated system of sources and sinks, with the sucrose potential being the factor most intimately involved in determining the direction that the synthesis-breakdown processes take at any one time and under any one set of conditions. Work on *Saccharum officinarum* (Hartt and her associates, Hatch and Glasziou), *Beta vulgaris* (Geiger, Swanson, Anisimov, Joy, Mortimer, Butcher), *Triticum aestivum* (Evans, Wardlaw, Carr, Porter, Quinlan, Sagar, Stoy), forest trees (Huber, Zimmermann, Ziegler), and a host of other plants show that source–sink relations are complex and labile, and subject to influence by light, temperature, water, growth, storage, senescence, disease, and many other factors. With this complex, labile situation in view, it should be interesting to consider the nature and function of the source in plants.

Obviously, photosynthesis is the primary process that provides the green portions of a plant with the solutes responsible, through osmosis, for the driving force of pressure flow. What are some of the anatomical and physiological factors that enable the green tissues—be they leaves, stems or fruits, to activate the translocation process?

In chapter 16 of the 1965 revision of her *Plant Anatomy,* Esau (1965b) gave a comprehensive description of the leaf. With excellent illustrations and detailed description, Esau covered various leaf forms, leaf development, leaf histology, and leaf abscission. In her description of leaf histology, she considered venation patterns, pointing out the close spatial relation of vascular tissues to mesophyll. Measurements on six species of dicotyledons of various forms showed that the total length of veins averaged 102 cm per square centimeter of leaf blade. This thorough distribution of vascular tissue was further reflected in the value of 130 μ for the average interveinal spacing in dicotyledonous leaves.

Esau further pointed out a correlation between vein distribution and the structure of the mesophyll. The larger the volume of non-conducting tissue, such as palisade parenchyma, the closer together are the vascular bundles. Thus, it has been observed that sun leaves, having strong palisade development, contain greater lengths of vein tissue than shade leaves.

However, despite the thorough distribution of veinal tissue in leaves, diffusion alone could not account for the rapid export of labeled assimilate from leaves reported by Kursanov (1963) and Webb and Gorham (1964). Protoplasmic streaming undoubtedly accelerates the movement of molecules from chloroplasts to the phloem. Cyclosis, such as that shown in *Elodea canadensis*, is an active process that carries the plastids around the cells in such a way that the cytoplasm is never very far from a pit area through which export may take place. In most higher plants, such movement of plastids is lacking; but strands on the inner surface of the cytoplasm, and often extending across the vacuole, show streaming motion of varying intensity.

Esau also illustrated the structure of small vascular bundles of leaves, showing the close relationship of tracheids and protophloem sieve tubes. Often these vascular elements are in direct contact, a fact that emphasizes the semipermeable nature of the sieve-tube plasmolemma; were this layer not impervious to assimilate molecules, the pressure-flow mechanism would not operate. Esau described the bundle sheaths, the parenchymatous cells that completely surround the small vascular bundles in the leaf. The bundle-sheath cells of dicotyledonous plants are called border parenchyma. These cells usually contain chloroplasts, although the chloroplasts may be less numerous than in palisade and spongy parenchyma cells. The bundle-sheath cells must function in the passage of molecules between palisade cells and the sieve tubes by symplastic movement; their protoplasts are also bathed by transpiration water, and are thereby provided with inorganic nutrients supplied by the roots. In bundle terminals, the differences in size and contents between border parenchyma cells and companion cells may be minor or lacking. These cells undoubtedly play important roles in the translocation process. The transfer cells of minor veins are in this same category (Gunning, Pate, and Briarty, 1968; Pate and Gunning, 1969).

Many workers have insisted that movement of assimilates from border parenchyma cells into sieve tubes is an active process; some term it secretion (Kursanov, 1963). Most work on this process, however, has been done by mass analysis of the tissues, and this affords no measure of the actual concentrations present in the symplast cytoplasm. Sucrose potential in chloroplasts may reach a point, minutes after illumination, such that starch is synthesized: concentrations must be high. It is unfortunate that it is impossible yet to measure cytoplasmic sugar concentration so that this matter can be settled. Very probably, sugars move along this route in phosphorylated forms, and since the ratio of sugars to phosphorus in the phloem sap does not show a stoichiometric relation (Wanner, 1953a, 1953b), the phosphorus must recycle to the palisade cells, possibly via the transpiration stream.

At the other end of the route (where sugars may be condensed to starch in phloem and xylem parenchyma) metabolic energy must be required, because the phloem sap, as collected by means of aphid stylets, usually runs between 10 and 20 percent sucrose (Mittler, 1958; Weatherley et al., 1959). The sugar potential here is hardly high enough to cause starch formation without a supply of energy.

Morretes (1962) described the vascular bundles of leaves in *Capsicum annuum* and *Phaseolus vulgaris*. In *C. annuum*, both the principal and the lateral vein endings contain phloem, and the xylem and phloem generally end together. In some bundles, however, the terminal xylem element extends slightly beyond the last sieve tube. In *P. vulgaris*, the main and lateral vein endings, according to Morretes, terminate with both phloem and xylem (or, in some places, with xylem only). Some lateral endings contain no sieve elements; in those endings, parenchyma may occupy the position of the phloem. The sieve elements of vein endings contain slime bodies, and the companion cells of terminal phloem in both species are wider than sieve elements. These transition forms have dense cytoplasm and rich, granular contents; in this, they contrast with sieve tubes, which, Morretes noted, are characterized by a clear, empty appearance. This is an expression of the mature state illustrated so well in root-tip sieve tubes in Figure 2.5 (page 27) and in minor veins in Figure 9.5. (Pate and Gunning, 1969).

Morretes pointed out that it is often difficult to determine whether a given cell in a vein ending is a companion cell or a phloem parenchyma cell. Some of the nucleated cells stain more densely and have more granular contents than others; there was no sharp demarcation between the different nucleated cells. This may reflect a similarity in function of all of the nucleated cells in this position. The association of sieve tubes, companion cells, and phloem parenchyma in the small veins suggests a cell system concerned with transfer of sugars from mesophyll to the phloem conduits. In this sytem, the companion cells are commonly visualized as metabolically active, and engaged in the synthesis of sucrose and its secretion into the sieve elements; possibly, all nucleated cells in small veins engage in this process.

Trip and Gorham (1968b) suggested that the tritiated sugar used in their histoautoradiographic studies is translocated in these "presumed companion cells," rather than in sieve tubes. The relation between protophloem sieve tubes and surrounding parenchyma cells illustrated in Figures 2.4 (page 26), 2.5 (page 27), and 9.5 indicates the difficulty involved in Trip and Gorham's interpretation. Morretes

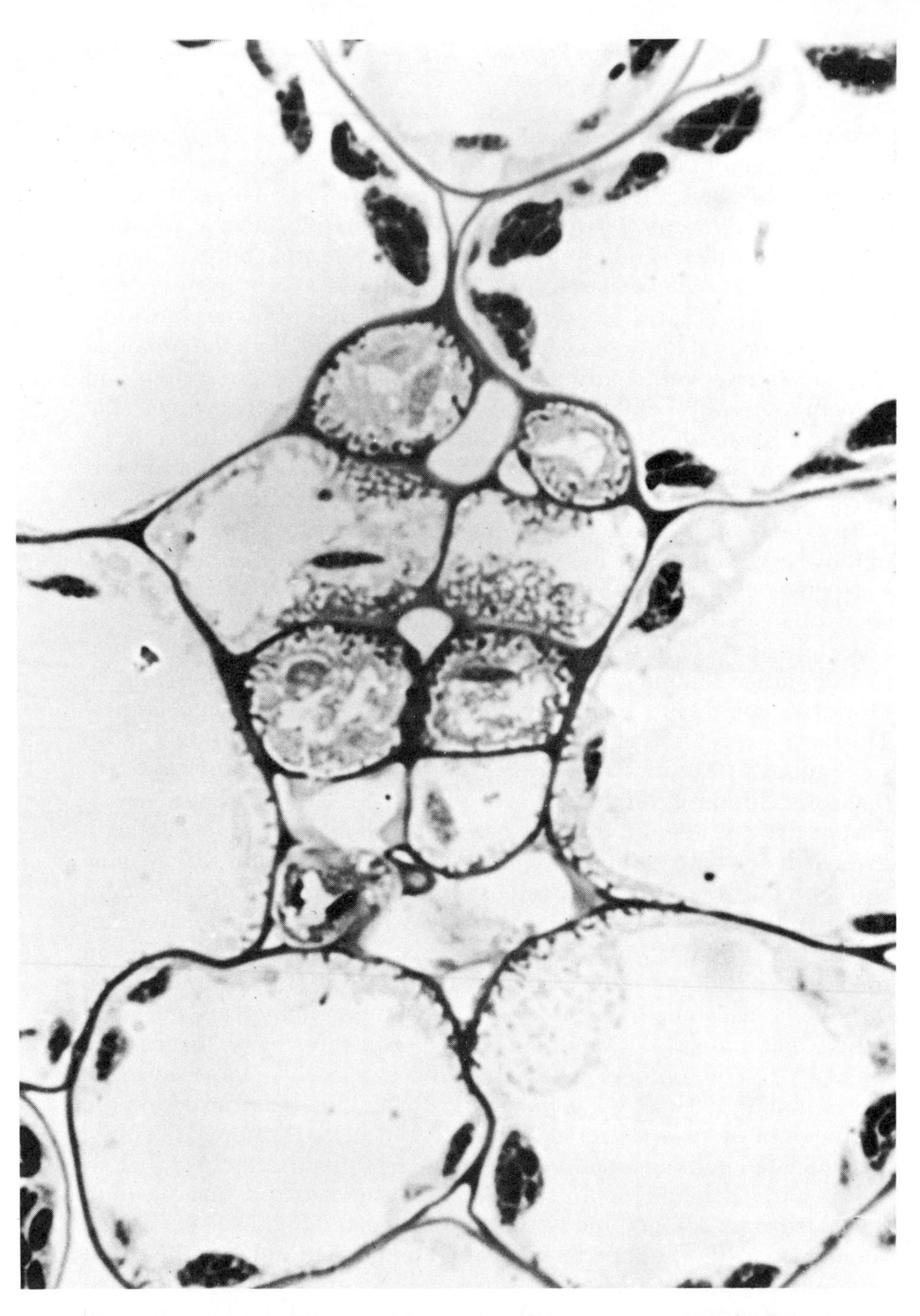

Figure 9.5. Functioning protophloem sieve tubes in a minor vein of a mature leaf of *Anacyclus pyrethrum*. Each sieve tube is flanked by two transfer cells with dense contents. The highly vacuolated cells are phloem parenchyma. From Pate and Gunning (1969).

(1962) noted the clear, empty appearance of functioning protophloem sieve tubes in bundle endings.

Esau (1967a) studied the minor veins in leaves of *Beta vulgaris*, considering structure–function relations. In this study, she used both the light microscope and the electron microscope, and provided excellent illustrations by both techniques. The netted type of venation was shown in a view of a cleared leaf, and interveinal distances were given; three typical values were 140 μ, 240 μ, and 310 μ. Trip (1969) cited a distance of 100 μ. Because the parenchyma cells in this tissue have diameters in the range of 30–80 μ, it is obvious that solutes would have to move through only a few cells in order to reach a vein.

Esau showed the intimate relation between sieve tubes and tracheids. She explained that the primary walls of the mature tracheary element may be largely disintegrated, so that the walls of the adjacent parenchyma cells and sieve elements are virtually in contact with the contents of the tracheary element, making for rapid and easy apoplastic movement of solutes moving in the transpiration stream.

Esau noted that the spatial relations of the veins in the leaves of *Beta vulgaris* provide an effective system for exchange of materials between vascular tissues and photosynthetic cells. Accepting the common concept that the movement of sugars into and out of the phloem depends on active metabolic transfer that involves enzyme function, she pointed out the strong phosphatase activity of phloem parenchyma, and the abundance of ribosomes shown in these cells by the electron microscope. Thus, it seems that the nucleate cells of phloem may have an active carrier system bringing about the transfer of sugars across membranes by a succession of phosphorylations and dephosphorylations resulting in their secretion into the sieve tubes.

She further noted the facility with which turgor may be established in the sieve tubes of the small veins when water moves in by osmosis from neighboring tracheids. The sensitivity of transport rates in the phloem to water deficits (Wardlaw, 1967) may be a reflection of this association.

Esau's light microscope preparations (her figures 10–14) show the clear, empty appearance of mature sieve tubes, and her electron microscope views (her figures 20, 23, and 24) emphasize this apparent paucity of contents. These relatively open elements are the structures that are capable of conducting the assimilate stream at rates of 100–300 cm per hour.

Geiger and Cataldo (1968) made histoautoradiographic studies of the leaves of *Beta vulgaris*. On cleared leaves they found 70 ± 9 cm of veins in each square centimeter of leaf, and they stated that a 20 μ

length of minor vein serves about 40 mesophyll cells. They found that ^{14}C fed to leaves in $^{14}CO_2$ is concentrated in the cytoplasm of mesophyll cells. They suggested that the structural specialization of phloem in minor veins leads to sugar accumulation, causing movement of the assimilate stream by actively establishing a concentration gradient in the phloem.

In their complete paper Geiger and Cataldo (1969) provided convincing evidence for the presence of soluble carbohydrates in the minor veins (their figure 8B), and for insoluble carbon compounds in mesophyll cells (their figure 12). Sieve tubes in minor veins are relatively free of organelles, adjacent border parenchyma cells which exceed the sieve tubes many times in transverse sectional area are rich in organelles. Sieve tubes of petiole veins are much larger than those of minor veins of the leaf. Geiger and Cataldo presented some interesting calculations correlating structural features of the leaf with the process of vein loading. Details of this process however must await methods whereby the identification and quantification of carbon compounds in individual cells are developed.

These descriptions of leaf structure point up the futility of mass analysis of mesophyll and veinal tissues of leaves. Such analysis is futile because the bulk of the minor veins, embedded in the mesophyll, are so masked that they cannot be isolated, and are not discernable by the naked eye.

This pattern applies to cereals, except that the leaves constitute a smaller part of the dry weight at flowering. In *Beta vulgaris*, the absence of an extended stem results in a modified pattern in which a large portion of exported assimilates goes to the root. The distance moved by assimilate from a given leaf relates to its position, which, by the availability of accessible channels, directs the flow to the nearest sink. Removal of leaves results in diversion of materials, rather than inhibition of movement. The amount of material exported from a leaf does not seem to depend on any particular sink. Geiger, Saunders, and Cataldo (1969) used *Beta vulgaris* plants, pruned to include a single source leaf and a single expanding sink leaf, in studies on translocation and accumulation of labeled assimilates in petiole tissues. During an 8-hour period, 2.7 percent of the labeled assimilate was accumulated per centimeter of petiole. The assimilate was moved mainly as sucrose and accumulated as sucrose and as ethanol-insoluble compounds. The minimum peak of translocation velocity averaged 54 cm per hour. The ratio of cross-sectional area to sieve-tube area required for transport was 1.2, indicating that the total sieve-tube capacity was being utilized. The ability of the parenchyma of the petiole to act as sink tissue for storage is consistent with the concept that storage along the transport route serves to buffer sucrose concentration during periods of fluctuating assimilation, with sucrose potential apparently determining the direction of movement. The correlation ($r = 0.76$) between sieve-tube area and rate of mass transfer in morphologically similar plants indicates that the cross-sectional area of the conducting system limits translocation through the petiole in *Beta vulgaris.*

Swanson (1959) described leaf structure in relation to function as a source for transport of assimilates.

Porter (1966) reviewed current literature covering the subject of leaves as distributing agents of carbon. From many studies, including her own on *Nicotiana tabacum,* she reached the following conclusions. In *N. tabacum* plants at flowering, leaves constitute 50 percent of the dry weight of the plant. If these provide 70 percent of the assimilates, about 40 percent is incorporated at the site of fixation. Around 10 percent moves acropetally, and 25 percent is incorporated in the stem; 25 percent moves from lower leaves to roots. Thus, the plant is built up, tier by tier, by lateral withdrawal of the sucrose lost from the leaves.

In his 1963 review, Kursanov gave a fine treatment of the transport of assimilates from mesophyll to phloem. During the first few seconds, triose phosphates, hexose phosphates, organic acids, and amino acids may be formed. The first nonphosphorylated sugar is sucrose, followed rapidly by starch. The number and amounts of these assimilates vary with the age of the leaf, conditions of illumination, mineral nutrition, and the plant species.

Kursanov found ^{14}C-labeled assimilates in the fine veins of *Rheum rhaponticum* at a distance of 1–2 cm from the point of application of $^{14}CO_2$ within a few minutes of the onset of photosynthesis; the major portion of the radioactivity was in sucrose. Apparently the portion of sugars most recently synthesized moves most rapidly, while those in storage in vacuoles are only gradually drawn into the channels of transport. Because movement is regulated by metabolism, some assimilates are rapidly transported to the phloem, while others remain behind; sugars apparently move freely; organic acids, in *R. rhaponticum,* tend to accumulate in vein tissues; only small quantities of malic and citric acids were transported in phloem. Kursanov found amino acids to enter conducting tissues selectively: threonine was particularly mobile in *R. rhaponticum;* serine and alanine were freely mobile, whereas aspartic acid and proline were less mobile.

Migration from mesophyll to phloem was accelerated when the

leaves were given ATP. Half leaves receiving ATP had 2800 count/min, as compared with water treated controls that had 1,500 count/min in their glucose-6-phosphate fraction; fructose-6-phosphate and glucose-1-phosphate were increased from 1400 to 2200 count/min, and triose phosphates were increased from 1700 to 2400 count/min. Movement from leaves to roots in *Beta vulgaris* was also enhanced by ATP; young plants responded more than old ones. Phosphorylation seems to be essential to the rapid movement of sugars to the sieve tubes, but Ziegler (1956) found the hexose phosphate concentration of phloem exudate to be low, compared with the quantity of sugars under transport. This would seem to imply a recycling of the phosphorus, and its role would be that of carrier.

Kursanov and Brovchenko (1962) enriched *Beta vulgaris* leaves with ATP (Table 9.3), but found no sucrose phosphate in them. Kursanov concluded that transport of sucrose from mesophyll to phloem was preceded by its transformation into two molecules of hexose monophosphate, which, upon entering the phloem, were resynthesized into sucrose. He explained that an additional reserve of free energy would be required by this system; such mutual transformations are reversible, are catalyzed by transferase-type enzymes, and proceed with little change in the free energy reserve; transport is selective and requires ATP.

Crafts and Yamaguchi (1958), using a variegated variety of *Tradescantia fluminensis*, showed by autoradiography that photosynthesis is required for translocation of ^{14}C-labeled 2,4-D. The tracer was readily exported and translocated from green leaves; export from variegated leaves was proportional to the amount of green tissue in them, and export from chlorotic leaves failed completely. In 1964, Crafts and Yamaguchi demonstrated movement of 2,4-D, amitrole, and maleic hydrazide from green leaves, and failure of export of the same substances from chlorotic leaves. Considering that all of the compounds listed in Table 8.1 (page 176) as phloem mobile are capable of diffusing across the cuticle, migrating from epidermis to sieve tubes, and translocating to distant plant parts throws some doubt on the requirement for an active transport system in source leaves. It is difficult to rationalize active transport with the array of different types of compounds that are moved into and translocated along the phloem. Many of these compounds are exogenous materials of varying physiological activity, and some of them are highly toxic.

Quinlan's work (1965, 1966) on assimilate distribution in *Malus* brought its source–sink relations into sharp focus. Labeling single leaves at different positions he found the familiar pattern of export

Table 9.3. Distribution of ^{14}C in various phosphate esters in leaves of *Beta vulgaris* enriched with ATP* and ^{14}C-labeled sucrose.

	Radioactivity [(count/min)/g dry wt]				
	In leaf blade				Total radioactivity of phosphate esters in conducting bundles of petiole
	Total radioactivity of phosphate esters	Glucose-6-phosphate	Fructose-6-phosphate and Glucose-1-phosphate	Phosphoglyceric acid and phosphoglycerol	
Treated plants	7452	2810	2258	2384	395
Controls	4626	1520	1418	1688	233
Difference	2826	1290	840	696	162

Source: Kursanov and Brovchenko (1961).

* 0.006 M.

reported by Crafts (1956b) for *Convulvulus arvensis* and by many workers with tracers since. A young expanding leaf at the top of the stem did not export; the fourth and sixth youngest leaves exported predominantly in an acropetal direction; from leaf 8, assimilate moved both acropetally and basipetally; and all leaves below leaf 8 exported to the roots. Removal of leaves shifted the competition of source and sink. When a leaf near the top of the plant that would normally export only to the young growing shoot tip had all leaves below it removed, thus reducing the supply to the roots, the stream divided and went both to tip and to roots. Likewise, removal of leaves above a mature leaf midway on the stem increased acropetal movement. Thus, source and sink can be manipulated experimentally to shift distribution to fit the needs of existing sinks. In fact, the direction of flow can be completely reversed, as shown by Crafts and Yamaguchi in *Zebrina* in 1958.

Ziegler's work (1964) on storage, mobilization, and distribution of food reserves in forest trees has direct bearing on the effectiveness of phloem parenchyma and ray cells as channels of movement from phloem to xylem, and vice versa, in providing for the spring flush of growth. At least for a period, starch and fat reserves constitute the source that provides for this rapid growth; after expansion of leaves and differentiation of new phloem, the normal leaf-stem-root source–sink relation is restored.

In studies on the physiology of the wood rays of *Populus*, Sauter (1966) found that cell complexes connected with vessels, and those that are isolated from vessels, exhibit different physiological behavior. Those cells that actually connect rays and vessels have special cytological differentiation; they also have high physiological activity. During dormancy, the ray parenchyma cells have high acid phosphatase activity. In early spring, Sauter found there was a great increase in starch mobilization. During summer, the starch deposition rate decreased, with a second period of starch dissolution in late autumn and winter. Variations in acid-phosphatase activity seemed to be associated with deposition and mobilization of starch. The parenchyma cells that connect rays and vessels had a higher acid-phosphatase content than other ray cells, related to starch dissolution in these cells. During starch mobilization, phosphatase activity was localized on the starch grains; in spring, when starch was being mobilized, acid phosphatase was localized on the large pits between contact cells and vessels, apparently denoting active sugar transport into the vessels. These cytochemical studies seem to indicate that acid phosphatase participates actively in the intercellular and transcellular transport of carbohydrates by dephosphorylation–phosphorylation processes taking place at the ray-cell extremeties. The transcellular transport, Sauter assumed, is activated by a steep intracellular gradient in sucrose potential created enzymatically. This overall phenomenon involves not only symplastic sugar transport from phloem to xylem via the rays, but, eventually, release of sugar into the apoplast for upward movement in the xylem vessels.

Butcher (1965), using *Beta vulgaris* plants having a simple supply leaf as source and the root as sink, studied the kinetics of translocation of ^{14}C-labeled assimilate. Butcher supplied the leaf with $^{14}CO_2$ of constant and known specific activity during a 3-hour labeling period in light. After 2 hours, the rate of accumulation of activity became linear and remained so for the remainder of the light period; at the end of this period, the plant was darkened and the leaf was flushed with air. Accumulation during the dark period became linear after 20 minutes, and the rate was 43 ± 7 percent of that in the light.

In studies on sugar uptake and translocation in *Ricinus communis* seedlings, Kriedmann and Beevers (1967a) found that movement of materials through the cotyledons into the embryonic axis exceeded 2 mg per hour after 5–6 days of germination. Sucrose was the predominant sugar undergoing transport. The cotyledons were able to absorb sucrose after removal of the endosperm. Experiments in which cotyledons were immersed in ^{14}C-labeled sucrose showed increasing rates of sugar absorption up to 0.5 M. Removal of the embryonic axis (sink) drastically reduced sugar uptake. As observed in steady uptake and pulse-chase experiments, more than 80 percent of the sugar entering the cotyledons from endosperm is moved on into the embryonic axis. The cotyledons of excised seedlings absorbed sucrose against apparently high concentration gradients. Absorption was stimulated by phosphate, had a *p*H optimum at 6.4, and was inhibited by arsenate, azide, and DNP. In a second paper, Kriedmann and Beevers (1967b) showed that uptake of sucrose by *R. communis* cotyledons exceeded their uptake of glucose; it exceeded their uptake of glucose and fructose at concentrations above 0.1 M. Only minute amounts of labeled hexoses were recovered from cotyledons after exposure to the ^{14}C-labeled compounds; sucrose contained virtually all of the ^{14}C in the sugar fraction. When cotyledons were supplied with sucrose whose fructose portion had been labeled with ^{14}C, 90 percent of the radioactivity was retained in the fructose residue. Kriedmann and Beevers concluded that the greater part of sucrose absorbed by cotyledons is not hydrolyzed prior to or during uptake.

In contrast to Kriedmann and Beever's work, Brovchenko (1967) found, in leaves of *Beta vulgaris*, that glucose was absorbed from solution more intensively than sucrose. This he attributed to competition between glucose and fructose resulting from hydrolysis of sucrose at the interface of the cell wall and the cytoplasm. The addition of cold glucose to a solution of ^{14}C-labeled sucrose decreased absorption of the latter. Glucose and a mixture of hexoses penetrating the cells of mesophyll and fine bundles were rapidly isomerized into fructose; they were also utilized in the synthesis of ionic and higher molecular substances. Such a mixture, penetrating the cells of mesophyll and fine bundles, was used in sucrose synthesis, which took place most rapidly in fine bundles. Strongly asymmetrical molecules of ^{14}C-labeled sucrose with a predominance of ^{14}C-labeled glucose were formed from ^{14}C-labeled glucose and uniformly ^{14}C-labeled sucrose that entered mesophyll and fine bundle cells. Brovchenko was convinced by his experimental data that sucrose synthesized in green mesophyll cells is split in them by invertase before entering the cells of fine conducting bundles; and that, in the form of activated hexoses, it surmounts the cytoplasmic barriers of the mesophyll cells and adjacent conducting cells.

While studying the origin of carbon and nitrogen sources for protein synthesis and growth of *Beta vulgaris* leaves, Joy (1967) found that protein synthesis in young leaves utilizes carbon from photosynthesis and from translocated sucrose; nitrogen is imported via both xylem and phloem. Joy found that the carbon of young leaf protein comes mainly from assimilated CO_2; translocated sucrose contributes more of its carbon to insoluble carbohydrate. Most protein amino acids become labeled from $^{14}CO_2$ assimilated by the young leaf, but glutamine is an exception. Glutamine, or glutamate, is synthesized from sucrose in roots, and is translocated to young leaves. Joy suggested that a small but significant proportion of the nitrogen requirement of the young leaf is translocated from the root in the phloem as glutamine; inorganic nitrogen is translocated in the xylem. Joy defined "young leaf," for the purposes of his experiments, as a leaf that has not yet reached compensation; that is, as a leaf that is still importing assimilates via the phloem. Under such conditions, young leaves acquire a considerable portion of their organic nitrogen from the root, including storage tissue. Bennett (1956) showed that shading or defoliating a plant of *B. vulgaris* induced import of curly top virus from infected mature leaves. If virus moves in on the assimilate stream, it seems logical that organic nitrogen could do so as well. Joy offered evidence that this imported organic nitrogen is in the form of glutamine, glutamic acid, or both. Removal of mature leaves had little effect on growth of young leaves over a 4-day period; soluble nitrogen in the remaining leaves was double that in normal plants, a result that Joy attributed to the removal of competition.

Leonard and Glenn (1968a) studied the translocation of certain herbicides in detached leaves of *Phaseolus vulgaris*. Because such leaves are sinks for absorption of exogenous tracers and sources for export into petioles, this work is relevant to the present discussion. Labeled 2,4-D, 2,4,5-T, dicamba, picloram, maleic hydrazide, ^{14}C-labeled assimilates, and ^{32}P were all found to move basipetally out of leaves and to build up to high concentrations in petioles. In leaves that had been detached for 3 days before treatment, there was accumulation in callus tissue at the cut surface of the petiole. When the petioles were standing in water, there was some acropetal movement along veins. Labeled diuron, by contrast, moved only acropetally. When leaves received a preliminary treatment with endothall, vein loading and transport of assimilate, 2,4-D, dicamba, maleic hydrazide, and ^{32}P were greatly reduced. Where plants were treated three days after detachment, there was some leakage of tracers from the callus tissue into the ambient water: ^{32}P leaked a bit, assimilates somewhat more, and 2,4-D and dicamba to the greatest extent. Picloram is known to leak from roots following transport there from tops.

Leonard and Glenn (1968b) extended their studies with labeled assimilates and $^{32}PO_4^{---}$. Vein loading and petiole accumulation in detached leaves of *Phaseolus vulgaris* were not greatly affected by mild moisture stress, the age of mature leaves, or a dark period prior to exposure to $^{14}CO_2$; both, however, were greatly reduced by oxygen deficiency. Endothall at 5×10^{-3}M stopped the basipetal movement of $^{32}PO_4^{---}$.

Esau (1968) provided an excellent treatment of cell-to-cell movement of viruses in chapter 4 of her recent book. She ascribed rapid, long-distance transport in sieve tubes to mass flow, and discussed the possibility that the same mechanism may account for intercellular movement between parenchyma cells and sieve tubes. This aspect of assimilate and virus movement needs much research.

The Sink

Esau (1965b) gave excellent descriptions of all of the tissues and organs that constitute sinks for assimilates in plants, (roots, stems, meristems, flowers, fruits, and seeds). Growth, metabolism, and storage by these structures demand organic substrates that are synthesized in leaves: these are delivered by the phloem to meet the various requirements. Truly, the phloem is a distribution system structured to provide every living cell of the plant with nutrients necessary to its well-being. Only when viewed in this broad light can the diverse functions of this system be properly appreciated (Crafts, 1961b).

In *Fraxinus americana*, Zimmermann (1959) found a gradient in total molar concentration down the trunk, which disappeared upon defoliation. Because the parietal layer of cytoplasm in the functioning sieve tube is semipermeable, Zimmermann proposed that entry and exit of solute molecules is metabolically controlled; this lateral pumping may establish and maintain a longitudinal turgor gradient along which a mass flow may occur. Defoliation experiments suggest that removal of sugars from sieve tubes is a reversible process, which operates under remote control from the leaves (source).

Bieleski (1962), having found that about 90 percent of the sugar taken up by tissue of *Saccharum officinarum* is accumulated as sugar, whereas 10 percent is utilized in other processes, proposed that a true, active sugar transport process is operating in the immature storage tissues of that plant. Sucrose labeled with ^{14}C was found to be taken up against an existing sucrose gradient; and the phloem accumulated sugar at a rate about 20 times that of the storage parenchyma, with the sugar accumulation ratio paralleling the comparative respiration rates of the two tissues. Sugar accumulation, Bieleski proposed, follows the kinetics of a reaction between an enzyme and its substrate. Potassium chloride had little effect, its relationship with sucrose being noncompetitive in nature. Summing up, Bieleski suggested (1) that the rate-limiting reaction in accumulation has the characteristics of an enzyme reaction; (2) that, up to and including the rate-limiting step, sucrose, glucose, and fructose behave individually, each having its own dissociation constant of the enzyme-substrate complex, and each interacting competitively with the others; and (3) that a subsequent step is such that sucrose is the major or sole accumulation product, regardless of the sugar supplied (Glasziou, 1960).

Bieleski proposed a scheme involving (1) a permease step; (2) a second enzyme-catalysed step, in which the activated sugars may be interconverted; (3) a transport step, with a subsequent transfer to a different position on the inner surface of the membrane; and (4) the release, finally, of sucrose into the vacuole, where it might form glucose and fructose by hydrolysis. The first step involves the linking of sugars to an activated receptor, thereby forming a compound that behaves like UTP, a type compound known to be involved in carbohydrate metabolism in *Saccharum officinarum.*

Kursanov (1963) proposed that the uptake of sucrose, glucose, and fructose by young parenchymatous tissues involves two processes. The first is the apoplastic uptake of the sugars by diffusion into the free space or cell wall phase, attended by their being weakly bound to the cell surfaces. The second stage proceeds more slowly and for a longer time; it requires aeration, and is attended by increased respiration. This process is undoubtedly an active one, because assimilate is taken from a region of low metabolic activity (the cell wall or, in phloem, the sieve tube) and stored or utilized in active cells. This uptake process is inhibited by metabolic inhibitors, and often results in internal concentrations many times greater than those from which the solutes are absorbed.

In studies with *Saccharum officinarum,* Glasziou (1961) showed that the ratio of sucrose to fructose does not change in the cell vacuoles; it does not change even when the proportion of one of them in the external system is changed by a factor of several hundred. Apparently, the living cells of this plant contain the enzyme system required to transform glucose to fructose and vice versa, and to combine these in the synthesis of sucrose. Considering the variety of compounds synthesized from sugar and simple amino acids by meristem cells and storage parenchyma (and even sclereids, in some instances), it seems that sink tissues are even more diverse in their biochemical capabilities than are cells of the source.

Because ^{14}C-labeled sucrose may be formed in cells of *Saccharum officinarum* from ^{14}C-labeled glucose, and may turn out to be uniformly labeled, Kursanov (1963) concluded that sucrose synthesis from imported glucose is spatially separated from the sucrose storage pool, and that it takes place at an early stage of sugar uptake. A synthesis of this type should be preceded by phosphorylation of glucose and its isomerization into fructose phosphate, after which there can occur an enzymatic synthesis of sucrose. Kursanov presented the following scheme for this process: UDPG + fructose $\longrightarrow$ sucrose + UDP. For sucrose phosphate synthesis he proposed the following: UDPG + fructose-6-phosphate $\longrightarrow$ sucrose phosphate + UDP. Similar reactions were noted in Kursanov's laboratory in *Beta vulgaris* vascular bundles. He concluded that the hexose required for sucrose synthesis is not taken from a pool, but is formed from sugar entering the cells. Because no free ^{14}C-labeled glucose was found, Kursanov reasoned that conversions of sugars associated with their penetration into cells do not reach free hexoses, but are carried on at the level of their phosphoric esters. Thus, the active transport of sugars through the cytoplasmic membrane is accompanied by their mutual transformations that apparently take place only in the phosphorylated condition of sugars being translocated through the mesophyll. Kursanov presented the following scheme from Glasziou (1960) to illustrate these processes:

OUTER SPACE

$$\text{sugar} + [\text{R-X}] \longrightarrow [\text{sugar-X}] + \text{R} \quad (1)$$

$$[\text{sugar-X}] + \text{carrier} \longrightarrow [\text{sugar-carrier}] + \text{X} \quad (2)$$

$$[\text{sugar-carrier}] \longrightarrow \text{sugar} + \text{carrier} \quad (3)$$

INNER SPACE

He proposed that R-X may be ATP, by which the sugar is activated and converted into a phosphate ester (sugar-X). A second, more specific acceptor involved in penetration of the protoplasm is termed "carrier." Finally, upon entering the inner space, the sugar releases the carrier. Crafts proposed (1961b) that the endoplasmic reticulum might constitute the carrier; its membranelike properties seem suited to such a role. Apparently, in *Saccharum officinarum*, there is evidence that various sugars may compete for carrier, in that an excess of one inhibits uptake of a second. Because Bielski found that absorption of glucose and sucrose was inhibited by chloramphenicol, a protein-synthesis suppressant, Kursanov suggested that sugar transport is bound up with proteins—a logical conclusion, because enzymes seem to be involved. Kursanov concluded his discussion of sugar absorption by living cells by again emphasizing the metabolic nature of the process.

Tammes, Vonk, and Die (1967) found, when they allowed young inflorescence stalks of *Yucca flaccida* to absorb ^{14}C-labeled glucose or ^{14}C-labeled fructose through their cut ends for about $2\frac{1}{2}$ hours, that if they cut the opposite ends of the stalks and collected the phloem exudate, it contained ^{14}C-labeled sucrose almost exclusively. This indicates that hexoses absorbed from the apoplast (xylem, in this case) are converted to sucrose by the time they arrive in the sieve-tube lumina. Tammes and his co-workers concluded that (1) the concentration of sucrose in the sieve tubes and surrounding parenchyma determines the direction of sucrose movement between them; and (2) that the direction of assimilate flow in intact plants is determined by the secretion-absorption balance of all the nucleate phloem cells along the whole sieve-tube system.

Wardlaw (1968b), by decapitating plants of *Pisum sativum*, was able to divert the movement of ^{14}C-labeled assimilate to their axillary buds within 24 hours. During the early stages of bud development, starch was detected within the bud tissue. Normal vascular tissues developed 2–3 days after decapitation, and assimilate movement into the bud increased. The bud continued to import ^{14}C-labeled assimilate from leaves of the main shoot until after the first bud leaf was fully expanded.

Crafts and Yamaguchi (1958) pointed out that a root of *Zebrina* must be growing actively in order to serve as a sink for the import of 2,4-D. A root system in which there was no growth would not import this tracer; one in a nominal state of growth absorbed a nominal amount; and one in which growth was vigorous became very heavily labeled. Because translocation in these plants was obviously from the source of assimilates to a sink for assimilates, and not general throughout the plant, it seems conclusive that the ^{14}C-labeled 2,4-D tracer was accompanying assimilates in their normal distribution—riding, as it were, on the assimilate stream. Subsequent experiments have shown that the direction of movement can be reversed by manipulation of the source and sink; and that success in accomplishing a thorough labeling of the roots with a tracer of restricted phloem-mobility (2,4-D, for example) depends upon having a very active source and a very active sink.

Turkina (1961) studied the absorption of sucrose by vascular tissues of plants in comparison with absorption by the parenchyma of petioles. Using *Beta vulgaris* and *Heracleum sosnovski*, she separated the vascular bundles from petioles of mature leaves and treated pieces 1.5 cm long with 0.05 M ^{14}C-labeled sucrose having an activity equivalent to 10^5 count/min per milliliter. For comparative purposes, she treated similar quantities of the parenchyma tissues that surrounded the vascular bundles in the petioles. The rates of absorption by these two types of tissue differed greatly, as shown in Table 9.4. Turkina tried using sucrose, glucose, and glycine in her studies. As shown in Table 9.5, the conducting bundles absorbed seven times as much sucrose per gram of fresh weight as the parenchyma, and four times as much per gram of dry weight. Glucose was also absorbed, but less than twice as much by bundles as by parenchyma; and 1.6 times as much glycine was absorbed by bundles as by parenchyma. Absorption of glucose and glycine were both calculated on the basis of fresh weight; reduced to dry weight, the differences were not significant.

Table 9.4. Sucrose absorption by conducting bundles and leaf petiole parenchyma of *Beta vulgaris*.

Sucrose concentration	Experiment number	Absorption time (hr)	Amount absorbed (μmole/g dry wt) (*a*) Conducting bundles	(*b*) Petiole parenchyma	*a*/*b*
0.05 M	1	1	69.0	16.8	4.1
	2	1	59.3	16.2	3.6
	Mean	*1*	*64.1*	*16.5*	*3.9*
	3	2	120.8	27.8	4.3
	4	2	51.5	9.8	5.2
	5	2	160.8	31.9	5.0
	6	2	122.8	14.3	8.5
	7	2	90.6	12.4	7.3
	Mean	2	*118.2*	*19.2*	*6.1*
0.01 M	8	2	106.7		
	9	2	104.1		
	10	2	70.5		
	11	2	105.3		
	12	2	78.1		
	13	2	75.7		
	14	2	96.2		
	15	2	67.4		
	Mean	2	*88.0*		

Source: Turkina (1961).

Table 9.5. Absorption of ^{14}C-labeled sucrose, glucose, and glycine by petiole tissues of *Beta vulgaris*.

Substance	Amount absorbed (μmole/hr) Per gram fresh weight			Per gram dry weight		
	(*a*) Conducting bundles	(*b*) Petiole parenchyma	*a*/*b*	(*a*) Conducting bundles	(*b*) Petiole parenchyma	*a*/*b*
Sucrose	9.9	1.4	7.1	66.0	16.0	4.1
Glucose	3.4	1.8	1.8	22.6	20.9	1.1
Glycine	2.2	1.4	1.6	14.6	15.7	0.9

Source: Turkina (1961).

The tests with *H. sosnovski* also showed that phloem and xylem both have a greater affinity for sucrose than parenchyma.

That absorption against a gradient is an active process in these experiments is shown by the fact that potassium cyanide at 10^{-3} M and DNP at 10^{-4} M inhibited uptake by 67.7 percent and 59.3 percent, respectively, in one experiment, and by 59.6 percent and 41.0 percent in a second experiment. Turkina considered this ability of the conducting tissues to accumulate sucrose as evidence for their metabolic role in the translocation process. Its physiological significance, she proposed, may consist in eliminating diffusion of the transported sucrose to the surrounding tissues, thereby limiting the course of sucrose transport to the conducting bundles. Considering the anatomical relations of phloem tissues, it seems that the very nature of the sieve tubes serves to limit long-distance transport to their lumina. The great affinity of phloem parenchyma, then, would seem to provide for the distribution of sucrose to those active sinks in which it is an essential energy source.

Ziegler (1964) made researches on storage, mobilization, and distribution of reserve materials in trees. Cells in stems and roots of woody plants capable of accumulating reserve materials are living parenchyma; dead cells soon lose their stored reserves. Because storage cells often store starch, they must contain leucoplasts, the organelles that serve in that capacity. In *Castanea, Fagus, Betula, Alnus, Quercus,* and *Tilia,* cells of the pith remain alive for a long time and are able to store reserves. In *Populus, Acer, Robinia,* and *Sambucus,* the pith cells are short-lived, but cells of the medullary sheath remain alive and serve in storage. Here, the symplast concept applies: these living storage cells are interconnected by plasmodesmata; hence, transport of soluble carbohydrate materials can proceed laterally from phloem to xylem and medulla along continuous chains of living cells.

The arrangement of other storage cells in the xylem shows varying characteristics. Examples are deciduous gymnosperms, such as *Ginkgo* and *Larix,* in which there is no xylem parenchyma; in these, there is a much denser ray system. In those gymnosperms that possess xylem parenchyma, there is a tendency for localization of storage cells at the end of the growth rings; thus, materials must traverse only a short distance when filling or emptying these cells. Among the angiosperms, such terminal parenchyma is developed in many species; another tendency is for development of paratracheal parenchyma around vessels or vessel groups. Evidently, different plants have developed different arrangements of storage tissues, depending upon their specific requirements.

In bark, storage parenchyma is often concentrated in tangential layers that form an undisrupted network with the storage cells and translocating ray cells. The gymnosperms may be arrayed in three groups, according to the different types of cell arrangement in their bark: (1) those having a four-cycle rhythm, consisting of fibers, sieve cells, phloem parenchyma, sieve cells, and so forth (examples are *Ginkgo, Taxus,* and *Cephalotaxus*); (2) those having albuminous cells in the phloem parenchyma, combined with a four-cycle rhythm (examples are *Taxodium* and *Thuja*); and (3) those having no fixed rhythm, and with albuminous cells only on the margins of rays (the Pinaceae belong to this group).

The most important reserve materials in the stems and roots of trees are starch, lipids, and nitrogenous compounds. Foresters distinguish between starch trees and fat trees, a classification based upon histochemical analysis of wood samples. Starch is a common reserve material in many plant species, and apparently the sucrose $\rightleftarrows$ starch transformation proceeds quickly and easily. Apparently, fats in storage cells are derived from transport sucrose; oxygen is released in the storage process, resulting in a high respiratory quotient. The reciprocal transformation at the time the fat is utilized demands additional exogenous oxygen and the RQ falls below 1.0.

In starch trees, RQ values at all times of the year run around 1.0. In *Pinus sylvestris,* a fat tree, values determined in the autumn were 1.79 at 20°C, 1.80 at 10°C, and 2.69 at 2°C. In the spring, a 2-year-old stem of *P. sylvestris* with a fat content of 2.58 percent had an RQ of 0.7. The roots of both fat trees and starch trees store mainly starch; however, when exposed to freezing temperatures, they may store fat.

Although little is known about the form in which assimilates move into and out of sink storage tissues, it seems that they must pass by way of the companion cells (or albuminous cells) and the phloem parenchyma on their way to the rays. Cambium is supplied with assimilates and growth substances predominantly via the rays. Meristems of root and shoot tips must receive their nutrients by symplastic movement from terminal protophloem sieve tubes (see Figure 2.4, page 26).

Kluge and Ziegler (1964) have examined the ATP content of the sieve-tube sap of some broad-leaved trees in an attempt to clarify the role of this energy-rich compound in long-distance transport. By thin-layer chromatography they found ATP in significant quantities in *Robinia pseudoacacia, Tilia platyphyllos* and *Fraxinus americana.*

UTP and ITP were not found; the presence of GTP in large amounts was improbable. Determination of the components of the adenylic-acid systems of *T. platyphyllos* and *R. pseudoacacia* revealed predominances of ATP and adenine, respectively. This reflects the activity of an ATP-synthesizing enzyme system in *R. pseudoacacia* that is not present in *T. platyphyllos*. ATP occurred in 17 of the 18 species studied in from very different to very similar concentrations. Only in *Quercus borealis* could its presence not be demonstrated. The ATP concentration in the sieve-tube sap of *T. platyphyllos* and *R. pseudoacacia* showed annual trends, supposedly in correlation with transport output of the phloem. The concentration of ATP in the sieve-tube sap of *T. platyphyllos* stem showed no longitudinal gradient. Kluge and Ziegler felt that ATP should not be considered to be a typical phloem-mobile substance. The concentrations of ATP in the phloem, xylem, and parenchyma of the petiole of *Heracleum mantegazzianum* were compared. Although the xylem held significantly less ATP than the phloem and parenchyma, the concentrations in the latter differed only on the basis of dry weights. Because Kluge and Ziegler found no gradient of ATP along the stem of *T. platyphyllos*, they concluded that the exchange of components of the adenylic-acid system between sieve tube and companion cell is comparable throughout the length of the conduit system. In view of the enucleate condition of sieve tubes, this suggests that companion cells may be the site of ATP synthesis, and that the amounts found in sieve-tube sap may represent the influence of the companion cells in providing for maintenance of the integrity of sieve tubes.

Ziegler (1965) studied the distribution in *Fagus sylvatica* and *Prunus avium* of $^{35}SO_4^{--}$ from leaves, through phloem, to the rays, xylem parenchyma, and medullary sheath. Silver-grain counting showed the highest concentration of tracer in the phloem; the next highest concentration was in the medullary sheath, followed in order by the xylem rays, the xylem parenchyma, and the pith. These results are diagrammed in Figure 9.6. Because the rays constitute the most active symplast route from phloem to the medula, Ziegler's experiments implicate them in the distribution of $^{35}SO_4^{--}$ in this sink region.

Ziegler, making certain assumptions, calculated a probable rate of lateral movement along rays by diffusion; his value for the 20-hour transport period was 2.9 cm. The actual distance moved was up to 10 cm in some of his preparations, however. Ziegler felt that a completely unknown mechanism was responsible for this movement. Protoplasmic streaming, however, is common in such cells; though often difficult to detect in slides of this material, it probably goes on

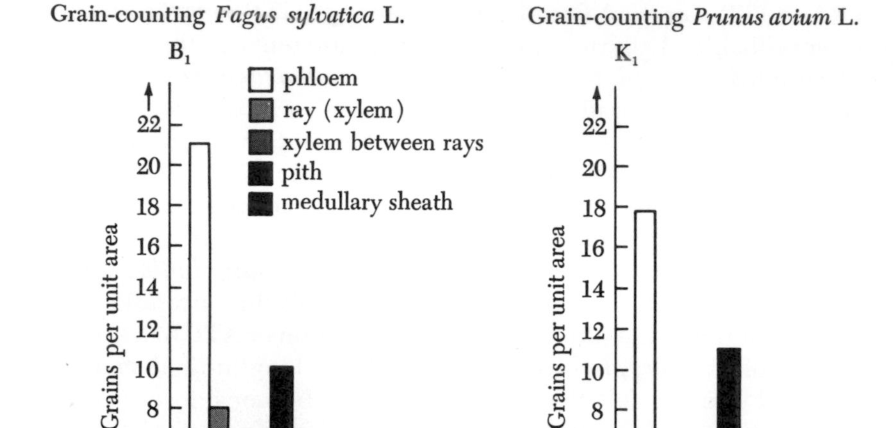

Figure 9.6. The distribution of $^{35}SO_4$ in different tissues of the stems of *Fagus sylvatica* and *Prunus avium*. From Ziegler (1965).

in the intact plant. There is no other known physical mechanism that could bring about this result.

Zimmermann (1964a) proposed three possible relationships between transport and growth in deciduous forest trees: (1) that assimilates are transported directly from active leaves to meristems via the phloem; (2) that new growth depends on retranslocation of mobilized reserves to regions of growth via phloem or xylem; and (3) that reserves stored very near the regions of growth are utilized, and that transport involves only short distances. Possibly all three of these are to be found in some plant species. In some species, phloem undergoing differentiation in late fall is arrested in development by the onset of winter. These partially differentiated sieve tubes become established as functioning conduits in spring, and serve in the supply of mobile nutrients during early growth (Alfieri and Evert, 1968). Anatomical studies and aphid feeding have reconfirmed these tissue relations of plants during inception and breaking of dormancy.

In some provocative work on the effects of gibberellin on translocation and on dry matter and water content of plant parts, Halevy et al. (1964) found that gibberellin increased the movement of dry matter from the cotyledons to the hypocotyl at concentrations as low as 10^{-6} M in *Cucumis sativus* seedlings; the optimum effects were at 10^{-3} M in light and at 10^{-4} M in dark. Four different gibberellins gave similar results in increasing translocation from cotyledons; but GA_4 was most active in promoting hypocotyl elongation, and in directing the internal distribution of dry matter to the hypocotyl, rather than to the radicle. Gibberellin increased water absorption at 3×10^{-6} M in the dark; 10^{-4} M was required in light. Gibberellin enhanced the depletion of dry matter from leaves of *Citrus aurantifolia* and *Gladiolus* sp. It did not affect translocation from young leaves of *Gladiolus* that were still importing assimilates. Gibberellin decreased the translocation of applied ^{14}C-labeled sucrose from leaves of *Phaseolus vulgaris* during the first 6 hours following treatment, but increased it after 20 hours. If gibberellin increases transport by affecting hydrolytic degradation of starch, radioactive sugar must first by incorporated into starch, and subsequently hydrolyzed for transport; thus, the time lag in transport from *P. vulgaris* leaves may be required for these transformations. Gibberellin always increased the acropetal translocation of ^{14}C in stems, and decreased the downward transport. If, as Zweig showed, gibberellin is readily phloem-mobile, it may accompany assimilates in their movement, and bring about a preferential action in stimulating growth of meristems of shoot apices. Thus, it may activate both source and sink.

Weaver, Shindy, and Kliewer (1969) studied the movement of assimilates into the berries of *Vitis vinifera* 'Black Corinth' as affected by treatment with CPA, gibberellin, and BAP. All three growth regulators caused the weight of the berries to increase. Gibberellin caused an eightfold increase in ^{14}C in 6 hours; ninefold and sixfold increases in radioactivity in tartaric and malic acids, respectively, after 12 hours; and strong increases also in γ-aminobutyric acid, pipecolic acid, and valine. Radioactivity in fructose was increased 70 percent in clusters treated for 12 hours with gibberellin; a thirty-one-fold increase in 96 hours.

When a whole shoot with treated berries and with leaves that had been fed $^{14}CO_2$ was sprayed with gibberellin, less assimilates reached the berries than in unsprayed controls. Weaver and his co-workers attributed this to the enhanced mobilizing power of the treated foliage, a deduction that seems logical in terms of sink activity. The concentration of DNA, RNA, and protein in berries treated with gibberellin and CPA increased during the initial 24-hour period after treatment; there was little change during the subsequent 24 hours. Gibberellin, CPA, and BAP all altered the percentage of ^{14}C in different constituents.

Fujiwara and Suzuki (1961b) studied the relation between respiration and translocation of assimilates in *Hordeum vulgare*. Their first experiment explored the effect of the *p*H of the root medium—culture solution, in this case—on the translocation of ^{14}C from leaves to roots. There were no differences between *p*H 6 and *p*H 10; at *p*H 4, there was less transport, presumably because of injury to the roots. If they had tried *p*H values greater than 10, they would undoubtedly have encountered the same lowering of transport rate as a result of root injury.

In a test using the inhibitors monoiodoacetic acid and sodium azide, they found that transport to roots declined with time at inhibitor concentrations of 10^{-3} M monoiodoacetic acid and 10^{-2} M sodium azide. No effect was recorded at 10^{-6} M concentrations. Removal of root tips had no effect on accumulation by the fully differentiated roots; accumulation was highest in the terminal 10 mm of the roots, and it was eliminated from roots having their tips removed. Excretion of $^{14}CO_2$ to the culture solution was almost doubled by the removal of the tips. These experiments show the effects of injury, inhibition of respiration, and elimination of the meristems on the uptake of assimilates by root sinks.

Source–Sink Relations

Münch (1930) studied the relations of growth to food reserves in trees. Fruits isolated by girdles failed to grow; needles of conifers isolated and darkened fell off. Evidently, substances other than food reserves are required for the normal growth and development of plant organs.

Weatherley et al. (1959) found that girdles some distance above and below exuding aphid stylets failed to stop exudation; evidently, foliage supplies were replaced by materials from storage cells. The switchover in *Salix* was fairly rapid; the exudation rate recovered within an hour when a girdle was made 15 cm above exuding stylets; a girdle made subsequently below the stylets reduced the exudate slightly, but it recovered within three days. Following these tests, Weatherley and his co-workers isolated segments of stem of varying length and determined that a piece of stem 16 cm long was sufficient to maintain a normal flow of exudate; from shorter pieces, exudation was reduced. Roughly 800–1000 sieve elements were found in a piece of stem 16 cm long.

When Peel and Weatherley (1962) used ^{14}C as a tracer in translocation studies with *Salix*, they found that exudation (through aphid stylets) from the sieve tubes of intact plants was affected by a girdle below the leaves that were acting as the source. The proportion of labeled sugar in the exudate fell, while the total quantity of sugar remained the same. Evidently, a switchover from the leaves to stored sucrose takes place rapidly, with the sink within the storage tissue of the stem becoming a source; sugar potential within the phloem apparently determines the direction of movement between the sieve tubes and the storage parenchyma.

In order to further study the energy relations of assimilate transport, Ullrich (1962) treated leaf veins of *Pelargonium zonale* simultaneously with ATP and fluorescein-K. He found that the energy-rich phosphate increased the transport of the dye. When he made separate applications at different loci on the same vein, there was no effect on the transport rate. Ullrich concluded that ATP stimulated the uptake of fluorescein-K into the sieve tubes, but that, when applied from the outside, it had no influence upon the transport process per se. He believed that ATP has great difficulty entering sieve tubes. This is in agreement with the concept that phosphorus compounds serve as carriers of assimilates across mesophyll up to the sieve tubes, but that they do not move in the assimilate stream during long-distance transport.

In a series of three papers, Hatch, Sacher, and Glasziou (1963), Hatch and Glasziou (1963), and Sacher, Hatch, and Glasziou (1963), described work on the sugar accumulation cycle in *Saccharum officinarum*. In the first paper, Hatch and his co-workers stated that, though they were unable to demonstrate the activity of sucrose phosphorylase, they did find UDPG-fructose glucosyltransferase; and some preparations of storage tissue converted UDPG and fructose-6-phosphate to a compound having the properties of sucrose phosphate. Enzymes for the synthesis, interconversion, and breakdown of hexose phosphates were identified in young and mature storage tissue. A soluble acid invertase was isolated from immature storage tissue. A soluble neutral invertase was present in mature storage tissue.

Hatch and Glasziou (1963), using 3-month-old plants of *Saccharum officinarum* under 5 temperatures and 3 watering regimes, measured growth rates, sugar contents, and invertase activities of storage tissues. Immature storage tissue contained an acid invertase that disappeared as the tissue matured. A neutral invertase was present in mature tissue. The acid-invertase content and the rate of elongation of immature internodes were linearly related; both increased with increasing temperature, provided the plants had an ample water supply. Changes in the total sugar content of mature tissue were related to neutral invertase activity. The neutral invertase activity and sugar content of mature internodes increased as the temperature was decreased, a result that Hatch and Glasziou attributed to slowing of growth.

Sacher et al. designed a schematic representation of the sugar-accumulation cycle in the storage tissue of *Saccharum officinarum*. Figure 9.7 presents this scheme. Sacher and his co-workers stated that there are three distinct compartments through which sugars move when accumulated from the ambient medium into storage tissues. These are the outer space (cell walls), the metabolic compartment (cytoplasm) and the storage compartment (vacuole). They found that sucrose is inverted in passing from the medium into the storage compartment where it reappears as sucrose. Inversion is mediated by an acid invertase in the outer space, and is apparently an integral step in the accumulation process; under some conditions, it may be rate-limiting. There is also invertase activity in the storage compartment.

In continued studies, Hatch (1964) showed that enzymes that catalyze the synthesis of sucrose phosphate (UDPG-fructose-6-phosphate glucosyltransferase) and the breakdown of sucrose phosphate (phosphatases) are present in the stem and leaf tissue of *Saccharum officinarum*. A compound having the properties of sucrose phosphate was formed from uniformly ^{14}C-labeled glucose in slices of stem tissue; and when radioactive sugars and sugar phosphates—including sucrose and sucrose phosphate, both fructosyl-labeled with ^{14}C—were supplied to slices of storage tissue, there was evidence for sucrose phosphate as an intermediate in sugar accumulation.

If the symplast concept is acceptable, then the insertion of an outer-space phase between the conducting tissue and the metabolic compartment is not necessary; in fact, many observations would indicate that transport from sieve tube to storage parenchyma is symplastic. If free sucrose were present in the apoplast system of vascular tissues, it would be reasonable to expect to find it crystallized out wherever transpiration water was being evaporated (in hydathodes, for example, or in stomatal chambers, or upon any cut or injured surface); this follows from the fact that the apoplast is an uncompartmentalized continuum.

Because the authors of this series of papers soaked small pieces of tissue in their labeled tracer solutions, uptake had to take place

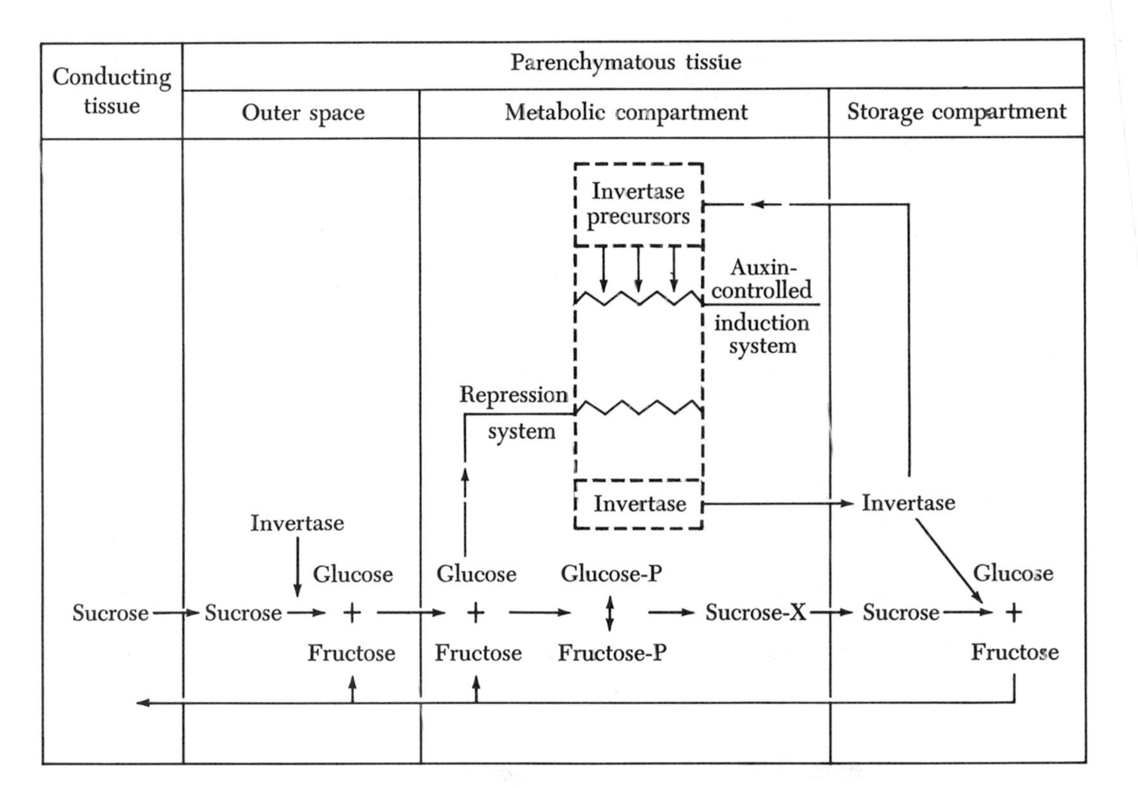

Figure 9.7. Schematic representation of the sugar-accumulation cycle in the storage tissue of *Saccharum officinarum*. From Sacher et al. (1963).

from apoplast into the symplast. Their results, however, do not prove that movement from sieve tubes takes place by this route. Inversion in their cultures could have been the result of contact of sucrose with invertase located on the outer surface of the plasmolemmas of storage cells. Crafts (1960) and Szabo (1963) found esters of 2,4-D to be hydrolyzed in the apoplast phase of plant leaves.

Despite this possible weakness in the interpretation of the work, this series of papers presents a wealth of valuable evidence on the biochemical processes relating to sugar storage in *Saccharum officinarum.* They reflect the labile state of assimilates as they are utilized in growth and storage; and they point up the possible role of invertases in the partitioning of available carbohydrate from photosynthesis between the processes of sugar storage and growth.

In studies on the mechanism of sugar storage in mature stem tissue of *Saccharum officinarum,* Hawker and Hatch (1965) found that two invertases were involved: a neutral one, and an acid invertase with an optimum at *p*H 3.8. When radioactive sucrose, glucose, or fructose was supplied to mature storage tissue, more than 50 percent of the radioactivity taken up was located in stored sucrose; less than 5 percent appeared as respired $^{14}CO_2$, and the remainder was recovered in the nonsugar components. Hydrolysis of sucrose by invertases was both a prerequisite and the rate-limiting process for the storage of sucrose. Mature tissue had an acid invertase, which was firmly bound to cell walls, and a neutral invertase, possibly located in the cytoplasm. Immature tissue had a soluble acid invertase. The storage of sucrose involved an active transport. Hawker and Hatch found that the concentration of sucrose in the intact plant may reach 23 percent w/v, where uptake is from a solution of 1 percent w/v; high concentrations of sucrose in the free space (apoplast) of storage tissues may contribute to the levels of stored sugar recorded.

Die and Tammes (1966) found that phloem exudate will move out from both ends when an inflorescense of *Yucca flaccida* is cut. By recutting both surfaces to remove plugged sieve plates, exudation was maintained for 48 hours. By comparing the exudates from the distal and proximal ends, Die and Tammes concluded that the cells that surround the sieve tubes can accumulate or secrete sugars reversibly, thus maintaining the sucrose potential between certain specific limits. Again the concept is illustrated that the phloem is an integrated system that can serve as sink or source. Work by Tammes, Vonk and Die (1967) emphasizes this point.

In studies on the length of an isolated stem segment of *Salix* that contributes solutes to the sap exuded through excised aphid stylets,

Ford and Peel (1966) demonstrated that when a gradient of $^{32}PO_4^{---}$ activity is established, the contributory length may be much greater than the 16 cm established by Weatherley et al. (1959). Upon cooling a portion of a stem segment between an aphid colony and the high-activity end, the pattern of specific activity indicated that cooling had increased the length of the contributory portion. Ford and Peel concluded that this provides evidence that lateral movement of solutes in phloem is directly or indirectly dependent upon metabolic energy, whereas longitudinal movement is by mass flow down a gradient of turgor pressure. They felt that the transcellular streaming hypothesis of Thaine (1962) could not possibly account for their results.

Ford and Peel (1967a) continued their studies on the effect of temperature on the movement of the assimilate stream using $^{14}CO_2$ to mark the assimilates. Using young shoots (3–5 weeks old) and mature stems (2–3 years old) of *Salix,* they measured translocation rate by determining the slope of the rise in activity of honeydew collected from aphid colonies. They found a marked difference in the response of young shoots and mature stems. In the former, low temperature from a cooling jacket brought about a lowering of the transport rate, whereas, in mature stems, cooling increased mass transport of the ^{14}C (Figure 9.8). Ford and Peel concluded that the difference in behavior between young and mature stems did not indicate a fundamental difference in transport mechanism. They reasoned that low temperature increased the viscosity of the stream, and appreciably increased resistance to flow through the sieve tubes of young stems, which had sieve-tube radii of 12.6 μ. However, in old stems, which had sieve-tube radii of 19.0 μ, resistance was not altered so drastically, but uptake in cooled sink tissues was inhibited, increasing the specific activity of the exudate—which, in their terms, indicated accelerated mass flow. Because resistance to flow through small conduits, according to Poiseuille's law, is proportional to the fourth power of the radius, the explanation of Ford and Peel seems reasonable.

Ford and Peel (1967b) performed experiments on the movement of sugars from the inner surface of bark strips into sieve tubes, either collecting honeydew from feeding aphids, or collecting phloem exudate by means of the aphid-mouthparts technique. Sucrose, glucose, and fructose, all labeled with ^{14}C, were applied to the cambial surface of bark strips of *Salix,* and their movement into the sieve-tube sap and their subsequent distribution were then determined. Sucrose, free hexoses, and sugar phosphates were found to contain the ^{14}C label. Rapid interconversions occurred between these compounds somewhere along the entry route. The uptake of sucrose was accom-

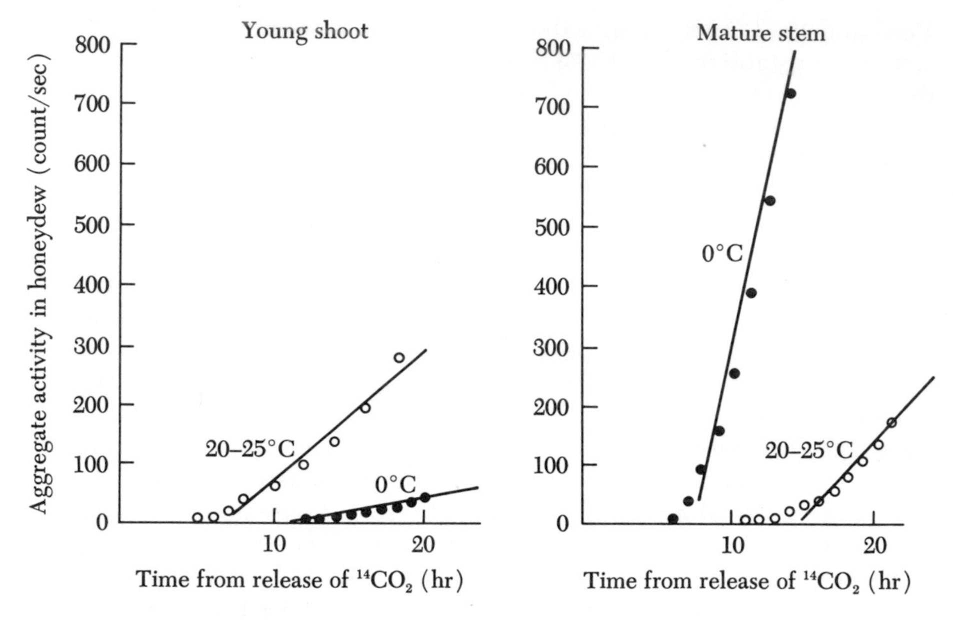

Figure 9.8. Effect of temperature on translocation of ^{14}C-labeled assimilate in young and mature *Salix* stems. From Ford and Peel (1967a).

panied by at least a partial breakdown into its component hexoses, and the involvement of a sugar-phosphate pool was found.

In continued studies on the movement of solutes into the sieve tubes of *Salix*, Peel and Ford (1968) bathed bark strips with sugar solutions, and determined the identity and concentration of sugars appearing in honeydew or in exudate from cut stylets of aphids. They found that when ^{14}C-labeled sucrose, or its constituent hexoses, are applied to the cambial surface of a bark strip, the sugars may move into the sieve tubes by two pathways: either by a direct transport via companion cells, or by an indirect movement involving storage parenchyma of the phloem. Labeled sugars entering by the first route do not mix with pools of unlabeled sugars before entering the sieve tubes; moving via this direct route, the sugars undergo metabolism leading to activity in a wide range of compounds in the sieve-tube exudate. Although movement via the indirect route involving storage parenchyma results in considerable metabolic change in the applied sugars, products of these changes are unable to enter the sieve elements and the exudate contains only labeled sucrose.

Peel and Ford were unable to determine whether or not sucrose molecules entering by the indirect pathway are split and reconstituted between the storage cells and the sieve elements; Kursanov's (1963) work would imply that they are. The activity distribution in sucrose, however, appeared to change between these two systems. In seeking the barrier that excluded hexoses from phloem exudate following their movement via the indirect route, Peel and Ford suggested that, once inside the symplast, sugars might move along the direct (companion cell) route, without encountering a tonoplast barrier; vacuoles in companion cells are very small. Via the indirect route, it would seem likely that movement across tonoplasts into vacuoles of storage cells would be probable; export would require recrossing the tonoplast, and sucrose might possibly have a preferred passage along this route. Peel and Ford criticized work on *Saccharum officinarum* tissues in which free-space movement of sucrose is implicated; such movement could not possibly have produced the results they found in their study. They cited Sacher et al. (1963) as showing that inversion of sucrose is an integral step in its accumulation in disks of *S. officinarum* tissue; the invertase responsible appeared to be present in the free space of the cell walls. Hawker (1965) expressed the same idea in a recent paper.

Ford and Peel felt that the rapidity with which fructose appears in the bathing solution indicates that a metabolic compartment responsible for the hydrolysis of sucrose is present between the bathing solution and the sieve-element lumen: they suggested the free space. A much more reasonable locus for this process would seem to be the exterior surface of the symplast: this is a living surface having the properties of a membrane; it contains proteins that possibly have exoenzymatic properties; and it constitutes a very large surface in the proper position to carry on this function. Crafts (1960) and Szabo (1963) have shown that esters of 2,4-D may be hydrolyzed in this region of leaves.

In continued studies on the movement of ions between tissue systems in stems of *Salix*, Peel (1967), using $^{14}CO_2$, ^{86}Rb, ^{22}Na and $^{32}PO_4^{---}$, determined movement into the xylem stream from phloem (with ^{14}C), and from filter-paper strips inserted between bark and wood (with ^{86}Rb, ^{22}Na, $^{32}PO_4^{---}$). When $^{14}CO_2$ was applied to leaves of a shoot on a stem segment, the xylem perfusate contained a portion of the activity, mainly in the form of sucrose; there was a time lag of about 30 hours between the peak of activity of xylem-sap and that of honeydew from the phloem. In shorter time experiments, Peel found that this lag could be reduced to 10–16 hours. Experiments in which ^{86}Rb, ^{22}Na and $^{32}PO_4^{---}$ were applied to the cambial surface of a raised portion of bark showed that these solutes could also pass into the xylem stream. The strongest passage from cambium to xylem took place during February, March and April; in June and July, little activity from these ions reached the wood; practically none showed up in the xylem perfusate.

It is well known that nutrient ions can move readily from the xylem into the phloem (Stout and Hoagland, 1939). This is apparently a reflection of the normal distribution function of the xylem system. It is also widely recognized that assimilates moving in the phloem readily pass via the symplast into the xylem; ray cells are definitely implicated in this movement (Ziegler, 1965). During the spring and early summer, the phloem normally supplies large quantities of assimilates to cambium and young growing xylem cells. It is not surprising that nutrient ions applied to cambium are absorbed into living cells and passed, in this same way, into the xylem; in Peel's experiments, the xylem was being flushed with distilled water, and the ionic concentrations of growing cells, therefore, were low.

The slow but ready movement of maleic hydrazide, picloram, TBA, sodium, rubidum, and cesium from phloem to xylem to the culture medium has been observed (Table 8.1, page 176). Potassium may also belong in this group. Phosphorus is known to circulate in plants—hence, to move from phloem to xylem, and vice versa. Considering the distribution functions of phloem and xylem (Crafts, 1961b), it should be apparent that this type of behavior is essential to the normal growth characteristics of plants. Plants that did not provide for such exchange and circulation would undoubtedly run short of essential elements for growth at certain times of the year.

Work by Gunning, Pate, and Briarty (1968) shows that the parenchyma cells immediately adjacent to phloem and xylem may be specialized in such a way as to facilitate the exchange inherent in distribution systems. Gunning and his co-workers called these specialized cells "transfer cells," and their specialization characteristic is their protuberant wall structure. The protuberances form an irregular layer of wall material deposited secondarily on the primary wall; their composition is of loose microfibrils. The result of this specialization is an increase in the surface-to-volume ratio; this increase may reach values of greater than 10, in some instances.

Protuberances may be found on all walls of a transfer cell, but they are often less abundant on the sister cell. They are found in opposition to sieve tubes, xylem elements, phloem parenchyma, bundle-sheath cells, and other transfer cells. It seems evident that the greatly increased surface presented by these specialized structures should facilitate uptake of solutes from the cell-wall fluids, and promote ready exchange between phloem and xylem conduits and the adjoining parenchymatous tissues. Experiments with isotopic tracers have shown that transfer cells of *Pisum arvense* can absorb and incorporate radioactivity derived from $^{14}CO_2$ applied to a leaf or labeled amino acid supplied through the transpiration stream. Histochemical tests have shown that the protuberance walls are the site of intense acid phosphate activity. Possibly, many enzymatic transformations may take place within these walls or at the cell wall–cytoplasm interfaces. The retrieval of solutes from the transpiration stream is an obvious function of these structures, as is the ready movement between the cytoplasm and the vacuole of the transfer cells. Further, the greatly expanded cytoplasmic surface should facilitate and accelerate symplastic movement between parenchyma cells and the sieve tubes.

In a taxonomic and morphologic survey of perivascular transfer cells in angiosperm leaves, Pate and Gunning (1969) reviewed studies on the fine structure of minor veins of mature leaves of 975 species from 242 families of angiosperms. They distinguished four types of transfer cells, two in phloem, one in xylem parenchyma, and one in bundle sheath. A most interesting discovery of the survey is the very definite association of vein transfer cells and the herbaceous habit.

Pate and Gunning found many species that lack transfer cells, but the differences between species possessing and not possessing such cells are relative and are differences of kind, not of function. In both kinds, the vein parenchyma must handle a large mass of photosynthate, and its ultrastructure suggests that some sort of active process takes place prior to the final movement of solutes into the sieve elements. In both, also, the leaves must recycle mineral ions and organic solutes arriving in the xylem (see Wanner, 1953a). Where transfer cells are lacking, interveinal recycling of solutes would be less efficient, involving loss via cell walls to mesophyll and return to the vein with other assimilates to the sieve elements via the symplasm. In the presence of transfer cells, this longer pathway might be short circuited by a massive transfer of solutes from xylem to phloem within the confines of the vein. In finer veins, xylem and sieve elements may be connected by transfer cells.

The effect of DNP on translocation in phloem has been investigated by Shulamith and Reinhold (1966) using ^{14}C-labeled sucrose applied through a flap in a leaf of *Glycine max;* steam girdling was used to check on possible xylem movement; DNP was supplied through the same flap as the labeled sucrose. Under these conditions, DNP drastically inhibited the transport of ^{14}C-labeled sucrose.

Transport was also inhibited if DNP was applied along the translocation path while the ^{14}C-treated terminal leaflet was still in position on the plant.

When DNP was applied through the cut petioles of primary leaves after removal of the ^{14}C-treated terminal leaflet of trifoliate 1, no inhibition was observed; in fact, transport appeared to have been promoted. Inhibition of basipetal assimilate movement by DNP was interpreted as resulting from hindered uptake from chlorenchyma to sieve tubes. The increased long-distance transport in plants receiving DNP through cut petioles after removal of the treated leaflet is explained by a similar inhibition of uptake out of the sieve tubes by parenchyma cells along the translocation route. Thus, DNP inhibition acts to reduce transport when it is allowed to act on the source, and to enhance it when applied along the sink tissues. Since the effects of DNP are not apparently upon the sieve tubes themselves, the authors consider their results compatible with the Münch pressure-flow hypothesis.

Lee, Whittle, and Dyer (1966) determined the effect of boron deficiency on the translocation profile of ^{14}C in *Helianthus annuus*. The ^{14}C-labeled assimilates were provided by $^{14}CO_2$; and both boron-deficient and boron-sufficient plants were used. In the boron-sufficient plants, the advancing front of radioactivity was always ahead of that in the boron-deficient plants; the general shape of the profile was the same in the two. Lee and his co-workers concluded that velocity of transport is reduced by the boron deficiency. A logical interpretation of this would be that the activity of apical leaf cells in the boron-deficient plants is reduced by the lack of sufficient boron for normal growth, making for a weak sink in the shoot tip.

In studies on the time-course of translocation of ^{14}C-labeled sucrose in *Beta vulgaris* during darkness, Geiger and Batey (1967) examined the source–sink relations of this plant. After a 4-hour light period, the arrival of ^{14}C-labeled assimilate in an immature sink leaf following darkening of the mature source leaf dropped rapidly. Within 5 minutes of the cessation of photosynthesis, the arrival of ^{14}C declined; after 1 hour, it was 50 percent of the value during the light period; after $2\frac{1}{2}$ hours, it was 25 percent.

After 2–3 hours, polysaccharide of the supply leaf reserves started to replenish the sucrose pool, which remained at around 60 percent of the light value. At the same time, the root and vascular parenchyma contributed to the sucrose supply. As occurred in the case of *Salix* in the experiments of Weatherley et al. (1959), the sucrose potential tends to maintain a steady level, with stored reserves serving as source when photosynthesis has stopped. This probably explains why it proved difficult to reverse the direction of flow of the assimilate stream in tracer experiments conducted by Crafts and Yamaguchi (1964), Forde (1966a, 1966b) and others.

Shiroya (1968), using seedlings of *Helianthus annuus*, found that when single leaves were fed $^{14}CO_2$ for 30 minutes, young plants showed a higher proportion of upward translocation than did older plants. Upward translocation decreased and downward increased with increasing age when one leaf at the second node was treated. ATP treatment stimulated downward transport; DNP inhibited it; neither chemical affected upward movement in these seedlings. Shiroya's results can be explained on the basis of altered source-sink relations, and do not seem to demand different translocation mechanisms. His suggestion that upward movement might take place through the xylem ignores the role of the transpiration stream in supplying water to the treated leaf.

Quinlan (1969) has shown that movement of labeled assimilates from leaves of one-year-old *Malus* rootstocks, treated with $^{14}CO_2$ in the autumn, was mainly downward to the root system; some ^{14}C was retained along the stem. All leaves and shoots formed the following spring contained ^{14}C, the level of radioactivity being highest in the newly formed leaves. Over the winter, there was a loss of radioactivity from the root system and old stem, and this was accompanied by a reduction in the dry weight of the root. Part of this may have been the result of respiration; the major amount was the result of spring utilization of assimilates in growth. During blossoming and the following 2–3 weeks, there was severe competition between meristematic regions, the early supply of substrates coming from both stored reserves and primary leaves. In autumn, the basal stem and roots constituted the principal sinks.

Most studies on assimilate movement have been conducted on young plants with rapidly developing tissues and organs. Kreidmann (1968b) determined translocation patterns of ^{14}C in shoots of *Prunus persica* and *P. armeniaca*, continuing into the ripening stages of fruits. Import of assimilates by fruits continued during ripening. Seeds imported during growth, but movement terminated during the ripening of fruit; the flesh of fruits continued to import until the fruits were fully ripe. After removal of the fruit, *P. armeniaca* leaves imported assimilates in limited quantitity, but *P. persica* leaves entered dormancy when export is normally expected.

Edelman et al. (1969), in studies on the pathway and control of translocation from the tuber of *Solanum tuberosum*, found that the breakdown of starch starts when buds begin to grow; it stops if the sprouts are removed. The sprout controls the translocation and utilization of food reserves by the tuber. Movement of reserves may occur over the whole cross section of the tuber, and is not restricted to the vascular shell. The presence of a growing sprout does not affect the permeability of the tuber tissue to sugar or amino acids. Edelman and his co-workers found that translocation was not interrupted by experimental limiting of the pathway of movement to a small channel of pith parenchyma between the apical and basal halves of a tuber, cutting off the main vascular connections. The aluminum sheet they used to restrict the translocation pathway did not limit the movement of ^{14}C-labeled sucrose or ^{14}C-labeled leucine injected into one segment of the tuber. Because these experiments involved periods of 6–14 weeks, it seems possible that phloem regeneration across the region of separation could have provided adequate conducting phloem to account for these results.

By means of analyses of the absorptive capacity, the auxin contents, and the respiration rates of stem segments, vascular bundles, and the tubers of *Solanum tuberosum*—using plants fertilized with ammonium sulfate, with unfertilized plants as controls—Anisimov (1968) concluded that nitrogen influences both the physiological-biochemical activity of the conductive tissues and the terminal mechanisms of the transport system. Nitrogen shifts the enzymatic carbohydrate transformations in the sources and sinks and increases the auxin content of the foliage and tubers to different extents. Thus, assimilate transport is governed partly by the absorptive capacity of the conducting tissues and their energy reserves, and partly by auxin-induced growth of foliage and tubers; the larger increment in foliage tends to lower the transport rate to tubers. A third factor is the enhanced enzymatic starch decomposition in leaves, and a decrease in activity of starch-synthesizing enzymes in the tubers. Complex source–sink relations, involving growth, storage, and enzyme activities, determine the course of assimilate transport in these plants; and inorganic nutrition may play an important role in tipping the balance between these processes.

Peel et al. (1969) tested the movement of water and solutes in sieve tubes of *Salix* in response to aphid-stylet puncture. They used ^{14}C-labeled sucrose, $^{32}PO_4^{---}$, and $^{3}H_2O$. Bark strips were sealed over windows cut in polythene tubing, two separate windows being used. Their set-up allowed determination of transport from one compartment to a stylet located in the back over the other compartment. Activity of ^{14}C and ^{32}P was detected in the stylet exudate usually within 1 hour; HTO activity was lacking or slight even after 8 hours; in some cases, low activity was found after 4 hours. When stylets were located over both windows, HTO activity moved from the phloem surface into a punctured sieve element more rapidly than did ^{14}C or ^{32}P activity; however, it did not move longitudinally to the stylet in the receptor compartment in appreciable quantity in 8 hours. In experiments using live and dead bark strips, HTO activity moved by diffusion along the whole length of a bark strip (9 cm per compartment), within 4 hours. Thus, after 4 hours, HTO activity should have been able to move into stylet exudate from one compartment to the other by a diffusional process. Peel and his co-workers concluded that these experiments indicate a transport process in the phloem other than mass flow.

There are several possible explanations for their results. First is the obvious exchange of ^{3}H for H in all hydrogen-containing compounds, of which water is present in the greatest quantity. The second is the low efficiency (35 percent) of ^{3}H counting. Third is the fact that, in their experimental setup, the only source for mass flow was the total phloem parenchyma over both compartments, and the only sinks were the stylet punctures. Whereas ^{14}C and ^{32}P compounds diffused from the bark surface into phloem parenchyma, and were probably secreted selectively into the sieve tubes below the donor compartment, water would have moved nonselectively from all of the phloem into the punctured sieve tube; because of the surface relations, much of the water would have entered from the receptor compartment, rather than from the donor compartment. A fourth factor is the relation of HTO to total water in the system. Because there was no active water movement, HTO would have diffused in all directions, and there would have been great dilution within the phloem tissue. Fifth, in the experiments where two stylets were used, the stylet under the donor compartment would have been the most available sink for receiving the tracer solution; there would have been no reason for water to move from donor to receptor regions along the punctured sieve tube; ^{14}C and ^{32}P activities were about twice as high in exudate from the donor region as from the receptor.

If, as was probably the case, a different sieve tube was punctured in the receptor region, the immediate reaction to lowered sugar and water potential would have been secretion of cold sugar and diffusion of water into that immediate vicinity. Although some ^{32}P and ^{14}C would have arrived from the donor region, there is little reason to expect HTO to have arrived in the exudate, because water would have been easily available in the immediate receptor region. Thus, although mass flow from the donor to the receptor region would have gone on slowly, there is little reason to expect HTO to have been present in the exudate in measurable quantity under the experimental conditions used by Peel and his co-workers. They offered no alternative mechanism to explain phloem transport in plants.

Pate (1966), studying the metabolism of *Pisum sativum*, found that the amides and amino acids normally supplied to the shoot via the transpiration stream transfer carbon to protein through the amino acids threonine, lysine, arginine, proline, aspartic acid, and glutamic acid. Carbon from CO_2 enters the protein of photosynthesizing tissues through a complementary set of amino acids, including glycine, alanine, serine, valine, tyrosine, phenylalanine, and histidine. Young tissues of the shoot synthesize certain amino acids by metabolizing sugars supplied by assimilating leaves.

Pate found that each mature leaf on a shoot contributes carbon to the current synthesis of protein at the shoot apex. Sucrose accounts for more than 90 percent of the labeled carbon leaving a leaf of any age fed with $^{14}CO_2$. Upper leaves supply labeled assimilates directly to the shoot apex, and the radiocarbon from these assimilates is soon incorporated into a wide range of amino acid units of protein. Most of the labeled assimilates from a lower leaf move downward to the root and nodules: thus, the amino acids and amides of root metabolism are strongly represented among compounds labeled in the shoot apex. Nitrogen must circulate in appreciable quantities in *Pisum sativum.*

Milthorpe and Moorby (1969) gave an excellent discussion of source–sink relations with respect to transport in leaf growth, ear development in cereals, tuber growth in *Solanum tuberosum*, and assimilate storage in the root of *Beta vulgaris*.

King et al. (1967) found that, in *Triticum aestivum*, photosynthesis in the flag leaf is regulated by the demand for assimilates from the flag leaf by the ear. Removal of the ear reduced photosynthesis in the flag leaf by about 50 percent; darkening of the remainder of the plant resulted in recovery of the photosynthetic rate.

Neales and Incoll (1968) reviewed the source-sink aspects of phloem transport in relation to the effects of assimilate movement on photosynthesis.

Some Constituents of the Assimilate Stream

Sucrose has long been recognized as the principal solute in the assimilate stream. Glucose and fructose were also mentioned in the early analyses, but there was uncertainty about whether they occurred normally in the assimilate stream or were present due to hydrolysis of sucrose during collection. With the introduction of chromatography, it became possible to identify much smaller amounts of solutes, and Mittler (1957, 1958), Zimmermann (1958a, 1958b, 1960b), Weatherley et al. (1959), and many others, have since contributed to our knowledge on this subject.

Kursanov et al. (1958) used double-grafted *Helianthus* plants (*H. tuberosus H. annuus–H. tuberosus* and *H. annuus–H. tuberosus–H. annuus*) to study sugar translocation. In such plants, of course, the phloem of the three components is fused into a common system. Kursanov and his co-workers showed that the grafted parts retained the carbohydrate content characteristic of their species, as shown by a predominance of monosaccharides in the cortex of *H. annuus*, and in the predominance of oligosaccharides in the cortex of *H. tuberosus*. The latter differed from the former, too, in that it had a relatively higher content of high polymer glucofructosans. Sucrose was found in both species in about the same amounts.

Feeding $^{14}CO_2$ to the leaves of the upper components of such plants proved that sucrose was the predominent radioactive sugar in the phloem along the entire length of the stem: 80–85 percent of the total carbohydrate radioactivity was in sucrose. Sucrose was the predominent sieve-tube sugar in either of the grafted components, regardless of whether the assimilation took place in the leaves of *Helianthus tuberosus* or of *H. annuus*. As the assimilates passed through the conducting system, the radioactivity ratio of sucrose to monosaccharides increased; however, this ratio did not change radically when the assimilates passed from one component to another. The radioactivity ratio of sucrose to oligosaccharide, on the contrary, showed a marked variation with the tissue metabolism. This corresponded to the specific carbohydrate composition of the component. Kursanov and his co-workers assumed that synthesis of oligosaccharides does not take place in the conducting elements, but in the surrounding tissues, into which sucrose may be exported enroute along the sieve tubes of the phloem.

Meyer-Mevius (1959) collected phloem exudate from plants of some 23 species and determined, by chromatography, the sugar constituents that were moving in their assimilate streams. Sucrose was the most prevalent sugar; it proved to be lacking in three species of *Allium* and in an undetermined species of *Clivia*. Glucose and fructose were found in most species, and galactose and raffinose in a few; there were roughly 10 unidentified sugars present in one or more species. Meyer-Mevius listed three plant groups on the basis of the sugar composition of their phloem exudates: (1) plants in which sucrose makes up the total sugar content; (2) plants in which monosaccharides and oligosaccharides accompany sucrose; and (3) plants in which only reducing sugars can be found. The three species of

Allium mentioned above fell into the third group. *Rubus idaeus* and *Wisteria sinensis* were the ones having only sucrose.

Twelve amino acids were identified in phloem exudate, and six plant species had one or more of these present. Feeding experiments with exogenous sugars (xylose, galactose, rhamnose, and lactose) revealed no relation between molecular size and transport velocity. Lactose, applied to single leaves of different plants, passed through the stems and into untreated leaves, in which it was immediately digested. Respiration measurements on leaf and stem tissues of *Tropaeolum majus* proved that lactose increased oxygen utilization, but to lesser extent than sucrose. The absorbed lactose was only partially oxidized; the main portion was converted to other compounds.

In reinvestigating the report that glucose was a major translocation sugar in *Malus*, Webb and Burley (1962) found that the report was the result of a misinterpretation of chromatographs. By careful technique, they were able to show that sorbitol is the principal compound translocated in *Malus*, and that sucrose is second in concentration among ^{14}C-labeled assimilates; they also found small amounts of raffinose and traces of stachyose in sieve-tube exudate.

In order to avoid repetition, the various contributions to this subject have been presented in tabular form in Table 6.1 (pages 89–99).

In addition to the constituents listed in Table 6.1, Eschrich and Kating (1964) identified purine bodies and a reducing oligosaccharide in *Cucurbita ficifolia;* they found additional compounds which they were unable to identify. Trip et al. (1963) found a number of phloem-mobile carbohydrates, as noted in Table 6.1; they also identified melibiose, galactose, glucose, fructose, and pentose as reducing sugars that did not move in the phloem.

Helga Kluge (1967), a student of Ziegler, conducted a survey to determine, by chromatography, the presence and relative amounts of various sugars and sugar alcohols in the sieve-tube sap of woody species; some 270 species were examined. The compounds identified were ajugose, verbascose, stachyose, raffinose, sucrose, sorbitol, dulcitol, and mannitol; myoinositol was also determined. Sucrose was found in all of the plants examined—in trace amounts in some, and in large concentrations in others. A few species contained only sucrose. All species of the Oleaceae that were examined contained mannitol; hexoses were suspected in a few species. Helga Kluge's work brings to over 500 the number of plant species in which phloem exudation has been demonstrated.

Nelson (1963) discussed the seasonal variations in distribution and translocation of assimilates. Figure 9.9 gives his results for 3-year-old seedlings of *Pinus strobus*. The measure of translocation was the

	April	May	July	Aug	Oct	
S	79	89	94	86	88	^{14}C in sugars in roots (%)
R	16	4	0	2	6	
G	2	3	2	6	3	
F	3	3	4	5	3	

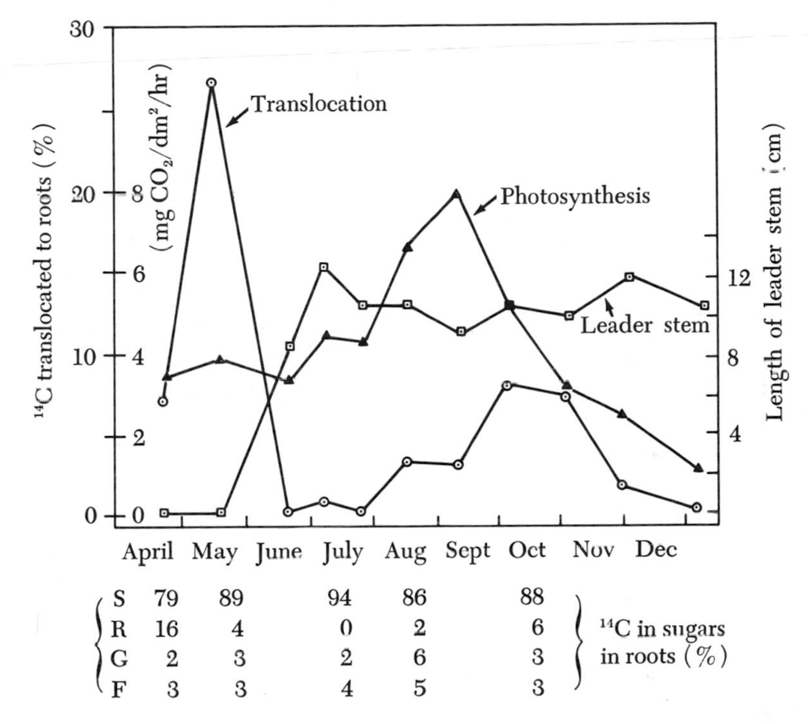

Figure 9.9. Changes, with season, in the translocation of ^{14}C to the roots, the rate of apparent photosynthesis of the shoots, the length of the new leader stem, and the percentage of ^{14}C in sugars in the roots in *Pinus strobus*. *S*, sucrose; *R*, raffinose; *G*, glucose; *F*, fructose. From Nelson (1964).

amount of ^{14}C recovered from the roots 8 hours after assimilation of $^{14}CO_2$ by the shoots. The greatest translocation was in May, before growth of the new leader and while photosynthesis was still low. Nelson gave data on the relative amounts of sucrose, raffinose, glucose, and fructose in the roots. Although the evolution of $^{14}CO_2$ by roots virtually stopped in June and July, this function probably reflects root growth, and not storage of reserves. Undoubtedly, some translocation into storage was going on during this period of very active photosynthesis.

The mineral composition of the assimilate stream of *Yucca flaccida* is shown in Table 5.3 (page 80). Peel, in 1963, noted that potassium and sodium are natural constituents of the phloem exudate of *Salix*.

Figure 9.10. [See following two pages.] Enzymes (in boldface type) that have been identified in sieve-tube sap from *Robinia pseudoacacia*. From Kennecke (1969).

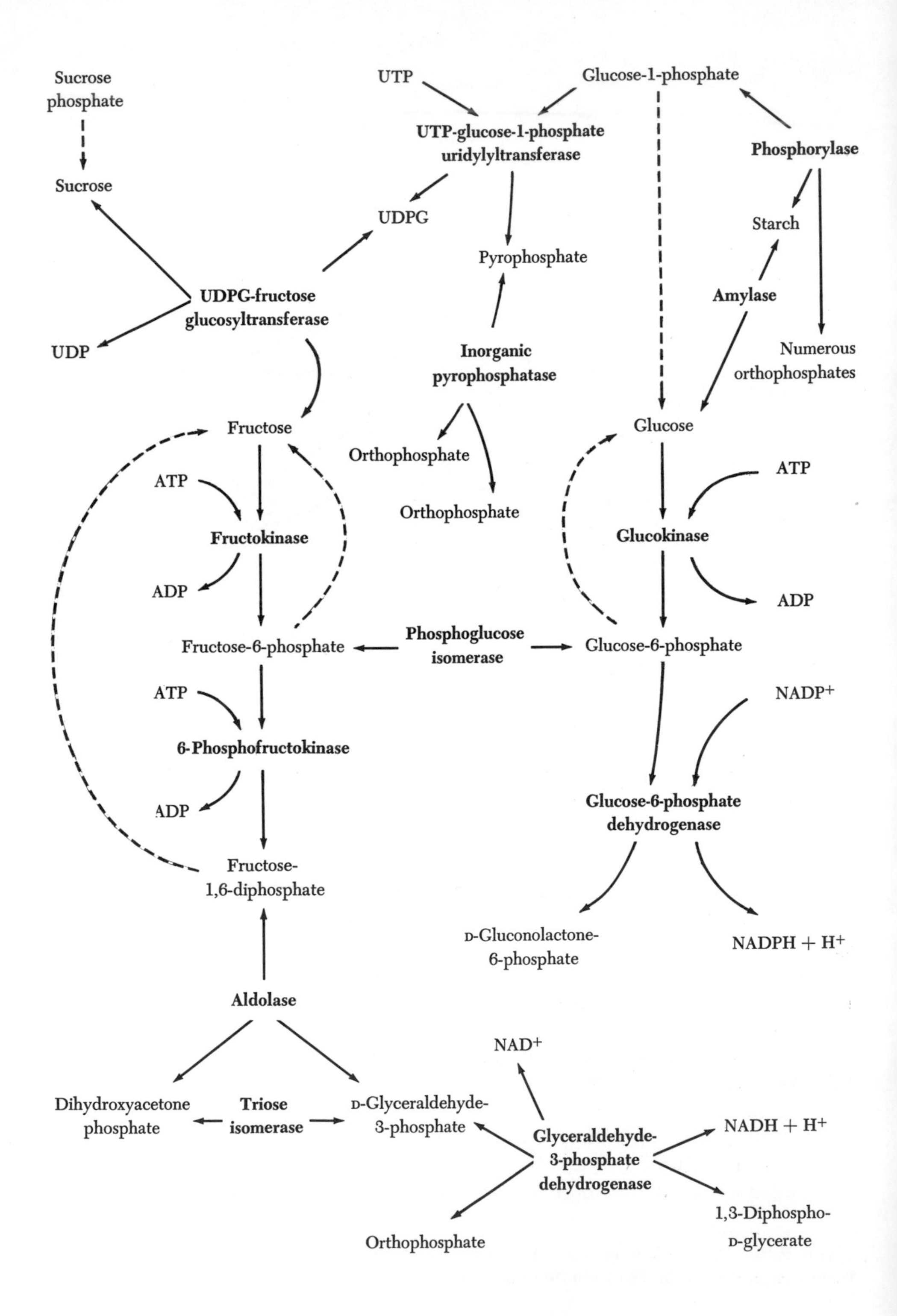

Sucrose phosphate
Sucrose
UTP
Glucose-1-phosphate
UTP-glucose-1-phosphate uridylyltransferase
Phosphorylase
UDPG
Starch
Pyrophosphate
UDPG-fructose glucosyltransferase
Amylase
UDP
Inorganic pyrophosphatase
Numerous orthophosphates
Fructose
Glucose
Orthophosphate
ATP
ATP
Orthophosphate
Fructokinase
Glucokinase
ADP
ADP
Fructose-6-phosphate
Phosphoglucose isomerase
Glucose-6-phosphate
ATP
NADP+
6-Phosphofructokinase
ADP
Glucose-6-phosphate dehydrogenase
Fructose-1,6-diphosphate
D-Gluconolactone-6-phosphate
NADPH + H+
Aldolase
NAD+
Dihydroxyacetone phosphate
Triose isomerase
D-Glyceraldehyde-3-phosphate
Glyceraldehyde-3-phosphate dehydrogenase
NADH + H+
Orthophosphate
1,3-Diphospho-D-glycerate

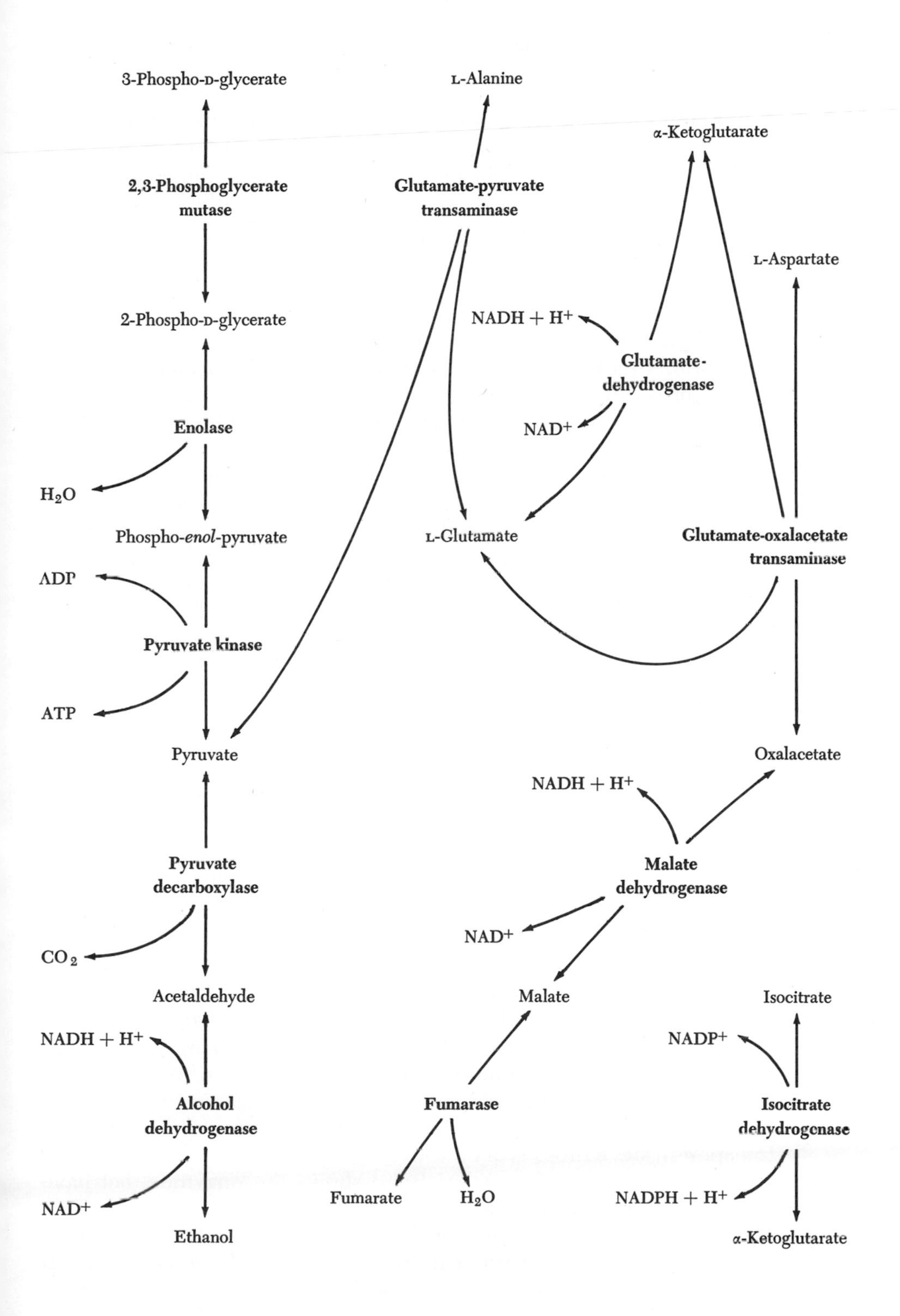

3-Phospho-D-glycerate
2,3-Phosphoglycerate mutase
2-Phospho-D-glycerate
Enolase
H2O
Phospho-enol-pyruvate
ADP
Pyruvate kinase
ATP
Pyruvate
Pyruvate decarboxylase
CO2
Acetaldehyde
NADH + H+
Alcohol dehydrogenase
NAD+
Ethanol
L-Alanine
Glutamate-pyruvate transaminase
L-Glutamate
α-Ketoglutarate
NADH + H+
Glutamate-dehydrogenase
NAD+
L-Aspartate
Glutamate-oxalacetate transaminase
Oxalacetate
NADH + H+
Malate dehydrogenase
NAD+
Malate
Fumarase
Fumarate
H2O
Isocitrate
NADP+
Isocitrate dehydrogenase
NADPH + H+
α-Ketoglutarate

Kluge and Ziegler (1964) found ATP in the phloem sap of broad-leaved trees; GTP was present in small quantities. Bieleski (1969) found a great variety of phosphorus compounds in phloem exudate of *Cucurbita maxima*, and suggested that inorganic phosphorus is the primary translocation form in the phloem.

Ziegler and Ziegler (1962) studied the content of the principal water-soluble vitamins in the sieve-tube sap of some 37 species of trees and shrubs by microbiological and chemical methods. They found thiamine in considerable amounts, almost exclusively in the free form. Riboflavine occurred in very small quantities, partly in bound form. Nicotinic acid appeared in abundance; often were large amounts recognized in bound form (with DPN or TPN).

Pantothenic acid was of general occurrence; a part of this had first to be hydrolyzed in order to be obtained in a free form. Pyridoxine was generally present, and always occurred in the free form. Only in *Robinia pseudoacacia*—and possibly in *Populus nigra*—did the sieve-tube sap contain no pyridoxine, although the roots require this vitamin. Ziegler and Ziegler suggested that, in the case of *R. pseudoacacia*, the nodule bacteria may provide this essential compound. For *P. nigra*, the quantity of sieve-tube sap was limited. The presence of vitamin B_{12} could not be accurately determined. So far as we know, at the time of their work, Ziegler and Ziegler knew of no influence of vitamin B_{12} on the growth of higher plants.

All tested sap samples contained biotin in such small amounts that its transport in the phloem could be of no significance; hydrolysis did not increase the activity of this vitamin. Folic acid—in *Robinia pseudoacacia*, for example—could be determined only in large samples of sieve-tub sap. The concentrations were small, and were not increased by incubation in conjugase. It seems that all organs of higher plants can synthesize folic acid and its biologically active derivatives: hence, its mobility in phloem would be of no significance.

Small amounts of unconjugated biopterin (Crithidia-factor) were found in sieve-tube sap of *Robinia pseudoacacia*. Myoinositol was found in important quantities in all tested samples of sieve-tube sap; hydrolysis increased yields significantly. Also, ascorbic acid was demonstrable in concentrations. In addition, both dihydroascorbic acid and diketogulonic acid were found. Because ascorbic acid is found in large quantities in leaves, fruits, and roots of plants, its transport in sieve tubes is of doubtful significance.

Figure 9.10 (after Kennecke, 1969) shows the enzymes that have been identified in sieve-tube sap of *Robinia pseudoacacia*. This figure points up two significant facts: that a tremendously complex biochemical system is present in the sieve tubes of one species; and that chromatography is of great value for identifying very small quantities of organic substances.

Apparently, as shown by work with tissue cultures, coconut-milk factor is one supplement needed by cells growing on synthetic media: it seems to be essential to root growth. On the other hand, considering the phloem-mobility of a wide variety of compounds, it seems that any such compounds synthesized in cells in excess of local needs would migrate to the sieve tubes and ride along on the assimilate stream. Possibly, ascorbic acid is such a compound.

One problem concerning the composition of phloem exudate of *Cucurbita* is that the ratio of carbon to nitrogen in the exudate, as found by Crafts and Lorenz (1944a), averages around 3:1, whereas that of the fruit averages 13:1. Because the carbon content of the exudate differs little from that of the fruit, the difference apparently occurs in the nitrogen: it is about 4 times as high in the exudate as in the fruit. The situation shown in Figure 2.2 and explained on page 32 indicates that the highly proteinaceous slime of *Cucurbita* may be retained in an adsorbed state on the filamentous reticulum. When a stem is cut and the exudate flows out at around 1000 cm per hour, this slime is washed free, and it flows along with the assimilate stream. Thus, the exudate is abnormally high in protein, an artifact caused by cutting. Crafts (1954) reported that 15 amino acids had been identified in the acid hydrolysate of phloem exudate of *Cucurbita maxima*. Eschrich (1963b) found phloem sap from the apical portion of a stem to be much higher in "lipoprotein" than sap from the basal end.

In collecting phloem exudate from stems of cucurbits, the custom has been to start near the tip of the stem and cut successive slices off in a basipetal direction. This has been done because the xylem vessels and cortical intercellular spaces at the tip are smaller and more easily clogged; exudate flows more copiously from cuts near the tip than from basal cuts. Thus, the usual collection of sap from cucurbit stems is inherently high in protein or amino acids. Crafts and Lorenz (1944a) decided that phloem exudate from *Cucurbita maxima* was not a typical sample of the assimilate stream. This is true: it contains large amounts of slime that normally remain in the sieve tubes of young stems.

In addition to the naturally occurring constituents of the sieve-tube sap, many exogenous materials are phloem-mobile, as shown in Table 8.1 (page 176). Crafts and Yamaguchi (1964) and Yamaguchi and Islam (1967) found the ^{14}C from eight amino acids to be phloem-mobile. These are listed in Table 8.4 (page 197). Thus, a great many compounds will move in the assimilate stream in sieve tubes. As for the distribution of pesticidal chemicals, penetration into the sieve-tube conduits seems more critical than mobility within them.

10

Effects of Environmental Factors

Given the conclusion that the phloem tissue in plants includes a complex, ramifying, inflated, elastic, osmotic system operating at a positive pressure and capable of conducting assimilates in solution at velocities of over 300 cm per hour (Hatch and Glasziou, 1964), it seems obvious that a number of environmental factors must determine the rates and efficiency of its function. Among the obvious factors are light, temperature, availability of water, mineral nutrition, composition of the assimilate stream, the structure of the plant as it affects frictional resistance and, therefore, velocity, and many others. This chapter will consider each of these in detail.

Water Potential

The water balance in the plant has obvious effects upon the functions of vascular tissues. Because the water in a plant exists as a continuum (Crafts, 1968b), addition or subtraction of water in any part of a plant affects the total water balance; and the sieve-tube system, being a complex osmometer, is regulated in its function by the availability of water. Photosynthesis, and hence the provision of osmotically active solutes, is very much subject to the water supply (Crafts, 1968b); and the partition of assimilates to growth or to storage depends upon water potential in meristems and storage cells. Under conditions of ample water, photosynthesis proceeds at an optimum, concentrations in sieve tubes are low, velocities are high, and growth of stems and roots provides active sinks. Under these conditions, translocation proceeds at the optimum rate for the species.

With lowering water potential, photosynthesis may slow down, solute concentration in sieve tubes increases, velocity decreases, and storage often replaces growth in stem and root tissues. Translocation processes under these conditions may be retarded, but they continue, even under severe water stress: vegetative growth gives way to fruiting and storage. Thus, the future needs of the plant are taken care of. Crafts reported in 1936 that phloem exudation occurred from stems having wilted leaves.

Weatherley et al. (1959) found that, in the case of phloem exudate collected from aphid stylets, wilting of the leaves caused a marked decrease in exudation rate with an accompanying increase in concentration. By passing a solution of 25 percent sucrose through the xylem vessels, they were able to shift the osmotic balance to the extent that the exudate contained almost 50 percent sucrose (see Figure 9.3, page 307).

Pallas (1960), testing the effects of humidity on translocation of ^{14}C-labeled 2,4-D and benzoic acid in *Phaseolus vulgaris,* found lower rates of transport under relative humidities of 34–48 percent than when humidities were 70–74 percent. Hartt and Kortschak (1963) listed decreasing moisture percentage in *Saccharum officinarum* plants as a cause of reduced transport of sugar to the upper portions of stems.

Measuring the movement of ^{14}C-labeled sucrose from leaves to roots of *Ipomoea batatas,* Ehara and Sekicka (1962) found that an RH of 70 percent favored transport more than an RH of 100 percent. This result differs from those obtained by Clor et al. (1962, 1963, 1964), Pallas and Williams (1962) and many current workers.

Using *Phaseolus vulgaris* plants growing in an air-conditioned room in which the light intensity was 1500 ft-c, Pallas and Williams (1962) studied the effects of varying levels of soil moisture on the foliar absorption and subsequent translocation of ^{32}P and ^{14}C-labeled 2,4-D. They found, from autoradiographs, that there was a marked decrease in the movement of these tracers as soil moisture became depleted. More ^{32}P was absorbed when the soil moisture stress was

$\frac{1}{3}$ atm than when it was 3 atm; eight times as much was translocated. Absorption of 2,4-D was not affected by moisture stress, probably reflecting absorption via the lipoid route (Crafts, 1961a). Twice as much 2,4-D was translocated at a low level of soil moisture stress ($\frac{1}{3}$ atm) as at a high level (4 atm). The inhibition of ^{32}P under soil moisture stress probably results from the fact that the polar phosphate ion enters the leaf by way of the water-filled microcapillaries of the leaf cuticle (Crafts, 1961a). Under stress, the menisci recede in the microcapillaries of the cuticle, leaving air blocks that prevent ready entry of liquids or polar solutes.

In a series of three papers, Clor et al. (1962, 1963, 1964) described their studies of the effects of high humidity on the uptake and distribution of ^{14}C-labeled 2,4-D and urea by *Gossypium hirsutum* and seedlings of *Quercus wislizenii*, *Q. douglasii*, and *Q. suber*. It has been the experience of many users of herbicides that high RH greatly increases the effectiveness of the chemicals. Smith et al. (1959) found that high humidity greatly enhanced the effects of maleic hydrazide, and Pallas (1960) obtained similar results with 2,4-D. These workers had used closed chambers in which the RH was 95 percent or more. Clor and his co-workers enclosed each of their plants in a polyethylene bag, in which the atmosphere became supersaturated so that the transpiration water condensed on the inside of the bag and ran down. Under these conditions, there was greatly enhanced absorption of treatment solutions that were applied as droplets. Not only was translocation via the assimilate stream increased, but the tracer accumulated to high concentration in the opposite, untreated cotyledon—a result indicating a cross-transfer in the xylem. Apparently, under these conditions of a supersaturated atmosphere, moisture condensed on the treated spot which then served as a source for uptake and movement via both the phloem and the xylem. Evaporation continues into this very moist atmosphere because the leaves receive light, which is transformed to heat, and so the leaf is warmed above the ambient temperature. Under these conditions, not only does the opposite, untreated cotyledon become highly labeled, the untreated portion of the treated cotyledon also becomes labeled. This effect is shown in Figure 10.1. Bagging just the treated cotyledon did not produce this effect, because the remainder of the plant maintained the water deficit throughout, and the atmosphere about the treated cotyledon did not become supersaturated. When all of the plant except the treated cotyledon was bagged, the treated plants produced autographs that were just like those of plants in the open greenhouse; there was no condensation on the treated spot. When ^{14}C-labeled urea was used,

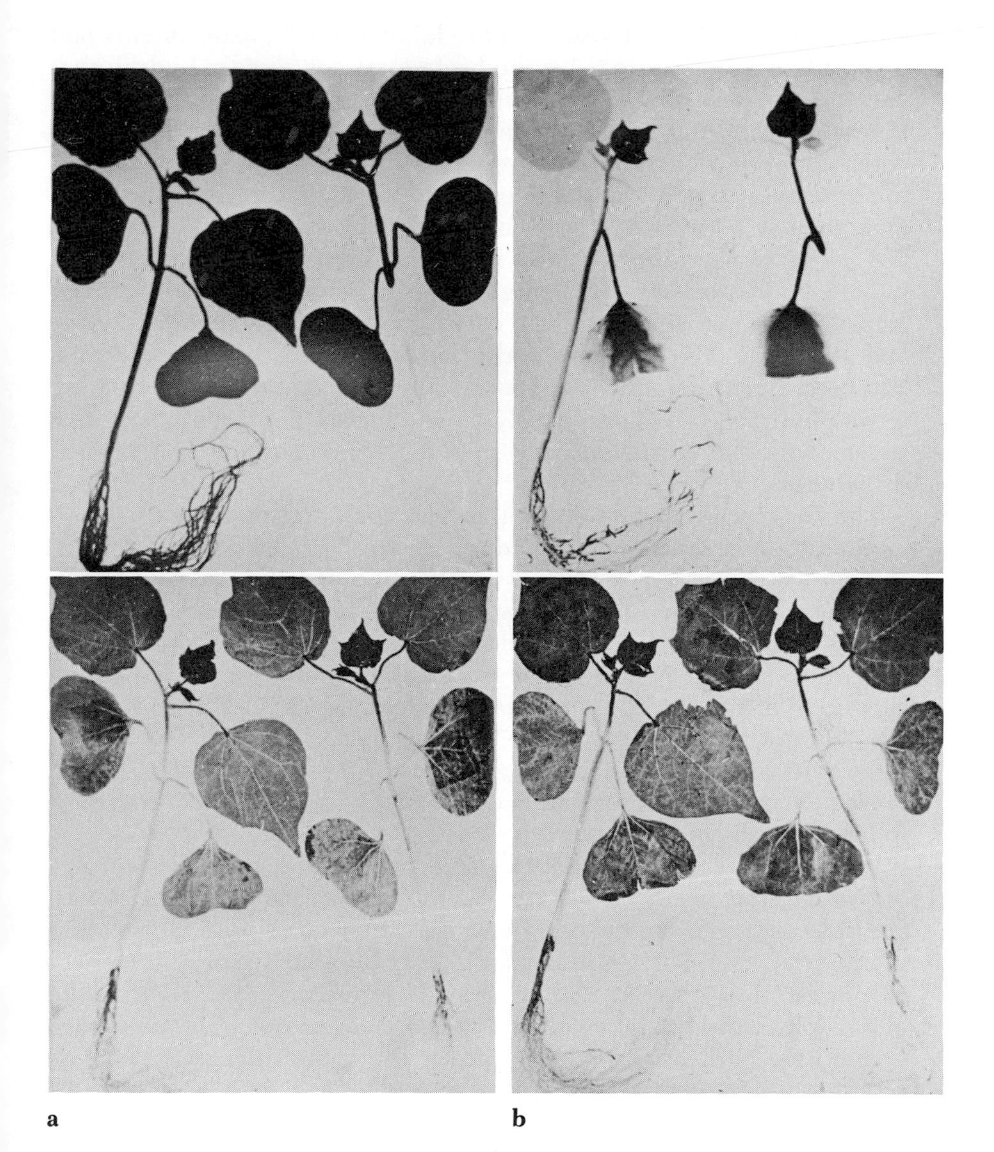

Figure 10.1. Effects of humidity and ringing on distribution of ^{14}C-labeled sucrose in seedlings of *Gossypium hirsutum*. Seedlings were kept (*a*) enclosed in polyethylene bags, or (*b*) in the open greenhouse. The righthand plant of each pair was steam-ringed below the cotyledons before treatment.

the results resembled those with the labeled 2,4-D; experiments had proved that labeled urea absorbed into leaves is rapidly hydrolysed, and that the resulting $^{14}CO_2$ is synthesized to sugars, which receive normal distribution in the assimilate stream. A steam ring below the cotyledonary node stopped movement to the roots; a similar ring in the epicotyl had no perceptible effect on tracer distribution; basipetal movement via the phloem was inhibited, but acropetal flow through the xylem was unimpaired by ringing. Ringing the petiole of the treated cotyledon was like ringing the epicotyl: the tracer moved across to the opposite, untreated cotyledon and up to expanded leaves and shoot tip. The roots were also labeled, indicating that the urea had been drawn into the other cotyledon and expanded leaves where it was hydrolysed, yielding $^{14}CO_2$, which was then synthesized to sugars, which, in turn, were moved to roots and shoot tip as in normal distribution.

The hygroscopic agent Tween 20 was included in all of the treatment solutions used by Clor and his co-workers. Possibly, by maintaining a wet surface, it favored condensation on the treatment spot; there was no visible film of moisture on the untreated surface. These results have important implications in the field of pesticide usage.

In their second paper, Clor et al. (1963) reported experiments on seedlings of *Gossypium hirsutum,* some of which had been held in the dark until their carbohydrate reserves were depleted. They found phloem transport to be greatly reduced in the plants with starved leaves. Starvation, however, had no effect on the xylem-transport induced by high humidity.

In their third paper, Clor et al. (1964) reported work with seedlings of *Quercus wislizenii, Q. douglasii,* and *Q. suber* using labeled 2,4-D, amitrole, and urea. In these experiments, they confirmed the fact that distribution of 2,4-D, amitrole, and sugar depends upon the activity of plants: those having active growth translocated tracers to the growing parts; dormant seedlings failed. The use of polythene bags to produce high humidity resulted in the same effects in *Quercus* seedlings.

The work of Ursino and Krotkov (1968) on translocation of recent photosynthate to the roots of young plants of *Pinus strobus,* as affected by water stress is marred by an apparent lack of understanding of the basic principles of irrigation. They state that their plants were maintained for 4 months on their specific soil moisture regimes of either 7 percent or 14 percent. When plants are watered by application to the soil, the water moves down and wets a certain depth of soil to field capacity and stops there. If twice as much water is added, then twice as much soil is wetted. Thus, under their experimental condition Ursino and Krotkov had two sets of plants: those at the 7 percent level growing in half as much soil as those at the 14 percent level; all of the roots were wet to field capacity at each irrigation. Their finding, therefore, that the magnitude of translocation to roots during the 8 hours after the assimilation of $^{14}CO_2$, was only slightly lower in the plants having the lower (7 percent) moisture application has little relevance to the effect of moisture stress on translocation.

Hatch and Glasziou (1963) found a direct relationship between the acid invertase content of *Saccharum officinarum* and the plant's water supply. Their results appear in Figures 10.2 and 10.3, and they show that, in their immature storage tissues, low water supply drastically inhibited invertase content. Because Hatch and Glasziou have concluded that invertases have a key role in regulating the movement of sucrose from conducting tissue and its subsequent utilization for growth or storage, it is apparent that the water regime of *S. officinarum* is important in the functioning of the plant in sugar production.

The effects of water stress on the translocation of ^{14}C-assimilates in 3-year-old trees of *Liriodendron tulipifera* was studied by Roberts (1964). The seedlings were kept under conditions favorable to normal translocation, and supplied with $^{14}CO_2$ in gaseous form. Water stresses up to a water deficit of 60 percent were used prior to exposure to $^{14}CO_2$. Photosynthesis proceeded for 1 hour under 2550 ft-c of light and at a temperature of 26 ± 1°C. The plants were sampled after a 4-hour translocation period.

Increasing water stress decreased both the uptake of $^{14}CO_2$ by leaves and the amount of ^{14}C moved out of leaves. Because reductions in translocation and photosynthesis were parallel, Roberts felt that a lowering of transport resulted from stomatal closure, and was therefore secondary to photosynthesis at the source. The amount of ^{14}C translocated out of leaves, the rate of translocation, and the distance to which translocate moved, were drastically reduced as leaf water stress increased from a stress of 5 percent to one of 20 percent. High water stress resulted in a larger percentage of acropetal ^{14}C movement than did low water stress, a finding substantiated by Wiebe and Wihrheim (1962).

The work of Plaut and Reinhold (1965) on water stress in *Phaseolus vulgaris* again shows that, under certain conditions, water deficiency reduces assimilate movement. When samples were taken 45 minutes or more after application of ^{14}C-labeled sucrose as droplets on leaves, transport of the ^{14}C was much reduced. Very little tracer moved acropetally in the stems, presumably as a result of water deficiency in

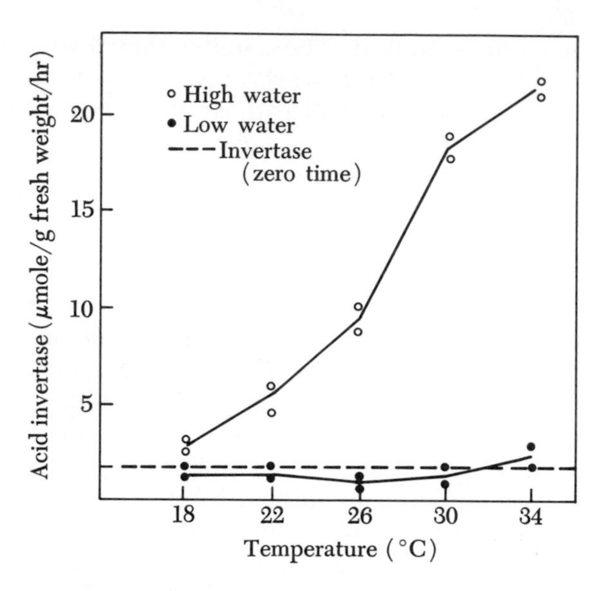

Figure 10.2. The effects of temperature and water supply on acid-invertase activity in *Saccharum officinarum*. The experimental time period was 10 days. High-water plants were irrigated twice daily, and low-water plants twice weekly. From Hatch and Glasziou (1963).

the growing region. The concentration of sugars in lower stems was reduced during the first hours, but later increased as a result of water stress on sugar incorporation by growth. Watering the plants accelerated basipetal transport, even before recovery from wilting. Plaut and Reinhold cited pressure flow as the mechanism responsible for their results.

Wardlaw (1967), working with *Triticum aestivum,* found that lack of water reduced photosynthesis (see Figure 6.8, page 133) and prolonged transport into conducting tissues; the mass rate of translocation was not seriously affected. Conversion of sugar to starch in ripening ears of *T. aestivum* may maintain the transport gradient, so that even though the velocity of flow is reduced, the concentration is increased in compensation. Thus *T. aestivum,* a plant that has evolved under conditions of reducing soil moisture during the fruiting period, seems adapted to drought, at least during the final ripening stages (Wardlaw, 1968a).

The role of soil moisture in growth and sugar production was studied by Hartt (1967). Moisture stress lowered the velocity and

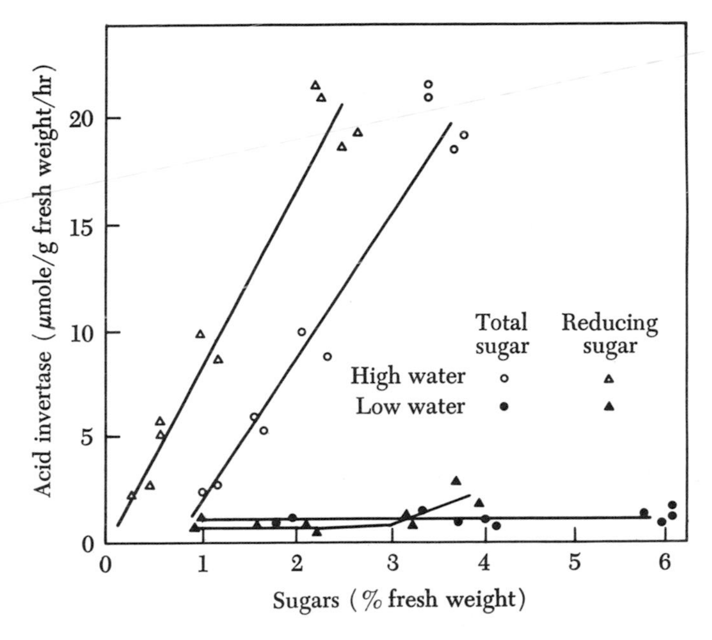

Figure 10.3. Relation between acid-invertase activity and sugar content of immature storage tissue after 10 days under high-water and low-water conditions. From Hatch and Glasziou (1963).

percentage rate of sugar transport. Sugar not used in growth moved only slowly in the phloem; most of it was stored in the stalk. In her experiments, Hartt found that moisture deficiency reduced transport more than it did photosynthesis, and so proposed that the water supply has a primary effect upon translocation.

Merkle and Davis (1967) studied the effect of moisture stress on absorption and translocation of picloram and 2,4,5-T in *Phaseolus vulgaris*. They used a thermoelectric method to determine the velocity of the transpiration stream in the stems, and found a good correlation between the velocity of sap and the moisture supply. They measured tracer concentrations in treated leaf, stem, apex, and central stem by gas chromatography. Table 10.1 gives their results, the leaf wash values being given in milligrams and the tracer concentrations in nanograms. It is apparent, from these data, that moisture deficits resulting in sap velocities of 57 and 22 percent of normal greatly reduced the amounts of picloram reaching the stem regions; sap velocities of 37 and 17 percent of normal correlated with reduced transport of 2,4,5-T. Thus, using two highly accurate methods—one

Table 10.1. Effect of moisture stress on the absorption and translocation of picloram and 2,4,5-T applied to the primary leaf of *Phaseolus vulgaris.* (Sap velocity was found to correlate with moisture supply. Values are averages of 12 replications consisting of 9 plants each.)

Herbicide	Xylem sap velocity (% of control)	Herbicide recovered			
		Leaf wash (mg)	Treated leaf (ng/mg fresh wt)	Shoot apex (ng/mg fresh wt)	Central stem (ng/mg fresh wt)
Picloram*	100	185.1	2.511	1.191	1.082
	57	180.5	2.592	0.983	1.068
	22	185.8	2.255	0.773	0.662
2,4,5-T†	100	351.8	7.063	1.680	0.787
	37	371.5	8.313	1.081	0.565
	17	364.8	7.384	0.781	0.344

Source: Merkle and Davis (1967).

*Data taken 4 hr after application of 225 mg.

†Data taken 8 hr after application of 450 mg.

to detect moisture deficit, the other to measure tracers—Merkle and Davis were able to give a clear picture of the role of water availability in herbicide transport.

Prasad et al. (1967), using methods similar to those of Clor et al. (1962, 1963, 1964), studied the effects of humidity on translocation. They used ^{14}C- and ^{36}Cl-labeled dalapon as the tracers. Count data and autoradiographic evidence showed that greater amounts of dalapon were absorbed and translocated at a post-treatment RH of 88 ± 3 percent than at a medium RH of 60 ± 5 percent or a low RH of 28 ± 3 percent. Results for *Hordeum vulgare* are presented in Table 10.2, and those for *Phaseolus vulgaris* in Table 10.3. Post-treatment temperature also determined the amount of dalapon absorbed and translocated in *P. vulgaris,* as shown in Table 10.4.

The rate of droplet drying and the number and behavior of stomata also played roles in dalapon distribution. Table 10.5 presents data on the effects of rewetting and bagging on dalapon uptake and distribution; Table 10.6 shows the effects of humidity and rewetting on dalapon uptake by leaf disks of *Zebrina pendula.* Thus, rewetting of the application droplets, or high ambient RH, promoted uptake through astomatous, as well as stomatous, surfaces. Absorption and translocation were greater in *Phaseolus vulgaris* at 95 ± 3 percent RH than in plants grown at 28 ± 3 percent RH, both of which were treated under high humidity conditions. Thus, cuticle development and hydration were implicated in the phenomenon of the enhancement of uptake by humidity.

Table 10.2. Effect of post-treatment relative humidity on distribution of ^{14}C-labeled dalapon in *Hordeum vulgare* at 26 ± 1°C and 800 ft-c, 6 and 18 hours after treatment. (Dalapon, at 1500 mg/liter, applied as a 10 μliter drop on the first true leaf. All plants grown at 26 ± 1°C, 60 ± 5% RH, and 800 ft-c prior to treatment and transfer.)

Relative humidity (%)	Radioactivity [(count/min)/mg dry wt]*			
	After 6 hr		After 18 hr	
	In shoots	In roots	In shoots	In roots
28 ± 3 (low)	104	51	904	130
60 ± 5 (medium)	130	62	1270	262
88 ± 3 (high)	148	76	1552	361

Source: Prasad et al. (1967).

*Counts are total net radioactivity in dalapon translocated out of a 2-inch section of leaf centered over the treated spot. The radioactivity of a standard planchet (equivalent to the applied dosage) was 14,286 count/min.

Table 10.3. Effect of post-treatment relative humidity on distribution of ^{14}C-labeled dalapon in *Phaseolus vulgaris* at 26 ± 1°C and 800 ft-c, 6 and 18 hours after treatment. (Dalapon, at 1500 mg/liter, applied as a 20 μliter drop on one leaflet of the first trifoliate leaf. All plants grown at 26 ± 1°C, 60 ± 5% RH, and 800 ft-c prior to treatment and transfer.)

Relative humidity (%)	Radioactivity [(count/min)/mg dry wt]*							
	After 6 hr				After 18 hr			
	Treated leaf	Stem	Terminal bud	Root	Treated leaf	Stem	Terminal bud	Root
28 ± 3 (low)	67	16	54	19	172	69	79	67
60 ± 5 (medium)	101	54	88	21	792	251	340	81
88 ± 3 (high)	117	57	81	21	864	301	740	85

Source: Prasad et al. (1967).

*Counts are total net radioactivity in dalapon translocated out of a 20-mm diameter section of leaflet centered over the treated spot. The radioactivity of a standard planchet (equivalent to the applied dose) was 29,477 count/min.

Several solution additives were studied in an attempt to find a substitute for high humidity. An effective surfactant almost doubled the counts in leaves of *Phaseolus vulgaris* plants treated with ^{14}C-labeled dalapon for 6 hours under conditions of low humidity (28 ± 3

Table 10.4. Effect of post-treatment temperature on distribution of ^{14}C-labeled dalapon in *Phaseolus vulgaris* at 60 ± 5% RH and 800 ft-c, 8 hours after treatment. (Dalapon, at 1500 mg/liter, applied as a 20 μliter drop on one primary leaf. All plants grown in the greenhouse initially, then held at 60 ± 5% RH, 26 ± 1°C, and 800 ft-c for 3 days prior to treatment and transfer.)

Temperature (°C)	Radioactivity [(count/min)/mg dry wt]*			
	Treated leaf	Stem	Terminal bud	Root
26 ± 1	181	72	33	88
43 ± 1	756	96	91	110

Source: Prasad et al. (1967).

*Counts are total net radioactivity in dalapon translocated out of a 20-mm diameter section of leaf centered over the treated spot. The radioactivity of a standard planchet (equivalent to the applied dose) was 29,578 count/min.

percent). These results help to explain both the greater effectiveness of dalapon against grasses in regions of high RH, and the benefits derived from use of a surfactant.

Uptake and distribution of the growth retardant Alar was studied by Sachs et al. (1967). One of the major factors involved in the movement of this compound in plants proved to be RH; transport was primarily basipetal in this work.

From these many studies on the effects of moisture supply on translocation, there are indications of a fairly consistent enhancement by adequate moisture, and a great increase under conditions of supersaturation of the atmosphere. Both absorption and translocation are increased, and often it is difficult to separate the two. Theoretically, a pressure-flow mechanism should not be sensitive to nominal differences in water potential because, by osmotic adjustment, velocity and concentration should compensate each other, with the result that mass transfer should remain constant. Under severe conditions of stress, the increased viscosity of the assimilate stream should slow the flow and reduce transport.

Under most conditions, the effects of water stress on photosynthesis and growth probably greatly outweigh those on transport per se. Thus, in the case of the harvesting of *Beta vulgaris* and *Saccharum officinarum*, withholding irrigation during the last days probably has little effect on sugar translocation; what deleterious effects there are are probably on photosynthesis; beneficial effects must relate to the inhibition of growth.

Table 10.5. Distribution of ^{14}C-labeled dalapon in *Phaseolus vulgaris* after 30 hours as affected by retardation of droplet drying (prolongation of exposure to dalapon solution). (Dalapon, at 7500 mg/liter, applied as a 20 μliter drop on one primary leaf. The droplet normally disappeared completely before 4 hours in the open greenhouse.)

Treated plants	Radioactivity [(count/min)/mg dry wt]*			
	Treated leaf	Stem	Terminal bud	Root
Left in open greenhouse†	185	60	205	51
Same, rewetted continually with distilled water for 4 hours	320	81	250	74
Placed in polyethylene bags (saturated atmosphere)	1020	324	507	124

Source: Prasad et al. (1967).

*Counts are total net radioactivity in dalapon translocated out of a 20-mm diameter section of leaf centered over the treated spot.

†Environmental conditions in the greenhouse during the day were 29 ± 2°C, 2500 ± 500 ft-c, and 60 ± 10% RH.

Table 10.6. Effect of post-treatment relative humidity and repeated rewetting on the uptake of ^{36}Cl-labeled dalapon by leaf disks of *Zebrina pendula* at 26 ± 1°C and 800 ft-c, 6 hours after treatment. (Dalapon, at 1500 mg/liter and containing 1.0 ml/liter of X-77 surfactant, applied as a 0.1 ml drop on either the upper or lower surface of each 13-mm diameter leaf disk. Disks were supported on moist filter paper in petri dishes and held at 26 ± 1°C, 800 ft-c, and 55 ± 5% RH prior to treatment and transfer.)

Relative humidity (%)	Radioactivity [(count/min)/leaf disk]*	
	Adaxial surface (astomatous)	Abaxial surface (stomated)
28 ± 3 (low)	82	230
28 ± 3 (low), rewetted	107	328
88 ± 3 (high)	121	411

Source: Prasad et al. (1967).

*Counts are total net radioactivity per leaf disk after thorough washing to remove dalapon adhering to the surface. Each value represents the average of 20 leaf disks.

Temperature

Curtis reported in 1929 that chilling the petioles of *Phaseolus vulgaris* retards removal of carboydrates. This he attributed to a cessation of protoplasmic streaming. Crafts (1932) found that chilling of the stems of *Cucumis sativus* reduced phloem exudation by about 56 percent. Cooling the entire plant reduced exudation rate even more. Crafts attributed this reduction to increased viscosity.

Research on the effects of temperature on translocation has continued through the years. Table 10.7 from Burr et al. (1958) shows that cool air around the tops of *Saccharum officinarum* plants reduced transport appreciably; cooling the roots reduced transport by 50 percent in 24 hours, and by 82 percent in 80 hours. Whitehead et al. (1959) found that cooling the stem to 5°C reduced transport in *Lycopersicon esculentum*. Pallas (1960), studying the effects of temperature and humidity on the movement of tracers in *Phaseolus vulgaris*, found increases in both uptake and translocation with increases in temperature between 20° and 30°C. Movement of ^{14}C-labeled 2,4-D followed the assimilate stream, and it moved as free 2,4-D.

Fujiwara and Suzuki (1961a) studied the effects of temperature and light on translocation of ^{14}C in *Hordeum vulgare*. They found a temperature of 25°C to be the optimum for photosynthesis under their conditions, and 30°C gave the maximum respiration rate of excised roots. In their translocation study, they used temperatures of 15°, 20°, 25°, and 30°C around the tops of the plants, and temperatures of 15°, 20°, and 30°C around the roots. Results of their study are shown in Figure 10.4. Because osmotic pressure is proportional to absolute temperature, the writers concluded that the larger the difference in

Table 10.7. Effect of air and root temperatures on growth and translocation of ^{14}C-labeled photosynthate from leaves of *Saccharum officinarum.*

Plant number	Temperature (°C)		Relative growth rate (dry wt as % of plant no. 1)	Relative photosynthesis (specific activity of lamina as % of plant no. 1)	Loss of ^{14}C from fed blade (%)	
	Air	Root			After 6 hr	After 24 hr
1	23.1	22.2	100.0	100.0	78	93
2	23.1	16.7	24.3	83.0	25	50
3	13.6	22.2	25.3	84.1	57	86

Source: Nelson (1963), based on data from Burr et al. (1958).

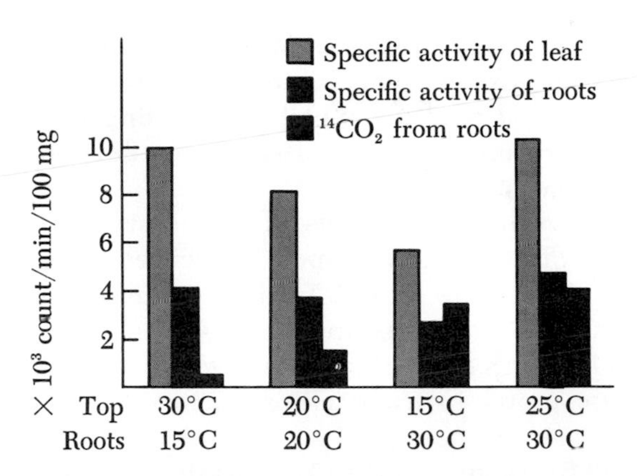

Figure 10.4. Effect of temperature on the translocation of ^{14}C-labeled assimilate in *Hordeum vulgare*. From Fugiwara and Suzuki (1961a).

temperature between source and sink, the higher the translocation rate in a pressure-flow system. This, of course, assumes a constant concentration, and ignores the activities of source and sink. As shown in Figure 10.4, they recorded their maximum transport when top temperature was 25°C and root temperature was 30°C. However, the excretion of $^{14}CO_2$ from roots increased severalfold between 15°C and 30°C, so that the benefits of high root temperature to translocation were more than offset by the respirational loss of $^{14}CO_2$. They found a direct correlation between the length of exposure to light and the translocation rate.

In 1962, Whitehead found that chilling and local killing of the stems of *Lycopersicon esculentum* reduced or stopped movement of sugars to the roots; the movement of ^{32}P was also inhibited by chilling to 0.5°C. Webb and Gorham (1965b) showed that chilling greatly reduced the export of ^{14}C-labeled sugar from leaf blades of *Cucurbita maxima* seedlings; temperatures above 25°C also reduced transport; movement was stopped at 55°C. The slowing of translocation by low temperature proved to be reversible: when the chilled node was warmed to 25°C, export from the blade was resumed.

Geiger (1966) and Swanson and Geiger (1967) studied the effects of low temperature upon sugar transport in *Beta vulgaris;* when a 2-cm portion of the petiole was cooled to 1°C, the translocation rate was cut 60–95 percent. The inhibition was temporary, for when the treatment was continued, the translocation rate gradually returned to near normal. This would seem to indicate that when the sugar potential built up in the source leaf, turgor pressure increased enough to overcome the viscous resistance, and flow was resumed. Swanson and Geiger proposed that the metabolic loading of the sieve-tube system results in resumption of a passive mass-flow of the assimilate stream in the sieve tubes.

Hartt (1965a) studied the effects of temperature on translocation in *Saccharum officinarum.* Transport velocities were 1.40 cm/min when the air temperature was 20°C; 1.56 cm/min at 24°C; and 2.00 cm/min at 33°C. This latter temperature was inhibitory to transport in Webb and Gorham's (1965b) *Cucurbita maxima* plants, but *Saccharum officinarum* grows in warm climates, in which temperatures of 35°C and above are common.

Thrower (1965), in her studies on translocation in *Glycine max,* tested the effects of chilling petioles. Chilling a petiole to 1°C prevented translocation past the chilled region; transport resumed 5–20 minutes after the petiole was warmed.

The effects of temperature extremes on translocation continue to interest plant physiologists (Hartt, 1965a; Webb and Gorham, 1965b; Swanson and Geiger, 1967). Webb (1967) performed a series of provocative experiments on the effects of localized temperature treatments on translocation rates in *Cucurbita melopepo.* Treatments consisted of sealing small troughs around selected loci on young plants having one fully expanded leaf. Treatments were made on petiole, node, stem, and hypocotyl; temperatures varied between 0° and 55°C.

In all cases temperatures of 0° and 5°C severely inhibited translocation of ^{14}C-labeled assimilate across the chilled region. A temperature of 10°C partially inhibited transport; 15°C gave extremely variable results; and between 15° and 35°C, temperature was not limiting. At 45°C, partial inhibition was observed, and 55°C caused almost complete cessation; 45°C causes temporary callosing of sieve plates that has been shown to block translocation (Webster, 1965; Currier, McNairn and Webster, 1966; McNairn, 1967; McNairn and Currier, 1967); 55°C causes almost instantaneous coagulation of the phloem sap of cucurbits.

The localized temperature treatment did not disturb the rate of assimilation of $^{14}CO_2$ or the export of ^{14}C-labeled compounds by the leaf blade. Translocates unable to pass a temperature-inhibited zone were diverted toward other sinks. Figure 10.5 shows the maximum and minimum amounts of ^{14}C translocated through the temperature-controlled regions. Webb proposed that these results indicate a uniform mechanism throughout the plant that controls the major portion

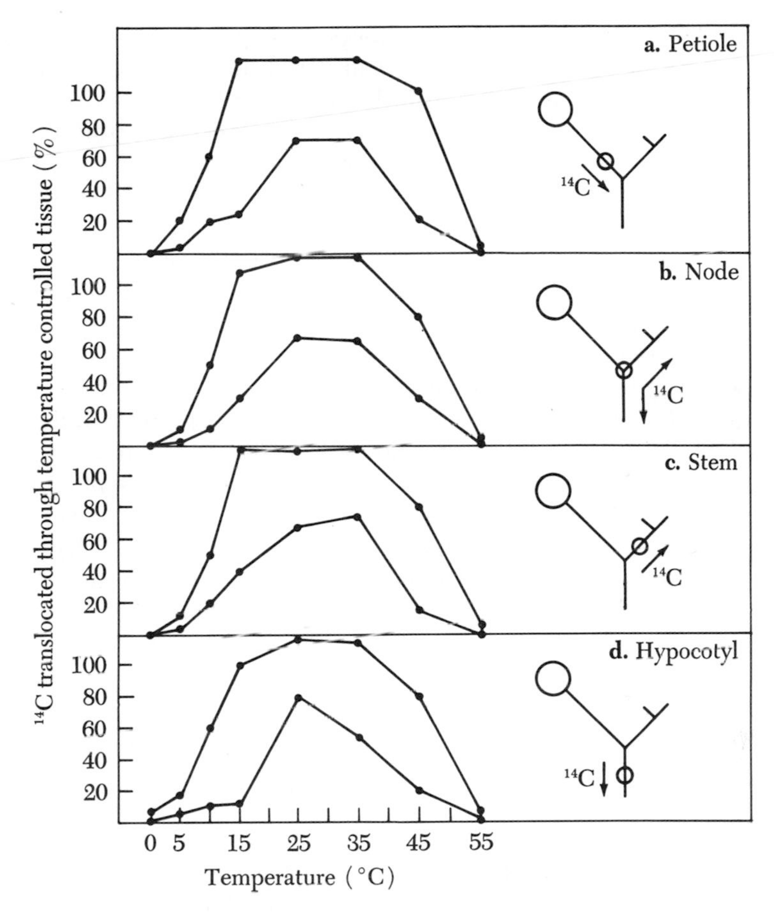

Figure 10.5. Maximum and minimum amounts recorded for the translocation of ^{14}C through *Cucurbita* tissues after 45 minutes of exposure to various temperatures. Results are expressed as percentages of an experimentally determined mean total of ^{14}C translocated in 45 minutes at 25°C. The tissues treated were (*a*) petiole, (*b*) node, (*c*) stem, and (*d*) hypocotyl. Each region of treatment is indicated in a diagram adjacent to its respective graph. From Webb (1967).

of the translocated compounds, and that the characteristics of the mechanism resemble those of protoplasmic streaming in chill-sensitive plants.

Ford and Peel (1967a) studied the effects of low temperature upon the movement of ^{14}C-labeled assimilates through phloem of young

(3–5 weeks) and old (2–3 years) shoots of *Salix*. Using water jackets 11 cm long, they compared cooled and normal stems by collecting and counting honeydew from aphids located at the distal ends of the jackets. Figure 9.8 (page 342) presents their results. Cooling of young shoots resulted in the expected lowering of mass transfer across the cooled region. Cooling the old stems gave the opposite result, that is, a greatly accelerated mass transport rate. Ford and Peel explained this on the basis of the difference in size of the sieve tubes in the young and old stems; sieve tubes in the young stems had radii of 12.6 μ with a standard deviation of 2.8 μ; in the old stems, the radii were 19.0 μ with a standard deviation of 3.9 μ. Increased viscosity upon cooling would retard movement much more in the younger shoots. In the older shoots, cooling might inhibit lateral movement to the xylem sink, and thereby actually cause an increase in mass transport of the ^{14}C-labeled assimilate. It is unfortunate that Ford and Peel did not measure the size of the tubular strands in the sieve plates. Resistance to flow through these is much more directly related to the transport capability of the phloem; and, because of the r^4 factor in the Poiseuille formula as applied to mass transport, resistance increases very rapidly with decreasing size of the conduits. On the basis of the radii of the sieve tubes, resistance per unit of length would be 2.3 times as great through the tubes of the young shoot as through those of the older one. Through the sieve pores, the difference could be very much greater.

In a recent communication, Swanson (1969) informed the authors of this book that in his *Beta vulgaris* test (Geiger, 1966; Swanson and Geiger, 1967), cooling a 4-cm length of petiole was no more effective in blocking transport than cooling a 1-cm length. This seems to indicate that, in such tests, the block is in the first range of sieve tubes, closest to the leaf. This recalls Crafts' (1936) exudation experiments with *Cucurbita maxima:* in these, cutting off a 1-mm slice from the end of the stem brought about resumed exudation; evidently, only a short range of sieve tubes becomes blocked.

Tammes et al. (1969) studied the effects of chilling on phloem exudation from *Yucca flaccida*. Cooling the tips of bleeding inflorescenses that were still attached to the plant prolonged the exudation period from a normal time of 7–8 hours to a period of 24–36 hours. Apparently, the plugging of the sieve tubes proceeds more slowly at 0°C than at 20–25°C. There was an appreciable increase in the dry-matter content of the sap collected at low temperatures, an effect that Tammes and his co-workers attributed to a cessation of normal export from sieve tubes to phloem parenchyma.

When isolated parts of the inflorescence stalk were cooled, exudation gradually decreased; exudation resumed when the parts were warmed to room temperature; it started again in 2 minutes when the parts were placed in water at 35°C. Tammes and his co-workers explained the slowing of exudation by low temperatures as an effect on the viscosity of the assimilate stream; at 20°C, a 20 percent sucrose solution has a viscosity of 2.0 cP; at 0°C, it has a viscosity of 3.8 cP. They assumed plugging to be a chemical or a metabolic process that is retarded at low temperatures. They attributed the driving force of the assimilate stream to turgor pressure developed osmotically and resulting from sugar secretion into the sieve tubes.

Wardlaw (1968a) cited literature describing the effects of temperature on patterns of assimilate distribution in plants. Brouwer and Levi (1969) found that altering the root temperature of *Phaseolus vulgaris* plants changes foliar uptake and subsequent transport of ^{22}Na, ^{32}P and ^{134}Cs. Uptake of ^{22}Na was slower than that of ^{134}Cs, but faster than that of ^{32}P. Root temperature greatly influenced uptake with an optimum range between 20° and 25°C. Penetration was found to respond more to root temperature during a pretreatment period than during the experimental period. Translocation during the first 6 hours of the experimental period was also influenced by the pretreatment temperature of the roots: pretreatment at 23°C resulted in much greater transport than did 8°C. Leakage of sodium and cesium into the culture medium occurred at all temperatures, provided transport had carried these elements into the roots.

Brouwer and Levi stressed the point that transport of foliarly applied substances follows the flow of carbohydrates in the assimilate stream. Low root temperatures that slowed basipetal flow resulted in accumulation in the stem and little or no loss from the roots.

In a subsequent paper, Levi (1969) reported experiments on uptake and distribution of ^{134}Cs by primary leaves of *Phaseolus vulgaris* under conditions of varying temperature, humidity, and air flow. Penetration was enhanced by increasing the humidity of the air or the moisture content of the leaves. Accumulation and transport were similarly influenced, and were affected more by temperature than by humidity. Levi (1966) had found in previous experiments that, under constant conditions of 23°C and 40 percent RH, a high absorption of cesium could take place for periods as long as 35 days, by which time fruits would be maturing. This has significance with respect to the hazards of ^{134}Cs residues in plants.

When considering the phenomenon of sieve-plate blocking, two factors should be borne in mind. The first factor is that in mass flow of a solution of elongated molecules, as in such movement of elongate virus rods, the elongated particles orient themselves parallel to the direction of flow. This is illustrated by stream double-refraction of rods of tobacco mosaic virus. The second factor is that such a stream, in approaching a sieve plate, is divided into a series of smaller, faster-flowing streams within the sieve pores, and that the elongate particles must become rearranged as they approach the ends of the pores. At normal temperatures, the viscosity allows such rearrangement; but with lowering temperature, a critical value must be reached at which a "log-jam" of the converging particles takes place, and the first sieve plate of each sieve tube becomes blocked with icelike sludge (Hodges, 1969). Thus, the length of chilled petiole behind this block is immaterial.

As pressure builds up in front of such a block, the effect of pressure upon the molecular structure of water would come into play (Crafts, 1968c): pressure is destructive of structure: a greater number of unbonded free water molecules would be formed. These molecules, filtering among the assimilate molecules, would act like a lubricant to bring about a reduction in frictional resistance. Viscosity would be lowered, and the "jam" would break. Such would seem to be a rational physical explanation for the blocking effect of temperatures around the freezing point of water, and the subsequent resumption of flow without a change in temperature. In those species in which flow is not resumed, the length and form of assimilate molecules, presumably, may be such that the quasi-crystaline block remains intact until the temperature is raised.

Light

Because light is required for photosynthesis, and photosynthesis provides the osmotically active substances for the osmotic pump that provides the driving force for flow of the assimilate stream, it seems logical to study the relation of light to translocation. Early work with $^{14}CO_2$ applied to leaves proved that ^{14}C-labeled assimilates move out of treated leaves within minutes of illumination, and that translocation is greater during the light period than in the dark (Rohrbaugh and Rice, 1949). When ^{14}C-labeled 2,4-D is applied to the leaves of *Phaseolus vulgaris,* it moves readily from photosynthesizing leaves, but very little or not at all from leaves in the dark. In such studies, leaves treated soon after exposure to light exported tracers while in the dark; it often proved necessary to hold leaves in the dark until the starch reserves were depleted, in order to prevent export; and such leaves would export tracer if sugar was included in the treatment.

More recently, Stoy (1963) fed $^{14}CO_2$ to the leaves of *Triticum aestivum* for 30 minutes in the light, and then placed them in the dark for study of distribution patterns. Translocation from flag leaves was preferentially to the inflorescences. During flowering, after growth had ceased, there was intense export to storage cells in stems; later, this stored assimilate was remobilized and moved into the developing seeds. Carr and Wardlaw (1965) found that, in low light, stem growth in *T. aestivum* may compete with grain for assimilates.

Butcher (1965), studying the kinetics of translocation of ^{14}C-labeled assimilate in *Beta vulgaris*, observed that 2 hours after supplying $^{14}CO_2$, the mass-rate of transport became linear, and remained so until the plants were darkened, whereupon the mass-rate decreased, becoming linear after 20 minutes at a level of less than half the light value. Decrease in accumulation was not caused by decrease in specific activity, but, rather, by a slowing of the stream. Concentration at the source seemed to determine the export rate; this, in turn, was a reflection of the synthesis of sucrose from starch reserves, and reflected sucrose potential at the source.

Nelson (1963), in a report on the environmental control of plant growth, described experiments on *Triticum aestivum* conducted in a phytotron, showing increasing root-to-shoot ratios with increasing light intensity. The increases in weight of roots he attributed to translocation. Table 10.8 illustrates this relation. Experiments on young seedlings of *Pinus* also showed a relation between light and translocation. Low light during the translocation period had little effect; low light during the growth period prior to the translocation test appreciably reduced transport of assimilates to roots. Table 10.9 shows these results. This experiment illustrates the point that the influence of

Table 10.8. The effect of light intensity on the ratio of root dry weight to shoot dry weight in *Triticum aestivum* 'Marquis' after 5 weeks growth. (Plants were grown at 20°C for a 24-hour photoperiod at constant light intensities, at 70% RH, and in sand culture. Values are means of 60 plants.)

Light intensity (ft-c)	200	500	1000	1750	2500	5000
Root/shoot ratio	0.14	0.17	0.27	0.32	0.32	0.43

Source: Nelson (1963).

Table 10.9. The effect of light intensity during growth and during translocation on the translocation of ^{14}C-labeled photosynthate from shoots to roots of *Pinus strobus*.

Light intensity			Distribution of ^{14}C (%)		
During growth	During translocation	$\left(\frac{\text{Root wt}}{\text{shoot wt}}\right)$	Shoot	Stem	Root
Full sunlight	2500 ft-c	3.7	87.0	0.5	12.5
Full sunlight	250 ft-c	2.7	84.0	0.5	15.5
6% of full sunlight	2500 ft-c	2.3	95.6	0.4	4.0
6% of full sunlight	250 ft-c	2.4	97.6	0.3	2.1

Source: Nelson (1963).

light on translocation is not direct, but is expressed indirectly through the supply of osmotically active solutes that drive the osmotic pump.

Brady (1969) found that light intensity influences translocation of 2,4,5-T in woody plants. Absorption of the isooctyl ester of 2,4,5-T varied by more than 20 percent when light intensity was increased from 40 ft-c to 4000 ft-c. Variation in translocation reached 45–50 percent in *Quercus stellata* and *Q. nigra;* the former showed a steady decline with increasing intensity, the latter a steady increase.

Hartt (1965b) ran tests on the effects of light on assimilate movement in detached leaves of *Saccharum officinarum.* In leaves detached in the morning with low reserve supply, basipetal translocation in the light increased about 16 times over blades in the dark. Blades taken in the afternoon gave transport values about three-quarters those from morning collection; transport was only twice that of afternoon blades in the dark. Thus, the assimilate pool built in the early part of the day provided solute for export even in the dark. Darkening the fed portion of a leaf and the tip above it brought about a reversal of flow from basipetal to acropetal in about 2 hours.

Hartt found that photosynthetic assimilation of CO_2 in *Saccharum officinarum* was saturated at 6000 ft-c; compensation was reached at around 125 ft-c. Because of this low value, Hartt concluded that translocation comes under photocontrol that does not involve pressure flow.

In 1966, Hartt reported work using lights of different colors. She used red, green, blue, and cool white fluorescent lamps, far-red sunlight, and incandescent bulbs in tests on detached leaves of *Saccharum officinarum.* She found transport of ^{14}C-labeled assimilate to be stimulated by red and blue more than by green or cool white illumination. Because the lights used had wide emission characteristics, Hartt could not correlate phototranslocation with any specific pigment system, but she concluded that transport of sucrose in phloem is influenced by the quality of illumination.

Thrower (1962), using plants of *Glycine max*, found that illumination in the range of 1000–2000 ft-c resulted in greater translocation of assimilate than did illumination in the range of 500–700 ft-c. Illumination of plants previous to exposure to $^{14}CO_2$ in the light resulted in reduced transport of ^{14}C, presumably because the sinks were saturated. This contrasts with Nelson's (1963) results, in which pretreatment of *Pinus* seedlings in full sunlight resulted in greater transport to the roots. Nelson proposed that the intensity of light has its effect not on the translocation process, but on the physiological state of the source. This is a logical conclusion, because there is no way known by which the energy of light can be applied directly to the translocation process as it goes on in sieve tubes of the phloem. By conversion to heat, it can raise the temperature of plant parts.

Wardlaw (1968a) reported on light in his review on translocation.

Inhibitors

Metabolic inhibitors, like temperature treatments, have been used to study the relation of metabolism to translocation. Curtis (1929) showed that anoxia and hydrogen cyanide inhibited the movement of assimilates from leaves. Schumacher (1930) proved that a very dilute eosin Y solution applied to leaves caused callose formation on sieve plates at some distance from the region of application. Such callose formation inhibited loss of nitrogen and dry weight from a treated leaf.

Much work on metabolic inhibitors has been done since these early tests. Kendall (1955) found that DNP and sodium fluoride inhibited the translocation of ^{32}P from *Phaseolus vulgaris* leaves by roughly 75 percent. With sodium arsenite, fluoroacetic acid, IAA, 2,4,-D, and TIBA results were indefinite. Kursanov (1956b) reported that poisoning of the cytochrome oxidase of the leaf petiole of *Beta vulgaris* impedes the passage of assimilates.

After careful measurements on respiration of vascular bundles of *Pelargonium zonale* petioles, Willenbrink (1957a) made detailed studies on the effects of enzyme inhibitors on translocation through such petioles. He determined the movement of exogenous indicators, of fluorescein-K, of endogenous nitrogen and phosphorus compounds, as well as of a single example of a ^{14}C-labeled compound from leaf laminae. Willenbrink found the respiration intensity of the isolated vascular bundle of *P. zonale* to be significantly higher than that of surrounding petiolar tissues, even if determined on the basis of protein nitrogen. The height of the RQ signified that not all plants studied have aerobic respiration in their vascular bundles, so that only the strikingly high metabolic intensity seems unique. The surrounding of freed vascular bundles of *P. zonale* by molecular nitrogen or hydrogen did not inhibit the transport of nitrogen compounds or fluorescein-K; an atmosphere of carbon monoxide had no effect on transport.

By contrast, hydrogen cyanide, volatilized from a 10^{-3} M solution, fully blocked the transport of fluorescein-K and compounds of nitrogen and phosphorus; such blocking could be counteracted by removal of hydrogen cyanide. Obstruction of transport differed according to the substances used. By simultaneous determination of the transport of fluorescein-K and the transport of compounds of nitrogen and phosphorus, it was shown that the movement of the first two is more sensitive to hydrogen cyanide than is that of compounds of phosphorus. Willenbrink took this as an indication of independent movement of these materials.

Other known enzyme inhibitors (DNP, sodium arsenite, potassium azide, and iodoacetic acid, for example) were found to hinder transport; sodium fluoride, however, acted only weakly. Although Willenbrink was unable to prove the reversibility of these inhibitions, he found that, at the end of the experiments, the treated cells were mostly plasmolysable. Fluorescein-K transport was hindered only by use of high concentrations of inhibitors, but the transport of phosphorus and nitrogen compounds was sensitive to all inhibitors used, and at lower concentrations. A test in which $^{14}CO_2$ was applied to the leaf showed that transport of ^{14}C-labeled sugar was reversibly hindered by an atmosphere of hydrogen cyanide. Willenbrink concluded that long-distance transport in sieve tubes is dependent upon respiration, in a manner similar to that shown by Arisz (1964) for transport in parenchyma. He further concluded that the respiratory poisons blocked the individual sieve tubes, rather than impairing the mobilization process in the leaf.

Currier (1957) found that boron and aniline blue promote the synthesis of callose, which tends to plug the sieve plates of plants and hinders translocation. Dunning (1959) found that callose could be induced to form on sieve plates of *Cucurbita* by allowing cut petioles to absorb solutions of endothall, eosin Y, maleic hydrazide,

2,4-D, monuron, boric acid, dalapon, amitrole, or phloridzin. Eschrich (1961) reported callose formation stimulated by barium acetate; and Lerch (1960) noted callose synthesis promoted by treatment with ammonia, IAA, and TIBA.

Bieleski (1960) conducted extended experiments with potassium cyanide on *Saccharum officinarum*, and he reported complete inhibition of sucrose accumulation in stems. Turkina (1961)—reasoning that the accumulation of sucrose against a concentration gradient that she had shown requires expenditure of metabolic energy—used potassium cyanide at 10^{-3} M concentration, and showed that sucrose absorption in *Beta vulgaris* was reduced by 60 percent for a period of 90 minutes (see Table 10.10). After inhibiting oxidative phosphorylation with DNP at 10^{-4} M concentration, Turkina found sucrose absorption to be depressed by 40–60 percent in 2-hour experiments. She also found a 35–40 percent increase in the respiration of vascular bundles of *Beta vulgaris* in an atmosphere of molecular oxygen after treatment with DNP. She reasoned that DNP, by suppressing ATP synthesis, lowered the energy level of the cells, causing inhibition of sugar accumulation. There is an inconsistency here in interpretation: cells having high respiration should have an ample supply of energy. Possibly, the injury incurred in isolating the vascular bundles was such that the sucrose tended to leak from storage cells, resulting in reduced retention, rather than inhibited absorption. Crafts and Yamaguchi (1964) found that DNP at 10^{-4} M reduced retention of ^{14}C-labeled 2,4-D by the roots of *Glycine max;* the tracer moved to the tops in sufficient quantity to produce visible labeling in autoradiographs.

Table 10.10. Effect of potassium cyanide and DNP on the absorption of ^{14}C-labeled sucrose by vascular bundles of *Beta vulgaris*.

Absorption time (min)	Inhibitor	Experiment 1		Experiment 2	
		Sucrose absorbed (μmole/g dry wt)	Inhibition (%)	Sucrose absorbed (μmole/g dry wt)	Inhibition (%)
90	None	126.6		100.1	
	KCN*	40.9	67.7	40.9	59.6
120	None	177.0		146.6	
	DNP†	72.0	59.3	86.6	41.0

Source: Turkina (1961).
* 10^{-3} M.
† 10^{-4} M.

Concerning the site of action of translocation inhibitors applied to petioles, Duloy et al. (1961) suggested that they may migrate to the leaves and hinder movement into phloem; or, alternatively, that they move into sink tissues and lower metabolic activity, so that assimilates are not readily absorbed out of the phloem.

Nelson (1963) reported unpublished results of Mortimer to the effect that localized application of hydrogen cyanide to *Beta vulgaris* petioles inhibits translocation. Using labeled cyanide, Mortimer found that the inhibitor entered the petiole, translocated to the leaf blade, and inhibited passage of sucrose from the mesophyll tissue to the veins. These suggestions of Duloy and his co-workers and Nelson controvert Willenbrink's conclusion that respiratory poisons have a local effect upon conduits of the petiole, rather than upon the source or the sink.

Crafts and Yamaguchi (1964) gave evidence that 2,4-D will inhibit translocation of amitrole in *Zebrina pendula* and *Phaseolus vulgaris*.

Moorby (1964) found that prometryne, an inhibitor of photosynthesis, produced the same changes in uptake and translocation of ^{137}Cs as did darkening the leaf. His results are shown in Table 10.11. The uptake of this tracer, and the amount translocated from the leaf, were appreciably reduced by the prometryne treatment; distribution was also affected.

Moorby also used DNOC, an uncoupler of phosphorylation, in his studies with ^{137}Cs. This inhibitor reduced the movement of the tracer to basal parts of the plant; it increased transport to the apical parts and to the untreated portion of the treated leaf (Table 10.12). This indicates that it reduced uptake into the symplast and freed the tracer for uptake and movement in the apoplast. These results resemble those of Crafts and Yamaguchi (1964) on the effects of DNP on the transport of 2,4-D from root application.

Harel and Reinhold (1966) used DNP as a metabolic inhibitor in an attempt to distinguish between direct and indirect effects of such inhibitors on sugar translocation. They used *Glycine max* seedlings and applied ^{14}C-labeled sucrose in capillary tubes to narrow leaf flaps involving the midribs near the leaf tips. DNP was supplied to cut petioles of primary leaves; and ^{14}C-labeled sucrose was applied to the first trifoliate leaf, or to the opposite primary leaf. DNP severely inhibited the transport of ^{14}C-labeled sucrose, if it was applied to the ^{14}C-treated leaf immediately before or during the ^{14}C treatment. Transport was also inhibited if DNP was applied along the translocation path during treatment with ^{14}C-labeled sugar. However, when DNP was applied through cut petioles of primary leaves after removal of

Table 10.11. The effect of prometryne and darkening on the uptake and translocation of ^{137}Cs by the third leaf of *Pisum sativum*.

	Amount absorbed (%)	Amount translocated (as % of amount absorbed)	Distribution (as % of amount translocated)		
			To rest of treated leaflet	To apical parts	To basal parts
Control (light, no prometryne)	86.9	29.4	5.6	33.5	60.9
Prometryne applied to illuminated third leaf	40.5	10.4	6.0	65.9	23.1
Prometryne applied to darkened third leaf	67.3	13.8	8.5	77.3	14.3
Third leaf in dark (no prometryne)	44.8	10.6	12.1	74.9	13.0
Least significant differences					
$P = 0.05$	21.0	3.8	2.2	10.0	8.9
$P = 0.01$	36.8	6.6	3.9	17.5	15.5

Source: Moorby (1964)

Table 10.12. The effect of DNOC on the uptake and translocation of ^{137}Cs by the third leaf of *Pisum sativum*.

	Amount absorbed (%)	Amount translocated (as % of amount absorbed)	Distribution (as % of amount translocated)		
			To rest of treated leaflet	To apical parts	To basal parts
Control	97.2	25.5	8.3	21.8	69.9
DNOC applied to third leaf	96.0	14.1	16.2	38.2	46.7
Least significant differences					
$P = 0.05$	3.1	2.1	7.2	10.8	4.0
$P = 0.01$	5.1	3.4	12.7	19.0	7.1

Source: Moorby (1964).

the ^{14}C-treated terminal leaflet of the first trifoliate leaf, no inhibition occurred. On the contrary, transport appeared to have been accelerated; ^{14}C disappeared from the upper regions and accumulated in the lower stem and roots.

Harel and Reinhold proposed that if ^{14}C circulates in the plant, the DNP treatment may retard redistribution from roots to tops, citing the work of Biddulph and Cory (1965) for evidence of such circulation. Biddulph and Cory, however, did not demonstrate circulation from roots to tops via the phloem. Harel and Reinhold concluded that the inhibitory effect of DNP on downward phloem transport was probably due to a retardation of uptake and secretion into sieve tubes, rather than to a direct effect on the sieve tubes themselves.

Eschrich et al. (1965) studied the effects of callose formation on assimilate transport. They induced callose synthesis in sieve tubes by injecting boric acid and calcium chloride solutions into hollow petioles of *Cucurbita;* as many as 34 percent of the sieve plates were callosed, but long-distance transport of fluorescein-K was not inhibited.

Willenbrink (1966b) used hydrogen cyanide to inhibit transport through *Pelargonium zonale* and *Phaseolus vulgaris* petioles from leaves treated with $^{14}CO_2$. In *Pelargonium zonale,* the application of hydrogen cyanide to the bared central bundle of the petiole brought about a localized inhibition of transport. Even when the hydrogen cyanide treatment preceded the $^{14}CO_2$ feeding by 3 hours, the inhibition was in evidence. Of the ^{14}C-labeled substances identified in the untreated portions of plants, only alanine was affected: its movement was increased. Sucrose was the principal assimilate that moved and blocking with hydrogen cyanide did not cause starch storage; amylase treatment showed little increase in ^{14}C. Hydrogen cyanide treatment was reversible: 90 minutes after removal of the inhibitor, translocation across the treated region was resumed. Willenbrink interpreted his experiments as indicating that long-distance transport via the phloem is not only determined by a metabolically regulated import and export of assimilates, but that an undisturbed functioning of cyanide-sensitive enzyme systems is essential to a free flow of materials through the phloem.

Webster (1965) found that local heat treatment at 45°C induced callose formation in the phloem of *Gossypium hirsutum* seedlings. McNairn (1967) demonstrated that such treatment causing callose in the hypocotyl retarded the translocation of ^{14}C-labeled assimilate out of the cotyledons. This response was discussed further by McNairn and Currier (1968).

From this review of work with inhibitors, it is obvious that two sorts of chemicals have been used: (1) the traditional metabolic inhibitors, used principally in efforts to prove that translocation results from the activity of living cells; and (2) callose inducers, used in a search for the causes of transport blockage. Although many studies have involved the use of metabolic inhibitors, and a good many examples of transport inhibition have been reported, little effort has been put to discovering a realistic mechanism to account for the action of inhibitors. Are they xylem-mobile, and therefore capable of movement into leaves where they inhibit transfer from mesophyll to phloem; are they phloem-mobile, and therefore able to migrate to sinks, and to hinder uptake from the sieve tubes; or do they act locally to block flow by plugging the sieve plates with callose or filaments? The conclusions of DuLoy et al. (1961) and Nelson (1963) would favor the first, Willenbrink's the second; work by Currier and his students points to callose as a possible mechanism. Before a satisfactory rationalization of inhibitor action can be obtained, these questions must have answers: we need to know how and where metabolic energy is brought to bear on the translocation process.

11

Some Quantitative Aspects of Translocation

Mass Transfer

The form in which translocation data are expressed has given rise to much discussion in the literature (Canny, 1960b, 1962b). Thus, concerning rate of translocation, Canny (1960b) remarked:

> It is natural to picture the transport taking place by the movement of a solution of organic substances in water, and to imagine that this solution has a concentration to which some value may be assigned. . . . This solution has, presumably, some average velocity, which is related to the volume transfer thus:
>
> $$\underset{(\mathrm{cm^3/hr})}{\text{volume transfer}} = \text{area} \times \underset{(\mathrm{cm^2})(\mathrm{cm/hr})}{\text{velocity.}}$$

This is the expression commonly used to express measured velocities of flow of phloem exudation (Crafts, 1931, 1932; Crafts and Lorenz, 1944a; Weatherley et al. 1959; Weatherley and Johnson, 1968).

In contrast to this is the expression of the same data in the form of specific mass transfer. For example, to quote Canny (1960b) again:

> A favorite system has been a large developing fruit, which though green, grows in dry weight mainly by virtue of the organic substance imported through the peduncle, and, though respiring, is gaining dry weight much faster than losing it. In such a system it is possible to measure a rate of arrival of organic substance at the sink which unequivocally implies a certain time rate of transfer of mass through the connecting tissue.

This specific mass transfer Canny expresses thus:

$$\underset{(\text{g/cm}^2/\text{hr})}{\text{specific mass transfer}} = \underset{(\text{cm/hr})}{\text{velocity}} \times \underset{(\text{g/cm}^3)}{\text{concentration}}$$

Table 11.1, from Canny, gives a number of such values. Those quoted from Crafts (1933) were for tubers of *Solanum tuberosum*, which obviously could not photosynthesize; they were, therefore, minimum values needing correction for respiration. The values of Crafts and Lorenz (1944a) were for a crookneck variety of *Cucurbita*, the fruits of which are yellow and, therefore, nonphotosynthetic.

Canny (1960a) stated:

> Direct measurements of the mass transfer of sucrose in phloem are few and have usually been made on systems selected to demonstrate the highest rates attainable rather than the rates normally realized in the parts of growing shoots. All those that give results in velocities are useless to us for the present purpose, and only those from which we can extract a mass-transfer rate are helpful.

To understand Canny's statement, it is important to realize that he viewed translocation in the phloem as a "diffusion analogue"; that is, that the molecules in transit are, presumably, moving independent of one another and of the water with which they are associated. In this view, the rate of translocation means the number of molecules moving from point *a* to point *b* in a given time. The velocity at which the fastest molecule moves over the specified distance is irrelevant; the important datum is the number of molecules delivered per unit of time.

For those who accept the mass-flow concept, velocity has an entirely different significance, because all solutes in the assimilate stream are viewed as moving along en masse with the included water; and although the first molecule to move from point *a* to point *b* in a given time may not indicate the true average velocity of flow, very soon a steady state of flow is established, and the velocity of this flow

Table 11.1. Translocation rates as measured by mass transfer of dry weight.

Plant system	Specific mass transfer (g dry wt/cm^2 phloem/hr)	Reference
Solanum tuberosum tuber stem	4.5	Dixon and Ball (1922)
Dioscorea alata tuber stem	4.4	Mason and Lewin (1926)
Solanum tuberosum tuber stem	2.1	Crafts (1933)
Kigelia africana fruit peduncle	2.6	Clements (1940)
Cucurbita maxima fruit peduncle	3.3	Crafts and Lorenz (1944a)
Cucurbita pepo fruit peduncle	4.8	Colwell (1942)
Gossypium hirsutum bark flaps (probably damaged)	0.14–0.64	Mason and Maskell (1928b)
Phaseolus multiflorus petiole	0.56	Birch-Hirschfeld (1920)
Phaseolus multiflorus petiole	0.7	Crafts (1931)
Tropaeolum majus petiole	0.7	Crafts (1931)

Source: Canny (1960b).

times the concentration will give the mass transfer, as shown in the equation from Canny cited above.

Many values for the concentration of phloem exudate have been given in the literature (Crafts, 1931, 1932, 1936; Huber et al., 1937; Zimmermann, 1957a, 1957b, 1958a, 1960a). Although all such measurements are subject to the criticism that they do not represent the normal condition of the functioning phloem, the more recent values obtained by the aphid-stylet technique are much more accurate, particularly when obtained over periods of several hours; values given by Weatherley et al. (1959) in their figure 3 and tables I and II are good examples.

Given the concentration of sucrose in the assimilate stream, it is simple to convert mass-transfer values to velocities; errors due to amino acids, polypeptides, and other constituents are minor. Table 11.2 gives specific gravities of various sucrose solutions within the range commonly found in phloem exudates.

As an example, using the average value of 3.6 g dry weight per square centimeter of phloem per hour from Table 11.1, if the concentration were 10 percent, the velocity of flow would be $36 \times 1/1.040 = 34.6$ cm per hour.

It is obvious, in all such calculations, that only minimum values are obtained, because there is bound to be a loss of solutes from the

Table 11.2. Specific gravity of sucrose solutions of different concentrations.

Percent sucrose by weight	Specific gravity (d^{20}_{20})	Percent sucrose by weight	Specific gravity (d^{20}_{20})	Percent sucrose by weight	Specific gravity (d^{20}_{20})
7.0	1.028	16.0	1.065	25.0	1.105
7.5	1.030	16.5	1.067	25.5	1.108
8.0	1.032	17.0	1.070	26.0	1.110
8.5	1.034	17.5	1.072	26.5	1.112
9.0	1.036	18.0	1.074	27.0	1.115
9.5	1.038	18.5	1.076	27.5	1.117
10.0	1.040	19.0	1.078	28.0	1.119
10.5	1.042	19.5	1.080	28.5	1.122
11.0	1.044	20.0	1.083	29.0	1.124
11.5	1.046	20.5	1.085	29.5	1.126
12.0	1.048	21.0	1.087	30.0	1.129
12.5	1.050	21.5	1.089	30.5	1.131
13.0	1.052	22.0	1.092	31.0	1.134
13.5	1.054	22.5	1.094	31.5	1.136
14.0	1.057	23.0	1.096	32.0	1.138
14.5	1.059	23.5	1.098	32.5	1.141
15.0	1.061	24.0	1.101	33.0	1.143
15.5	1.063	24.5	1.103	33.5	1.145

Source: Data from Bates et al. (1942).

assimilate stream. First, if one is using a labeled tracer, such as ^{14}C-labeled sucrose, there will be an exchange of labeled for unlabeled sucrose throughout the length of the conduits. Second, there may be fixation by adsorption or by incorporation along the channels of transit: the phloem is, after all, a distribution system. Third, there is storage in many tissues, including phloem parenchyma, xylem and phloem rays, cortex, pith, and so forth. Finally, respiration going on in all living cells accounts for a certain percentage of the translocated molecules (Canny and Markus, 1960; Canny 1962b).

The root of the problem of expression is that there exist different views of the translocation mechanism. If the mechanism is seen as one of independent movement—be it activated diffusion, protoplasmic streaming, surface migration, or some sort of metabolic acceleration—then specific mass transfer is relevant. If, on the other hand—because of evidence on phloem anatomy, information about phloem exudation, and an appreciation of the significance of tracer distribution—it is seen as a mechanism of mass flow, then velocity, expressed in centimeters per hour, is a true measure of translocation rate, for it can be a direct measure of the amount of solute delivered in a given time. Concentration of the solute in the moving stream must be known.

Advancing Radioactive Fronts

Much importance has been attached by some to the fact that a front of radioactivity in a moving solute mass in the phloem is approximately logarithmic in form (Vernon and Aronoff, 1952; Canny, 1962a, 1962b; Canny and Phillips, 1963; Horwitz, 1958; Spanner, 1958, 1963; Spanner and Prebble, 1962). Because such a front is common to a system involving transport by diffusion, it has been reasoned that the transport mechanism must be a diffusion analogue (Canny and Askham, 1967). A number of physiologists have argued that such fronts constitute evidence against mass flow.

When a tracer is applied to a leaf, be it as $^{14}CO_2$ or as some tracer in solution, it enters the leaf by diffusion. Phloem-mobile tracers rapidly enter the symplast, where they move by diffusion, possibly accelerated by streaming, to the vascular bundles; ^{14}C-labeled sucrose is synthesized within the symplast, and it moves to the vascular channels in the same way. Having arrived in the border parenchyma cells that surround the sieve tubes, tracers enter the sieve tubes by diffusion. Hence, the advancing front of any phloem-mobile tracer, whether it moves in the sieve tubes by diffusion or in a stream by mass flow, is necessarily logarithmic in form. The critical matter is not the form of the advancing front, but the velocity of movement. It is difficult indeed to visualize movement at a velocity of 100–300 cm per hour in sieve tubes as diffusional in nature.

Velocity of Movement

The velocity of flow of the assimilate stream has been determined in a number of ways. Dixon (1922) calculated a velocity of 50 cm per hour for movement of organic nutrients into the tuber of *Solanum tuberosum,* and he considered that the small size of phloem elements and the viscous nature of phloem sap precluded the possibility that this tissue provided a route for the movement of nutrients. Mason and Maskell (1928a, 1928b) found that assimilates moved rapidly in *Gossypium hirsutum,* that movement was from source to sink, and that the velocity of transport was up to 20,000 times that of diffusion. Many rates of translocation have been published since these early works; a few examples of contemporary values are given in Table 11.3.

Of the values cited, those of Canny (1961), calculated from curves showing the rate of advance of tracer fronts with respect to time and distance, do not take into consideration the time required for penetration of the tracer, and they are based on the mass delivered per unit of time, not on the velocity of flow. In his paper on measurements of velocity, Canny (1961) proposed that transport rates should be expressed on the basis of mass per unit of time per unit of distance, rather than on the basis of velocity. This followed from his belief that translocation is a diffusionlike process, with each solute species moving independently along its own individual gradient. With a pressure-flow system, all solutes move along together, and except for preferential distribution along the route, all solutes in the stream should be delivered at the same time. Many experiments with exogenous tracers indicate that this is approximately what takes place in plants.

One other value in Table 11.3 seems out of line. This is the high rate of 7200 cm per hour cited from Nelson. According to Nelson, this very rapid translocation of ^{14}C occurred out of leaves of *Glycine max* treated with $^{14}CO_2$. The tracer was found within seconds in the roots, with voids of tracer in the stem. He proposed that this rapid transport took place in living cells and in xylem, while a mass flow of around 100 cm per hour was going on in the phloem. It is hard to visualize movement through living cells at velocities of this magnitude, where the only moving phase is the streaming of protoplasm at rates not exceeding 6 cm per hour. It is even more difficult to visualize it through xylem, for there the transpiration stream is moving at rates of hundreds of centimeters per hour in the opposite direction.

Calculations of translocation rates based upon daily increments of increase of dry weight have been made on tubers of *Solanum tuberosum* (Crafts, 1933), cucurbit fruits (Crafts and Lorenz, 1944a), and other growing plant organs. Such calculations are valid tests of the physical principles upon which translocation mechanisms are based. However, before the anatomical work proving that the sieve-plate pores are open, calculations attempting to determine the resistance to flow through sieve plates (Crafts, 1932, p. 212) were confused by the

Table 11.3. Some samples of translocation velocities quoted in the literature.

Velocity (cm/hr)	Plant	Translocate	Reference
1.5–2.0	*Salix* sp.	Assimilate	Canny (1961)
2.0	*Lolium temulentum*	Flowering stimulus	Evans and Wardlaw (1964)
5–6	*Abies concolor*	Assimilate	Leonard and Hull (1965)
10–12	*Phaseolus vulgaris*	2,4-D; 2,4,5-T	Little and Blackman (1963)
13–40	*Populus* sp.	^{32}P	Vogl (1964)
17	*Glycine max*	Assimilate	Thrower (1965)
20–24	*Phaseolus vulgaris*	IAA	Little and Blackman (1963)
25–33	*Salix* sp.	Assimilate	Peel and Weatherley (1962)
30–70	*Cucurbita pepo*	Assimilates	Kursanov (1956b)
35–40	*Gossypium hirsutum*	Assimilates	Kursanov (1956b)
35–70	*Heracleum mantegazzianum*	Assimilate	Ziegler and Vieweg (1961)
39–57	*Triticum aestivum* (leaf sheath)	Assimilate	Wardlaw (1965)
48–60	*Metasequoia glyptostroboides*	Assimilate	Willenbrink and Kollmann (1966)
50–104	*Phaseolus vulgaris*	2,4-D	Day (1952)
50–135	*Beta vulgaris*	Assimilates	Mortimer (1965)
60–80	*Phaseolus vulgaris*	Assimilates	Kursanov (1956b)
60–120	*Saccharum officinarum*	Sucrose	Hartt and Kortschak (1963)
70–100	*Beta vulgaris*	Assimilates	Kursanov (1956b)
87–109	*Triticum aestivum* (internode)	Assimilate	Wardlaw (1965)

Table 11.3.–*Continued*

Velocity (cm/hr)	Plant	Translocate	Reference
88	*Cucurbita pepo*	Assimilate	Hendrix (1968)
100	*Salix* sp.	Assimilate	Weatherley et al. (1959)
150	*Saccharum officinarum*	Sucrose	Hartt and Kortschak (1963)
250–300	*Cucurbita melopepo torticollis*	Assimilate	Webb and Gorham (1964)
300	*Saccharum officinarum*	Sucrose	David North Research Center (1964)
360	*Saccharum officinarum*	Sucrose	Hatch and Glasziou (1964)
7200	*Glycine max*	^{14}C	Nelson et al. (1958)

fact that rapid callose formation on sieve plates was not recognized. The pore dimensions used in these early calculations were all too small. Calculations aimed at estimating intermicellar spaces in phloem walls (Crafts, 1932, p. 213; Crafts, 1933, table V, p. 92) were equally futile. The average value of 130 cm per hour for flow of a 10 percent solution through sieve-tube lumina given by Crafts and Lorenz (1944a) should be accurate; the velocity through sieve-plate pores would be at least twice this, but the plates occupy only around 1 percent of the total length of the sieve tube. Thus, calculations of the resistance to flow of the assimilate stream through sieve tubes do not produce impossible values. Weatherley et al. (1959) arrived at a similar conclusion, although, at the time of their publication, plugging of sieve plates presented a serious problem.

Webb and Gorham (1964) made a detailed study of translocation of ^{14}C-labeled assimilates in young *Cucurbita* plants which are known to move materials at high rates (Crafts and Lorenz, 1944a). By refined, modern methods of isotope analysis, data on translocation velocity, on assimilate distribution, and on assimilate metabolism were obtained. The time of initial exposure to $^{14}CO_2$ was 15 seconds, the radiation dosage was 60 μc, and the translocation times were from 10 seconds to 14 hours.

Webb and Gorham reported that ^{14}C-labeled assimilates from mature leaf blades appear in the petiole within 5 minutes, and may move at a maximum velocity of 290 cm per hour. They recognized that this high velocity may involve movement of gaseous $^{14}CO_2$, with fixation along the route of flow. The rate of export of ^{14}C from the blade reached a maximum within 10–15 minutes, and diminished during the following 30 minutes, and more rapidly thereafter. There was no further export after 3 hours, at which time 55 percent of the total assimilated ^{14}C had been exported. Stachyose and a trace of raffinose and sucrose were the principal compounds translocated; they moved into all parts of the plant, except other mature leaves. Stachyose labeled with ^{14}C was almost completely exported from a blade within 45 minutes. In roots and stem, it was slowly metabolized; in young growing tissues of stem and leaves, it was rapidly metabolized to sucrose and other compounds. Labeled sucrose was only partially exported following a lag period of about 30 minutes.

Moorby et al. (1963) used the short-lived isotope ^{11}C in studies on assimilate movement from the leaves of *Glycine max* down the main plant axis. This isotope emits positrons having an energy of 0.96 Mev, which, upon annihilation, give two annihilation-photons in opposite directions; each of these has an energy of 0.51 Mev, and they can be detected by scintillation counters positioned near the stem in which the tracer is moving, thus making possible counting *in vivo.* Using this equipment, Moorby and his co-workers studied the rate of accumulation of radioactivity at several points on the stem during continuous illumination. This rate quickly became constant, and remained so for several hours. There was a period of about 30 minutes between the switching on of the light and arrival of ^{11}C at a point about 20 cm along the stem. When the light was turned off, a similar period expired before the rate of movement slowed down. A similar effect followed replacing the ^{11}C-labeled gas with inactive gas. By measuring the attainment of steady-state transport at different positions along the stem, it was possible to calculate an average velocity of transport; this turned out to be 60 cm per hour. Moorby and his co-workers visualized a pressure-flow type of mechanism to explain their results. The loss of assimilate along the channels of movement, which they term "leakage," is more probably normal distribution.

In a second attending paper, Evans, Ebert, and Moorby (1963) visualized a model to account for the results of the experiments described above. They discussed this in terms of a pool or reservoir from which flow takes place. By mathematical analysis of their model, they concluded that the mixing compartment in the plant foliage is virtually of constant dimensions during illumination, and that it contains a few milligrams of exchangeable carbon. This compartment may not constitute the sole source of sucrose entering the sieve tubes, because by calculation, it would provide a sucrose concentration of only about 3 percent or less, whereas the usual concentration of the assimilate stream should lie between 10 and 25 percent. The rate of migration ("leakage") from the sieve tubes turned out to be about 0.8 percent of the sieve tube contents per centimeter. The translocation velocity under their conditions was about 60 cm per hour.

It is unfortunate that these workers did not utilize the existing knowledge of leaf anatomy in their analysis of the transport function in their plants. By applying the symplast concept, they surely could have arrived at a more satisfying picture of assimilate movement in *Glycine max.* Their experimental methods are beyond criticism, and their results should prove to be highly reliable.

As noted in Chapter 6 and in Table 11.3, Hartt and Kortschak (1963) measured translocation velocities in *Saccharum officinarum* and found values lying between 60 and 150 cm per hour. Hatch and Glasziou (1964) found sucrose to be the principal component of the translocated assimilates. Translocation velocities 240–360 cm per hour were recorded.

Mortimer (1965) measured translocation velocities in *Beta vulgaris* petioles. Using $^{14}CO_2$ in a 5-minute exposure period he found that ^{14}C moved through the length of the petiole at velocities ranging from 50 to 135 cm per hour. This is comparable with the velocity of 150 cm per hour found by Bennett (1934) for the movement of curly top virus in the same organ.

Wardlaw (1965), using labeled assimilates from $^{14}CO_2$ as a measure of translocation velocity and distribution pattern, found that assimilates move through the top internode of *Triticum aestivum* at rates of 89–109 cm per hour, and through *T. aestivum* leaf sheaths at 39–57 cm per hour. Removal of grains from the developing heads reduced the velocity of upward movement and increased that of basipetal flow; the rate of movement out of the leaf was unaffected. With full grain development, assimilates from the flag leaf moved directly from the node of insertion of the leaf up the stem to the ear.

Huber et al. (1937) published a paper describing a sucrose concentration wave that they found to descend the bark of *Quercus borealis* at velocities of 150–450 cm per hour. This phenomenon has become widely accepted as an indication of the velocity of phloem transport in *Q. borealis*. Zimmermann (1969), observing ratios of sugar concentrations (sucrose to stachyose, raffinose to stachyose, and sucrose to raffinose) in the sieve-tube exudate from *Fraxinus americana*, found that these ratios moved down the bark at velocities of 30–70 cm per hour. These are independent of the absolute exudate concentration, and thus are unaffected by hydrodynamic pressure changes in the xylem. Zimmermann did not doubt that the wave of sugar concentration observed by Huber et al. was a real one. He suspected, however, that the velocity values were distorted by the effects of xylem pressure upon the concentration in the sieve tubes. In view of the results of Peel and Weatherley (1963), this interpretation seems logical.

From these data, it is evident that many substances—including viruses, assimilates, and exogenous tracers—are able to move through the sieve tubes of functioning phloem at rapid velocities. In addition to providing evidence on the mode of supply of foods, hormones, and vitamins in plants, this information is an indication of the nature of the translocation mechanism and the structure of the sieve plate. Only perforate sieve plates could conduct the assimilate stream at the cited velocities, and only a mass-flow mechanism would seem to be capable of delivering foods at the velocities that studies of tuber formation (Crafts, 1933) and fruit growth (Crafts and Lorenz, 1944a) imply.

Mathematical Models

Many attempts have been made to set up mathematical models in order to test hypotheses about the mechanism of translocation. Most of these have been oversimplifications that have not dealt realistically with phloem anatomy, sieve tube structure, the relation of phloem exudation to mechanism, and the specific results of tracer studies. Dixon (1923), Mason and Maskell (1928b), Mason and Phillis (1937), Crafts (1931, 1932, 1933), and Curtis (1935) were among the earlier workers who used the mathematical approach. More recently, Spanner (1958, 1962, 1963) and Spanner and Prebble (1962) carried out more detailed analyses; Canny (1960a, 1960b, 1961, 1962a, 1962b), Canny and Phillips (1963), and Canny and Askham (1967) examined the diffusion-analogue concept of translocation. Horwitz (1958) took an unbiased approach, analysing a number of proposed mechanisms. N. T. S. Evans et al. (1963) provided a clear treatment, but failed to consider the details of phloem anatomy and, particularly, the symplast concept. Weatherley and Johnson (1968) made mathematical analyses of several models.

Swanson (1959), using data from Crafts and Lorenz (1944a), calculated a series of flow velocities for assimilate in *Cucurbita pepo,* based on the assumption of a 20 percent concentration. These are shown graphically in Figure 11.1. His maximum mass rate of transport turned out to be 1.70 g per hour. This, divided by 0.036 cm^2 (the transverse area of sieve tubes at the time of maximum transport velocity), gives a delivery velocity of 47.2 cm per hour for pure assimilate. To multiply this by 5 (to account for the 20 percent dillution), and to divide by 1.083 (the density of a 20 percent sucrose solution), yields a velocity of 217.9 cm per hour. Swanson used a density value of 1.5 (the density of pure sucrose), which lowers the velocity to 157.3 cm per hour—a false value, because the assimilate stream does not have such a density. To calculate on the basis of a 10 percent solution, which is more nearly that found in phloem exudate (Crafts and Lorenz, 1944b), yields velocities approximately twice as great. They represent maximum values for translocation at its highest velocity; and are not out of line, by comparison to the values derived from tracer movement determined by Webb and Gorham (1964) and Hatch and Glasziou (1964).

Swanson went a step farther, and multiplied his value of 157.3 by 20, to find the velocity required to provide transport through the space occupied by the cytoplasm; the value, 3147 cm per hour, illustrates the difficulty in rationalizing any mechanism requiring transport by protoplasmic streaming, or any metabolic mechanism. Swanson's

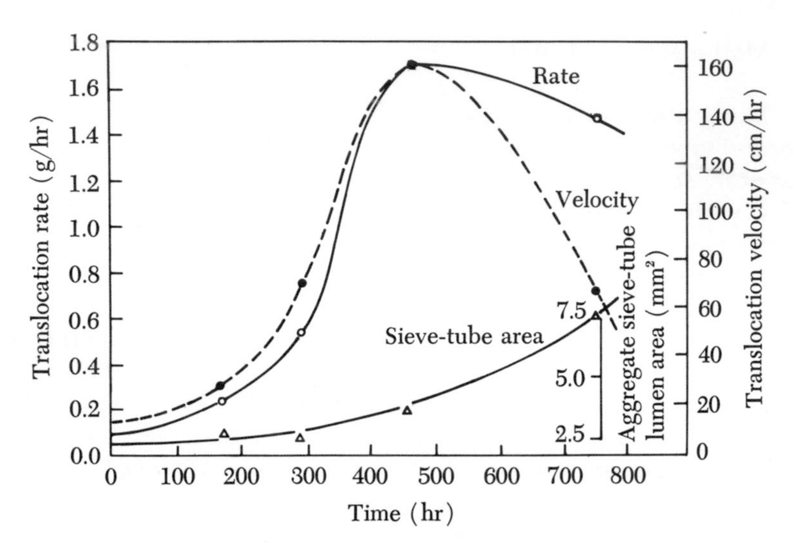

Figure 11.1. Rate and velocity of translocation of dry-weight materials into the fruits of *Cucurbita pepo* in relation to time. From Swanson (1959).

further reference to an additional twentyfold increase for transit through the connecting strands is questionable: if the pores of the sieve plate are blocked by solid cytoplasm, all such calculations are futile, for we do not know the permeability of such structures. If, on the other hand, the pores are open, as indicated by Figure 2.10 (page 41); and if strands occupy around 40 percent of the transverse cell area, as concluded by Esau and Cheadle (1959); then the value of 157.3 should be multiplied by 2.5, not by 400 (20×20) as Swanson suggested.

Webb and Gorham (1965a) concluded:

> A mathematical analysis of such a complicated physiological system must extend beyond the relatively simple treatments of Canny (1962a), Evans et al. (1963), and Horwitz (1958). With the insufficient and conflicting quantitative information on translocation that is presently available it is doubtful whether the process can yet be usefully expressed in mathematical terms.

This is certainly a realistic reaction to the various attempts to fit translocation data to mathematical formulations.

Some Calculations

That mathematics may be effectively used to test the validity of a hypothesis on translocation is shown by the following calculations. If we grant that movement along sieve tubes occurs through the lumina and across the sieve plates via tubular protoplasmic strands, then values for the required pressure gradients may be calculated using the Poiseuille formula. In terms of velocity of flow, this formula states that

$$P = \frac{8R_1\eta l}{r^2},$$

where P = pressure in dynes per square centimeter, R_1 = velocity in centimeters per second, η = viscosity of the liquid in poises, l = length of the tube or gradient in centimeters, and r = radius of the conducting element in centimeters. Using the figure obtained by Crafts and Lorenz (1944a) for phloem transport in *Cucurbita melopepo* 'Early Prolific Straightneck' of 130 cm per hour for the velocity of flow of a 10 percent solution through the phloem, the velocity through the sieve plates may be calculated as follows. An average-sized sieve plate of this plant with area of 5485 μ^2 proved, by actual count, to have 120 pores with an average diameter of 4.82 μ. The average radius would be 2.41 μ, and $r^2 = 5.8\ \mu^2$.

Taking the value of Esau and Cheadle (1959, table 8) of 49.3 percent for that part of the transverse cell area occupied by strands, and discounting the 9.3 percent for the parietal layer and plasmatic filaments, the velocity of flow through the open pores, R_1, would be 130/0.40 = 325 cm per hour, or 0.09 cm per second; the viscosity of a 10 percent sugar solution, η, is 0.015; and the thickness of a sieve plate, l, averages around 5 μ, or 5×10^{-4} cm. Then

$$\begin{aligned} P &= \frac{8 \times 0.09 \times 0.015 \times 5 \times 10^{-4}}{5.8 \times 10^{-8}} \\ &= 93 \text{ dyn/cm}^2 \\ &= 0.000092 \text{ atm/sieve plate.} \end{aligned}$$

With sieve tubes averaging 0.25 mm in length, there are 4000 of them per meter, and the value of P required to maintain flow through open sieve-plate pores for a total length of $4000 \times 5\ \mu$ or 0.2 cm in that meter of stem is 0.37 atm per meter.

Using the same formula, the resistance to flow through the open lumina of those sieve-tube members devoid of sieve plates may be calculated. Where $R_1 = 130/3600$, or 0.036 cm/sec; and $r = 40\ \mu$, or 0.004 cm; P proves to be 0.00027 atm per centimeter, or 0.027 atm per meter. Adding the value for the sieve plates (0.37) to that for the open lumina (0.027), the sum is 0.397 atm per meter. At this value, 1.0 atm of turgor pressure would account for movement at the rate of 130 cm per hour through 2.5 meters, and assuming a 10 percent concentration for the assimilate stream, the pressure should take care of movement through any shoot of *Cucurbita*. The nitrogenous compounds and potassium present in the phloem exudate might result in additional turgor sufficient to compensate for resistance to flow through the roots.

A study of the conformation of sieve-plate pores (Figure 11.2) reveals two features that would seem to demand a formula different than that of Poiseuille. First, the openings of the pores are not angular but curving; and flow, instead of being laminar for the total length, converges at the entrance and diverges at the exit—laminar flow, if present at all, would occur for only a very short distance. An attempt was made to set up a new formula to cover this type of flow. Solution of the equations was carried out on a computer, and the formula was approximately as follows:

$$P = \frac{4R_1\eta l}{r^2}.$$

This means that the calculation of required pressure are all twice too high, and it appears that the osmotic pressures of the assimilate stream are thoroughly adequate to account for flow through the length of the tallest trees. This is most gratifying, at first glance, but there is an additional condition that may prove to be even more pertinent. The diameters of sieve-plate pores are equal to, or greater than, the sieve-plate thickness; that is, $d > l$. It has long been known that, when this is true—taking into account the Reynolds numbers for the flow of the assimilate stream at measured velocities (see Table 11.3)—flow may be creeping rather than laminar. Under these conditions, the uniform liquid accommodates itself to the constrictions of the sieve pores with little or no turbulence, and resistance is at a minimum. The formula for such flow (Happel and Brenner, 1965), when expressed in terms of velocity, is

$$P = \frac{3\pi R_1\eta}{r}.$$

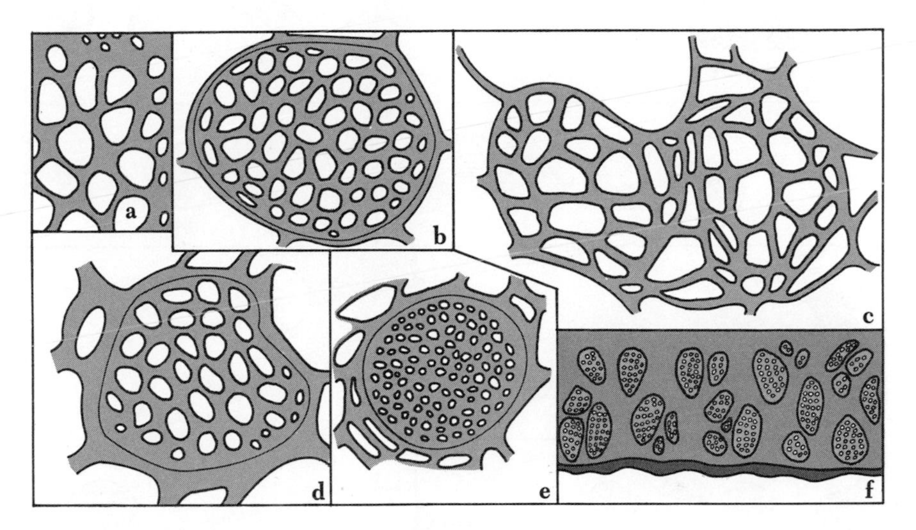

Figure 11.2. Transverse sections showing sieve plates with open pores: (*a*) *Tetracera* sp.; (*b*) *Cucurbita* sp.; (*c*) *Fraxinus excelsior;* (*d*) *Robinia pseudoacacia;* (*e*) *Macrocystis pyrifera;* (*f*) *Smilax rotundifolia.* The thin plasmolemma lining has been omitted in these drawings; some pore space has been lost in reproduction, particularly in *e* and *f*.

Substituting the values given on page 399,

$$P = \frac{9.42 \times 0.09 \times 0.015 \times 5.0 \times 10^{-4}}{2.41 \times 10^{-4}}$$

$$= 51.21 \text{ dyn/cm}^2$$

$$= 0.000051 \text{ atm/sieve plate}$$

The pressure P, then, is 0.202 atm per meter for flow through the open sieve-plate pores for a total length of 0.2 cm in the 1 meter of stem. Flow through the lumina would be laminar, and the value of 0.027 atm per meter would hold. The total, then, would be 0.229 atm per meter, and an osmotic pressure of 1.0 atm would be sufficient to cause flow at 130 cm per hour along a stem 4.4 meters in length.

The sieve pores upon which these calculations are based (see Figure 11.3) occur mainly in primary phloem. Sieve plates in the secondary phloem of trees are thicker; pores through thicker plates may be tubular; and flow through tubular pores is laminar. Thus, the analysis of the function of the total flow of the assimilate stream in a plant requires that both creeping and laminar flow be taken into

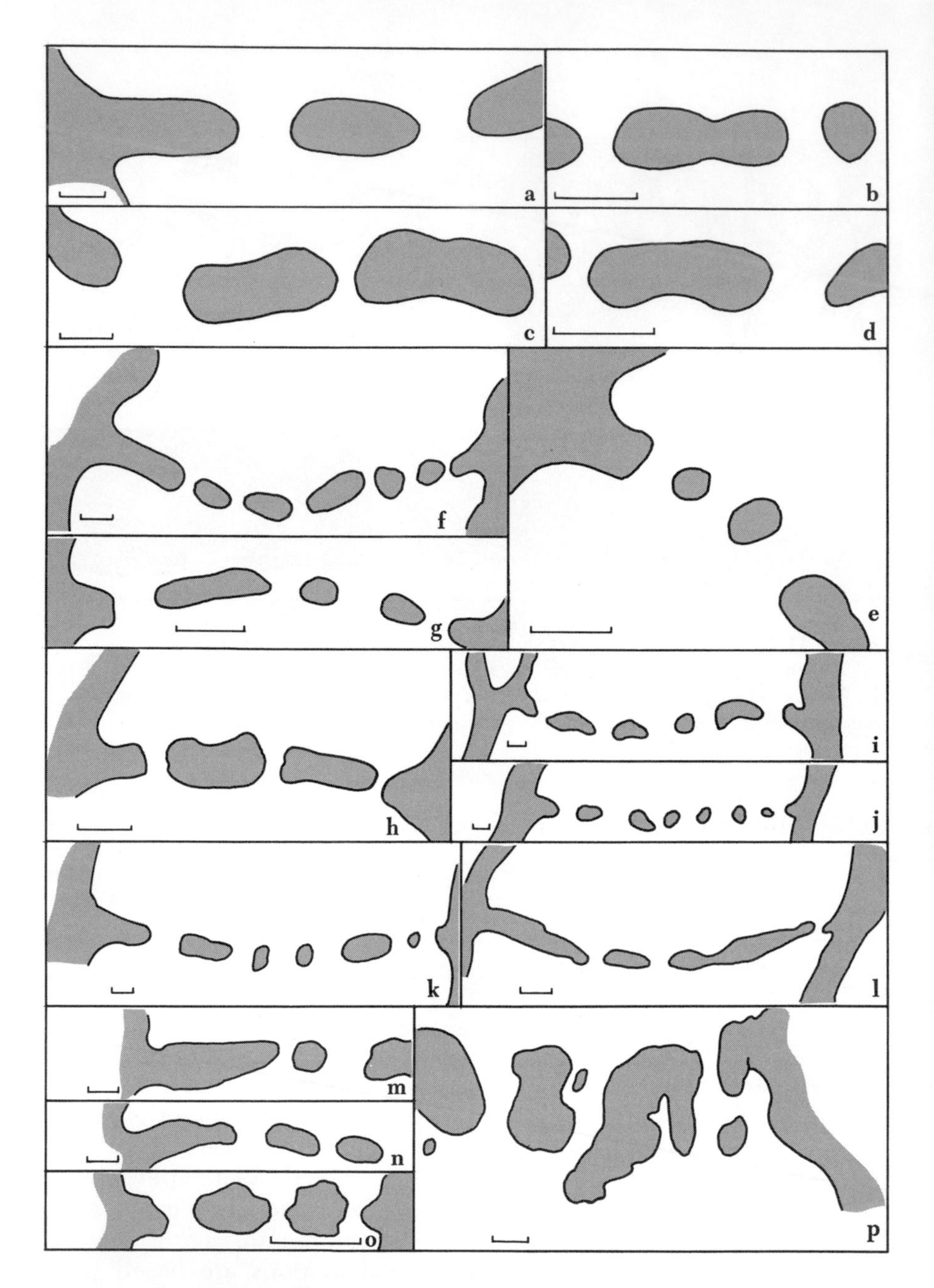

Figure 11.3. Longitudinal sections of phloem showing sieve plates with open pores: (*a, b, c, d, e*) *Cucurbita maxima;* (*f*) *Impatiens sultanii;* (*g*) *Cucurbita ficifolia;* (*h*) *Impatiens sultanii;* (*i, j, k*) *Cucurbita maxima;* (*l, m, n*) *Cucurbita pepo;* (*o*) *Beta vulgaris;* (*p*) *Pinus strobus.* The thin plasmolemma lining has been omitted in these drawings. The bar in each figure represents one micron.

account. The formula for creeping flow will certainly apply in the case of leaves and fine roots. In petioles, twigs, and small branches, the modified formula

$$P = \frac{4R_1\eta l}{r^2}$$

may more nearly meet requirements. In the trunks of tall trees, flow may be more strictly laminar through the sieve plates. Mittler (1967), studying the relations between suction and flow rate and viscosity and flow rate for water and sugar solutions through the stylet bundle of a large milkweed bug (*Oncopeltus fasciatus*), found that these fit the Poiseuille concept of laminar flow through small tubes. There was a positive correlation between pressure and flow rate, and a negative correlation between viscosity of the moving solution and flow rate. Figure 11.4 shows the relation between sucrose concentration, viscosity, and relative flow rate.

The authors of this book made a set of calculations for *Pinus strobus*. Using Münch's value for rate [20 cm per hour for flow through the conducting phloem (Münch, 1930, table 2, p. 81)], measuring the numbers and dimensions of strands, and counting sieve areas in the common cross-walls between sieve elements in slides of *P. strobus*, the following values were obtained. There were 24 strands per field and 30 fields per common cross-wall, or 720 strands per common cross-wall. If the average value of 0.7 μ is used for pore diameter (Murmanis and Evert, 1966, figure 18), then $r = 0.35\ \mu$; $r^2 = 0.1225\ \mu^2$, and $\pi r^2 = 0.3845\ \mu^2$. Then the total area of strands is $0.3845 \times 720 = 276.8\ \mu^2$.

The average sieve element was $25\ \mu \times 15\ \mu$ in area, or $375\ \mu^2$; and $276.8/375 \times 100 = 74$ percent; the acceleration through the strands would be 100/74, or 1.36 times. If the sieve tube lumina occupy 50 percent of the total phloem area in *Pinus strobus*, then the linear velocity through the lumina would be 40 cm per hour; and $40 \times 1.36 = 54.4$ cm per hour, or 0.015 cm per second, through the strands. Therefore, where $R_1 = 0.015$, $\eta = 0.018$, $l = 5.0 \times 10^{-4}$ cm, and $r^2 = 0.1225 \times 10^{-8}$ cm^2, then

$$\begin{aligned} P &= \frac{8 \times 0.015 \times 0.018 \times 5.0 \times 10^{-4}}{0.1225 \times 10^{-8}} \\ &= 881.6 \text{ dyn/cm}^2 \\ &= 0.00087 \text{ atm/sieve plate} \end{aligned}$$

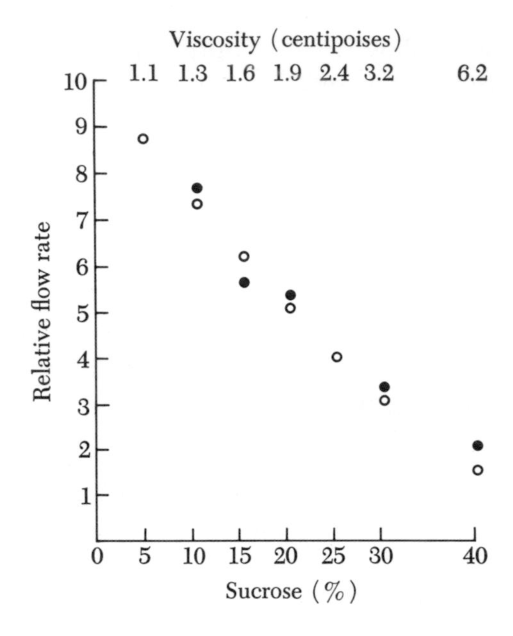

Figure 11.4. Relation between viscosity and relative flow rate of sucrose solutions through the stylet bundle of the milkweed bug *Oncopeltus fasciatus*. From Mittler (1967).

The sieve cells reported in *Pinus strobus* in Crafts (1961b, table 6.1) were 1.58 mm in length. There were, therefore, 6.3 end walls per centimeter, or 630 per meter. The pressure gradient required to bring about flow through pores at the rate of 54.4 cm per hour would be 0.00087×630, or 0.55 atm per meter, as compared with 0.37 atm per meter for *Cucurbita melopepo* for a velocity of 325 cm per hour.

The resistance to flow through the open lumina of sieve cells of *Pinus strobus* was calculated by the authors of this book. Thus, where $R_1 = 0.0056$ cm per second, and $r = 10.9\ \mu$, P proves to be 0.055 atm per meter. The total resistance is $0.55 + 0.055 = 0.61$ atm per meter. In the tree, gravity accounts for a positive pressure of approximately 0.1 atm per meter, so that the actual resistance would be 0.51 atm per meter. If the concentration of sugars in the assimilate stream of *Pinus strobus* should reach 20 percent, equivalent to 20 atm of turgor pressure, then flow from a height of 20/0.51 meters, or 39 meters (154 feet) could be accounted for. Usually, there are branches below this level that provide a supply of organic nutrients to the trunk and roots; and creeping flow may occur in foliage and roots.

In their recent review, Weatherley and Johnson (1968) calculated the pressure required to move the assimilate stream through the sieve tubes of a stem of *Salix* at 100 cm per hour. Thus, where $R_1 = 100$ cm per hour, and $r = 12\ \mu$, the pressure $P = 0.24$ atm per meter. Weatherley and Johnson calculated that the pressure required for flow through the sieve pores is 0.32 atm per meter. Thus, the sum 0.56 atm per meter is the total required pressure; gravity would reduce it to 0.46 atm per meter. Assuming a total turgor pressure of 15 atm in the phloem, transport should occur over 15/0.46, or 33 meters; this equals 108 feet, a height seldom reached by *Salix;* usually, many branches occur below this height. Weatherley and Johnson concluded from these calculations that the concentration of solutes in the phloem sap would need to be very high to account for transport in the tallest trees up to 100 meters. There are two objections to this conclusion. First, *Salix* trees don't attain a height of 100 meters, and those species that do have sap concentrations in the range of 20–27 percent and phloem with larger sieve tubes than those of *Salix* examined by Weatherley and Johnson (see, for example, Ford and Peel, 1967a). If the value of $r = 19\ \mu$ given by Ford and Peel is substituted for 12 in the above calculation, P turns out to be 0.31 atm and the height to be 156 feet. Second, it is becoming very evident that the trunk and roots of trees are nourished by assimilates synthesized in the lower branches, not in the leaves in the upper canopy, and that gravity is always in favor of downward flow. Creeping flow in foliage and roots reduces the resistance even further.

Although the preceding calculations would seem to indicate that turgor gradients in the phloem of plants are adequate to account for known rates of food transport, there are still a number of conditions that have not yet been given sufficient consideration. For example, we do not know, with sufficient certainty, the actual dimensions of the tubular strands traversing the pores in sieve plates. It is certain that most of the old measures and estimates (Dixon, 1922; Crafts, 1931, 1932, 1933) were too small, because the rapid formation of sieve-plate callose was not recognized. Before 1961 and recognition of the open condition of these strands, there could have been no rational use of the Poiseuille formula to calculate sieve-plate resistance, because it was almost universally agreed that the pores were blocked by slime or cytoplasm.

Figure 11.3 shows the condition of sieve plates as illustrated in various publications. From these views, it is evident that the sieve pores occupy a major portion of the transverse area of the sieve plate; the callose may take up varying amounts of space. From the electron microscope views of sieve plates presented by Duloy et al. (1961), Esau and Cheadle (1961), Esau et al. (1962), Engleman (1965a, 1965b),

Eschrich (1965), Esau (1966), Evert et al. (1966), and others (as shown in Figure 11.3), it is evident that, in the natural state in intact plants, the callose may be present as only a very thin layer and so present little resistance to solution flow—a situation indicated for years by phloem exudation. Plant physiologists generally were too much impressed with the closed condition of the sieve plates shown in slides prepared from cut material processed in killing fluids.

The data presented by Esau and Cheadle (1959), as shown in Table 11.4, give an accurate evaluation of the percentage of the transverse area of the sieve tube occupied by pores and strands. Because of the particulate material that adheres to the plasmolemma lining of the sieve-plate strands shown in electron micrographs, it is difficult to measure the thickness of this lining. Where the discrete plasmolemma can be seen (in, for example, Esau and Cheadle, 1961, figures 9 and 10) it appears to be around 0.005 μ in thickness. Hence, for a pore having a transverse area of 12.69 μ^2 ($r = 2.01$, according to Esau and Cheadle, 1959, table 1), a layer 0.005 μ thick would occupy an area equivalent to 0.5 percent of the 12.69 μ^2 value, and the inner diameter of the pore would be 4.01 μ, and the area 12.63 μ^2; the plasmolemma area = 0.06 μ^2. If the strands of the plasmatic reticulum occupied an equal volume, the transverse area of the pore would still not be severely restricted.

From the values in Table 11.4, it is apparent that the transverse areas of strands are from one-third to one-half those of the sieve plates, and consequently, the velocities of flow are 2–3 times the velocities in the open lumina. Using a value of 2.5 times, as represented by the calculation on page 399, the resistance of 0.397 atm per meter does not prove to constitute a serious hindrance to flow.

Table 11.4. Percent of transverse cell area occupied by pores and strands of sieve plates in ten species in each of three sieve-plate groups.

Sieve-plate group	Mean area per sieve plate (μ^2)		Percent of transverse area of the sieve tube occupied by	
	Pores	Strands	Pores	Strands
Simple	793.1	469.6	83.2	49.3
Simple and scalariform	821.9	364.7	104.8	46.5
Scalariform	1111.0	315.1	112.9	32.0

Source: Esau and Cheadle (1959).

The actual form of plants—particularly of trees—may serve to reduce resistance to flow. Because the branches converge to form the trunk, there must be a decrease in the total volume per unit length of phloem and, hence, an increase in velocity of flow along the trunk. The converse applies to the root system. In the leaves and fine roots, the phloem is made up primarily of protophloem having very thin sieve plates, so that strictly laminar flow occurs over extremely short distances, if at all.

Huber et al. (1937) stressed two additional factors that favor mass flow in the phloem of plants. First, as water stress heightens during the day the volume of the assimilate stream is reduced and the concentration is increased. This tends to reduce velocity and, hence, frictional resistance, and to increase mass transport. Although the reverse effect takes place during the night, the accelerating influence takes place during the day, after initial photosynthesis in the morning has loaded the phloem in the foliar regions. The second factor has to do with phloem function as a distribution system: assimilates are constantly lost from the phloem during the flow of the assimilate stream, particularly in the trunk and roots of trees. Huber and his co-workers showed the decreasing concentration values at lower levels on the trunk of *Quercus borealis* (their figures 1–7). Hammel (1968) confirmed the validity of such gradients (his table 1).

The preceding considerations would seem to indicate that the values calculated for pressures required to drive the assimilate stream in various plants are maximum values, and that true values may be appreciably lower. Yet the calculated values appear to fall well within reasonable limits, in view of the concentrations of sugars reported in phloem exudate. Huber et al. (1937, tables 1 and 2) report values for the percentage of sugar ranging from 14.8 to 27.1. The latter value would account for a turgor pressure of at least 29 atm; and if potassium were present in its usual concentration, a turgor of 30 atm or more might be expected. Given a resistance equivalent to 0.61 atm per meter, as calculated above for *Pinus*, 30 atm turgor pressure would account for flow through 49.2 meters, or 194 feet. This is well within the height of the lower branches of most big trees.

Thus, it seems that we do not have to account for transport at 100–300 cm per hour through 100 meters in order to substantiate the mass-flow mechanism. In fact, if use of the Poiseuille formula is justified at all, transport would be along channels not to exceed 50 meters in tall trees; resistance to flow in the branches, the upper one-quarter of the trunk, and the roots must be appreciably lower because of the factors enumerated above. The realities of plant form and structure would seem to indicate that there is an excess of pressure in the phloem over and above that required for normal growth. Possibly, this accounts for the fact that Mason and Maskell (1928b), in partial ringing experiments, had to reduce the phloem to about 33 percent of normal before they materially hindered assimilate movement into the roots of *Gossypium hirsutum;* and that callosing of sieve plates in one-third of the phloem did not slow the transport of labeled assimilate (Eschrich et al., 1965).

12

In Retrospect

With the overwhelming evidence in favor of a mass-flow mechanism to account for phloem transport in plants, attention should now be turned to filling in the multitude of details needed for a complete understanding of assimilate distribution. Most of the responses to environmental factors seem to follow the principles of physical chemistry. There are, apparently, few exceptions to predicted effects of temperature, of light, of the availability of water, and of nutritional and regulatory factors.

Although the experience gained by the expanding use of systemic pesticides has provided a mass of evidence that corroborates the mass-flow mechanism, we still have a lot to learn. We need to study the many differences that exist among the array of plants from brown algae to forest trees and desert succulents. From the literature, there is a suggestion that fast-growing, herbaceous plants generally move assimilates faster than do large trees. How general is this? It is reported that some woody monocots have functional phloem 50–100 years old. If this is true, then what is the significance?

Reports have come from xerophytic areas that protracted spells of temperatures running above 100°F have lowered yields of fiber by

Gossypium hirsutum. Can this be attributed to blocking of sieve-plate pores by callose, with an attendant reduction in assimilate movement?

Much of the early research on the use of hormone-type herbicides revolved around the problem of obtaining optimum physiological responses from these growth regulators. The time of application was an important aspect. After much empirical testing, it turned out that two factors determine the optimum time of application: first, there is the problem of obtaining sufficient leaf maturity to provide ample absorption without excessive contact injury; second, there is the necessity of making the application before diminishing soil moisture limits the growth activity of roots. In the western United States, for many perennial weeds, this is early summer—when the soil moisture is still adequate for root growth—or middle to late fall, after vegetative activity and root growth have been revived by early rains. With *Prosopis juliflora* in Texas, it turns out to be a short period in early summer, before the soil moisture supply has become exhausted. With amitrole, which does not depend upon root growth for toxicity to roots, a longer period is available (Crafts, 1961d).

The distribution patterns cited in this book would seem to indicate that, in many mature plants, the roots receive the bulk of their organic nutrients from the lower leaves. This fact is significant with respect to a number of the new crop varieties that are producing phenomenal yields *Oryza sativa* '1R8,' for instance. Most such high-yielding varieties have a more erect stature than the older varieties. In addition to being less prone to lodging, this leaf arrangement allows some light to penetrate to the lower leaves; they remain green, and they provide enery-rich nutrients to the roots. This enables the roots to continue to absorb nitrogen for a longer period; the plants can take advantage of a higher level of soil nitrogen, and this in turn enables them to carry and mature a greater crop. Thus, the physiological principle that enables the pesticide user to obtain optimum results with systemic materials also makes possible high yielding crops of *Oryza sativa, Triticum aestivum, Hordeum vulgare,* and *Zea mays.* A necessary corollary of this principle is that, in order to obtain the higher yields, the grower must use a higher level of nitrogen fertilization. Without this, the new varieties may underyield the old.

The widely confirmed observation that assimilates exported from basal leaves move to the roots; that from young mature leaves they move to shoot tips; that young, growing leaves fail to export; and that, in the process of translocation, mature, exporting leaves are by-passed; seems reconcilable only with a mass-flow concept in which solutes and water move en masse. Furthermore, the fact that a host of tracers, including many pesticides, assume this same distribution pattern indicates that their movement takes place along with that of assimilates. The numbers of exogenous materials, including many phytotoxic substances, that assume not an independent pattern but the same one taken by assimilates, would seem to prove that the driving force is not a diffusion gradient but an osmotic gradient that activates a pressure flow.

Because a survey of the array of phloem-mobile substances in Table 8.1 (page 176) shows that these vary from common nutrients to highly toxic herbicides, it is difficult to visualize a unique role for metabolism in their long-distance transport in sieve tubes. Metabolism certainly serves in the exporting sources, in the importing sinks, and, nominally, in maintaining the sieve tubes in a functioning condition.

What, then, is the significance of the great array of enzymes, vitamins, growth regulators, nucleotides, amino acids, and assorted organic residues, which, in addition to assimilates, have been identified in phloem exudate (Table 6.1, page 89). Are they simply end products of the processes of degradation that mark the maturation of sieve tubes? From Figures 2.4–2.6 (pages 26, 27, and 29), it is evident that much of this residual material must be present in sieve-tube sap, particularly in the rapidly growing parts of the plant. It must be reduced to a state in which it can migrate from the sieve cell in which it originated to the surrounding parenchyma cells.

Obviously, the sugars, sugar alcohols, and amino acids serve in the organic nutrition of the plant. The organic acids produced in the roots may balance the excess of cations commonly absorbed; the sugar phosphates may be intermediates in the absorption, transport, and utilization of sugars.

The organic phosphates and enzymes are components related to many stages of metabolism. In the enucleate sieve element, are they simply leftovers from a previous active stage, or do they serve in some way to maintain the functional integrity of these unique cells? With respect to the growth regulators, nucleic acids, and vitamins, it seems difficult to accept that they occur simply as a chance product of the processes of differentiation. There is much yet to be learned about phloem.

Advances in the field of translocation physiology find ready application in production agriculture. For example, with the discovery that herbicides move with assimilates in plants, recommendations were made for the effective use of 2,4-D and 2,4,5-T in the control of *Convolvulus arvensis* (Crafts, 1956b) and of brushy plants on range lands (Leonard and Crafts, 1956). The discovery and development of amitrole, TBA, dicamba, dalapon, and picloram (see Table 8.1, page 176) as translocated herbicides has greatly expanded the arsenal of chemicals that are available for controlling weeds. The distribution patterns of these chemicals all indicate that they follow the assimilate stream in its flow from foliar sources to underground sinks in roots and other storage tissues.

In Chapter 8, pages 180–264 give detailed information on the translocation of viruses, chemotherapeutants, pesticides, and inorganic nutrients. Much of the research described in that chapter may find use in practical agriculture. The redistribution patterns of phosphorus, potassium, and sulfur indicate the roles played by these compounds in plant growth and metabolism. Similar patterns of distribution for picloram, TBA, dalapon, and maleic hydrazide indicate the reasons for their particular effectiveness as weed killers. Some of the organic phosphorus insecticides and growth retardants follow similar patterns, which explains their systemic effects.

Knowing the nature of the processes of translocation enables those engaged in pesticide testing to include isotope techniques among their methods of providing information on penetration and distribution, as well as on toxicity. Thus, progress in the discovery and adoption of new chemicals for pest control is hastened; and detailed information on fate and breakdown, gained by tracer techniques, makes possible a quick evaluation of the residue problems to be met.

Given the wealth of experience that has been gained in the field of agricultural chemicals within the past two decades, it is possible to see more clearly the needs of the future. New and more specific chemicals are needed to control such pests as *Cyperus rotundus* and *Eichhornia crassipes*, weeds that handicap agriculture and transportation throughout the semitropical and tropical world. Cheaper and more effective materials are needed in the fields of range and forest management. *Convolvulus arvensis, Cardaria draba, Centaurea repens, Cirsium arvense, Euphorbia esula, Sonchus arvensis,* and a host of other perennial weeds still interfere with crop production on millions of acres. Detailed research on translocation of assimilates and labeled herbicides in these species is needed for solution of the perennial weed problem.

Systemic fungicides, insecticides, and nematocides that move as freely in plants as phosphorus, potassium, TBA, and picloram would be of inestimable value in the control of pests. Great advances have been made in the formulation of inorganic fertilizer elements to enable the farmer to use them with maximum effectiveness. This work must be continued; and the design of experiments must take into consideration the penetration and distribution of the elements, in order that they arrive at the target sites at the proper time and in adequate concentrations. Tracer techniques are of tremendous value in all research of this type; and translocation physiology has an essential role to play.

There have been remarkable advances in the use of translocated pesticides in recent times, notably the application of 2,4-D in the control of perennial weeds, the introduction of 2,4,5-T in the control of brush and weedy trees, and the discovery and development of picloram for handling those species that the chlorophenoxy herbicides fail to kill. As our knowledge of the means by which these phenomenal materials act becomes widespread we will continue to progress in the technology of plant control.

New and more specific reagents are needed in the fields of range and forest management. Means must be found for handling the vigorous growth of unwanted plants in the tropics in order to salvage the millions of acres of jungle for the production of food and fiber. And safer and more convenient chemicals are needed for handling domestic pest problems. In all of these fields translocation physiology has an important role to play.

We obviously do not yet have all of the answers to translocation problems. Although mass flow seems best able to explain the multitude of observations recorded in this book, there are innumerable variations on this theme. We still need to determine specific responses that will enable us to make maximum use of translocation physiology in crop production. We need to work out the great number of aspects of competition as they relate to range and forest management. If we are to have success in the use of systemic herbicides, insecticides, and fungicides in agriculture, we must comprehend the mechanics of their distribution in plants. From the experience of reviewing the great amount of literature covered in this book, it seems imperative that we carry on more realistic research. Instead of using metabolic inhibitors in search of "metabolic pumps," we should take a more thorough look at the obvious osmotic pumps that occur in the leaves of all plants. We need to understand the nature and source of the turgor pressure that is common to all sieve-tube systems. We need information on the size, distribution and structure of the fine veins and bundle terminals that serve as the generators of the hydrostatic pressure in phloem systems. Morrettes (1962), Esau (1967a), Trip (1969), and Geiger and

Cataldo (1969) have made valuable contributions along this line.

Instead of wasting energy attempting to analyze mathematically the logarithmic form of tracer fronts in phloem, where the very function of the tissues involves irreversible loss, efforts should be applied to careful analyses of the distribution patterns of tracers. The consistent patterns of export by mature leaves, import by young leaves, unidirectional and bidirectional movement, bypassing, and accumulation in meristems all preclude the independent movement of solute species as a mechanism of long-distance transport. Thus, attempts to prove the diffusional nature of translocation can only involve processes at the sources and sinks: rapid flow of the assimilate stream cannot be explained as a diffusion analogue.

As for using metabolic inhibitors, we know that some of them hinder translocation. If we wish to know how, we should label them, trace them, and determine their site, or sites, of action: if xylem-mobile and penetrable, they should move into leaves via the transpiration stream, and affect loading at the source; if phloem-mobile, they should move to sinks, and alter processes of accumulation, growth, and storage; and if they induce callosing of sieve plates, they should act locally to block transport, as well as affecting loading or unloading at source or sink. Tools are available to make a realistic approach to these problems. It seems futile to search for metabolic pumps in sieve tubes until someone comes up with a physical explanation of how such mechanisms might possibly work.

The symplast-apoplast concept originated by Münch (1930) has been slow to take hold. Finally, among people working on translocation, its obvious logic is becoming recognized; and the use of these terms to describe transport phenomena is becoming increasingly more common in recent publications. In contrast to the idea of free space, which is described as a mathematical concept, symplast and apoplast are recognizable plant structures that plant physiologists work with all of the time. The coordinating, unifying, and simplifying nature of the ideas of symplast and apoplast should be more widely used. In spite of the common familiarity with the use of these terms, there are still plant physiologists who think of the higher plants as assemblages of individual cells wrapped up in a bag of cuticle; and to make the matter worse, they call upon energy from ATP and other such compounds to overcome the barriers to intercellular transport. If we could only devise methods for determining the concentrations of solutes within living cytoplasm, we might find that many of the imagined barriers to transport do not exist. Only where concentrations of dissolved solutes in sink tissues exceed those in the assimilate stream is it necessary to postulate active, energy-requiring steps in the translocation process.

With the recognition of the existence of open pores in sieve plates, of the composition and nature of phloem exudate, and of common distribution patterns of tracers in plants, pressure flow or mass flow seems to be the inescapable answer to the translocation enigma. It will be well when this mechanism is as widely recognized as is the cohesion mechanism of water movement in plants. When the various untenable ideas on translocation have been abandoned, more time and energy will be available to study the real problems of this fascinating subject.

Common Names of Plants and Plant Pests

Plants

Abies concolor—White fir
Abies grandis—Grand fir
Acacia farnesiana—Huisache; sweet acacia
Acer circinatum—Vine maple
Acer macrophyllum—Bigleaf maple
Acer negundo—Box elder
Acer pseudoplatanus—Sycamore maple
Acer rubrum—Red maple
Aesculus sp.—Horse-chestnut
Agropyron repens—Couch grass
Allium sp.—Onion
Alnus sp.—Alder
Amyema pendula—Eucalyptus mistletoe
Anacyclus pyrethrum—Alexander's-foot
Apium graveolens—Celery
Arceuthobium sp.—Dwarf mistletoe
Arenga pinnata—Sugar palm
Asparagus officinalis—Asparagus
Atropa belladonna—Belladonna
Avena sativa—Cultivated oat
Avena fatua—Wild oat

Beta vulgaris—Sugar beet; garden beet
Betula sp.—Birch
Brassica capitata—Cabbage
Brassica napus—Rape
Brassica oleracea—Cole
Brassica rapa—Turnip

Calendula officinalis—Calendula
Capsicum annuum—Chili pepper; bush red pepper
Cardaria draba—Hoary-cress
Carpinus caroliniana—American hornbean

Carya sp.—Hickory
Castanea dentata—American chestnut
Castanea sativa—Spanish chestnut
Celtis milbraedii—Hackberry
Centaurea repens—Russian knapweed
Cephalotaxus sp.—Plum-yew
Ceratonia siliqua—Carob
Chenopodium capitatum—Blite goosefoot
Chondrilla juncea—Gum succory; skeleton weed
Chrysanthemum morifolium—Florist's chrysanthemum
Cirsium arvense—Canada thistle
Citrus aurantifolia—Lime
Citrus limon—Lemon
Citrus paradisi—Grapefruit
Citrus sinensis—Orange
Clivia sp.—Kafir-lily
Coffea arabica—Coffee
Coleus blumei—Common coleus
Convallaria majalis—Lily-of-the-valley
Convolvulus arvensis—Field Bindweed
Cornus sp.—Dogwood
Cornus mas—Cornelian-cherry
Cucumis sativus—Cucumber
Cucurbita ficifolia—Gourd
Cucurbita maxima—Squash
Cucurbita melopepo—Summer squash
Cucurbita melopepo torticollis—Crook-neck squash
Cucurbita pepo—Pumpkin
Cuscuta sp.—Dodder
Cynodon dactylon—Bermuda grass
Cyperus rotundus—Purple nutsedge

Dactylis glomerata—Orchard grass
Datura stramonium—Jimson-weed
Dioscorea alata—Winged yam
Dioscorea obconica—Yam
Dioscorea reticulata—Greater yam
Dioscorea sansibarensis—Zanzibar yam

Ecballium elaterium—Squirting-cucumber
Echinocystis lobata—Wild-cucumber
Eichhornia crassipes—Water-hyacinth
Elodea canadensis—Waterweed
Elodea densa—Dense-leafed elodea
Euphorbia esula—Leafy spurge

Fagus grandiflora—American beech
Fagus sylvatica—European beech
Fraxinus americana—White ash
Fraxinus excelsior—European ash

Ginkgo biloba—Ginkgo; maidenhair tree
Gladiolus sp.—Gladiolus
Glycine max—Soybean
Gossypium hirsutum—Highland cotton

Helianthus annuus—Sunflower
Helianthus tuberosus—Jerusalem artichoke
Heracleum mantegazzianum—Bigleaf cow-parsnip
Heracleum sosnovski—Cow-parsnip
Heracleum sphondylium—Cow-parsnip
Hippophae salicifolia—Sea-buckthorn
Hordeum vulgare—Barley
Humulus lupulus—Hops

Ilex vomitoria—Yaupon
Impatiens sultanii—Snapweed
Ipomoea batatas—Sweet-potato

Juniperus communis—Common juniper

Kigelia africana—Sausage tree
Kigelia pinnata—Sausage tree

Larix sp.—Larch
Lens culinaris—Lentil
Linum usitatissimum—Flax
Liquidambar styraciflua—Sweet gum

Liriodendron tulipifera—Tulip tree; yellow poplar
Lolium multiflorum—Italian ryegrass
Lolium perenne—Perennial ryegrass; English ryegrass
Lolium temulentum—Ryegrass; darnel
Lycopersicon esculentum—Tomato
Lygodesmia juncea—Skeleton weed

Macrocystis pyrifera—Giant kelp
Malus sp.—Apple
Malus sylvestris—Common apple
Medicago sativa—Alfalfa
Mentha piperita—Peppermint
Metasequoia glyptostroboides—Dawn redwood; metasequoia
Morus alba—Mulberry

Nicotiana tabacum—Tobacco
Nitella sp.—Nitella
Nymphoides peltatum—Floating heart

Olea europaea—Olive
Oryza sativa—Rice

Paspalum distichum—Knot grass
Passiflora coerulea—Passion flower
Pelargonium hortorum—Fish geranium
Pelargonium zonale—Horseshoe geranium
Persea americana—Avocado
Pharbitis nil—Morning glory
Phaseolus coccineus—Scarlet runner bean
Phaseolus vulgaris—Common bean
Phlox sp.—Phlox
Phoradendron sp.—Green mistletoe
Picea abies—Norway spruce
Picea sitchensis—Sitka spruce
Pinus banksiana—Jack pine
Pinus halepensis—Aleppo pine
Pinus pinea—Italian stone pine
Pinus ponderosa—Western yellow pine
Pinus resinosa—Red pine
Pinus strobus—Eastern white pine
Pinus sylvestris—Scots pine
Pisum arvense—Field pea
Pisum sativum—Garden pea
Platanus occidentalis—American sycamore; American plane-tree
Polygonum convolvulus—Black bindweed
Populus sp.—Poplar
Populus nigra—Black poplar
Populus regenerata—Poplar
Populus tremula—European aspen
Populus tremuloides—Quaking aspen
Potamogeton nodosus—Pondweed
Primula obconica—Primrose
Prosopis juliflora—Mesquite
Prunus armeniaca—Apricot
Prunus avium—Sweet cherry
Prunus persica—Peach
Prunus persica nectarina—Nectarine
Pseudotsuga menziesii—Douglas-fir

Quercus borealis—Red oak
Quercus douglasii—Blue oak
Quercus ellipsoidalis—Northern pin oak
Quercus marilandica—Blackjack oak
Quercus nigra—Water oak
Quercus robur—English oak
Quercus rubra—Southern red oak; Spanish oak
Quercus stellata—Post oak
Quercus suber—Cork oak
Quercus wislizenii—Coast live oak

Rheum rhaponticum—Rhubarb
Ricinodendron heudelotii—Manketti
Ricinus communis—Castor-bean
Robinia pseudoacacia—Black locust
Rubus idaeus—Raspberry
Rubus occidentalis—Blackcap blackberry

Sabal palmetto—Common palmetto
Saccharum officinarum—Sugar-cane
Salix sp.—Willow
Salix alba—White willow
Salix viminalis—Basket willow
Sambucus sp.—Elderberry
Sasa sp.—Sasa bamboo
Secale cereale—Rye
Sequoia sempervirens—Coast redwood
Sequoiadendron giganteum—Big tree; giant-sequoia
Sesamum indicum—Sesame
Sicyos angulatus—Wild cucumber
Silene armeria—Sweet-william catchfly
Smilax rotundifolia—Common greenbrier; horse brier
Solanum tuberosum—Potato
Sonchus arvensis—Perennial sow-thistle
Sorbus aucuparia—European mountain-ash; rowan
Sorghum halepense—Johnson grass
Sorghum vulgare—Sorghum
Streptanthus tortuosus—Mountain strepthanthus
Syringa vulgaris—Lilac

Tamus communis—No common name
Taraxacum officinale—Dandelion
Taxodium sp.—Baldcypress
Taxus sp.—Yew
Tetracera sp.—No common name
Tetragonia expansa—New-Zealand-spinach
Tilia americana—American basswood
Tilia cordata—Small-leaved linden
Tilia europaea—European linden
Tilia platyphyllos—Large-leaved linden
Tilia tomentosa—White linden; silver linden
Tradescantia albiflora—Spiderwort
Tradescantia fluminensis—Wandering Jew
Trifolium subterraneum—Subterranean clover
Triticum aestivum—Wheat
Tropaeolum majus—Garden nasturium

Ulmus carpinifolia—Smooth-leaved elm

Vallisneria sp.—Tape grass
Verbascum thapsus—Common mullein
Verbascum virgatum—Mullein
Vicia faba—Broad-bean; horse-bean
Vinca rosea—Madagascar periwinkle
Vitis sp.—Grape
Vitis labruscana—American vinyard grape
Vitis vinifera—European grape

Wisteria sinensis—Chinese wistaria

Xylopia quintasii—African pepper tree

Yucca flaccida—Yucca

Zantedeschia sp.—Calla-lily
Zea mays—Corn; maize
Zebrina pendula—Wandering Jew
Zinnia sp.—Zinnia

Insects

Choristoneura fumiferana—Spruce budworm
Cinara laricicola—Larch lachnid
Contarinia oregonensis—Douglas-fir cone midge
Cupressobium juniperi—Juniper aphid
Longistigma caryae—Giant bark aphid
Myzus persicae—Green peach aphid
Oncopeltus fasciatus—Milkweed bug
Tuberolachnus salignus—Willow aphid

Nematodes

Meloidogyne sp.—Root-knot nematode

Trichodorus christiei—Stubbyroot nematode

Fungi

Cytispora cincta—of cytospora canker of peaches

Erysiphe polygoni—of red clover powdery mildew

Fusarium moniliforme—of fusarium wilt

Sclerotinia fructicola—of brown rot

References

Abbe, L. B., and A. S. Crafts, 1939. The phloem of white pine and other coniferous species. *Bot. Gaz.* 100:695–722.

Addicott, F. T., and J. L. Lyon, 1969. Physiology of abscisic acid and related substances. *Ann. Rev. Plant Physiol.* 20:139–164.

Adedipe, N. O., D. P. Ormrod, and A. R. Maurer, 1968. Response of pea plants to soil and foliar applications of Cycocel. *Can. J. Plant Sci.* 48:323–325.

Alfieri, F. J., and R. F. Evert, 1968. Seasonal development of the secondary phloem in *Pinus. Am. J. Bot.* 55:518–528.

Anderson, O., 1957. The physiological action of 3-amino-1,2,4-triazole on nutgrass, with special emphasis on *Cyperus rotundus* L. Doctoral dissertation, University of California, Davis.

Anderson, R., and J. Cronshaw, 1969. The effects of pressure release on the sieve plate pores of *Nicotiana. J. Ultrastruct. Res.* 29:50–59.

——— and ———, 1970. Sieve plate pores in tobacco and bean. *Planta* 91:173–180.

Anisimov, A. A., 1959. The translocation of assimilates in wheat seedlings in connection with root nutrition conditions. *Plant Physiol.* (*USSR*). 6:149–153.

———, 1964. The effect of superphosphate on the translocation of assimilates in potato. *Plant Physiol* (*USSR*). 11:55–60.

———, 1965. Effect of ammonium sulfate on translocation of assimilates in sugar beet. *Plant Physiol* (*USSR*). 12:238–241.

———, 1968. Mechanisms by which nitrogen nutrition affects assimilate movement in potato. *Plant Physiol.* (*USSR*) 15:8–12.

Arisz, W. H., 1961. Symplasm theory of salt uptake into and transport in parenchymatic tissue. In *Recent Advances in Botany: from Lectures and Symposia presented to the IX International Botanical Congress, 1959*, pp. 1125–1128. Toronto: University of Toronto Press.

———, 1964. Translocation of labelled chloride ions in the symplasm of *Vallisneria* leaves. *Koninkl. Ned. Akad. Wetenschap. Proc.* (C)67:128–137.

Arnold, W. N., 1968. The selection of sucrose as the translocate of higher plants. *J. Theoret. Biol.* 21:13–20.

Bachofen, R., 1962a. Transport und Verteilung von markierten Substanzen, III. Einfluss der Fütterungs methode auf Absorption und Transport von ^{32}P bei *Phaseolus vulgaris. Vierteljahresschr. Naturforsch. Ges. Zürich* 107:31–39.

———, 1962b. Transport und Verteilung von markierten Substanzen, V. Über die Natur der transportierten Kohlenhydrate bei *Phaseolus multiflorus. Vierteljahresschr. Naturforsch. Ges. Zürich* 107:41–47.

——— and H. Wanner, 1962. Transport und Verteilung von markierten Assimilaten, II. Über die Transportbahnen von Assimilaten in Fruchtstielen von *Phaseolus. Planta* 58:225–236.

Badiei, A. A., E. Basler, and P. W. Santelmann, 1966. Aspects of movement of 2,4,5-T in blackjack oak. *Weeds* 14:302–305.

Balatinecz, J. J., D. F. Forward, and R. G. S. Bidwell, 1966. Distribution of photoassimilated $^{14}CO_2$ in young jack pine seedlings. *Can. J. Bot.* 44:362–365.

Bates, F., et al., 1942. Polarimetry, saccharimetry and the sugars. *U.S. Dept. Com. Natl. Bur. Std. Circ.* C440, table 109.

Bauer, L., 1949. Über den Wanderungsweg fluoreszierenden Farbstoffe in den Siebröhren. *Planta* 37:221–243.

———, 1953. Zur Frage der Stoffbewegungen in der Pflanze mit besonderer Berucksichtigung der Wanderung von Fluorochromen. *Planta* 42:367–451.

Beer, M., 1959. Fine structure of phloem of cucurbit as revealed by the electron microscope. *Proc. Internatl. Bot. Congr. 9th* 2:26.

Behnke, H., 1965a. Über das Phloem der *Dioscoreaceen* unter besonderer Berücksichtigung ihrer Phloembecken, I. Lichtoptische Untersuchungen zur Struktur der Phloembecken und ihrer Einordnung in das Sprossleitsystem. *Z. Pflanzenphysiol.* 53:97–125.

———, 1965b. Über das Phloem der *Dioscoreaceen* unter besonderer Berücksichtigung ihrer Phloembecken, II. Electronenoptische Untersuchungen zur Feinstruktur des Phloembeckens. *Z. Pflanzenphysiol.* 53:214–244.

———, 1968. Zum Aufbaugitterartiger Membranstrukturen im Siebelementplasma von *Dioscorea. Protoplasma* 66:287–310.

——— and I. Dörr, 1967. Zur Herkunft und struktur des Plasmafilimente in Assimilatleitbahner. *Planta* 74:18–44.

Belikov, I. F., 1955. Movement and distribution of photosynthesis products in soybean during vegetation period. *Plant Physiol.* (*USSR*) 2:345–357.

Bennet, S. H., 1957. The behavior of systemic insecticides applied to plants. *Ann. Rev. Entomol.* 2:279–296.

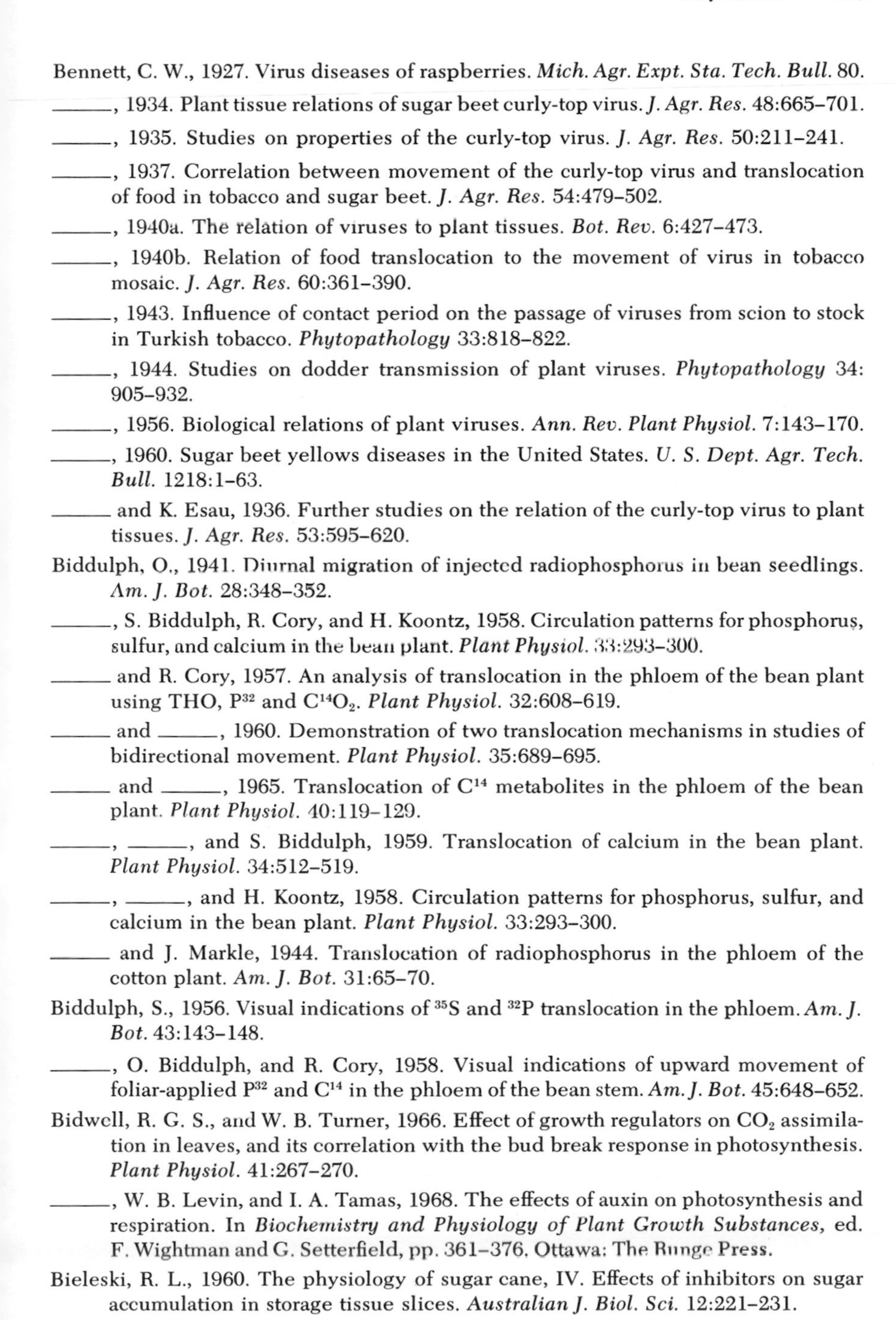

Bennett, C. W., 1927. Virus diseases of raspberries. *Mich. Agr. Expt. Sta. Tech. Bull.* 80.

———, 1934. Plant tissue relations of sugar beet curly-top virus. *J. Agr. Res.* 48:665–701.

———, 1935. Studies on properties of the curly-top virus. *J. Agr. Res.* 50:211–241.

———, 1937. Correlation between movement of the curly-top virus and translocation of food in tobacco and sugar beet. *J. Agr. Res.* 54:479–502.

———, 1940a. The relation of viruses to plant tissues. *Bot. Rev.* 6:427–473.

———, 1940b. Relation of food translocation to the movement of virus in tobacco mosaic. *J. Agr. Res.* 60:361–390.

———, 1943. Influence of contact period on the passage of viruses from scion to stock in Turkish tobacco. *Phytopathology* 33:818–822.

———, 1944. Studies on dodder transmission of plant viruses. *Phytopathology* 34: 905–932.

———, 1956. Biological relations of plant viruses. *Ann. Rev. Plant Physiol.* 7:143–170.

———, 1960. Sugar beet yellows diseases in the United States. *U. S. Dept. Agr. Tech. Bull.* 1218:1–63.

——— and K. Esau, 1936. Further studies on the relation of the curly-top virus to plant tissues. *J. Agr. Res.* 53:595–620.

Biddulph, O., 1941. Diurnal migration of injected radiophosphorus in bean seedlings. *Am. J. Bot.* 28:348–352.

———, S. Biddulph, R. Cory, and H. Koontz, 1958. Circulation patterns for phosphorus, sulfur, and calcium in the bean plant. *Plant Physiol.* 33:293–300.

——— and R. Cory, 1957. An analysis of translocation in the phloem of the bean plant using THO, P^{32} and $C^{14}O_2$. *Plant Physiol.* 32:608–619.

——— and ———, 1960. Demonstration of two translocation mechanisms in studies of bidirectional movement. *Plant Physiol.* 35:689–695.

——— and ———, 1965. Translocation of C^{14} metabolites in the phloem of the bean plant. *Plant Physiol.* 40:119–129.

———, ———, and S. Biddulph, 1959. Translocation of calcium in the bean plant. *Plant Physiol.* 34:512–519.

———, ———, and H. Koontz, 1958. Circulation patterns for phosphorus, sulfur, and calcium in the bean plant. *Plant Physiol.* 33:293–300.

——— and J. Markle, 1944. Translocation of radiophosphorus in the phloem of the cotton plant. *Am. J. Bot.* 31:65–70.

Biddulph, S., 1956. Visual indications of ^{35}S and ^{32}P translocation in the phloem. *Am. J. Bot.* 43:143–148.

———, O. Biddulph, and R. Cory, 1958. Visual indications of upward movement of foliar-applied P^{32} and C^{14} in the phloem of the bean stem. *Am. J. Bot.* 45:648–652.

Bidwell, R. G. S., and W. B. Turner, 1966. Effect of growth regulators on CO_2 assimilation in leaves, and its correlation with the bud break response in photosynthesis. *Plant Physiol.* 41:267–270.

———, W. B. Levin, and I. A. Tamas, 1968. The effects of auxin on photosynthesis and respiration. In *Biochemistry and Physiology of Plant Growth Substances*, ed. F. Wightman and G. Setterfield, pp. 361–376. Ottawa: The Runge Press.

Bieleski, R. L., 1960. The physiology of sugar cane, IV. Effects of inhibitors on sugar accumulation in storage tissue slices. *Australian J. Biol. Sci.* 12:221–231.

———, 1962. The physiology of sugar cane, V. Kinetics of sugar accumulation. *Australian J. Biol. Sci.* 15:429–444.

———, 1966a. Sites of accumulation in excised phloem and vascular tissues. *Plant Physiol.* 41:455–466.

———, 1966b. Accumulation of phosphate, sulfate and sucrose by excised phloem tissues. *Plant Physiol.* 41:447–454.

———, 1969. Phosphorus compounds in translocating phloem. *Plant Physiol.* 44: 497–502.

Birch-Hirschfeld, L., 1920. Untersuchungen über die Ausbreitungsgeschwindigkeit gelöster Stoffe der Pflanze, *Jahrb. Wiss. Bot.* 59:171–262.

Birecka, H., 1967. Translocation and distribution of ^{14}C-labeled (2 chloroethyl) trimethyl ammonium chloride (CCC) in wheat. *Bull. Acad. Polon. Sci. Ser. Sci. Biol.* 15:707–714.

Blinn, R. C., 1967. Plant growth regulant. Biochemical behavior of 2-chloroethyl trimethyl ammonium chloride in wheat and in rats. *J. Agr. Food Chem.* 15:984–988.

Bollag, J. M., and E. Galun, 1966. Distribution of labelled indolyl-3-acetic acid in intact cucumber plants. *Nature* 211:647–648.

Bolli, H. K., 1967. Absorption und Transport markierter Aminosauren durch Wurzeln von *Zea mays. Ber. Schweiz. Bot. Ges.* 77:61–102.

Bonnemain, M. J., 1965. Sur le transport diurne des produits d'assimilation lors de la floraison chez la Tomate. *Compt. Rend.* 260:2054–2057.

Bouck, G. B., and J. Cronshaw, 1965. The fine structure of differentiating sieve tube elements. *J. Cell. Biol.* 25:79–96.

Bovey, R. W., F. S. Davis, and M. G. Merkle, 1967. Distribution of picloram in huisache after foliar and soil applications. *Weeds* 15:245–249.

Bowen, M. D., and G. V. Hoad, 1968. Inhibitor content of phloem and xylem sap obtained from willow (*Salix viminalis*) entering dormancy. *Planta* 81:64–70.

Bowling, D. J. F., 1968. Translocation at 0° in *Helianthus annuus. J. Exptl. Bot.* 19: 381–388.

Brady, H. A., 1969. Light intensity and the absorption and translocation of 2,4,5-T by woody plants. *Weed Sci.* 17:320–322.

Braun, H. J., and J. J. Sauter, 1964. Phosphatase-Lokalisation in Phloembeckenzellen und Siebröhren der Dioscoreacene und ihr mögloche Bedeutung fur den aktiven Assimilat-Transport. *Planta* 60:543–557.

Brennen, H., J. S. Pate, and W. Wallace, 1964. Nitrogen-containing compounds in the shoot system of *Pisum arvense* L., I. Amino compounds in the bleeding saps from petioles. *Ann. Bot.* (*London*) (N.S.)28:527–540.

Brian, P. W. 1966. Uptake and transport of systematic fungicides and bactericides. In *Trasporto delle molecole organiche nelle piante,* pp. 97–122. Vienna: 6th International Symposium on Agrochemistry.

Brouwer, R., 1953. The arrangement of the vascular bundles in the nodes of the Dioscoreaceae. *Acta Bot. Neerl.* 11:66–73.

——— and E. Levi, 1969. Responses of bean plants to root temperatures. *Acta Bot. Neerl.* 18:58–66.

Brovchenko, M. L., 1965. On the movement of sugars from the mesophyll to the conducting bundles in sugar beet leaves. *Plant Physiol.* (*USSR*) 12:230–237.

______, 1967. Some proofs of splitting of sucrose during its translocation from the mesophyll to the thin bundles of sugar beet leaves. *Plant Physiol.* (*USSR*) 14:352–359.

Brown, A. G., and C. G. Greenham, 1965. Further investigations in the control of mistletoe by trunk injections. *Australian J. Exptl. Agr. Animal Husbandry* 5:305–309.

Brown, A. L., S. Yamaguchi, and J. Leal-Diaz, 1965. Evidence for translocation of iron in plants. *Plant Physiol.* 40:35–38.

Burley, J. W. A., 1961. Carbohydrate translocation in raspberry and soybean. *Plant Physiol.* 36:820–824.

Burr, G. D., C. E. Hartt, T. Tanimoto, T. Takahashi, and H. W. Brodie, 1958. The circulatory system of the sugarcane plant. *Radioisotopes Sci. Res. Proc. Internatl. Conf. Paris 1957* 4:351.

Butcher, H. C., 1965. The kinetics of carbon-^{14} translocation in sugar beet: an effect of illumination. *Dissertation Abstr.* 25:7350–7350.

Buvat, R., 1963a. Infrastructure et differenciation des cellules criblees de *Cucurbita pepo*. Evolution du tonoplast et signification du contenu cellulaire final. *Compt. Rend.* 256:5193–5195.

______, 1963b. Sur la presence d'acide ribonucleique dans les corpuscules muqueux des cellules criblees de *Cucurbita pepo*. *Compt Rend.* 257:733–735.

Bykhovskii, E. G., A. V. Sokolov, E. V. Fomina, and V. V. Kabanov, 1968. Use of radioactive isotopes for measuring the rate of translocation of resin and metabolites along the pine trunk. *Gidrolizn. Lesokhim. Prom.* 21:16–18. (Abstract in *Chem. Abstr.* 68:9896.)

Caldwell, J., 1934. The physiology of virus diseases in plants, V. The movement of the virus agent in tobacco and tomato. *Ann. Appl. Biol.* 21:191–205.

Canny, M. J., 1960a. The breakdown of sucrose during translocation. *Ann. Bot.* (*London*) (N.S.)24:330–344.

______, 1960b. The rate of translocation. *Biol. Rev. Cambridge Phil. Soc.* 35:507–532.

______, 1961. Measurements of the velocity of translocation. *Ann. Bot.* (*London*) (N.S.) 25:152–167.

______, 1962a. The translocation profile: sucrose and carbon dioxide. *Ann. Bot.* (*London*) (N.S.)26:181–196.

______, 1962b. The mechanism of translocation. *Ann. Bot.* (*London*) (N.S.)26:603–617.

______ and M. J. Askham, 1967. Physiological inferences from the evidence of translocated tracer: a caution. *Ann. Bot.* (*London*) (N.S.)31:409–416.

______ and K. Markus, 1960. The metabolism of phloem isolated from grapevine. *Australian J. Biol. Sci.* 13:292–299.

______ and O. M. Phillips, 1963. Quantitative aspects of a theory of translocation. *Ann. Bot.* (*London*) (N.S.)27:379–402.

Carr, J. D., and I. F. Wardlaw, 1965. The supply of photosynthetic assimilates to the grain from the flag leaf and ear of wheat. *Australian J. Biol. Sci.* 18:711–719.

Chang, F. Y., and W. H. Vanden Born, 1968. Translocation of dicamba in Canada thistle. *Weed Sci.* 16:176–181.

Chen, S. L., 1951. Simultaneous movement of ^{32}P and ^{14}C in opposite directions in phloem tissue. *Am. J. Bot.* 38:203–211.

Chin, T. Y., and J. A. Lockhart, 1965. Translocation of applied gibberellin in bean seedlings. *Am. J. Bot.* 52:828–833.

Choi, I. C. and S. Aronoff, 1966. Photosynthate transport using tritiated water. *Plant Physiol.* 41:1119–1129.

Chopowick, R. E., and D. F. Forward, 1970. Translocation of ^{14}C-sucrose when ^{14}C-alanine is applied to sunflower leaves. Private communication.

Clauss, H., D. C. Mortimer, and P. R. Gorham, 1964. Time-course study of translocation of products of photosynthesis in soybean plants. *Plant Physiol.* 39:269–273.

Clements, H. F., 1934. Translocation of solutes in plants. *Northwest Sci.* 8:9–21.

———, 1940. Movement of organic solutes in the sausage tree, *Kigelia africana*. *Plant Physiol.* 15:689–700.

Clor, M. A., 1959. Comparative studies on translocation of ^{14}C-labeled 2,4-D, urea, and amino triazole in cotton and oaks. Doctoral dissertation, University of California, Davis.

———, A. S. Crafts, and S. Yamaguchi, 1962. Effects of high humidity on translocation of foliar-applied labeled compounds in plants, I. *Plant Physiol.* 37:609–617.

———, ———, and ———, 1963. Effects of high humidity on translocation of foliar-applied labeled compounds in plants, II. Translocation from starved leaves. *Plant Physiol.* 38:501–507.

———, ———, and ———, 1964. Translocation of ^{14}C-labeled compounds in cotton and oaks. *Weeds* 12:194–200.

Colwell, R. N., 1942. The use of radioactive phosphorus in translocation studies. *Am. J. Bot.* 29:798–807.

Crafts, A. S., 1931. Movement of organic materials in plants. *Plant Physiol.* 6:1–41.

———, 1932. Phloem anatomy, exudation, and transport of organic nutrients in cucurbits. *Plant Physiol.* 7:183–225.

———, 1933. Sieve-tube structure and translocation in the potato. *Plant Physiol.* 8:81–104.

———, 1934. Phloem anatomy in two species of *Nicotiana*, with notes on the interspecific graft union. *Bot. Gaz.* 45:592–608.

———, 1936. Further studies on exudation in cucurbits. *Plant Physiol.* 11:63–79.

———, 1938. Translocation in plants. *Plant Physiol.* 13:791–814.

———, 1939a. The relation between structure and function of the phloem. *Am. J. Bot.* 26:172–177.

———, 1939b. The protoplasmic properties of sieve tubes. *Protoplasma* 33:389–398.

———, 1943. Vascular differentiation in the shoot apex of *Sequoia sempervirens*. *Am. J. Bot.* 30:110–121.

———, 1948. Movement of materials in phloem as influenced by the porous nature of the tissues. *Discussions Faraday Soc.* (3):153–159.

———, 1951. Movement of assimilates, viruses, growth regulators, and chemical indicators in plants. *Bot. Rev.* 17:203–284.

———, 1954. Composition of the sap of xylem and phloem and its relation to nutrition of the plant. In *Extrait de la Brochure Analyse des Plantes et Problemes des Fumures Minerales*, pp. 18–21. Paris: 8th International Botanical Congress.

———, 1956a. Translocation of herbicides, I. The mechanism of translocation. Methods of study with ^{14}C-labeled 2,4-D. *Hilgardia* 26:287–334.

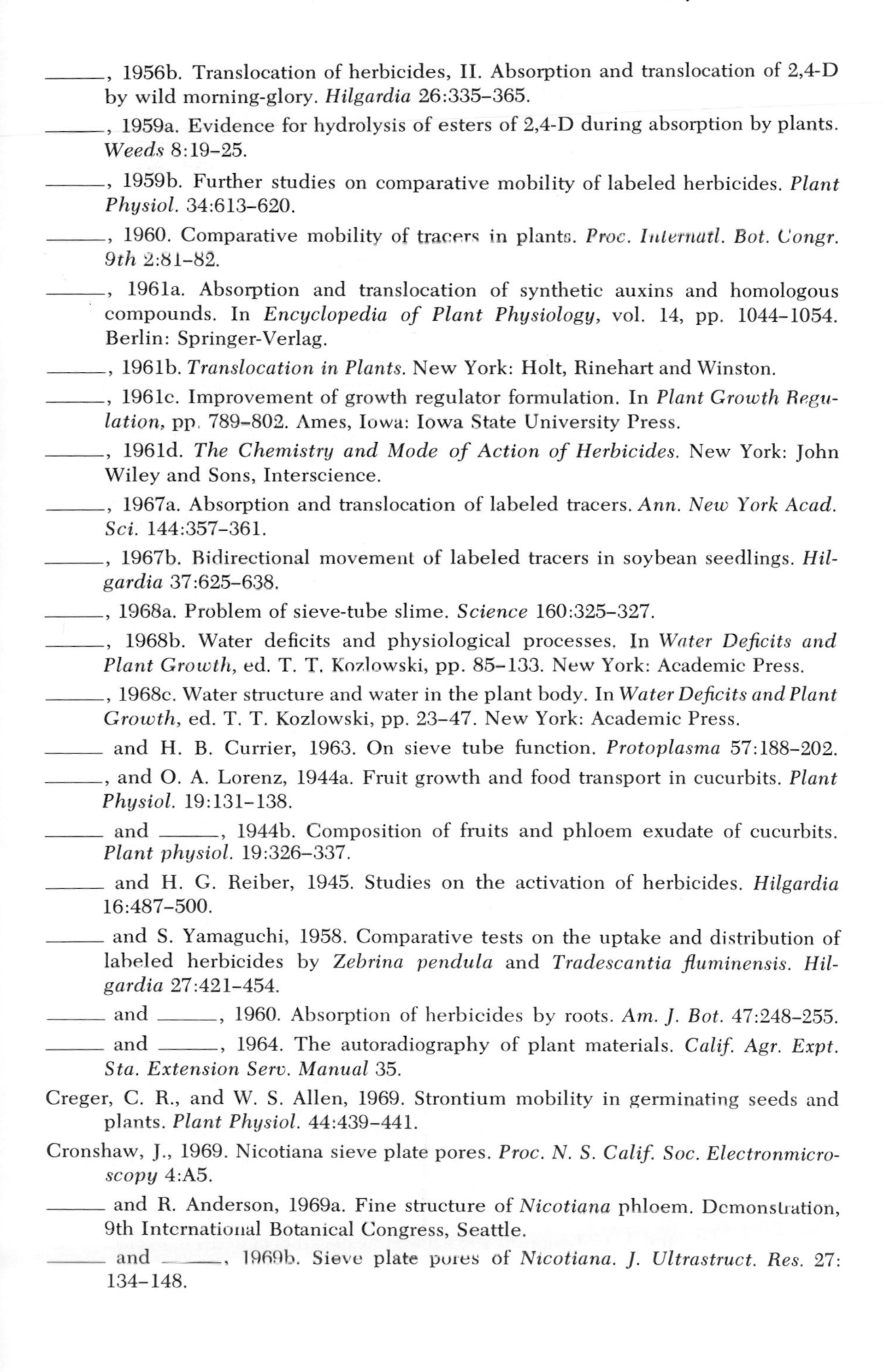

———, 1956b. Translocation of herbicides, II. Absorption and translocation of 2,4-D by wild morning-glory. *Hilgardia* 26:335–365.

———, 1959a. Evidence for hydrolysis of esters of 2,4-D during absorption by plants. *Weeds* 8:19–25.

———, 1959b. Further studies on comparative mobility of labeled herbicides. *Plant Physiol.* 34:613–620.

———, 1960. Comparative mobility of tracers in plants. *Proc. Internatl. Bot. Congr. 9th* 2:81–82.

———, 1961a. Absorption and translocation of synthetic auxins and homologous compounds. In *Encyclopedia of Plant Physiology,* vol. 14, pp. 1044–1054. Berlin: Springer-Verlag.

———, 1961b. *Translocation in Plants.* New York: Holt, Rinehart and Winston.

———, 1961c. Improvement of growth regulator formulation. In *Plant Growth Regulation,* pp. 789–802. Ames, Iowa: Iowa State University Press.

———, 1961d. *The Chemistry and Mode of Action of Herbicides.* New York: John Wiley and Sons, Interscience.

———, 1967a. Absorption and translocation of labeled tracers. *Ann. New York Acad. Sci.* 144:357–361.

———, 1967b. Bidirectional movement of labeled tracers in soybean seedlings. *Hilgardia* 37:625–638.

———, 1968a. Problem of sieve-tube slime. *Science* 160:325–327.

———, 1968b. Water deficits and physiological processes. In *Water Deficits and Plant Growth,* ed. T. T. Kozlowski, pp. 85–133. New York: Academic Press.

———, 1968c. Water structure and water in the plant body. In *Water Deficits and Plant Growth,* ed. T. T. Kozlowski, pp. 23–47. New York: Academic Press.

——— and H. B. Currier, 1963. On sieve tube function. *Protoplasma* 57:188–202.

———, and O. A. Lorenz, 1944a. Fruit growth and food transport in cucurbits. *Plant Physiol.* 19:131–138.

——— and ———, 1944b. Composition of fruits and phloem exudate of cucurbits. *Plant physiol.* 19:326–337.

——— and H. G. Reiber, 1945. Studies on the activation of herbicides. *Hilgardia* 16:487–500.

——— and S. Yamaguchi, 1958. Comparative tests on the uptake and distribution of labeled herbicides by *Zebrina pendula* and *Tradescantia fluminensis. Hilgardia* 27:421–454.

——— and ———, 1960. Absorption of herbicides by roots. *Am. J. Bot.* 47:248–255.

——— and ———, 1964. The autoradiography of plant materials. *Calif. Agr. Expt. Sta. Extension Serv. Manual* 35.

Creger, C. R., and W. S. Allen, 1969. Strontium mobility in germinating seeds and plants. *Plant Physiol.* 44:439–441.

Cronshaw, J., 1969. Nicotiana sieve plate pores. *Proc. N. S. Calif. Soc. Electronmicroscopy* 4:A5.

——— and R. Anderson, 1969a. Fine structure of *Nicotiana* phloem. Demonstration, 9th International Botanical Congress, Seattle.

——— and ———, 1969b. Sieve plate pores of *Nicotiana. J. Ultrastruct. Res.* 27: 134–148.

——— and K. Esau, 1967. Tubular and fibrillar components of mature and differentiating sieve elements. *J. Cell Biol.* 34:801–816.

——— and ———, 1968a. P-protein in the phloem of *Cucurbita,* I. The development of P-protein bodies. *J. Cell Biol.* 38:25–39.

——— and ———, 1968b. P-protein in the phloem of *Cucurbita,* II. The P-protein of mature sieve elements. *J. Cell Biol.* 38:292–303.

Crowdy, S. H., A. P. Green, J. F. Grove, P. McCloskey, and A. Morrison, 1959. The translocation of antibiotics in plants, 3. The estimation of griseofulvin relatives in plant tissue. *Biochem. J.* 72:230–241.

———, J. F. Grove, and P. McCloskey, 1959. The translocation of antibiotics in plants, 4. Systemic fungicidal activity and chemical structure in griseofulvin relatives. *Biochem. J.* 72:241–249.

Cruz-Perez, L. M., and D. Durkin, 1964. Sugar translocation in peppermint (*Mentha piperita* L.). *Proc. Am. Soc. Hort. Sci.* 85:414–418.

Currier, H. B., 1957. Callose substance in plant cells. *Am. J. Bot.* 44:478–488.

———, K. Esau, and V. I. Cheadle, 1955. Plasmolytic studies of phloem. *Am. J. Bot.* 42:68–81.

———, R. B. McNairn, and D. H. Webster, 1966. Blockage phenomena in axial phloem conduction. *Plant Physiol.* 41(Suppl.):xx.

——— and C. Y. Shih, 1968. Sieve tubes and callose in *Elodea* leaves. *Am. J. Bot.* 55:145–152.

——— and D. H. Webster, 1964. Callose formation and subsequent disappearance: Studies in ultrasound stimulation. *Plant Physiol.* 39:843–847.

Curtis, O. F., 1925. Studies on the tissues concerned in the transfer of solutes in plants. The effect on the upward transfer of solutes of cutting the xylem as compared with that of cutting the phloem. *Ann. Bot.* (*London*) (N.S.)39:573–585.

———, 1929. Studies on solute translocation in plants. Experiments indicating that translocation is dependent on the activity of living cells. *Am. J. Bot.* 26:154–168.

———, 1935. *The Translocation of Solutes in Plants.* New York: McGraw-Hill.

Czapek, F., 1897. Über die Leitungswege der organischen Baustoffe in Pflanzenkörper. *Sitzber. Kaiserl. Akad. Wiss.* 66:117–170.

Davey, C. B., and G. C. Papavizas, 1961. Translocation of streptomycin from coleus leaves and its effect on rhizosphere bacteria. *Science* 134:1368–1369.

David North Plant Research Center, 1964. Translocation of sugars and other compounds. *Rept. David North Plant Res. Ctr.* (Toowong, Brisbane, Queensland, Australia).

Davis, F. S., R. W. Bovey, and M. G. Merkle, 1968. Effect of paraquat and 2,4,5-T on the uptake and transport of picloram in woody plants. *Weed Sci.* 16:336–339.

Day, B. E., 1950. Absorption and translocation of 2,4-dichlorophenoxyacetic acid by bean plants. Doctoral dissertation, University of California, Davis.

———, 1952. The absorption and translocation of 2,4-dichlorophenoxyacetic acid by bean plants. *Plant Physiol.* 27:143–152.

DeKock, P. C., 1955. Iron nutrition of plants at high *p*H. *Soil Sci.* 79:167–175.

De la Fuente, R., and A. C. Leopold, 1965. Two translocation systems in excised bean petioles. *Plant Physiol.* 40(Suppl.):xivi.

de Pietri-Tonelli, P., 1965. Penetration and translocation of Rogor applied to plants. *Advan. Pest Control Res.* 6:31–84.

Derr, W. F., and R. F. Evert, 1967. The cambium and seasonal development of the phloem in *Robinia pseudoacacia. Am. J. Bot.* 54:147–153.

de Stigter, H. C. M., 1966. Parallelism between the transport of ^{14}C-photosynthates and the flowering response in grafted *Silene armeria* L. *Z. Pflanzenphysiol.* 55:11–19.

De Vries, H., 1885. Über die Bedeutung der Circulation und der Rotation des Protoplasma für das Stofftransport in der Pflanze. *Bot. Ztg.* 43:1–6, 16–26.

Dezsi, L., M. Barkoczi, and G. Palfi, 1967. Data on the translocation of amino acids of wheat, maize and rice, and the role of ornithine. *Acta Agron. Acad. Sci. Hung.* 16:17–24.

Dickmann, D. I., and T. T. Kozlowski, 1968. Mobilization by *Pinus resinosa* cones and shoots of ^{14}C-photosynthate from needles of different ages. *Am. J. Bot.* 55:900–906.

Die, J. van, 1968. The use of phloem exudates from *Yucca flaccida* Haw. in the study of translocation of assimilates. *Vorträge Gesamtgebiet Bot.* (N.S.)2:27–30.

______ and P. M. L. Tammes, 1964. Studies on phloem exudation from *Yucca flaccida* Haw., II. The translocation of assimilates. *Acta Bot. Neerl.* 13:84–90.

______ and ______, 1966. Studies on phloem exudate from *Yucca flaccida* Haw., III. Prolonged bleeding from isolated parts of the young inflorescense. *Koninkl. Ned. Akad. Wetenschap. Proc.* (C)69:648–654.

Dimond, A. E., 1965. Natural models for plant chemotherapy. *Advan. Pest Control Res.* 6:127–169.

______ and J. G. Horsfall, 1959. Plant chemotherapy. *Ann. Rev. Plant Physiol.* 10:257–276.

Dixon, H. H., 1922. Transport of organic substances in plants. *Nature* 110:547–551.

______, 1923. Transport of organic substances in plants. *Notes Bot. School Trinity Coll. Dublin* 3:207–215.

______ and N. G. Ball, 1922. Transport of organic substances in plants. *Nature* 109: 236–237.

Doi, Y., M. Terenaka, K. Yora, and H. Asuyama, 1967. Mycoplasma- or PLT group-like organisms found in the phloem elements of plants infected with mulberry dwarf, potato witches broom, aster yellows, or paulownia witches broom [In Japanese with English summary]. *Ann. Phytopathol. Soc. Japan* 33:259–266.

Donoho, C. W., Jr., A. E. Mitchell, and M. V. Bukovac, 1961. The absorption and translocation of ring-labeled ^{14}C naphthalene acetic acid in the apple and peach. *Proc. Am. Soc. Hort. Sci.* 78:96–103.

Duble, R. L., E. C. Holt, and G. G. McBee, 1968. The translocation of two organic arsenicals in purple nutsedge. *Weed Sci.* 16:421–424.

Duloy, M. D., and F. V. Mercer, 1961. Studies in translocation, I. The respiration of the phloem. *Australian J. Biol. Sci.* 14:391–401.

______, ______, and N. Rathgeber, 1961. Studies in translocation, II. Submicroscopic anatomy of the phloem. *Australian J. Biol. Sci.* 14:506–526.

Dunning, J. J., 1958. Factors influencing the formation of callose in plant cells. Doctoral dissertation, University of California, Davis.

Dyar, J. J., and K. L. Webb, 1961. A relationship between boron and auxin in ^{14}C translocation in bean plants. *Plant Physiol.* 36:672–676.

Eddings, J. L., and A. L. Brown, 1967. Absorption and translocation of foliar-applied iron. *Plant Physiol.* 42:15–19.

Edelman, J. T., T. G. Jefford, and S. P. Singh, 1969. Studies on the biochemical basis of physiological processes in the potato tuber. *Planta* 84:48–56.

Eden, W. G., and B. W. Arthur, 1965. Translocation of DDT and heptachlor in soybeans. *J. Econ. Entomol.* 58:161.

Edgerton, L. J., and W. J. Greenhalgh, 1967. Absorption, translocation and accumulation of labeled *N*-dimethyl amino-succinamic acid in apple tissues. *Proc. Am. Soc. Hort. Sci.* 91:25–30.

Edgington, L. V., and A. E. Dimond, 1964. The effect of adsorption of organic cations to plant tissue on their use as systemic fungicides. *Phytopathology* 54:1193–1197.

Ehara, K., and H. Sekicka, 1962. Effect of atmospheric humidity and soil moisture on the translocation of sucrose-^{14}C in the sweet potato plant. *Proc. Crop Sci. Soc. Japan* 31:41–44.

Eliasson, L., 1965. Interference of the transpiration stream with the basipetal translocation of leaf-applied chlorophenoxy herbicides in aspen (*Populus tremula* L.). *Physiol. Plantarum* 18:506–515.

El-Zayat, M. M., R. J. Lukens, A. E. Dimond, and J. G. Horsfall, 1968. Systemic action of nitrophenols against powdery mildew. *Phytopathology* 58:434–437.

Engleman, E. M., 1963. Fine structure of the proteinaceous substance in sieve tubes. *Planta* 59:420–426.

______, 1965a. Sieve element of *Impatiens sultanii,* I. Wound reaction. *Ann. Bot.* (*London*) (N.S.)29:83–101.

______, 1965b. Sieve elements of *Impatiens sultanii,* II. Developmental aspects. *Ann. Bot.* (*London*) (N.S.)29:103–118.

Esau, K., 1933. Pathologic changes in the anatomy of the sugar beet, *Beta vulgaris* L. affected by curly top. *Phytopathology* 23:679–712.

______, 1935. Ontogeny of the phloem in sugar beets affected by the curly top disease. *Am. J. Bot.* 22:149–163.

______, 1938. Ontogeny and structure of the phloem of tobacco. *Hilgardia* 11:343–424.

______, 1941. Phloem anatomy of tobacco affected with curly top and mosaic. *Hilgardia* 13:437–490.

______, 1948. Phloem structure in the grapevine and its seasonal changes. *Hilgardia* 18:217–296.

______, 1960. Development of inclusions in sugar beets infected with the beet yellows virus. *Virology* 11:317–328.

______, 1961. *Plants, Viruses, and Insects.* Cambridge, Mass.: Harvard University Press.

______, 1963. Ultrastructure of differentiated cells in higher plants. *Am. J. Bot.* 50: 495–506.

______, 1964. Aspects of ultrastructure of phloem. In *The Formation of Wood in Forest Trees,* ed. M. H. Zimmerman, pp. 51–63. New York: Academic Press.

______, 1965a. Fixation images of sieve element plastids in *Beta. Proc. Natl. Acad. Sci. U. S.* 54:429–437.

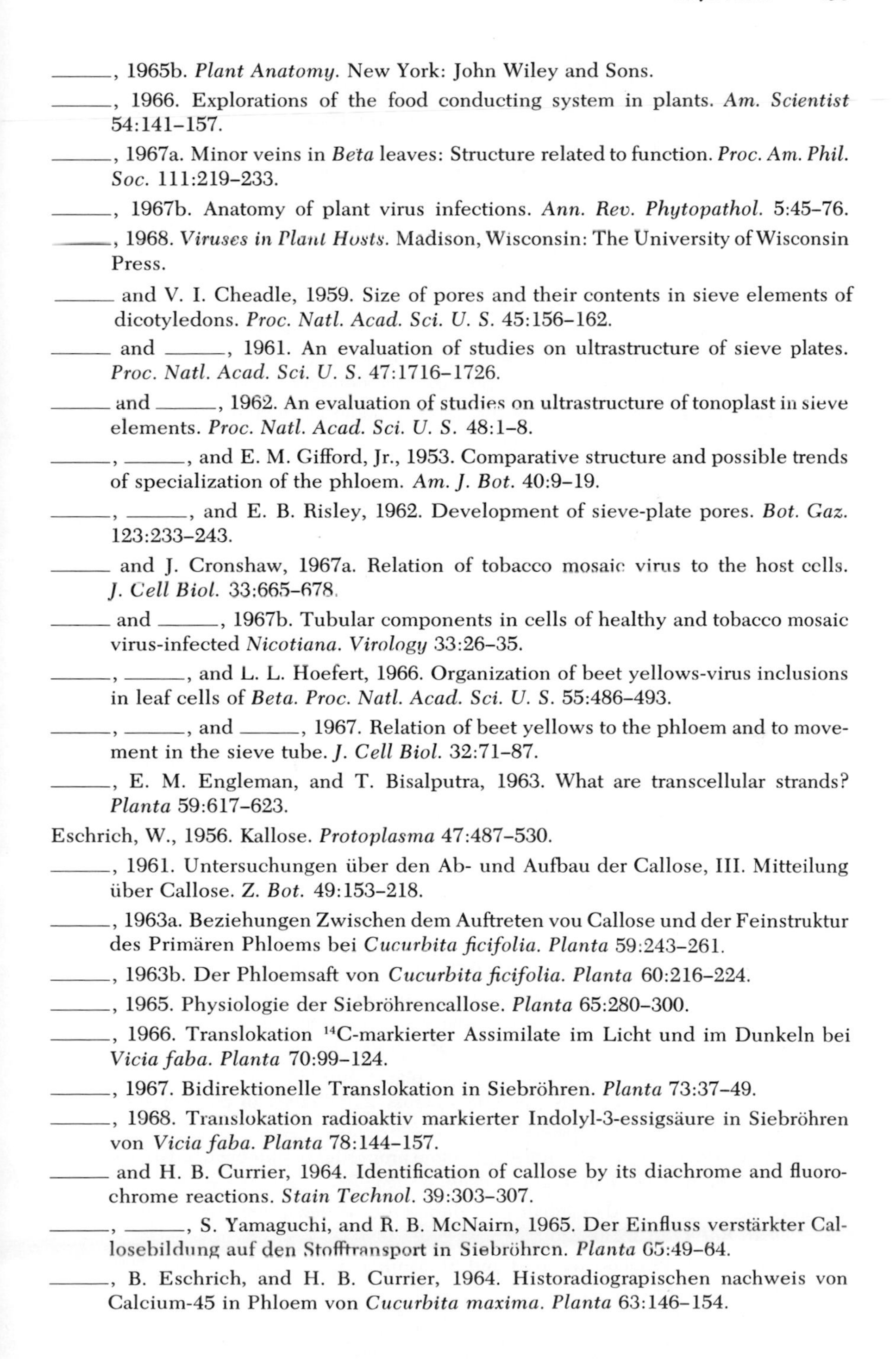

———, 1965b. *Plant Anatomy.* New York: John Wiley and Sons.

———, 1966. Explorations of the food conducting system in plants. *Am. Scientist* 54:141–157.

———, 1967a. Minor veins in *Beta* leaves: Structure related to function. *Proc. Am. Phil. Soc.* 111:219–233.

———, 1967b. Anatomy of plant virus infections. *Ann. Rev. Phytopathol.* 5:45–76.

———, 1968. *Viruses in Plant Hosts.* Madison, Wisconsin: The University of Wisconsin Press.

——— and V. I. Cheadle, 1959. Size of pores and their contents in sieve elements of dicotyledons. *Proc. Natl. Acad. Sci. U. S.* 45:156–162.

——— and ———, 1961. An evaluation of studies on ultrastructure of sieve plates. *Proc. Natl. Acad. Sci. U. S.* 47:1716–1726.

——— and ———, 1962. An evaluation of studies on ultrastructure of tonoplast in sieve elements. *Proc. Natl. Acad. Sci. U. S.* 48:1–8.

———, ———, and E. M. Gifford, Jr., 1953. Comparative structure and possible trends of specialization of the phloem. *Am. J. Bot.* 40:9–19.

———, ———, and E. B. Risley, 1962. Development of sieve-plate pores. *Bot. Gaz.* 123:233–243.

——— and J. Cronshaw, 1967a. Relation of tobacco mosaic virus to the host cells. *J. Cell Biol.* 33:665–678.

——— and ———, 1967b. Tubular components in cells of healthy and tobacco mosaic virus-infected *Nicotiana. Virology* 33:26–35.

———, ———, and L. L. Hoefert, 1966. Organization of beet yellows-virus inclusions in leaf cells of *Beta. Proc. Natl. Acad. Sci. U. S.* 55:486–493.

———, ———, and ———, 1967. Relation of beet yellows to the phloem and to movement in the sieve tube. *J. Cell Biol.* 32:71–87.

———, E. M. Engleman, and T. Bisalputra, 1963. What are transcellular strands? *Planta* 59:617–623.

Eschrich, W., 1956. Kallose. *Protoplasma* 47:487–530.

———, 1961. Untersuchungen über den Ab- und Aufbau der Callose, III. Mitteilung über Callose. *Z. Bot.* 49:153–218.

———, 1963a. Beziehungen Zwischen dem Auftreten vou Callose und der Feinstruktur des Primären Phloems bei *Cucurbita ficifolia. Planta* 59:243–261.

———, 1963b. Der Phloemsaft von *Cucurbita ficifolia. Planta* 60:216–224.

———, 1965. Physiologie der Siebröhrencallose. *Planta* 65:280–300.

———, 1966. Translokation ^{14}C-markierter Assimilate im Licht und im Dunkeln bei *Vicia faba. Planta* 70:99–124.

———, 1967. Bidirektionelle Translokation in Siebröhren. *Planta* 73:37–49.

———, 1968. Translokation radioaktiv markierter Indolyl-3-essigsäure in Siebröhren von *Vicia faba. Planta* 78:144–157.

——— and H. B. Currier, 1964. Identification of callose by its diachrome and fluorochrome reactions. *Stain Technol.* 39:303–307.

———, ———, S. Yamaguchi, and R. B. McNairn, 1965. Der Einfluss verstärkter Callosebildung auf den Stofftransport in Siebröhren. *Planta* 65:49–64.

———, B. Eschrich, and H. B. Currier, 1964. Historadiograpischen nachweis von Calcium-45 in Phloem von *Cucurbita maxima. Planta* 63:146–154.

——— and T. Hartmann, 1969. Translocation and biochemical behavior of D- and L-phenylalanine in *Vicia faba*. *Planta* 85:213–227.

——— and H. Kating, 1964. Aufnahme, Einbau und Transport von ^{14}C in *Cucurbita ficifolia*. *Planta* 60:523–539.

Evans, L. T., 1966. Abscisin II: Inhibitory effect on flower induction in a long-day plant. *Science* 151:107.

——— and I. F. Wardlaw, 1964. Inflorescence initiation in *Lolium temulentum* L., IV. Translocation of the floral stimulus in relation to that of assimilates. *Australian J. Biol. Sci.* 17:1–9.

——— and ———, 1966. Independent translocation of ^{14}C-labelled assimilates and of the floral stimulus in *Lolium temulentum*. *Planta* 68:310–326.

———, ———, and C. N. Williams, 1964. Environmental control of growth. In *Grasses and Grasslands*, ed. C. Barnard, pp. 102–125. London: Macmillan.

Evans, N. T. S., M. Ebert, and J. Moorby, 1963. A model for the translocation of photosynthate in the soybean. *J. Exptl. Bot.* 14:221–231.

Evert, R. F., 1962. Some aspects of phloem development in *Tilia americana*. *Am. J. Bot.* 49:659.

——— and F. J. Alfieri, 1965. Ontogeny and structure of coniferous sieve cells. *Am. J. Bot.* 52:1058–1066.

——— and W. F. Derr, 1964a. Callose substance in sieve elements. *Am. J. Bot.* 51: 552–559.

——— and ———, 1964b. Slime substance and strands in sieve elements. *Am. J. Bot.* 51:875–880.

———, W. Eschrich, J. T. Medler, and F. J. Alfieri, 1968. Observations on the penetration of linden branches by stylets of the aphid *Longistigma caryae*. *Am. J. Bot.* 55:860–874.

——— and L. Murmanis, 1965. Ultrastructure of the secondary phloem of *Tilia americana*. *Am. J. Bot.* 52:95–106.

———, ———, and I. B. Sachs, 1966. Another view of the ultrastructure of *Cucurbita* phloem. *Ann. Bot. (London)* (N.S.)30:563–585.

Falk, H., 1964. Zer Herkunft des Siebröhren schleimes bei *Tetragonia expansa* Murr. *Planta* 60:558–567.

Fensom, D. S., 1957. The bioelectric potentials of plants and their functional significance, I. An electrokinetic theory of transport. *Can. J. Bot.* 35:573–582.

———, 1958. The bioelectric potentials of plants and their functional significance, II. The patterns of bio-electric potential and exudation rate in excised sunflower roots and stems. *Can. J. Bot.* 36:367–383.

———, 1959. The bioelectric potentials of plants and their significance, III. The production of continuous potentials across membranes in plant tissue by the circulation of the hydrogen ion. *Can. J. Bot.* 37:1003–1026.

Fife, J. M., C. Price, and D. C. Fife, 1962. Some properties of phloem exudate collected from root of sugar beet. *Plant Physiol.* 37:791–792.

Fischer, A., 1885. Über den inhalt der Siebrohren in der unverletzter Pflanze. *Ber. Deut. Bot. Ges.* 3:230–239.

Fischer, H., 1967. Phloem transport and Stoffaufnahme. *Z. Pflanzenernär. Düng. Bodenk.* 118:100–111.

Fisher, D. B., 1970. Kinetics of C-[14] translocation in soybean, I. Kinetics in the stem. *Plant Physiol.* 45:107–113.

———, 1970. Kinetics of C-[14] translocation in soybean, II. Kinetics in the leaf. *Plant Physiol.* 45:114–118.

———, 1970. Kinetics of C-[14] translocation in soybean, III. Theoretical considerations. *Plant Physiol.* 45:119–125.

Fletcher, R. A., and S. Zalik, 1965. Effect of light of several spectral bands on the metabolism of radioactive IAA in bean seedlings. *Plant Physiol.* 40:549–552.

Ford, J., and A. J. Peel, 1966. The contributory length of sieve tubes in isolated segments of willow, and the effect on it of low temperatures. *J. Exptl. Bot.* 17:522–533.

——— and ———, 1967a. Preliminary experiments on the effect of temperature on the movement of ^{14}C-labelled assimilates through the phloem of willow. *J. Exptl. Bot.* 18:406–415.

——— and ———, 1967b. The movement of sugars into the sieve elements of bark strips of willow. *J. Exptl. Bot.* 18:607–619.

Forde, B. J., 1965. Differentiation and continuity of the phloem in the leaf intercalary meristem of *Lolium perenne*. *Am. J. Bot.* 52:953–961.

———, 1966a. Translocation in grasses, I. Bermuda grass. *New Zealand J. Bot.* 4:479–495.

———, 1966b. Translocation in grasses, II. Perennial ryegrass and couch grass. *New Zealand J. Bot.* 4:496–514.

Fox, J. E., 1966. Incorporation of a kinin, N,6-benzyladenine, into soluble RNA. *Plant Physiol.* 41:75–82.

Foy, C. L., 1961a. Absorption, distribution, and metabolism of 2,2-dichloropropionic acid in relation to phytotoxicity, I. Penetration and translocation of [36]Cl- and ^{14}C-labeled dalapon. *Plant Physiol.* 36:688–697.

———, 1961b. Absorption, distribution and metabolism of 2,2-dichloropropionic acid in relation to phytotoxicity, II. Distribution and metabolic fate of dalapon in plants. *Plant Physiol.* 36:698–709.

Fraenkel-Conrat, H., 1956. The role of the nucleic acid in the reconstitution of active tobacco mosaic virus. *J. Am. Chem. Soc.* 78:882–883.

Fujiwara, A., and M. Suzuki, 1961a. Effects of temperature and light on the translocation of photosynthetic products. *Tohoku J. Agr. Res.* 12:363–367.

——— and ———, 1961b. Relation between respiration and translocation of photosynthetic products. *Tohoku J. Agr. Res.* 12:369–373.

Gage, R. S., and S. Aronoff, 1960a. Translocation, III. Experiments with carbon 14, chlorine 36, and hydrogen 3. *Plant Physiol.* 35:53–64.

——— and ———, 1960b. Radioautography of tritiated photosynthate arising from HTO. *Plant Physiol.* 35:65–68.

Gardner, D. C. J., and A. J. Peel, 1969. ATP in sieve tube sap from willow. *Nature* 222:774.

Geiger, D. R., 1966. Effect of sink region cooling on translocation rate. *Plant Physiol.* 41(Suppl.):xx.

——— and J. W. Batey, 1967. Translocation of ^{14}C sucrose in sugar beet during darkness. *Plant Physiol.* 42:1743–1749.

——— and D. Cataldo, 1968. Relation of leaf structure to translocation. *Plant Physiol.* 43(Suppl.):35.

——— and ———, 1969. Leaf structure and translocation in sugar beet. *Plant Physiol.* 44:45–54.

——— and M. A. Saunders, 1969. Path parameters of a translocating sugar beet leaf. *Plant Physiol.* 44(Suppl.):22.

——— and C. A. Swanson, 1965a. Sucrose translocation in the sugar beet. *Plant Physiol.* 40:685–690.

——— and ———, 1965b. Evaluation of selected parameters in a sugar beet translocation system. *Plant Physiol.* 40:942–947.

Gierer, A., and G. Schramm, 1956. Infectivity of ribonucleic acid from tobacco mosaic virus. *Nature* 177:702–703.

Glasziou, K. T., 1960. Accumulation and tranformation of sugars in sugar cane stalks. *Plant Physiol.* 35:895–901.

———, 1961. Accumulation and transformation of sugars in stalks of sugar cane. Origin of glucose and fructose in the inner space. *Plant Physiol.* 36:175–179.

Goodman, R. N., 1962. The impact of antibiotics upon plant disease control. *Advan. Pest Control Res.* 5:1–46.

Gray, R. A., 1958. The downward translocation of antibiotics in plants. *Phytopathology* 48:71–78.

Greenham, C. G., 1962. Studies on translocation in skeleton weed (*Chondrilla juncea* L.). *Australian J. Agr. Res.* 13:624–637.

——— and A. G. Brown, 1957. The control of mistletoe by trunk injection. *J. Australian Inst. Agr. Sci.* 23:308–318.

Greenway, H., and A. Gunn, 1966. Phosphorus retranslocation in *Hordeum vulgare* during early tillering. *Planta* 71:43–67.

Gunning, B. E. S., and W. K. Barkley, 1963. Kinin-induced directed transport and senescence in detached oat leaves. *Nature* 199:262–265.

———, J. S. Pate, and L. G. Briarty, 1968. Specialized "transfer cells" in minor veins of leaves and their possible significance in phloem translocation. *J. Cell Biol.* 37:C7–C12.

Gustafson, F. G., and M. Darken, 1937. Further evidence for the upward movement of minerals through the phloem of stems. *Am. J. Bot.* 24:615–621.

Hale, V. Q., and A. Wallace, 1961. Translocation and retranslocation of C^{14}-labeled chelating agents in plants. *Proc. Am. Soc. Hort. Sci.* 78:597–604.

Halevy, A. H., S. P. Monselise, and Z. Plaut, 1964. Effects of gibberellin on translocation and on dry matter and water content in several plant species. *Physiol. Plantarum* 17:49–62.

Hammel, H. T., 1968. Measurement of turgor pressure and its gradient in the phloem of oak. *Plant Physiol.* 43:1042–1048.

Hanawa, J., 1968. Formation of double leaf by AMO 1,618 and 2,4-D in *Sesamum indicum. Shokubutsugaku Zasshi.* 81:545–555.

Hansen, P., 1967a. ^{14}C-studies on apple trees, I. The effect of the fruit on the translocation and distribution of photosynthates. *Physiol. Plantarum* 20:382–391.

———, 1967b. ^{14}C studies on apple trees, II. Distribution of photosynthates from top and base leaves from extension shoots. *Physiol. Plantarum* 20:720–725.

———, 1967c. ^{14}C studies on apple trees, III. The influence of season on storage and mobilization of labelled compounds. *Physiol. Plantarum* 20:1103–1111.

Happel, J., and H. Brenner, 1965. *Low Reynolds Number Hydrodynamics: With Special Applications to Particulate Media,* p. 153. Englewood Cliffs, N.J.: Prentice-Hall.

Harel, S., and L. Reinhold, 1966. The effect of 2,4-dinitrophenol on translocation in the phloem. *Physiol. Plantarum* 19:634–643.

Hartig, T., 1858. Über den Herbstsaft der Holzpflanzen. *Bot. Z.* 16:369–370.

Hartt, C. E., 1963. Translocation as a factor in photosynthesis. *Naturwissenschaften* 50:666–667.

———, 1965a. The effect of temperature upon translocation of ^{14}C in sugar cane. *Plant Physiol.* 40:774–781.

———, 1965b. Light and translocation of ^{14}C in detached blades of sugar cane. *Plant Physiol.* 40:718–724.

———, 1966. Translocation in colored light. *Plant Physiol.* 41:369–372.

———, 1967. Effect of moisture supply upon translocation and storage of ^{14}C in sugar cane. *Plant Physiol.* 42:338–346.

——— and H. P. Kortschak, 1963. Tracing sugar in the cane plant. *Proc. 11th Congr. Internatl. Soc. Sugar Chemists Technologists (Mauritius)*, pp. 323–334.

——— and ———, 1964. Sugar gradients and translocation of sucrose in detached blades of sugar cane. *Plant Physiol.* 39:460–474.

——— and ———, 1967. Translocation of ^{14}C in the sugar cane plant during the day and night. *Plant Physiol.* 42:89–94.

———, ———, and G. O. Burr, 1964. Effects of defoliation, deradication, and darkening the blade upon translocation of ^{14}C in sugar cane. *Plant Physiol.* 39:15–22.

Hatch, M. D., 1964. Sugar accumulation by sugar-cane storage tissue: the role of sucrose phosphate. *Biochem. J.* 94:521–526.

——— and K. T. Glasziou, 1963. Sugar accumulation cycle in sugar cane, II. Relationship of invertase activity to sugar content and growth rate in storage tissue of plants grown in controlled environments. *Plant Physiol.* 38:344–348.

——— and ———, 1964. Direct evidence for translocation of sucrose in sugar cane leaves and stems. *Plant Physiol.* 39:180–184.

———, J. A. Sacher, and K. T. Glasziou, 1963. Sugar accumulation cycle in sugar cane, I. Studies on enzymes of the cycle. *Plant Physiol.* 38:338–343.

Hawker, J. S., 1965. The sugar content of cell walls and intercellular spaces in sugar cane stems and its relation to sugar transport. *Australian J. Biol. Sci.* 18:959–969.

——— and M. D. Hatch, 1965. Mechanism of sugar storage by mature stem tissue of sugar cane. *Physiol. Plantarum* 18:444–453.

Hendrix, J. E., 1968. Labeling pattern of translocated stachyose in squash. *Plant Physiol.* 43:1631–1636.

Hepton, C. E. L., and R. D. Preston, 1960. Electron microscopic observations of the structure of sieve-connections in the phloem of angiosperms and gymnosperms. *J. Exptl. Bot.* 11:381–394.

———, ———, and G. W. Ripley, 1955. Electron microscopic observations on the structure of the sieve plates in Cucurbita. *Nature* 176:868–870.

Hew, C. S., C. D. Nelson, and G. Krotkov, 1967. Hormonal control of translocation of photosynthetically assimilated ^{14}C in young soybean plants. *Am. J. Bot.* 54:252–256.

Hewitt, A. A., 1963. The effect of boron sprays on the translocation and distribution of endogenous carbohydrates as measured by ^{14}C sucrose in the York Imperial apple tree. *Dissertation Abstr.* 23(a):3063.

Heyser, W., W. Eschrich, and R. F. Evert, 1969. Translocation in perennial monocotyledons. *Science* 164:572–574.

Hill, A. W., 1908. The histology of the sieve tubes of Angiosperms. *Ann. Bot.* (*London*) 22:245–290.

Hill, G. P., 1960. Seasonal variation in sieve-tube activity and sugar content of sieve-tube sap in *Tilia americana*. *Plant Physiol.* 35(Suppl):iv.

———, 1962. Exudation from aphid stylets during the period from dormancy to bud break in *Tilia americana* (L.). *J. Exptl. Bot.* 13:144–151.

———, 1963. The sources of sugars in sieve-tube sap. *Ann. Bot.* (*London*) (N.S.)27: 79–87.

Hoad, G. V., 1967. (+)-abscisin II, (+)-dormin in phloem exudate of willow. *Life Sci.* 6:1113–1118.

——— and M. R. Bowen, 1968. Evidence for gibberellin-like substances in phloem exudate of higher plants. *Planta* 82:22–32.

——— and A. J. Peel, 1965a. Studies on the movement of solutes between the sieve tubes and surrounding tissues in willow, I. Interference between solutes and rate of translocation measurements. *J. Exptl. Bot.* 16:433–451.

——— and ———, 1965b. Studies on the movement of solutes between the sieve tubes and the surrounding tissues in willow, II. Pathways of ion transport from the xylem to the phloem. *J. Exptl. Bot.* 16:742–758.

Hodges, H. F., 1969. The basis for resistance of plants to subfreezing temperatures. In *Physiological Limitations on Crop Production under Temperature and Moisture Stress* (Rept. Res. Planning Conf., San Jose, Costa Rico, 1967), p. 8. Washington, D.C.: National Academy of Sciences.

Holt, E. C., J. L. Faubion, W. W. Allen, and G. G. McBee, 1967. Arsenic translocation in nutsedge tuber systems and its effect on tuber viability. *Weeds* 15:13–15.

Honert, T. H. van den, 1932. On the mechanism of the transport of organic materials in plants. *Proc. Koninkl. Akad. Wetenschap. Amsterdam* 35:1104–1111.

Horwitz, L., 1958. Some simplified mathematical treatments of translocation in plants. *Plant Physiol.* 33:81–93.

Huber, B., 1937. Methoden, Ergebnisse und Probleme der neuen Baumphysiologie. *Ber. Deut. Bot. Ges.* 55:46–62.

———, 1942. Die Siebröhren der Pflanzen als Nahrungsquelle frem der Organismen und als Transport bahnen von Krankheitskeimen. *Biol. Gen.* 16:310–343.

——— and W. Liese, 1963. Studies on the ultrastructure of the secondary phloem. *Protoplasma* 57:429–439.

——— and E. Rouschal, 1938. Anatomische und zellphysiologische Beobachtungen am Siebröhrensystem der Bäume. *Ber. Deut. Bot. Ges.* 56:380–391.

———, E. Schmidt, and H. Jahnel, 1937. Untersuchungen über den Assimilatström,

I. Mitteilung aus der sächsischen forstlichen Versuchsanstalt Tharandt, Abteilung für Botanik. *Tharandter. Forstl. Jahrb.* 88:1017–1050.

Hull, H. M., 1960. A tabular summary of research dealing with translocation of foliar applied herbicides and selected growth regulators. *Weeds* 8:214–231.

Hull, R. J., 1969. Translocation of assimilates and dalapon in established Johnson grass. *Weed Sci.* 17:314–320.

——— and M. R. Weisenberg, 1967. Translocation and metabolism of dicamba in *Sorghum halepense* and *Phaseolus vulgaris*. *Plant Physiol.* 42(Suppl.):49.

Humphries, E. C., 1968. Responses of crop plants to growth regulators. *Soc. Chem. Ind. (London) Monograph* (31):251–258.

Husain, A., and D. C. Spanner, 1966. The influence of varying concentrations of applied sugar on the transport of tracers in cereal leaves. *Ann. Bot. (London)* (N.S.)30: 549–561.

Ie, T. A., P. M. L. Tammes, and J. van Die, 1966. Studies on phloem exudation from *Yucca flaccida* Haw., V. Electron microscopy of sieve plate pores. *Koninkl. Ned. Akad. Wetenschap. Proc.* (C)69:661–663.

Itai, C., and Y. Vaadia, 1965. Kinetin-like activity in root exudate of water-stressed sunflower plants. *Physiol. Plantarum* 18:941–944.

Jansen, L. L., W. A. Gentner, and W. C. Shaw, 1961. Effects of surfactants on the herbicidal activity of several herbicides in aqueous spray systems. *Weeds* 9:381–405.

Jeffrys, R. A., V. Q. Hale, and A. Wallace, 1961. Uptake and translocation in plants of labeled iron and labeled chelating agents. *Soil Sci.* 92:268–273.

Johnson, N. E., and J. G. Zingg, 1967. Effective translocation of four systemic insecticides following application to the foliage and cones of Douglas fir. *J. Econ. Entomol.* 60:575–578.

Johnson, R. P. C., 1966. Potassium permanganate fixation and the electron microscopy of sieve tube contents. *Planta* 68:36–43.

———, 1968. Microfilaments in pores between frozen-etched sieve elements. *Planta* 81:314–332.

———, 1969. Crystalline fibrils and complexes of membranes in the parietal layer in sieve elements. *Planta* 84:68–80.

Jones, H., and J. E. Eagles, 1962. Translocation of ^{14}Carbon within and between leaves. *Ann. Bot. (London)*(N.S.)26:505–510.

———, R. V. Martin, and H. K. Porter, 1959. Translocation of ^{14}Carbon in tobacco following assimilation of ^{14}Carbon dioxide by a single leaf. *Ann. Bot. (London)* (N.S.)23:493–510.

Joy, K. W., 1962. Transport of organic nitrogen through the phloem in sugar beet. *Nature* 195:618–619.

———, 1964. Translocation in sugar beet, I. Assimilation of $^{14}CO_2$ and distribution of materials from leaves. *J. Exptl. Bot.* 15:485–494.

———, 1967. Carbon and nitrogen sources for protein synthesis and growth of sugar beet leaves. *J. Exptl. Bot.* 18:140–150.

——— and A. J. Anticliff, 1966. Translocation of amino acids in sugar beets. *Nature* 211:210–211.

Kasprzyk, Z., Z. Wojceichowski, and K. Czerniakowska, 1968. The transport of oleanolic acid in *Calendula officinalis*. *Physiol. Plantanum* 21:966–970.

Kating, H., and W. Eschrich, 1964. Uptake, incorporation, and transport of ^{14}C in *Cucurbita ficifolia*, II. Application of bicarbonate ^{14}C to the roots. *Planta* 60:598–611.

Keen, N. T., and P. H. Williams, 1969. Translocation of sugars into infected cabbage tissues during clubroot development. *Plant Physiol.* 44:748–754.

Kendall, W. A., 1955. Effect of certain metabolic inhibitors on translocation of ^{32}P in bean plants. *Plant Physiol.* 30:347–350.

Kende, H., 1965. Kinetin-like factors in the root exudate of sunflowers. *Proc. Nat. Acad. Sci. U. S.* 53:1302–1307.

Kennecke, M., 1969. Über Enzyme in Siebröhren-säften. Doctoral dissertation, Darmstadt Technical University.

Kennedy, J. S., and T. E. Mittler, 1953. A method for obtaining phloem sap via the mouth-parts of aphids. *Nature* 171:528.

Kessler, B., and Z. W. Moscicki, 1958. Effect of triiodobenzoic acid and maleic hydrazide upon the transport of foliar-applied calcium and iron. *Plant Physiol.* 33: 70–72.

Key, J. L., 1969. Hormones and nucleic acid metabolism. *Ann. Rev. Plant Physiol.* 20: 449–474.

Khan, A. A., and G. R. Sagar, 1966. Distribution of ^{14}C-labelled products of photosynthesis during the commercial life of the tomato crop. *Ann. Bot. (London)* (N.S.)30: 727–743.

King, R. W., L. T. Evans, and I. F. Wardlaw, 1968. Translocation of floral stimulus in *Pharbitis nil* in relation to that of assimilates. *Z. Pflanzenphysiol.* 59:377–388.

———, I. F. Wardlaw, and L. T. Evans, 1967. Effect of assimilate utilization on photosynthetic rate in wheat. *Planta* 77:261–276.

Kluge, Helga, 1967. Untersuchungen über Kohlenhydrate und Myo-inosit in Siebröhren-saften van Holz. Doctoral dissertation, Darmstadt Technical University.

Kluge, M., 1967. Viruspartikel im Siebröhrensaft von *Cucumis sativus* L. nach infektion durch das Cucumis Virus 2A. *Planta* 73:50–61.

———, E. Reinhard, and H. Ziegler, 1964. Gibberellinaktivität von Siebröhren-säften. *Naturwissenschaften* 51:145–146.

———, and H. Ziegler, 1964. Der ATP-gehalt der Siebröhrensäfte von Laubbäumen. *Planta* 61:167–177.

Kollmann, R., 1960a. Untersuchungen über das Protoplasma der Siebröhren von *Passiflora coerulea*, I. Lichtoptische Untersuchungen. *Planta* 54:611–640.

———, 1960b. Untersuchungen über das Protoplasma der Siebröhren von *Passiflora coerulea*, II. Elektronenoptische Untersuchungen. *Planta* 55:67–107.

———, 1963. Fine structure in sieve tubes. *Progr. Biophys. Biophys. Chem.* 13:244–246.

———, 1964. On the fine structure of the sieve element protoplast. *Phytomorphology* 14:247–264.

———, 1965. Localization of the functioning sieve cells in secondary phloem tissue of *Metasequoia glyptostroboides*. *Planta* 65:173–179.

———, 1967. Autoradiographic evidence of the assimilate-conducting elements in the secondary phloem of *Metasequoia glyptostroboides*. *Z. Pflanzenphysiol.* 56: 401–409.

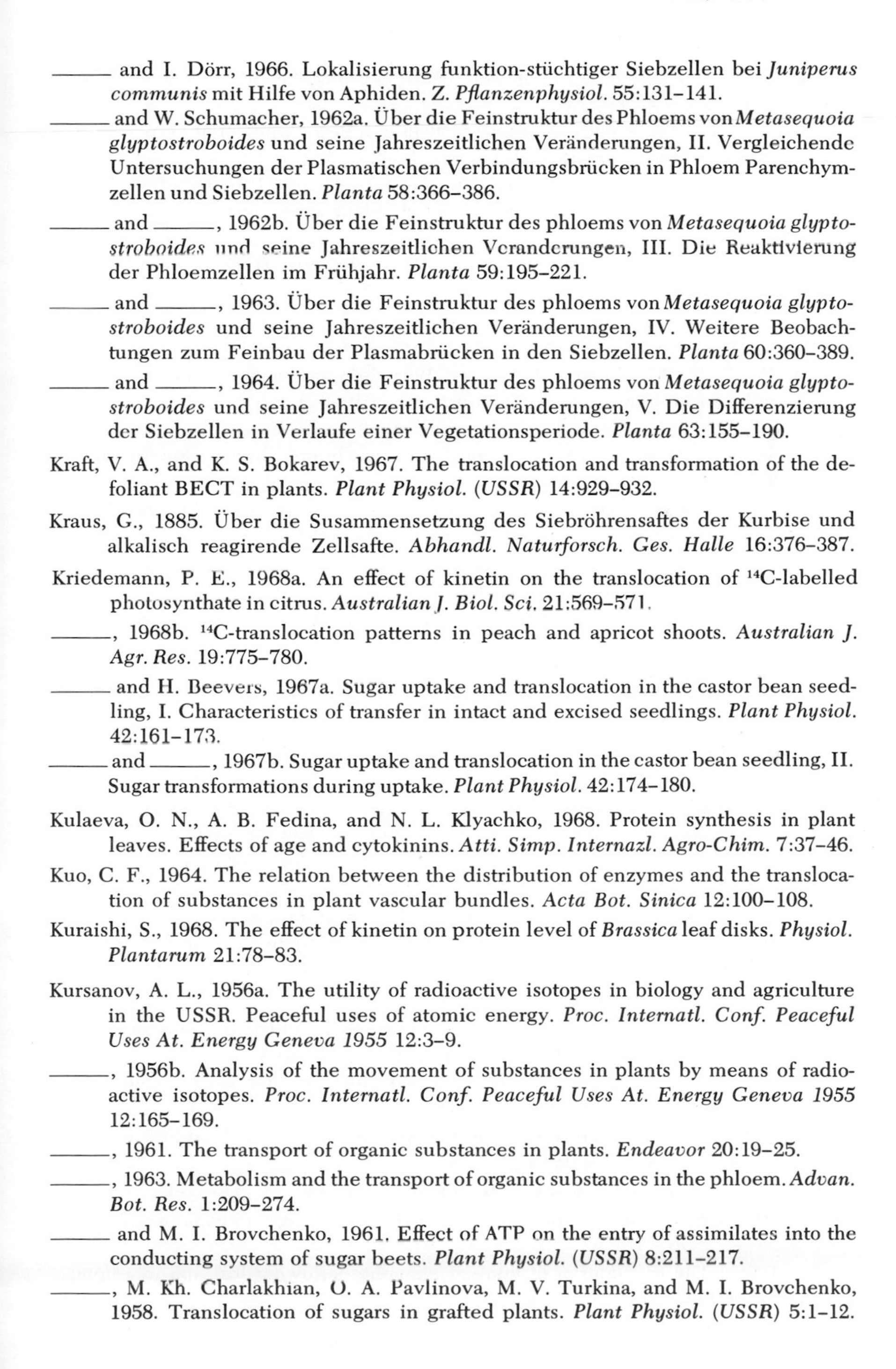

——— and I. Dörr, 1966. Lokalisierung funktion-stüchtiger Siebzellen bei *Juniperus communis* mit Hilfe von Aphiden. *Z. Pflanzenphysiol.* 55:131–141.

——— and W. Schumacher, 1962a. Über die Feinstruktur des Phloems von *Metasequoia glyptostroboides* und seine Jahreszeitlichen Veränderungen, II. Vergleichende Untersuchungen der Plasmatischen Verbindungsbrücken in Phloem Parenchymzellen und Siebzellen. *Planta* 58:366–386.

——— and ———, 1962b. Über die Feinstruktur des phloems von *Metasequoia glyptostroboides* und seine Jahreszeitlichen Verandcrungen, III. Die Reaktivierung der Phloemzellen im Frühjahr. *Planta* 59:195–221.

——— and ———, 1963. Über die Feinstruktur des phloems von *Metasequoia glyptostroboides* und seine Jahreszeitlichen Veränderungen, IV. Weitere Beobachtungen zum Feinbau der Plasmabrücken in den Siebzellen. *Planta* 60:360–389.

——— and ———, 1964. Über die Feinstruktur des phloems von *Metasequoia glyptostroboides* und seine Jahreszeitlichen Veränderungen, V. Die Differenzierung der Siebzellen in Verlaufe einer Vegetationsperiode. *Planta* 63:155–190.

Kraft, V. A., and K. S. Bokarev, 1967. The translocation and transformation of the defoliant BECT in plants. *Plant Physiol.* (*USSR*) 14:929–932.

Kraus, G., 1885. Über die Susammensetzung des Siebröhrensaftes der Kurbise und alkalisch reagirende Zellsafte. *Abhandl. Naturforsch. Ges. Halle* 16:376–387.

Kriedemann, P. E., 1968a. An effect of kinetin on the translocation of ^{14}C-labelled photosynthate in citrus. *Australian J. Biol. Sci.* 21:569–571.

———, 1968b. ^{14}C-translocation patterns in peach and apricot shoots. *Australian J. Agr. Res.* 19:775–780.

——— and H. Beevers, 1967a. Sugar uptake and translocation in the castor bean seedling, I. Characteristics of transfer in intact and excised seedlings. *Plant Physiol.* 42:161–173.

——— and ———, 1967b. Sugar uptake and translocation in the castor bean seedling, II. Sugar transformations during uptake. *Plant Physiol.* 42:174–180.

Kulaeva, O. N., A. B. Fedina, and N. L. Klyachko, 1968. Protein synthesis in plant leaves. Effects of age and cytokinins. *Atti. Simp. Internazl. Agro-Chim.* 7:37–46.

Kuo, C. F., 1964. The relation between the distribution of enzymes and the translocation of substances in plant vascular bundles. *Acta Bot. Sinica* 12:100–108.

Kuraishi, S., 1968. The effect of kinetin on protein level of *Brassica* leaf disks. *Physiol. Plantarum* 21:78–83.

Kursanov, A. L., 1956a. The utility of radioactive isotopes in biology and agriculture in the USSR. Peaceful uses of atomic energy. *Proc. Internatl. Conf. Peaceful Uses At. Energy Geneva 1955* 12:3–9.

———, 1956b. Analysis of the movement of substances in plants by means of radioactive isotopes. *Proc. Internatl. Conf. Peaceful Uses At. Energy Geneva 1955* 12:165–169.

———, 1961. The transport of organic substances in plants. *Endeavor* 20:19–25.

———, 1963. Metabolism and the transport of organic substances in the phloem. *Advan. Bot. Res.* 1:209–274.

——— and M. I. Brovchenko, 1961. Effect of ATP on the entry of assimilates into the conducting system of sugar beets. *Plant Physiol.* (*USSR*) 8:211–217.

———, M. Kh. Charlakhian, O. A. Pavlinova, M. V. Turkina, and M. I. Brovchenko, 1958. Translocation of sugars in grafted plants. *Plant Physiol.* (*USSR*) 5:1–12.

Lagerstedt, H. B., and R. G. Langston, 1967. Translocation of radioactive kinetin. *Plant Physiol.* 42:611–622.

Lawton, J. R. S., 1966. A note on callose distribution in the phloem of Dioscoreaceae. *Z. Pflanzenphysiol.* 55:287–291.

———, 1967a. Translocation in the phloem of *Dioscorea* spp., I. Distribution of radioactive photosynthates as shown by whole-plant autoradiographs. *Z. Pflanzenphysiol.* 58:1–7.

———, 1967b. Translocation in the phloem of *Dioscorea* spp., II. Distribution of translocates in the stem. *Z. Pflanzenphysiol.* 58:8–16.

——— and O. Biddulph, 1964. The accumulation of translocated material at the sieve plate of *Dioscorea sansibarensis* Pax. *J. Exptl. Bot.* 15:201–204.

Lee, K. W., C. M. Whittle, and H. J. Dyer, 1966. Boron deficiency and translocation profiles in sunflower. *Physiol. Plantarum* 19:919–924.

Lenton, J. R., M. R. Bowen, and P. F. Saunders, 1968. Detection of abscisic acid in the xylem sap of willow (*Salix viminalis*) by gas-liquid chromatography. *Nature* 220:86–87.

Leonard, O. A., D. E. Bayer, and R. K. Glenn, 1966. Translocation of herbicides and assimilates in red maple and white ash. *Bot. Gaz.* 127:193–201.

——— and A. S. Crafts, 1956. Translocation of herbicides, III. Uptake and distribution of radioactive 2,4-D by brush species. *Hilgardia* 26:366–415.

——— and R. K. Glenn, 1968a. Translocation of herbicides in detached bean leaves. *Weed Sci.* 16:352–356.

——— and ———, 1968b. Translocation of assimilates and phosphate in detached bean leaves. *Plant Physiol.* 43:1380–1388.

——— and R. J. Hull, 1965. Translocation relationships in and between mistletoes and their hosts. *Hilgardia* 37:115–153.

——— and ———, 1966. Translocation of ^{14}C-labelled substances and $^{32}PO_4$ in mistletoe-infected and uninfected conifers and dicotyledonous trees. In *Isotopes in Weed Research,* pp. 31–46. Vienna: International Atomic Energy Agency.

——— and D. L. King, 1968. Vein loading and transport in detached leaves. *Plant Physiol.* 43:460–463.

———, L. A. Lider, and R. K. Glenn. 1966. Absorption and translocation of herbicides by Thompson Seedless (Sultanina) grape, *Vitis Vinifera* L. *Weed Res.* 6:37–49.

——— and R. J. Weaver, 1961. Absorption and translocation of 2,4-D and amitrole in shoots of the Tokay grape. *Hilgardia* 31:327–368.

———, ———, and R. K. Glenn, 1967. Effect of 2,4-D and picloram on translocation of C^{14}-assimilates in *Vitis vinifera* L. *Weed Res.* 7:208–219.

Lerch, G., 1960. Untersuchungen über Wurzelkallose. In *Botanische Studien,* ed. W. Troll, vol. 2, p. 3. Jena: Gustav Fischer.

Lester, H. H., and R. F. Evert, 1965. Acid-phosphatase activity in sieve-tube members of *Tilia americana. Planta* 65:180–185.

Levi, E., 1962. An artifact in plant autoradiography. *Science* 137:343–344.

———, 1966. Uptake and distribution of ^{134}Cs applied to leaves of bean plants. *Radiation Bot.* 6:567–574.

———, 1968a. The distribution of mineral elements following leaf and root uptake. *Physiol. Plantarum* 21:213–226.

———. 1968b. Losses through roots of foliar applied mineral elements. *Naturwissen-.aften* 42:1–2.

———, 1969. The penetration and adsorption of cesium in bean leaves. *Acta Bot. Neerl.* 18:463–469.

Linder, P. J., J. W. Brown, and J. W. Mitchell, 1949. Movement of externally applied phenoxy compounds in bean plants in relation to conditions favoring carbohydrate translocation, *Bot. Gaz.* 110:628–632.

Little, E. C. S., and G. E. Blackman, 1963. The movement of growth regulators in plants, III. Comparative studies of transport in *Phaseolus vulgaris*. *New Phytologist* 62:173–197.

Livingston, L. G., 1964. The nature of plasmodesmata in normal (living) plant tissue. *Am. J. Bot.* 51:950–957.

Lucier, G. W., and R. E. Menzer, 1968. Metabolism of dimethoate in bean plants in relation to its mode of application. *J. Agr. Food Chem.* 16:936–945.

Luckwill, L. C., and C. P. Lloyd-Jones, 1962. The absorption, translocation and metabolism of 1-naphthaleneacetic acid applied to apple leaves. *J. Hort. Sci.* 37:190–206.

Lund-Høie, K., and D. E. Bayer, 1968. Absorption, translocation and metabolism of 3-amino-1,2,4-triazole in *Pinus ponderosa* and *Abies concolor*. *Physiol. Plantarum* 21:196–212.

Lupton, F. G. H., 1966. Translocation of photosynthetic assimilates in wheat. *Ann. Appl. Biol.* 57:355–365.

Macchia, F., 1967. Effect of growth retardants on the structure of pea seedlings cultured in nutrient solution. *Giorn. Bot. Ital.* 101:361–390.

Maestri, M., 1967. Structural and functional effects of endothall on plants. Doctoral dissertation, University of California, Davis.

Magalhaes, A. C., F. M. Ashton, and C. L. Foy, 1968. Translocation and fate of dicamba in purple nutsedge. *Weed Sci.* 16:240–245.

Maier, C. R., 1960. Streptomycin absorption, translocation and retention in hops. *Phytopathology* 50:351–356.

Mangham, S., 1917. On the mechanism of translocation in plant tissues. An hypothesis with special reference to sugar conduction in sieve tubes. *Ann. Bot.* (*London*) 31:293–311.

Maramorosch, K., E. Shikata, and R. R. Granados, 1968. Structures resembling mycoplasma in diseased plants and in insect vectors. *Trans. N. Y. Acad. Sci.* (2)30: 841–855.

Marshall, C., 1967. The use of radioisotopes to investigate organization in plants with special reference to the grass plant. In *Isotopes in Plant Nutrition and Physiology*, pp. 203–216. Vienna: International Atomic Energy Agency.

——— and G. R. Sagar, 1965. The influence of defoliation on the distribution of assimilates in *Lolium multiflorum* Lam. *Ann. Bot.* (*London*) (N.S.)29:365–370.

Mason, G. W., 1960. The absorption, translocation, and metabolism of 2,3,6-trichlorobenzoic acid in plants. Doctoral dissertation, University of California, Davis.

Mason, T. G., 1926. Preliminary note on the physiological aspects of certain undescribed structures in the phloem of greater yam *Dioscorea alata* Linn. *Sci. Proc. Roy. Dublin Soc.* 18:195–198.

——— and C. J. Lewin, 1926. On the rate of carbohydrate transport in greater yam, *Dioscorea alata* Linn. *Sci. Proc. Roy. Dublin Soc.* 18:203–205.

——— and E. J. Maskell, 1928a. Studies on the transport of carbohydrates in the cotton plant, I. A study of diurnal variation in the carbohydrates of leaf, bark, and wood and of the effects of ringing. *Ann. Bot.* (*London*) 42:189–253.

——— and ———, 1928b. Studies on the transport of carbohydrates in the cotton plant, II. The factors determining the rate and the direction of movement of sugars. *Ann. Bot.* (*London*) 42:571–636.

———, ———, and E. Phillis, 1936. Concerning the independence of solute movement in the phloem. *Ann. Bot.* (*London*) 50:23–58.

——— and E. Phillis, 1936. On the simultaneous movement of solutes in opposite directions through the phloem. *Ann. Bot.* (*London*) 50:161–174.

——— and ———, 1937. The migration of solutes. *Bot. Rev.* 3:47–71.

Mayer, A., and H. K. Porter, 1960. Translocation from leaves of rye. *Nature* 188:921–922.

McNairn, R. B., 1967. The influence of elevated temperature on axial translocation and callose formation in cotton hypocotyl. Doctoral dissertation, University of California, Davis.

——— and H. B. Currier, 1965. The influence of boron on callose formation in primary leaves of *Phaseolus vulgaris* L. *Phyton* (*Buenos Aires*) 22:153–158.

——— and ———, 1967. Sieve plate callose. A factor in blockage of axial phloem transport. *Naturwissenschaften* 22:591.

——— and ———, 1968. Translocation blockage by sieve plate callose. *Planta* 82:369–380.

Mehta, A. S., and D. C. Spanner, 1962. The fine structure of the sieve tubes of the petiole of *Nymphoides peltatum* (Gmel.) O. Kunze. *Ann. Bot.* (*London*) (N.S.)26:291–299.

Meikle, R. W., 1968. Metabolism of nellite nematocide (phenyl-*N,N'*-dimethyl phosphordiamidate) in cucumber plants. *J. Agr. Food Chem.* 16:928–935.

Merkle, M. G., and F. S. Davis, 1967. Effect of moisture stress on absorption and movement of picloram and 2,4,5-T in beans. *Weeds* 15:10–12.

Metcalf, R. L., 1966. Absorption and translocation of systemic insecticides. In *Transporto della molecole organiche nelle piante*, pp. 1–20. Vienna: 6th International Symposium on Agrochemistry.

———, 1967. Absorption and translocation of systemic insecticides. *Agrochemica* 11:105–123.

Meyer-Mevius, U., 1959. Vorkommen und Transport von Kohlenhydraten und Stickstoffverbindungen in den pflanzlichen Leitungsbahnen. *Flora* 147:553–594.

Milborrow, B. V., 1968. Identification and measurement of (+)-abscisic acid in plants. In *Biochemistry and Physiology of Plant Growth Substances,* ed. F. Wightman and G. Setterfield, pp. 1531–1545. Ottawa: The Runge Press.

Milthorpe, F. L., and J. Moorby, 1969. Vascular transport and its significance in plant growth. *Ann. Rev. Plant Physiol.* 20:1117–138.

Mitchell, J. W., and J. W. Brown, 1946. Movement of 2,4-dichlorophenoxyacetic acid stimulus and its relation to the translocation of organic food materials in plants. *Bot. Gaz.* 107:393–407.

———, I. R. Schneider, and H. G. Gauch, 1960. Translocation of particles within plants.

The translocation systems of plants can move particles that vary in size from the ionic to the macromolecular. *Science* 131:1863–1870.

______, B. C. Smale, and R. L. Metcalf, 1960. Absorption and translocation of regulators and compounds used to control plant diseases and insects. *Advan. Pest Control Res.* 3:359–436.

______ and J. F. Worley, 1964. Intracellular transport apparatus of phloem fibers. *Science* 145:409–410.

Mittler, T. E., 1953. Amino-acids in phloem sap and their excretion by aphids. *Nature* 172:207.

______, 1957. Studies on the feeding and nutrition of *Tuberolachnus salignus* (Gmelin), I. The uptake of phloem sap. *J. Exptl. Biol.* 34:334–341.

______, 1958. Studies on the feeding and nutrition of *Tuberolachnus salignus* (Gmelin), II. The nitrogen and sugar composition of ingested phloem sap and excreted honeydew. *J. Exptl. Biol.* 35:74–84.

______, 1959. Aphids and their supply of sieve tube sap. *Proc. Internatl. Bot. Congr. 9th* 2:266.

______, 1967. Flow relationships for hemipterous stylets. *Ann. Entomol. Soc. Am.* 60: 1112–1114.

Mokronosov, A. T., and N. B. Bubenshchikova, 1962. Translocation of assimilates in potato plants. *Plant Physiol. (USSR)* 8:447–454.

Moorby, J., 1964. The foliar uptake and translocation of caesium. *J. Exptl. Bot.* 15: 457–469.

______, M. Ebert, and N. T. S. Evans, 1963. The translocation of C^{11}-labelled photosynthate in the soyabean. *J. Exptl. Bot.* 14:210–220.

Moore, T. C., 1968. Translocation of the growth retardant *N,N*-dimethylaminosuccinamic acid-^{14}C. *Bot. Gaz.* 129:280–285.

Moose, C. A., 1938. Chemical and spectroscopic analysis of plants. *Plant Physiol.* 13: 365–380.

Morretes, B. L. de, 1962. Terminal phloem in vascular bundles of leaves of *Capsicum annum* and *Phaseolus vulgaris*. *Am. J. Bot.* 49:560–567.

Morris, D. A., and E. E. Thomas, 1968. Distribution of ^{14}C-labelled sucrose in seedlings of *Pisum sativum* L. treated with indoleacetic acid and kinetin. *Planta* 83:276–281.

Mortimer, D. C., 1965. Translocation of the products of photosynthesis in sugar beet petioles. *Can. J. Bot.* 43:269–280.

Morton, H. L., 1966. Influence of temperature and humidity on foliar absorption, translocation and metabolism of 2,4,5-T by mesquite seedlings. *Weeds* 14:136–141.

Mothes, K., and L. Engelbrecht, 1961. Kinetin-induced directed transport of substances in excised leaves in the dark. *Phytochemistry* 1:58–62.

______, ______, and H. R. Schütte, 1961. Über die Akkumulation von alpha-aminoisobuttersäure im Blattgewebe unter dem Einfluss von Kinetin. *Physiol. Plantarum* 14:72–75.

Muller, K., and A. C. Leopold, 1966a. Correlative aging and transport of ^{32}P in corn leaves under the influence of kinetin. *Planta* 68:167–185.

______ and ______, 1966b. The mechanism of kinetin-induced transport in corn leaves. *Planta* 68:186–205.

Münch, E., 1927. Dynamik der Saftströmungen. *Ber. Deut. Bot. Ges.* 44:68–71.

———, 1930. *Die Stoffbewegungen in der Pflanze.* Jena: Gustav Fischer.

Murmanis, L., and R. F. Evert, 1966. Some aspects of sieve cell ultrastructure in *Pinus strobus. Am. J. Bot.* 53:1065–1078.

Nageli, C., 1861. Über die Siebröhren von *Cucurbita. Sitzber. Königl. Bayer. Akad. Wiss. München* 1:212–238.

Nakata, S., and A. C. Leopold, 1967. Radioautographic study of translocation in bean leaves. *Am. J. Bot.* 54:769–772.

Nalewaja, J. D., 1968. Uptake and translocation of diallate in wheat, barley, flax and wild oat. *Weed Sci.* 16:309–312.

Neales, T. F., and L. D. Incoll, 1968. The control of leaf photosynthesis rate by the level of assimilate concentration in the leaf: a review of the hypothesis. *Bot. Rev.* 34:107–125.

Neeracher, H., 1966. Transportuntersuchungen an *Zea mays* mit Hilfe von THO und Mikroautoradiographie. *Ber. Schweiz. Bot. Ges.* 75:303–342.

Nelson, C. D., 1962. The translocation of organic compounds in plants. *Can. J. Bot.* 40:757–770.

———, 1963. Effect of climate on the distribution and translocation of assimilates. In *Environmental Control of Plant Growth,* ed. L. T. Evans, pp. 149–174. New York: Academic Press.

———, 1964. The production and translocation of photosynthate-^{14}C in conifers. In *The Formation of Wood in Forest Trees,* ed. M. H. Zimmerman, pp. 243–257. New York: Academic Press.

———, H. F. Perkins, and P. R. Gorham, 1958. Note on a rapid translocation of photosynthetically assimilated ^{14}C out of the primary leaf of the young soybean plant. *Can. J. Biochem. Physiol.* 36:1277–1279.

Newhall, W. F., and A. P. Pieringer, 1967. Derivatives of (+)-limonene. Detection and translocation of quaternary ammonium plant growth retardants in young grape fruit and bean seedlings. *J. Agr. Food Chem.* 15:488–491.

Norris, L. A., and V. H. Freed, 1966a. The absorption and translocation characteristics of several phenoxyalkyl acid herbicides in bigleaf maple. *Weed Res.* 6:203–211.

——— and ———, 1966b. The metabolism of a series of chlorophenoxyalkyl acid herbicides in bigleaf maple. *Weed Res.* 6:212–220.

——— and ———, 1966c. The absorption, translocation and metabolism characteristics of 4-(2,4-dichlorophenoxy) butyric acid in bigleaf maple. *Weed Res.* 6:283–291.

Northcote, D. H., and F. B. P. Wooding, 1966. Development of sieve tubes in *Acer pseudoplatanus. Proc. Roy, Soc. (London)* (B)163:524–536.

Ogawa, J. M., B. T. Manji, and E. Bose, 1968. Efficacy of fungicide 1991 in reducing fruit rot of stone fruits. *Plant Disease Reptr.* 52:722–726.

Othlinghaus, D., K. Schmitz, and J. Willenbrink, 1968. Zum Assimilattransport in wachsende Fruchte von *Phaseolus vulgaris. Planta* 80:89–95.

Pallas, J. E., Jr., 1960. Effects of temperature and humidity on foliar absorption and translocation of 2,4-dichlorophenoxyacetic acid and benzoic acid. *Plant Physiol.* 35:575–580.

———, 1963. Absorption and translocation of the triethylamine salt of 2,4-D and 2,4,5-T in four woody species. *Forest Sci.* 9:485–491.

——— and A. S. Crafts, 1957. Critical preparation of plant material for radioautography. *Science* 125:192–193.

——— and G. G. Williams, 1962. Foliar absorption and translocation of ^{32}P and 2,4-dichlorophenoxyacetic acid as affected by soil-moisture tension. *Bot. Gaz.* 123:175–180.

Palmquist, E. M., 1938. The simultaneous movement of carbohydrates and fluorescein in opposite directions in the phloem. *Am. J. Bot.* 25:97–105.

Panak, H., and R. Cz. Szafranck, 1967. Translocation of foliar and root applied sulfur-35 in rape and bean plants. *Acta Agrobot.* (*Warsaw*) 20:143–152.

Parker, J., 1964a. Transcellular strands and intercellular particle movement in sieve tubes of some common tree species. *Naturwissenschaften* 51:273–274.

———, 1964b. Sieve tube strands in tree bark. *Nature* 202:926–927.

———, 1965. Strand characteristics in sieve tubes of some common tree species. *Protoplasma* 60:86–93.

——— and D. E. Philpott, 1961. The ultrastructure of sieve plates of *Macrocystis pyrifera. Bull. Torrey Bot. Club* 88:85–90.

Parthasarathy, M. V., and P. B. Tomlinson, 1967. Anatomical features of metaphloem in stems of *Sabal, Cocos* and two other palms. *Am. J. Bot.* 54:1143–1151.

Pate, J. S., 1966. Photosynthesizing leaves and nodulated roots as donors of carbon to protein of the shoot of the field pea (*Pisum arvense* L.). *Ann. Bot.* (*London*) (N.S.)30:93–109.

——— and B. E. S. Gunning, 1969. Perivascular transfer cells in angiosperm leaves—a taxonomic and morphological survey. *Protoplasma* 68:135–156.

Peacock, F. C., 1966. Nematode control by plant chemotherapy. *Nematologica* 12:70–86.

Peel, A. J., 1963. The movement of ions from the xylem solution into the sieve tubes of willow. *J. Exptl. Bot.* 14:438–447.

———, 1967. Demonstration of solute movement from the extracambial tissues into the xylem stream in willow. *J. Exptl. Bot.* 18:600–606.

———, R. J. Field, C. L. Coulson, and D. C. J. Gardner, 1969. Movement of water and solutes in sieve tubes of willow in response to puncture by aphid stylets. Evidence against a mass flow of solution. *Physiol. Plantarum* 22:768–775.

——— and J. Ford, 1968. The movement of sugars into the sieve elements of bark strips of willow, II. Evidence for two pathways from the bathing solution. *J. Exptl. Bot.* 19:370–380.

——— and P. E. Weatherley, 1959. Composition of sieve-tube sap. *Nature* 184:1955–1956.

——— and ———, 1962. Studies in sieve-tube exudation through aphid mouthparts, I. The effects of light and girdling. *Ann. Bot.* (*London*) (N.S.)26:633–646.

——— and ———, 1963. Studies in sieve-tube exudation through aphid mouth parts, II. The effects of pressure gradients in the wood and metabolic inhibitors. *Ann. Bot.* (*London*) (N.S.)27:197–211.

Penot, M., 1965. Etude du rôle des appels moléculaires dans la circulation libérienne. *Physiol. Végétal* 3:41–89.

Pereira, J. F., A. S. Crafts, and S. Yamaguchi, 1963. Translocation in coffee plants. *Turrialba* 13:64–79.

Peterson, C. A., 1968. Studies on bidirectional translocation in the phloem using a fluorescent dye tracer. Doctoral dissertation, University of California, Davis.

Phelps, W. R., and J. E. Kuntz, 1965. Translocation and persistence of cycloheximide and oligomycin in northern pin oak. *Forest Sci.* 11:353–359.

Pilet, P. E., 1968. *In vitro* and *in vivo* auxin and cytokinin translocation. In *Biochemistry* and *Physiology of Plant Growth Substances*, ed. F. Wightman and G. Setterfield, pp. 993–1004. Ottawa: The Runge Press.

Plaut, Z., and L. Reinhold, 1965. The effect of water stress on ^{14}C sucrose transport in bean plants. *Australian J. Biol. Sci.* 18:1143–1155.

______ and ______, 1967. The effect of water stress on the movement of sucrose and tritiated water within the supply leaf of young bean plants, *Australian J. Biol. Sci.* 20:297–309.

Porter, H. K., 1966. Leaves as collecting and distributing agents of carbon. *Australian J. Sci.* 29:31–40.

Pozsár, B. I., M. E. Hammady, and Z. Király, 1967. Cytokinin effect of benzyladenine. *Nature* 214:273–274.

Prasad, R., C. L. Foy, and A. S. Crafts, 1967. Effects of relative humidity on absorption and translocation of foliarly applied dalapon. *Weeds* 15:149–156.

Pristupa, N. A., 1959. The transport form of carbohydrates in pumpkin plants. *Plant Physiol. (USSR)* 6:30–35.

Quinlan, J. D., 1965. The pattern of distribution of ^{14}Carbon in a potted apple rootstock following assimilation of ^{14}Carbon by a single leaf. *Ann. Rept. East Malling. Res. Sta. Kent 1964.*

______, 1966. The effects of partial defoliation on the pattern of assimilate movement in an apple rootstock. *Ann. Rept. East Malling Res. Sta. Kent 1965.*

______, 1969. Mobilization of ^{14}C in the spring following autumn assimilation of $^{14}CO_2$ by an apple root stock. *J. Hort. Sci.* 44:107–110.

______, and G. R. Sagar, 1962. An autoradiographic study of the movement of C^{14}-labelled assimilates in the developing wheat plant. *Weed Res.* 2:264–273.

______ and R. J. Weaver, 1969. Influence of benzyladenine, leaf darkening, and ringing on movement of ^{14}C-labelled assimilates into expanded leaves of *Vitis vinifera* L. *Plant Physiol.* 44:1247–1252.

Radwan, M. A., 1966. Absorption and distribution of ^{14}C-labeled tetramine in relation to its possible use in animal damage control. *U. S. Dept. Agr. Forest Serv. Pacific Northwest Forest Range Expt. Sta. Res. Paper* PNW-34.

______, 1967. Translocation and metabolism of ^{14}C-labeled tetramine by Douglas fir, orchard grass and blackberry. *Forest Sci.* 13:265–273.

Randall, A. P., 1962. A laboratory test of three systemic insecticides against the spruce budworm, *Choristoneura fumiferana.* (Clem.). *Can. Entomologist* 94:1156.

Rediske, J. H., and N. E. Johnson, 1965. The absorption and translocation of the systemic insecticide Schradan in Sitka spruce and grand fir. *Weyerhaeuser Forestry Paper* (5):1–9.

Reid, D. M., and W. J. Burrows, 1968. Cytokinin and gibberellin-like activity in the spring sap of trees. *Experientia* 24:189–191.

Resch, A., 1961. Cytology of Phloem. *Ber. Deut. Bot. Ges.* 74:55–58.

Ripper, W. E., 1957. The status of systemic insecticides in pest control practices. *Adv. Pest Control Res.* 1:305–352.

Roberts, B. R., 1964. Effect of water stress on the translocation of photosynthetically assimilated carbon-14 in yellow poplar. In *The Formation of Wood in Forest Trees*, ed. M. H. Zimmerman, pp. 273–288. New York: Academic Press.

Rohde, R. A., and W. R. Jenkins, 1958. Basis for resistance of *Asparagus officinalis* var. *altrilis* to the stubby-root nematode, *Trichadorus christiei* Allen 1957. *Maryland Agr. Expt. Sta. Bull.* A-97.

Rohrbaugh, L. M., and E. L. Rice, 1949. Effect of application of sugar on the translocation of sodium 2,4-dichlorophenoxyacetate by bean plants in the dark. *Bot. Gaz.* 110:85–89.

Rouschal, E., 1941. Untersuchungen über die Protoplasmatik und Funktion der Siebröhren. *Flora* 35:135–200.

Sacher, J. A., M. D. Hatch, and K. T. Glasziou, 1963. Sugar accumulation cycle in sugar cane, III. Physical and metabolic aspects of cycle in immature storage tissues. *Plant Physiol.* 38:348–354.

Sachs, R. M., Y. P. Shia, and R. G, Maire, 1967. Penetration, translocation, and metabolism of C^{14}-Alar (B-9), a plant growth retardant. *Plant Physiol.* 42(Suppl.):50.

Sagar, G. R., and C. Marshall, 1967. The grass plant as an integrated unit. Some studies of assimilate distribution in *Lolium multiflorum* Lam. *Proc. Internatl. Grassland Congr. 9th*, pp. 493–497.

Sasaki, S., N. Ota, J. Eguchi, Y. Furukawa, T. Akashiba, T. Tsuchiyama, and S. Suzuki, 1968. Polyoxins, antifungal antibiotics, VIII. Mechanism of action on sheath blight of rice plants. *Nippon Shokubutsu Byori Gakkaiho* 34:272–279.

Sauter, J. J., 1966. Investigations on the physiology of the wood rays of poplar, II. Seasonal changes in activity of acid phosphatase in woody ray parenchyma and its relation to the metabolism and the transport of carbohydrates. *Z. Pflanzenphysiol.* 55:349–362.

Schmitz, K., and J. Willenbrink, 1967. Histoautoradiographischer Nachweis ^{14}C-markierter Assimilate in Phloem. *Z. Pflanzenphysiol.* 58:97–107.

——— and ———, 1968. Zum Nachweis tritierter Assimilate in dem Siebrohren von Cucurbita. *Planta* 83:111–114.

Schneider, H., 1945. The anatomy of peach and cherry phloem. *Bull. Torrey Bot. Club* 72:137–156.

Schneider, I. R., 1964. Difference in the translocatability of tobacco ringspot and southern bean mosaic viruses in bean. *Phytopathology* 54:701–705.

———, 1965. Introduction, translocation, and distribution of viruses in plants. *Advan. Virus Res.* 11:163–221.

——— and J. F. Worley, 1958. Apparent movement of southern bean mosaic virus across steamed areas of bean stems. *Science* 127:1050–1051.

——— and ———, 1959a. Distribution of translocated particles of southern bean and tobacco mosaic viruses after intake into tracheary elements. *Proc. Internatl. Bot. Congr. 9th* 11:347.

——— and ———, 1959b. Upward and downward transport of infectious particles of southern bean mosaic virus through steamed portions of bean stems. *Virology* 8:230–242.

——— and ———, 1959c. Rapid entry of infectious particles of southern bean mosaic virus into living cells following transport of the particles in the water stream. *Virology* 8:243–249.

Schumacher, W., 1930. Untersuchungen über die Lokalization der Stoffwanderung in den Leitbündeln höherer Pflanzen. *Jahrb. Wiss. Bot.* 73:770–823.

———, 1933. Untersuchungen über die Wanderung des Fluoreszeins in den Siebröhren. *Jahrb. Wiss. Bot.* 77:685–732.

———, 1967. Der Stofftransport in der Pflanze. In *Handbuch der Pflanzenphysiologie*, ed. W. Ruhland et al., vol. 13, p. 1–334. Berlin: Springer-Verlag.

——— and R. Kollmann, 1959. Zur Anatomie des Siebröhrenplasmas bei *Passiflora coerulea*. *Ber. Deut. Bot. Ges.* 72:176–179.

Seth, A. K., and P. F. Wareing, 1967. Hormone-directed transport of metabolites and its possible role in plant senescence. *J. Exptl. Bot.* 18:65–77.

Shih, C. Y., and H. B. Currier, 1969. Fine structure of phloem cells in relation to translocation in the cotton seedling. *Am. J. Bot.* 56:464–472.

Shindy, W., and R. J. Weaver, 1967. Plant regulators alter translocation of photosynthetic products. *Nature* 214:1024–1025.

Shiroya, M., 1968. Comparison of upward and downward translocation of ^{14}C from a single leaf of sunflower. *Plant Physiol.* 43:1605–1610.

Shiroya, T., G. R. Lister, C. D. Nelson, and G. Krotkov, 1961. Translocation of ^{14}C in tobacco at different stages of development following assimilation of $^{14}CO_2$ by a single leaf. *Can. J. Bot.* 39:855–864.

———, ———, V. Slankis, G. Krotkov, and C. D. Nelson, 1962. Translocation of the products of photosynthesis to roots of pine seedlings. *Can. J. Bot.* 40:1125–1135.

Shulamith, H., and L. Reinhold, 1966. The effect of 2,4-dinitrophenol on translocation in the phloem. *Physiol. Plantarum* 18:634–643.

Sijpesteijn, A. K., and G. J. M. van der Kerk, 1965. Action of toxicants. Fate of fungicides in plants. *Ann. Rev. Phytopath.* 3:127–153.

Sjolund, R. D., 1968. Chloroplast development and cellular differentiation in tissue cultures of *Streptanthus tortuosus* Kell. (Cruciferae). Doctoral dissertation, University of California, Davis.

Skoog, F., H. Q. Hamzi, A. M. Szweykowska, N. J. Leonard, K. L. Carraway, T. Fugii, J. P. Helgeson, and R. N. Loeppky, 1967. Cytokinins: structure/activity relationships. *Phytochemistry* 6:1169–1192.

Slife, F. W., J. L. Key, S. Yamaguchi, and A. S. Crafts, 1962. Penetration, translocation and metabolism of 2,4-D and 2,4,5-T in wild and cultivated cucumber plants. *Weeds* 10:29–35.

Small, J. G. C., and O. A. Leonard, 1968. Translocation of ^{14}C-labelled photosynthate in nodulated legumes as influenced by nitrate nitrogen. *Am. J. Bot.* 56:187–194.

Smith, A. E., J. W. Zukel, G. M. Stone, and J. A. Riddell, 1959. Factors affecting the performance of maleic hydrazide. *J. Agr. Food Chem.* 7:341–344.

Smith, L. W., and P. J. Davies, 1965. The translocation and distribution of three labelled herbicides in *Paspulum distichum* L. *Weed Res.* 5:343–347.

______ and C. L. Foy, 1967. Interactions of several paraquat-surfactant mixtures. *Weeds* 15:67–72.

______, ______, and D. E. Bayer, 1967. Herbicidal enhancement by certain new biodegradable surfactants. *Weeds* 15:87–88.

Spanner, D. C., 1958. The translocation of sugar in sieve tubes. *J. Exptl. Bot.* 9:332–342.

______, 1962. A note on the velocity and the energy requirement of translocation. *Ann. Bot.* (*London*) (N.S.)26:511–516.

______, 1963. The mathematical pattern of tracer movement. *Prog. Biophys. Biophys. Chem.* 13:246–251.

______ and J. N. Prebble, 1962. The movement of tracers along the petiole of *Nymphoides peltatum*, I. A preliminary study with ^{137}Cs. *J. Exptl. Bot.* 13:294–306.

Sprent, J. I., 1968. The effects of benzyladenine on the growth and development of peas. *Planta* 78:17–25.

Srivastava, L. M., and T. P. O'Brien, 1966. On the ultrastructure of cambium and its vascular derivatives, II. Secondary phloem of *Pinus strobus* L. *Protoplasma* 61:277–293.

Steucek, G. L., and H. V. Koontz, 1970. Phloem mobility of magnesium. *Plant Physiol.* 46:50–52.

Stewart, I., 1963. Chelation in the absorption and translocation of mineral elements. *Ann. Rev. Plant Physiol.* 14:295–310.

Stout, P. R., and D. R. Hoagland, 1939. Upward and lateral movement of salt in certain plants as indicated by radioactive isotopes of potassium, sodium, and phosphorus absorbed by roots. *Am. J. Bot.* 26:320–324.

Stoy, V., 1963. The translocation of ^{14}C-labelled photosynthetic products from the leaf to the ear in wheat. *Physiol. Plantarum* 16:851–866.

Summers, L. A., 1968. Recent advances in synthetic organic fungicides for the control of plant pathogens. *Rev. Pure Appl. Chem.* 18:1–16.

Sundaram, A., 1965. A preliminary investigation of the penetration and translocation of 2,4,5-T in some tropical trees. *Weed Res.* 5:213–225.

Swanson, C. A., 1959. Translocation of organic solutes. In *Plant Physiology*, ed. F. C. Steward, vol. 2, pp. 48–551. New York: Academic Press.

______, 1969. Private communication.

______ and R. H. Böhning, 1951. The effect of petiole temperature on the translocation of carbohydrates from bean leaves. *Plant Physiol.* 26:557–564.

______ and E. D. H. El Shishiny, 1958. Translocation of sugars in grape. *Plant Physiol.* 33:33–37.

______ and D. R. Geiger, 1967. Time course of low temperature inhibition of sucrose translocation in sugar beets. *Plant Physiol.* 42:751–756.

______ and J. B. Whitney, 1953. Studies on the translocation of foliar-applied ^{32}P and other radioisotopes in bean plants. *Am. J. Bot.* 40:816–823.

Szabo, S. S., 1963. The hydrolysis of 2,4-D esters by bean and corn plants. *Weeds* 11: 292–294.

Tammes, P. M. L., 1933. Observations on the bleeding of palm trees. *Rec. Trav. Bot. Neerl.* 33:514–538.

———, 1951. Bleeding of and sieve tube transport in palm trees. *Koninkl. Ned. Akad. Wetenschap. Proc.* (C)54:30–32.

———, 1958. Micro- and macro-nutrients in sieve-tube sap of palms. *Acta Bot. Neerl.* 7:233–234.

———, and J. van Die, 1964. Studies on phloem exudation from *Yucca flaccida* Haw., I. Some observations on the phenomenon of bleeding and the composition of the exudate. *Acta Bot. Neerl.* 13:76–83.

——— and ———, 1966. Studies on phloem exudation from *Yucca flaccida* Haw., IV. Translocation of macro- and micro-nutrients by the phloem sap stream. *Koninkl. Ned. Akad. Wetenschap. Proc.* (C)69:656–659.

———, C. R. Vonk, and J. van Die, 1967. Studies on phloem exudation from *Yucca flaccida* Haw., VI. The formation of exudate-sucrose from supplied hexoses in excised inflorescense parts. *Acta Bot. Neerl.* 16:244–246.

———, ———, and J. van Die, 1969. Studies on phloem exudation from *Yucca flaccida* Haw., VII. The effect of cooling on exudation. *Acta Bot. Neerl.* 18:224–229.

Tamulevich, S. R., and R. F. Evert, 1966. Aspects of sieve element ultrastructure in *Primula obconica*. *Planta* 69:319–337.

Taylor, G. A., J. N. Moore, and W. O. Drinkwater, 1961. Influence of 2,3,5-triiodobenzoic acid and method of sap collection on translocation of foliar applied radiocalcium. *Plant Physiol.* 36:360–363.

Thaine, R., 1961. Transcellular strands and particle movement in mature sieve tubes. *Nature* 192:772–773.

———, 1962. A translocation hypothesis based on the structure of plant cytoplasm. *J. Exptl. Bot.* 13:152–160.

———, 1964a. Cytoplasm exudate from cut phloem. *Nature* 203:544–545.

———, 1964b. The protoplasmic-streaming theory of phloem transport. *J. Exptl. Bot.* 15:470–484.

———, 1965. Surface associations between particles and the endoplasmic reticulum in protoplasmic streaming. *New Phytologist* 64:118–130.

———, S. L. Ovenden, and J. S. Turner, 1959. Translocation of labelled assimilates in the soybean. *Australian J. Biol. Sci.* 12:349–371.

———, M. C. Probine, and P. Y. Dyer, 1967. The existence of transcellular strands in mature sieve elements. *J. Exptl. Bot.* 18:110–127.

Thomas, W. A., 1967. Dye and calcium ascent in dogwood trees. *Plant Physiol.* 42:1800–1802.

———, 1968. Calcium distribution in leaves: autoradiography versus chemical analysis. *Internatl. J. Appl. Radiation Isotopes* 19:544–545.

Thomas, T. M., and D. E. Seaman, 1968. Translocation studies with endothal-^{14}C in *Potamogeton nodosus* Poir. *Weed Res.* 8:321–326.

Thrower, S. L., 1962. Translocation of labelled assimilates in the soybean, II,. The pattern of translocation in intact and defoliated plants. *Australian J. Biol. Sci.* 15:629–649.

———, 1965. Translocation of labelled assimilates in the soybean, IV. Some effects of low temperature on translocation. *Australian J. Biol. Sci.* 18:449–461.

Tiffin, L. O., and J. C. Brown, 1961. Selective absorption of iron from iron chelates by soybean plants. *Plant Physiol.* 36:710–714.

———, ———, and R. W. Krauss, 1960. Differential absorption of metal chelate components by plant roots. *Plant Physiol.* 35:362–367.

Tomlinson, P. B., 1964. Stem structure in arborescent monocotyledons. In *The Formation of Wood in Forest Trees,* ed. M. H. Zimmerman, pp. 65–86. New York: Academic Press.

Treece, R. E., and J. G. Matthysse, 1959. Use of systemic insecticides on woody ornamental plants. *N.Y. State Agr. Expt. Sta.* (*Geneva, N. Y.*) *Bull.* 145:1–30.

Trip, P., 1969. Sugar transport in conducting elements of sugar beet leaves. *Plant Physiol.* 44:717–725.

——— and P. R. Gorham, 1967. Autoradiographic study of the pathway of translocation. *Can. J. Bot.* 45:1567–1573.

——— and ———, 1968a. Bidirectional translocation of sugars in sieve tubes of squash plants. *Plant Physiol.* 43:877–882.

——— and ———, 1968b. Translocation of sugar in minor veins of sugar beet leaves. *Plant Physiol.* 43(Suppl.):35.

——— and ———, 1968c. Translocation of sugar and tritiated water in squash plants. *Plant Physiol.* 43:1845–1849.

——— and ———, 1968d. Translocation of radioactive sugars in vascular tissues of soybean plants. *Can. J. Bot.* 46:1129–1133.

———, G. Krotkov, and C. D. Nelson, 1963. Biosynthesis of mannitol-^{14}C from $^{14}CO_2$ by detached leaves of white ash and lilac. *Can. J. Bot.* 51:828–835.

———, C. D. Nelson, and G. Krotkov, 1965. Selective and preferential translocation of ^{14}C-labeled sugars in white ash and lilac. *Plant Physiol.* 40:740–747.

Tso, T. C., and I. Fisenne, 1968. Translocation and distribution of lead-210 and polonium-210 supplied to tobacco plants. *Radiation Bot.* 8:457–462.

Tucker, C. M., and R. F. Evert, 1969. Seasonal development of the secondary phloem in *Acer negundo. Am. J. Bot.* 56:275–284.

Tuichibaev, M., and A. S. Kruzhilin, 1965. Translocation of labeled assimilates from individual cotton leaves. *Plant Physiol.* (*USSR*) 12:918–922.

Turkina, M. V., 1961. Sucrose absorption by plant conducting tissues. *Plant Physiol.* (*USSR*) 8:523–528.

Ullrich, W., 1961. Zur Sauerstoffabhangigkeit des Transportes in den Siebröhren. *Planta* 57:402–429.

———, 1962. Zurwirkung von Adenosintriphosphat auf den Fluorescein-transport in den Siebröhren. *Planta* 57:713–717.

Ursino, D. J., and G. Krotkov, 1968. The effect of water stress on the translocation of recent photosynthate to the roots of young plants of *Pinus strobus. Can. J. Bot.* 46:1327–1329.

———, C. D. Nelson, and G. Krotkov, 1968. Seasonal changes in the distribution of photo-assimilated ^{14}C in young pine plants. *Plant Physiol.* 43:845–852.

Vernon, L. P., and S. Aronoff, 1952. Metabolism of soybean leaves, IV. Translocation from soybean leaves. *Arch. Biochem. Biophys.* 36:383–398.

Vogl, M., 1964. Die Abwanderung von markiertem Phosphor aus dem Pappelblatt. *Flora* 154:94–98.

Wanner, H., 1953a. Die Zusammensetzung des Siebröhrensaftes: Kohlenhydrate. *Ber. Schweiz. Bot. Ges.* 63:162–168.

———, 1953b. Enzyme der Glykolyse im Phlöemsaft. *Ber. Schweiz. Bot. Ges.* 63:201–212.

Wardlaw, I. F., 1965. The velocity and pattern of assimilate translocation in wheat plants during grain development. *Australian J. Biol. Sci.* 18:269–281.

———, 1967. The effect of water stress on translocation in relation to photosynthesis and growth, I. Effect during grain development in wheat. *Australian J. Biol. Sci.* 20:25–39.

———, 1968a. The control and pattern of movement of carbohydrates in plants. *Bot. Rev.* 34:79–105.

———, 1968b. Carbohydrate movement in pea plants in relation to axillary bud growth. *Plant Physiol.* 43(Suppl.):35.

———, 1969. The effect of water stress on translocation in relation to photosynthesis and growth. *Australian J. Biol. Sci.* 22:1–16.

——— and H. K. Porter, 1967. The redistribution of stem sugars in wheat during grain development. *Australian J. Biol. Sci.* 20:309–318.

Ware, G. W., 1968. DDT-^{14}C translocation in alfalfa. *J. Econ. Entomol.* 61:1451–1452.

Wark, M. C., and T. C. Chambers, 1965. Fine structure of the phloem of *Pisum sativum*, I. The sieve element ontogeny. *Australian J. Bot.* 13:171–183.

Weatherley, P. E., 1962. The mechanism of sieve tube translocation: observation, experiment and theory. *Advan. Sci.* 18:571–577.

Weatherley, P. E., and R. P. C. Johnson, 1968. The form and function of the sieve tube. *Internatl. Rev. Cytol.* 24:149–192.

———, A. J. Peel, and G. P. Hill, 1959. The physiology of the sieve tube. Preliminary experiments using aphid mouth parts. *J. Exptl. Bot.* 10:1–16.

Weaver, R. J., W. Shindy, and W. M. Kliewer, 1969. Growth regulator induced movement of photosynthetic products into fruits of 'Black Corinth' grapes. *Plant Physiol.* 44:183–188.

———, J. van Overbeek, and R. M. Pool, 1966. Effect of kinins on fruit-set and development in *Vitis vinifera*. *Hilgardia* 37:181–201.

Webb, J. A., 1967. Translocation of sugars in *Cucurbita melopepo*, IV. Effects of temperature change. *Plant Physiol.* 42:881–885.

Webb, K. L., and J. W. A. Burley, 1962. Sorbitol translocation in apple. *Science* 137:766.

——— and ———, 1964. Stachyose translocation in plants. *Plant Physiol.* 39:973–977.

——— and P. R. Gorham, 1964. Translocation of photosynthetically assimilated ^{14}C in straight-neck squash. *Plant Physiol.* 39:663–672.

——— and ———, 1965a. Radial movement of ^{14}C-translocates from squash phloem. *Can. J. Bot.* 43:97–103.

——— and ———, 1965b. The effect of node temperature on assimilation and translocation of ^{14}C in the squash. *Can. J. Bot.* 43:1009–1020.

Webster, D. H., 1965. Heat-induced callose and lateral movement of assimilates from phloem. Doctoral dissertation, University of California, Davis.

——— and H. B. Currier, 1965. Callose: lateral movement of assimilates from phloem. *Science* 150:1610–1611.

——— and ———, 1968. Heat-induced callose and lateral movement of assimilates from phloem. *Can. J. Bot.* 46:1215–1220.

Weise, A. H., and J. D. Seeley, 1964. Translocation of the floral stimulus in *Chrysanthemum. Proc. Am. Soc. Hort. Sci.* 85:574–583.

Weisgerber, I., W. Klein, A. Djirsarai, and F. Korte, 1968. Insecticides in metabolism, XV. Distribution and metabolism of endrin-^{14}C in cabbage. *Justus Liebigs Ann. Chem.* 713:175–179.

Wellensiek, S. J., 1966. The flower forming stimulus in *Silene armeria* L. *Z. Pflanzenphysiol.* 55:1–10.

Wells, W. A., W. Hurtt, and C. P. P. Reid, 1969. Modes of translocation of ^{14}C-picloram in *Phaseolus vulgaris. Plant Physiol.* 44(Suppl.):24.

Went, F. W., and H. M. Hull, 1949. The effect of temperature upon translocation of carbohydrates in the tomato plant. *Plant Physiol.* 24:505–526.

Whitehead, C. W., 1962. Translocation in tomato. *Dissertation Abstr.* 23:411–412.

———, N. G. Sansing, and W. E. Loomis, 1959. Temperature coefficient of translocation to tomatoes. *Plant Physiol.* 34(Suppl.):xxi.

Whitehouse, R. L., and S. Zalik, 1968. Translocation of radioactively labelled 3-indoleacetic acid and tryptophane in seedlings of *Phaseolus coccineus* L. and *Zea mays* L. In *Biochemistry and Physiology of Plant Growth Substances,* ed. F. Wightman and G. Setterfield, pp. 977–991. Ottawa: The Runge Press.

Wiebe, H. H., and S. E. Wihrheim, 1962. The influence of internal moisture deficit on translocation. *Plant Physiol.* 37(Suppl.):l–li.

Wiersum, L. K., 1967. The mass-flow theory of phloem transport: a supporting calculation. *J. Exptl. Bot.* 18:160–162.

Willenbrink, J., 1957. Über die hemmung des Stofftransports in den Siebröhren durch lokale Inaktivierung verschiedener Atmungsenzyme. *Planta* 48:269–342.

———, 1966a. Transport ^{14}C markierter Assimilate im Phloem von *Pelargonium zonale* und *Phaseolus vulgaris. Planta* 71:171–183.

———, 1966b. Zur lokalen Hemmung des Assimilat transports durch Blausäure. *Z. Pflanzenphysiol.* 55:119–130.

——— and R. Kollmann, 1966. The translocation of assimilates in the phloem of *Metasequoia. Z. Pflanzenphysiol.* 55:42–53.

Williams, M. W., G. C. Martin, and E. A. Stahly, 1967. The movement and fate of sorbitol-^{14}C in the apple tree and fruit. *Proc. Am. Soc. Hort. Sci.* 90:20–24.

Williams, R. E., and A. W. Helton, 1967. Prevention and cure of *Cytospora* canker disease of peach trees with systemic chemicals. *Plant Disease Reptr.* 51:834–838.

Wilson, P. M. W., 1959. Evidence of downward transport of "solanaceous" alkaloids. *New Phytologist* 58:326–329.

Wooding, F. B. P., 1966. The development of the sieve elements of *Pinus pinea. Planta* 69:230–243.

Worley, J. F., 1965a. The distribution of Southern bean mosaic virus antigen in systemically- and locally-infected tissues as indicated by fluorescent antibody. Doctoral dissertation, The George Washington University.

———, 1965b. Translocation of Southern bean mosaic virus in phloem fibers. *Phytopathology* 55:1299–1302.

———, 1966. Injection of foreign particles and their intracellular translocation in living phloem fibers. *Planta* 68:286–291.

———, 1968. Rotational streaming in fiber cells and its role in translocation. *Plant Physiol.* 43:1648–1655.

——— and I. R. Schneider, 1963. Progressive distribution of Southern bean mosaic virus antigen in bean leaves determined with a fluorescent antibody stain. *Phytopathology* 53:1255–1257.

Wurster, R. T., and O. Smith, 1966. Translocation of iron-59 in the potato plant. *Am. Potato J.* 43:184–192.

Yamaguchi, S., 1961. Absorption and distribution of EPTC-^{35}S. *Weeds* 9:374–380.

———, 1965. Analysis of 2,4-D transport. *Hilgardia* 36:349–378.

——— and A. S. Crafts, 1958. Autoradiographic method for studying absorption and translocation of herbicides using C^{14}-labeled compounds. *Hilgardia* 28:161–191.

——— and ———, 1959. Comparative studies with labeled herbicides on woody plants. *Hilgardia* 29:171–204.

——— and A. S. Islam, 1967. Translocation of eight C^{14}-labeled amino acids and three herbicides in two varieties of barley. *Hilgardia* 38:207–229.

Yamamoto, T., 1967. The distribution pattern of carbon-14 assimilated by a single leaf in tobacco plant. *Plant Cell Physiol.* (*Tokyo*) 8:353–362.

Yoshida, D., 1967. Relations between the translocation of photosynthetic products and nicotine synthesis in tobacco plants. *Soil Sci. Plant Nutr.* (*Tokyo*) 13:63–70.

Yushkov, P. L., 1965a. Distribution of photosynthetic products in pine. *Tr. Inst. Biol. Akad. Nauk. SSSR Ural'sk. Filial* 43:17–23.

———, 1965b. Carbon nutrition of growing branches of young pines. *Tr. Inst. Biol. Akad. Nauk. SSSR Ural'sk. Filial* 43:25–32.

Zacharias, E., 1884. Über den Inhalt der Siebröhren von *Cucurbita pepo*. *Bot. Ztg.* 42:65–73.

Ziegler, H., 1956. Unterzuchungen über die Leitung und Sekretion der Assimilate. *Planta* 47:447–500.

———, 1958. Über die Atmung und den Stofftransport in den isolatierten Leitbundeln der Blattstiele von *Heracleum mantegazzianum* Somm. et Lev. *Planta* 51:186–200.

———, 1960a. Untersuchungen über die Feinstruktur des Phloems, I. Die Siebplatten bei *Heracleum mantegazzianum* Somm. et Lev. *Planta* 55:1–12.

———, 1960b. Über den Nachweis von Uridindiphosphatglukose (UDPG) im Phloem von *Heracleum mantegazzianum* Somm. et Lev. *Naturwissenschaften* 6:140.

———, 1963a. Verwendung von ^{45}Calcium zur Analyse der Stoffversorgung washsender Früchte. *Planta* 60:41–45.

———, 1963b. Untersuchungen über die Feinstruktur des Phloems, II. Die Siebplatten bei der Braunalge *Macrocystis pyrifera*. *Protoplasma* 57:786–799.

———, 1964. Storage, mobilization and distribution of reserve material in trees. In *The Formation of Wood in Forest Trees*, ed. M. H. Zimmerman, pp. 303–320. New York: Academic Press.

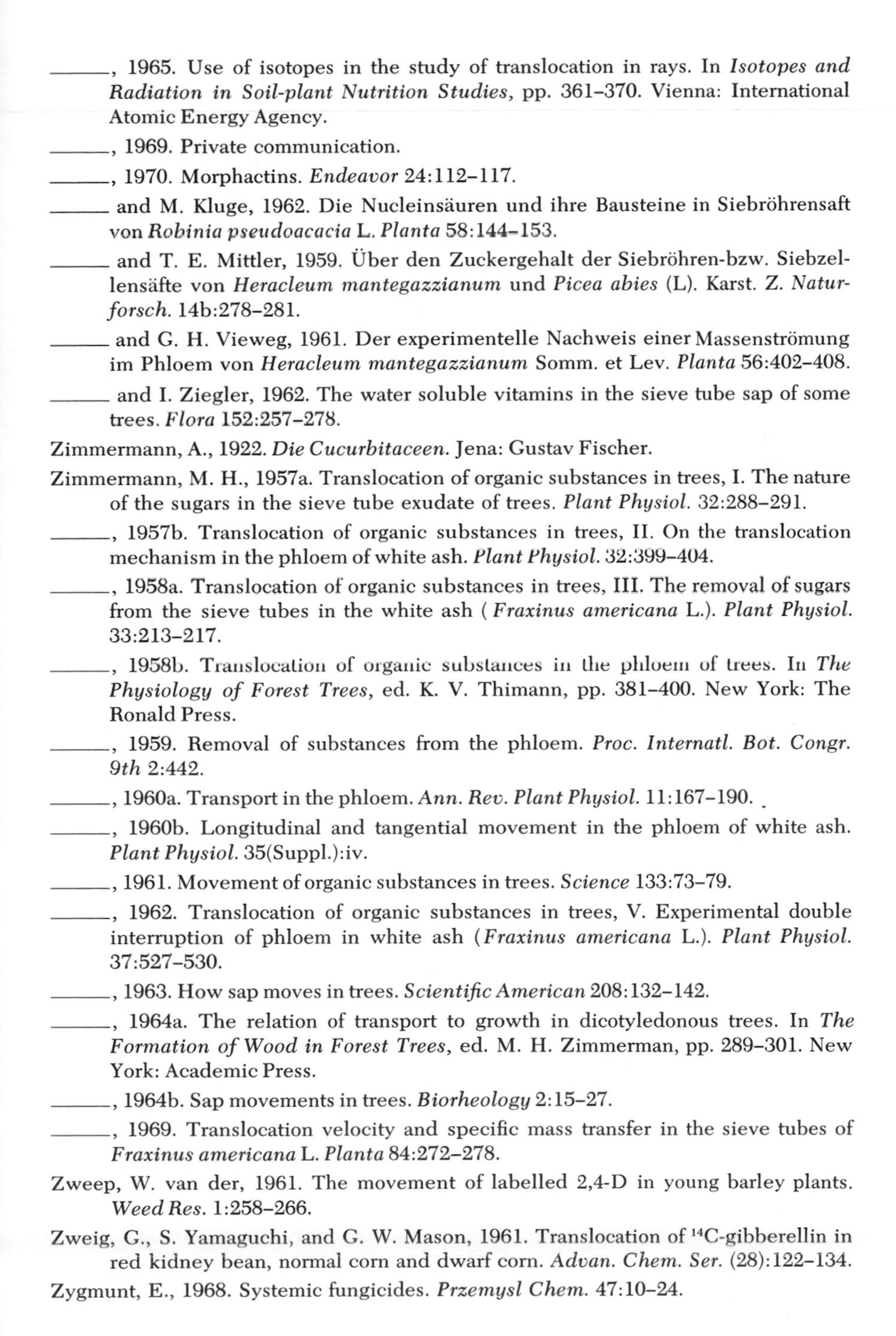

———, 1965. Use of isotopes in the study of translocation in rays. In *Isotopes and Radiation in Soil-plant Nutrition Studies,* pp. 361–370. Vienna: International Atomic Energy Agency.

———, 1969. Private communication.

———, 1970. Morphactins. *Endeavor* 24:112–117.

——— and M. Kluge, 1962. Die Nucleinsäuren und ihre Bausteine in Siebröhrensaft von *Robinia pseudoacacia* L. *Planta* 58:144–153.

——— and T. E. Mittler, 1959. Über den Zuckergehalt der Siebröhren-bzw. Siebzellensäfte von *Heracleum mantegazzianum* und *Picea abies* (L). Karst. *Z. Naturforsch.* 14b:278–281.

——— and G. H. Vieweg, 1961. Der experimentelle Nachweis einer Massenströmung im Phloem von *Heracleum mantegazzianum* Somm. et Lev. *Planta* 56:402–408.

——— and I. Ziegler, 1962. The water soluble vitamins in the sieve tube sap of some trees. *Flora* 152:257–278.

Zimmermann, A., 1922. *Die Cucurbitaceen.* Jena: Gustav Fischer.

Zimmermann, M. H., 1957a. Translocation of organic substances in trees, I. The nature of the sugars in the sieve tube exudate of trees. *Plant Physiol.* 32:288–291.

———, 1957b. Translocation of organic substances in trees, II. On the translocation mechanism in the phloem of white ash. *Plant Physiol.* 32:399–404.

———, 1958a. Translocation of organic substances in trees, III. The removal of sugars from the sieve tubes in the white ash (*Fraxinus americana* L.). *Plant Physiol.* 33:213–217.

———, 1958b. Translocation of organic substances in the phloem of trees. In *The Physiology of Forest Trees,* ed. K. V. Thimann, pp. 381–400. New York: The Ronald Press.

———, 1959. Removal of substances from the phloem. *Proc. Internatl. Bot. Congr. 9th* 2:442.

———, 1960a. Transport in the phloem. *Ann. Rev. Plant Physiol.* 11:167–190.

———, 1960b. Longitudinal and tangential movement in the phloem of white ash. *Plant Physiol.* 35(Suppl.):iv.

———, 1961. Movement of organic substances in trees. *Science* 133:73–79.

———, 1962. Translocation of organic substances in trees, V. Experimental double interruption of phloem in white ash (*Fraxinus americana* L.). *Plant Physiol.* 37:527–530.

———, 1963. How sap moves in trees. *Scientific American* 208:132–142.

———, 1964a. The relation of transport to growth in dicotyledonous trees. In *The Formation of Wood in Forest Trees,* ed. M. H. Zimmerman, pp. 289–301. New York: Academic Press.

———, 1964b. Sap movements in trees. *Biorheology* 2:15–27.

———, 1969. Translocation velocity and specific mass transfer in the sieve tubes of *Fraxinus americana* L. *Planta* 84:272–278.

Zweep, W. van der, 1961. The movement of labelled 2,4-D in young barley plants. *Weed Res.* 1:258–266.

Zweig, G., S. Yamaguchi, and G. W. Mason, 1961. Translocation of ^{14}C-gibberellin in red kidney bean, normal corn and dwarf corn. *Advan. Chem. Ser.* (28):122–134.

Zygmunt, E., 1968. Systemic fungicides. *Przemysl Chem.* 47:10–24.

Author Index

Subject Index